Chesterfield
Rowsley
Manch.r Buxton Matlock & Midl.d
Mansfield
Southwell
Newark
Nottingham
Midland
Railway
Great Northern
Sleaford
Boston
Boston Sleaford & Midl.d Co's
Ashbourne
Ripley
Ct. Northern
Midl.d Ry.
Ambercate Ry
Grantham
Uttoxeter
Derby
Willington
Staffordshire
Abbots Bromley
Burton
Edinham
Spalding
Bourne
Melton Mowbray
Ashby de la Zouch
Swannington
Mt. Sorrell
Rutland
Essendine
Stamford
Leicester Sh.
Leicester
Tamworth
Sutton Coldfield
L. & N. W.
S. Leicestershire
Nuneaton
North Western
Peterborough
Market Harborough
M. H. & M. M. Ry.
Northampton Sh.
Huntingdon
London & North Western
Coventry
Thrapstone
Kettering
Rugby
K & T
K & T. E.n
Warwick Sh.
Wellingborough
Henly
Warwick
Leamington
Daventry
Southam
Weedon
Stratford upon Avon
East and West Ry
Blisworth
Northampton
Bedford
Sandy
N & B
N. Pagnell
Banbury
Buckingham
Buckinghamsh. Ry.
Bedford Ry.
Midland
Bedford Sh.
Royston &
Chipping Norton
Hitchen
Bicester
L. & N. W.
Dunstable
Bourton
Oxf

MAPPING *the* RAILWAYS

Collins

MAPPING THE RAILWAYS

Collins
An imprint of HarperCollins Publishers
Westerhill Road, Bishopbriggs, Glasgow G64 2QT

First published as The Times Mapping the Railways 2011

This Edition 2013

ISBN 978 0 00 793178 1
Imp 001

British Library Cataloguing in Publication Data
A catalogue record for this book is available from the British Library

Printed in China

If you would like to comment on any aspect of this publication, please write to:
Collins Maps, HarperCollins Publishers, Westerhill Road, Bishopbriggs, Glasgow G64 2QT
e-mail: **collinsmaps@harpercollins.co.uk**
or visit our website at: **www.collinsmaps.com**

Follow us on twitter **@collinsmaps**

Historical maps available to view and buy at
www.mapseeker.co.uk

MAPPING the RAILWAYS

Julian Holland** and **David Spaven

With special thanks to

MAPSEEKER
ARCHIVE PUBLISHING
GATEWAY TO OUR HISTORIC PAST
www.mapseeker.co.uk

nls
National Library of Scotland

MIDLAND RAILWAY
WELSH HARP
Brent Reservoir
Kingsbury
Willesden Paddocks
Dollis Hill
GOLDERS GREEN
Bishops Wood
HIGHGATE
The Spaniards
North End
Ken Wood
Childs Hill
HAMPSTEAD
Hampstead Heath
Wembley Park
The Tower
Oakington Fm.
Wembley Ho.
Neasden
KINGSBURY NEASDEN
DUDDING HILL
Cricklewood
CHILDS HILL & CRICKLEWOOD
Hampstead Cemetery
Sherrick Green
Shoot up Hill
Finchley Road
West End
FINCHLEY RD
HAMPSTEAD HEATH
GOSPEL OAK
HIGHGATE ROAD
HAVERSTOCK HILL
KENTISH TOWN
CAMDEN ROAD
WILLESDEN GREEN
KILBURN & BRONDESBURY
WEST END LANE
FINCHLEY ROAD
WEST HAMPSTEAD
SWISS COTTAGE
CHALK FARM
MIDLAND & STH WESTERN JUN. RAILY
WILLESDEN
Willesden Green
EDGEWARE RD & BRONDESBURY
Brondesbury
STONEBRIDGE PARK
LOUDOUN ROAD
KILBURN & MAIDA VALE
ST JOHNS WOOD
MARLBOROUGH ROAD
Primrose Hill
Harlesden Green
HAMPSTEAD JUNCTION
KENSAL RISE
QUEENS PARK
WEST KILBURN
REGENTS PARK
Twyford
WILLESDEN JUNCTION
Kensal Green
Cemetery
Kilburn Park
ST JOHNS WOOD ROAD
EUSTON
ST PANCRAS
Acton Cemetery
Harrow Road
Westbourne
MARYLEBONE
Gt Central Terminus
METROPOLITAN RAIL
BAKER STR
PORTLAND ROAD
Wormwood Scrubs
ST QUINTON PARK
WEST BOURNE PARK
ROYAL OAK
BISHOPS ROAD
EDGEWARE RD
ACTON
WESTERN RAILWAY
NOTTING HILL
PADDINGTON
Gt Western Terminus
PRAED STR (Paddington)
East Acton
LATIMER RD
HAMMERSMITH BRANCH
BAYSWATER
HYDE PARK
MAYFAIR
Uxbridge Road
NOTTING HILL GATE
Kensington Gardens
Serpentine
ACTON
DISTRICT RAILWAY
UXBRIDGE ROAD
SHEPHERDS BUSH
Holland Ho.
Albert Memorial
Green Park
Gunnersbury
MILL HILL PARK
NORTH & SOUTH WESTERN JUNC RAILY
SOUTH ACTON
SHEPHERDS BUSH
KENSINGTON High St
Imperial Institute
BROMPTON
CHISWICK PARK & ACTON GREEN
Gunnersbury Park
Ravenscourt Park
ADDISON ROAD
Olympia
KENSINGTON
TURNHAM GREEN
HAMMERSMITH
VICTORIA
EARLS COURT
GLOUCESTER RD
SOUTH KENSINGTON
KEW BRI.
GUNNERSBURY
HAMMERSMITH & CHISWICK
RAVENSCOURT PARK
HAMMERSMITH
WEST KENSINGTON
SLOANE SQ
CHISWICK
KEW
Castelnau
WEST BROMPTON
CHELSEA
GROSVENOR RD
CHISWICK & GROVE PARK
WEST LONDON RAILY
Walham Green
WALHAM GREEN
CHELSEA
Battersea Park
BARNES
Barnes Elms
PARSONS GREEN
BATTERSEA PARK
KEW GARDENS
RIVER THAMES
FULHAM
BATTERSEA
QUEENS ROAD
Barnes Common
Palace
MORTLAKE
L. & S.W. RY
RICHMOND LINE
BARNES
PUTNEY BRIDGE FULHAM
WANDSWORTH ROAD
CLAPHAM JUNCTION
East Sheen
Upper Richmond Road
PUTNEY
Putney Park
WANDSWORTH
Wandsworth Road
CLAPHAM & NTH STOCKWELL
CLAPHAM
Sheen Common
EAST PUTNEY
NEW WANDSWORTH
Clapham Common
Wandsworth Common
Southfield
ROEHAMPTON
Putney Heath
Combe Park
Kingston Road
Surrey County Prison
WANDSWORTH COMMON
Balham Hill
RICHMOND PARK
SOUTHFIELDS
R. Wandle
CRYSTAL PALACE RAILY
WIMBLEDON PARK
EARLSFIELD
BALHAM & UPR TOOTING
Garrat
Halfway Ho.
Lunatic Asylum
Wimbledon Road
WIMBLEDON PARK
Streatham Cemetery
Robin Hood Inn
WIMBLEDON COMMON
Upper Tooting
Summers Town
Lambeth Cemetery
HAYDENS LANE
WIMBLEDON
Lowr Tooting

CONTENTS

	Introduction	6
Up to 1859	The Railway Revolution	10
1860–1900	The Railways Pre-eminent	64
1901–1922	The Zenith of the Network	130
1923–1947	The 'Grouping' – and a Quest for Speed	174
1948–1963	Nationalization and Modernization	202
1963–1970	Dr Beeching's Prescription	230
1970–1993	The Revival of the Passenger Railway	256
1993 onwards	Privatization and Beyond	276
	Conclusion: A Journey from the Past to the Future	298
	Index	300
	Acknowledgements and Bibliography	304

INTRODUCTION

Britain invented railways, and can also lay claim to being the best-mapped country in the world. Put the two together and you have a recipe for adventure and discovery. *Mapping the Railways* illustrates the history of Britain's railways through a carefully chosen selection of maps – from a wide variety of sources – that have never before appeared in a single publication. While maps are, of course, primarily functional, guiding us across the face of Britain, they can also be a visual feast – works of art with a very practical purpose – as this selection demonstrates.

The inherent characteristics of rail operation – steel wheel on steel rail, on a guided track, within a segregated and signalled right of way – give railways a particular strength, in that the great reduction in friction between wheel and land surface allows heavy loads to be moved at high speeds with a small power input. However, this lack of friction prevents railway locomotives from negotiating the steepness of gradients which road vehicles can readily tackle.

To minimize gradients, the construction of railways therefore demands careful choice of routes, and major engineering works on embankments, cuttings, bridges and tunnels – and this has had a distinctive impact on the landscape. The other major impact of topography on railways is through route curvature – in areas of more difficult terrain, the railway builders often faced choosing between sharper curves which would reduce the cost of earthworks (but would also limit train speeds), and gentler curves which would increase construction costs but allow faster speeds. The story of how railways overcame these challenges is well illustrated by maps, particularly those of larger scale.

Railways and maps have been closely linked since the first days of the 'iron road'. From 1803 – by which time the wooden 'waggonways' which preceded railways as we know

them were being replaced with iron rails – Parliament required the sponsors of new railways to submit detailed surveyed plans or large-scale maps of the intended routes. Once the successful schemes had received Parliamentary approval, and had been built and opened to traffic, the presence of these major new lines of communication in the landscape were inevitably marked on the various general topographic maps covering the country. In the era of early railways, lines would typically be marked on the privately produced 'county map' series, but from the mid-19th century onwards the publicly owned Ordnance Survey came to dominate the market for general maps – and the level of detail of railway features increased to reflect their central importance to Victorian economy and society.

The widespread, publicly available archive of general maps depicting railways and specialist railway maps allows us to trace the development of the rail network across Britain – from waggonways, through the pioneering public railways of the late 1820s and the subsequent railway mania, to the zenith of the system in the early 20th century. The mid-20th century saw great changes on the railways, with maps graphically illustrating the drastic reduction in the network following Dr Beeching's infamous report of 1963. More recently, maps now record the growing patchwork of lines re-opened in response to population changes, road congestion and environmental concerns, while two entirely new rail routes have been built – the Doncaster to York coalfield diversion of the East Coast Main Line in 1983 and the Channel Tunnel Rail Link (or HS1) completed in 2007.

Mapping the Railways charts the history of Britain's railways through a unique collection of 121 maps, with different origins, purposes, design features and content. General maps, such as the Ordnance Survey and the Bartholomew's series produced until the late 1980s, depict railways in the wider topographical scene, and have always been in the public domain. The very first specialist maps of railways were those showing planned alignments of proposed routes as part of Parliamentary depositions – but many of these either did not succeed or were amended to follow different alignments when actually built. Railway companies also produced maps and plans to guide the construction of their new lines, and after 1853 were obliged to produce formal 'demolition statements' – illustrated by large-scale plans – if they wished to demolish more than 30 houses in any one parish in the process of construction.

Once new lines had opened, the railway companies were soon producing network maps – together with timetables – for existing passengers and prospective customers. Other specialist maps were produced privately within the railway industry as working plans for legal, estate, regulatory, system development, commercial and operational purposes – and in the latter category detailed plans and diagrams of rail routes were eventually published more widely in response to public interest. Last but not least, there is now a long tradition of maps produced retrospectively to illustrate the history of railways, such as the Jowett atlases, excerpts from which appear in this volume. *Mapping the Railways* also features three maps specially created for the book: the first showing lines which were reprieved from the 'Beeching Axe' as well as those which later succumbed despite *not* being on the

A triumph of Victorian railway engineering through the West Yorkshire landscape. An 'A4' Pacific-hauled charter train crosses Milnsbridge Viaduct in 1984, on the steep trans-Pennine climb from Huddersfield to Marsden.

Beeching blacklist; the second and third setting out railways re-opened and new railways built since 1969.

One of the key characteristics of the history of maps depicting railways is *change* – both in the rail network as it grew, contracted and eventually revived in modern times, and in the surrounding developments which were stimulated by the arrival of railways, and in turn subsequently influenced the extent and complexity of railway infrastructure in the landscape. During the boom years of rail growth there was a massive demand for maps showing railways – it was a time of unprecedented urban and industrial change, and, as succinctly explained by Catherine Delano-Smith and Roger Kain in *English Maps: A History* (1999):

> *Railway companies required maps for planning new ventures, and the fast-developing railway network was itself transforming the pattern of economic activity and the way land was being used. Regional specialisation in agriculture was intensifying. Urban-based industry was developing in conjunction with the railways as the latter provided a means for the assembly of raw materials and the distribution of manufactured products. All these processes brought a demand for printed maps on the largest possible scale.*

The old and the new – Eurostar international passenger trains at rest below WH Barlow's vast trainshed at St Pancras International, terminus of the Channel Tunnel Rail Link (or HS1) since 2007.

Luckily, a large number of maps have survived to help us follow the development of our railway system over a period of nearly 200 years. In *Mapping the Railways* you will find various examples of 'before' and 'after' maps, showing just how much the pattern of railways changed in specific locales; while comparison of the national maps demonstrates the radical differences in the rail network from one era to another – notably the rapid expansion of the mid-19th century and the drastic reduction of the 1960s.

Mapping the Railways has been designed to appeal to both the general reader and those with a special interest in railways, history, geography or cartography. The maps are presented in chronological order within eight broad eras of railway development, except where different period snapshots of a particular locale are compared in a single map feature. While half of our map features are devoted to the first century of Britain's railways, ending in the 'Grouping' of 1923, we have not neglected the modern period – from the last years of private ownership, through nationalization and modernization, the Beeching era and subsequent passenger rail recovery, to the new world of late 20th-century privatization.

Each map feature covers two or four pages. Many of the maps are shown three times – in their entirety, but reduced in scale to fit the page; in an enlarged extract; and in a 100 per cent scale extract over two pages – allowing the fullest appreciation of particular areas of interest. The maps are interspersed and linked with 24 thematic 'mileposts' – illustrated with archive drawings, photographs, and railway ephemera; these explore in some detail a range of pivotal events, personalities and topics in Britain's railway history from the early railways through to the present day.

Every map feature is supported by a well-researched commentary, drawing out key aspects of railway history and geography, as well as any notable features of mapping design. While each nation and region of Britain is explored in the maps, we have deliberately increased the focus on those which particularly illustrate key aspects of the development of our railways – such as the role of the north of England in early railway building, Scotland in the development of railways for tourist travel through scenic areas, and London and the South East as the focus for one of the world's great commuter rail networks.

Of the 121 maps in this unique collection, some 20 are partially or fully 'diagrammatic', rather than topographically accurate 'geographic' representations – that is to say, lines of route are straightened and simplified on the map in order to ease understanding of the railway and its stations, or in some cases to fit into a particular format such as a brochure. A classic example is Harry Beck's London Underground map of 1933,

Right: an extract from the North British Railway's striking network map of 1896 (see pages 120–123)

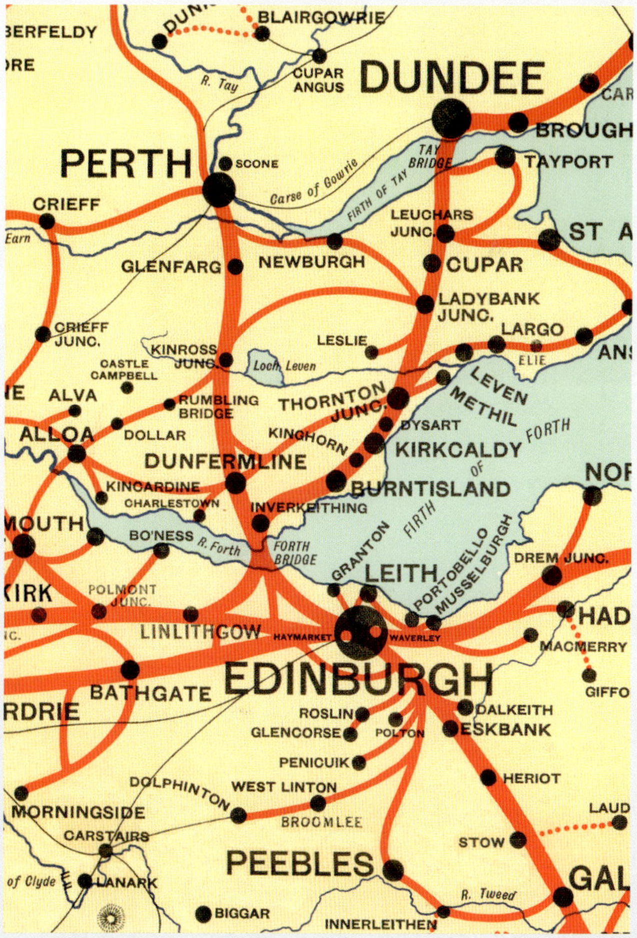

which eliminates virtually all geographical detail other than the rail network – but this is not just a 20th and 21st century phenomenon, the earliest of our selection of simplified maps dating back to 1849. Towards the end of the book, in a feature on Britain's railways today, we compare the relative success of fully diagrammatic, partially diagrammatic and geographic maps in representing the national rail network in the modern era.

In *Mapping the Railways* you will discover, or rediscover, that fascinating combination of railways and maps, helping you to understand better the development of the British railway system over two centuries, or simply to appreciate a collection of maps which for the most part appeal to that most basic of pleasures – they are pleasing to the eye. We hope this selection of railway maps will encourage you to explore further our modern rail system with maps. *Mapping the Railways* includes 15 Ordnance Survey maps dating from 1851 to the present day, and in one of our later map features we examine the way in which Ordnance Survey representation of railways has changed in the last 40 years, as an aid to interpreting railways in the landscape in the current 1:50,000 Landranger series. But now it is time to begin our journey with maps through the history of Britain's railways!

THE RAILWAY REVOLUTION

From origins as wooden waggonways conveying coal, railways as we know them swiftly developed into a nationwide network of steam-hauled train services carrying both passengers and goods. The railways transformed the Victorian economy and made their mark on landscapes the length and breadth of Britain.

MILEPOST 1 — Waggonways, Plateways and the Advent of Steam — 12

1819 — Robert Stevenson's Fife, Strathearn and Strathmore — 14

1820 — Greenwood's County Durham — 16

MILEPOST 2 — The First Public Railways — 20

1820 — Plan of Proposed Stockton & Darlington Railway — 22

MILEPOST 3 — The First Inter-City and Trunk Lines — 26

1836 — Grand Junction Railway — 28

1837 — Newcastle & Carlisle Railway — 32

MILEPOST 4 — George Bradshaw and Railway Timetables — 34

UP TO 1859

1839	Bradshaw's Great Britain	36
1840	Hobson's County Durham	40
MILEPOST 5	Railway Mania	42
1848	Cheffins's English & Scotch railways	44
c.1848–9	Reid's London to Edinburgh 'Railway Ride'	48
1851	York	52
1858	Bills of Session (England & Wales)	56
1858	Fife and the Lothians before the Forth Bridge	60

MILEPOST

1

The use of ruts in rocks for guiding carts and wagons along goes back into the mists of time. Certainly the Ancient Greeks exploited them and at various sites around the Mediterranean quite sophisticated layouts still exist where the early engineers carved out contour-hugging rutways, sidings and passing loops.

However, railways as we know them began in the infancy of the Industrial Revolution in Europe and by the mid-17th century horse-drawn wagons with wooden flanged wheels were running along oak rails laid on wooden ties between collieries and the River Tyne in the Newcastle area. It was a dangerous operation with loaded coal wagons running downhill by gravity, their progress slowed by a brakeman operating an extremely primitive brake. After the coal was loaded onto a sailing boat the wagons were returned empty to the colliery by horse power. The problem with this early railway technology was that the wooden wheels quickly wore out and cast iron wheels, introduced around 1760, soon wore out the wooden rails.

The first recorded use of iron rails in Britain came in 1767 when Richard Reynolds of Coalbrookdale in Shropshire cast rails for use on his ironworks waggonway (note the traditional spelling). These proved an immediate success with the reduced friction between wheel and rail allowing much heavier loads to be horse-drawn.

While the latter part of the 18th century saw the dawn of the short-lived canal age, the evolution of the railway continued with various developments, such as Benjamin Outram's L-shaped plate rail which found favour in the South Wales valleys and Lord Penrhyn's double-flanged wheels running on oval rails in his North Wales' slate quarries. Neither of these progressed much further, the latter being particularly over-complicated at crossings and points.

Meanwhile down in Cornwall, stationary steam engines had been in use to pump water out of mines since the early 18th century. Further developed by James Watt, an additional impetus came when the first crude steam-powered road vehicles appeared in the latter part of the century. The most successful of these was built by Richard Trevithick at Coalbrookdale in 1802 and he was soon commissioned to build a steam locomotive for the nine-mile 4 ft 2 in. gauge Penydarren plateway which served ironworks at Dowlais in South Wales. Despite the locomotive's successful maiden voyage with a loaded train in 1804, the cast iron plate rails broke under its weight and the experiment ended. However, the viability of steam haulage was proved beyond a shadow of doubt, although it was to be some years before it superseded traditional horse power.

With the Napoleonic Wars seeing ever-increasing prices for horse fodder, the next 10 years saw major strides forward in the development of the steam locomotive. The public's appetite was whetted by Trevithick when he ran his 'Catch-Me-Who-Can' circular demonstration steam-hauled train in Euston Square (London) in 1808. In Yorkshire, John Blenkinsop designed two steam locomotives that operated with cogs and rail-mounted racks ('rack and pinion') for use on the privately owned Middleton colliery railway near Leeds. These first ran in 1812 and continued in use for 30 years on the world's first commercially successful steam railway.

The horse-drawn Parkmore Waggonway can be seen in the foreground of this 1783 view of Newcastle.

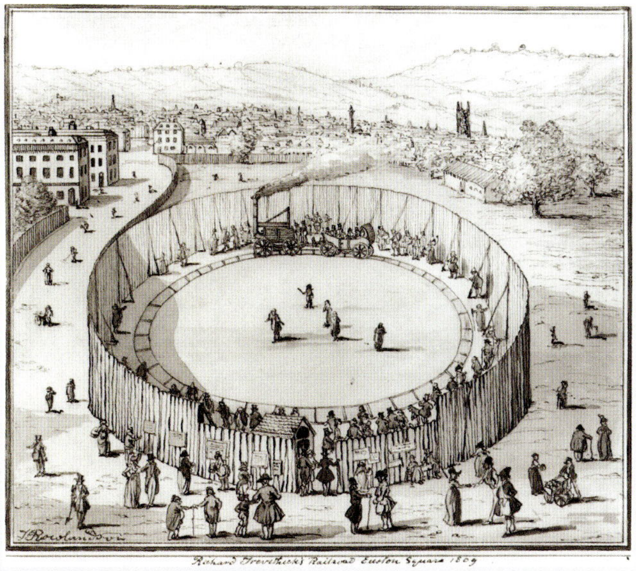

Richard Trevithick's 'Catch-Me-Who-Can' circular steam railway was exhibited at Euston Square, London, in 1808.

Near Newcastle, the Wylam colliery railway had already been operating as a wooden waggonway for nearly 200 years. The colliery's chief engineer, William Hedley, had the line converted to a 5 ft gauge plateway and successfully introduced his new steam locomotive, 'Puffing Billy', in 1813. Building on Trevithick's proven design of an efficient boiler with a single straight flue, his locomotive initially proved too heavy for the plate rails. To overcome this problem Hedley successfully rebuilt it as an articulated locomotive with two four-wheel bogies driven by gears.

At the same time as Hedley's success at Wylam a local self-taught engineer was about to launch himself onto the unsuspecting world. Born in Wylam in 1781, George Stephenson was soon to be known as the 'Father of Railways', the first great engineer of the Railway Age.

Built in 1813, William Hedley's 'Puffing Billy' was used at Wylam Colliery near Newcastle until 1862. The earliest surviving steam locomotive in the world, it was acquired by the Science Museum in London in 1862.

Title	Sketch of the Country from Stirling to Aberdeen. Referring to a Memorial regarding a communication by Canal or Railway through the Great Vallies of Strathearn and Strathmore.
By	Robert Stevenson, Civil Engineer
Date	1819

For a brief period before railways achieved their ascendancy as the most efficient form of inland transport, railways and canals could be planned almost interchangeably, as shown in this map setting out 1819 proposals for a new form of communication through Fife, Strathearn and Strathmore. Some 19 years earlier Dr James Thomson, minister of Markinch in his *General View of the Agriculture of the County of Fife: with Observations on the Means of its Improvement* had commented that:

> *There are no canals in Fife. The situation of the county, between two large navigable rivers, furnished with numerous convenient harbours, from which the greater part of the county is removed not more than three or four miles, and no part above eight or nine miles, renders this public accommodation less necessary here, than in many other parts of the kingdom.*

However, Robert Stevenson (better known for building lighthouses, and, of course, no relation to George Stephenson) here explored a number of possible routes for canals or railways over a large swathe of eastern Scotland, including the coastal town of 'Aberbrothock', better known

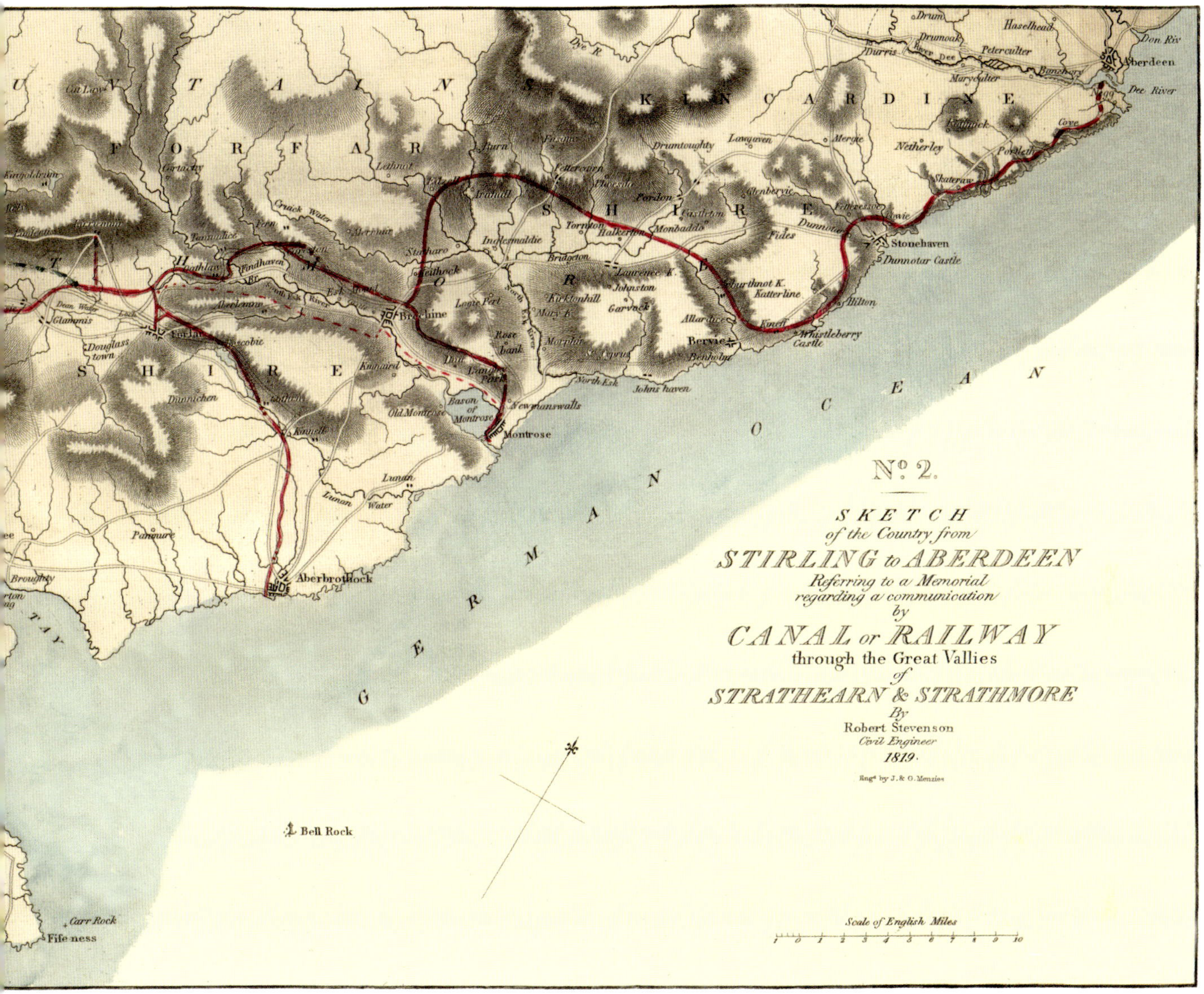

now as Arbroath. To avoid steeply graded ground, some circuitous routes following the contours were surveyed. The routes shown by red dashed lines were alternatives based on 'inclined planes' (cable haulage) or a series of locks, enabling railway or canal respectively to overcome steep gradients without meandering diversions. Interestingly the direct line that was eventually built between Perth and Dunfermline via Glenfarg (not planned by Stevenson on this map) required a six-mile climb at a gradient of 1 in 75 out of Strathearn. Very few stretches of the lines mooted on the map were built as railways on the alignments shown, not least because inclined planes or the alternative of meandering routes soon fell out of fashion.

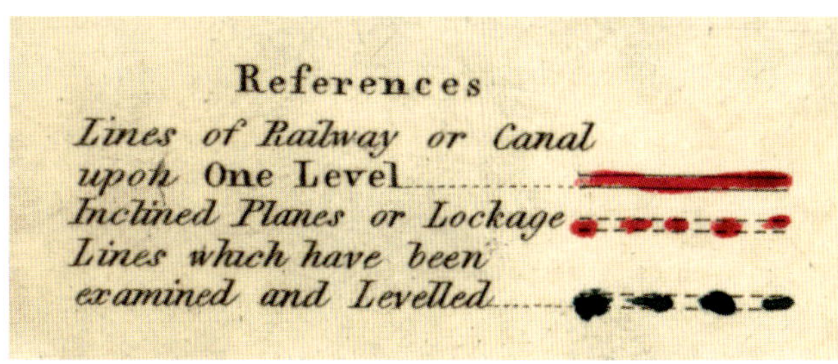

Several railways in Britain were built along earlier canal alignments, including parts of those from Glasgow to Paisley, Carlisle to Port Carlisle and the isolated narrow-gauge railway from Campbeltown to Machrihanish.

Title	Map of the County Palatine of Durham
Publisher	Greenwood, Pringle & Co.
Date	1820

The origins of the railway revolution lie in the waggonways of northeast England, linking collieries with the coast and with rivers. The Northumberland and Durham coalfield lay close to the coast, facilitating bulk sea shipments to London and other key markets, and wooden (later iron) waggonways evolved over a period of almost 200 years as an efficient overland alternative to poor roads.

The role of waggonways is shown in this map produced by Christopher Greenwood's long-established private mapmaking business, shortly before the Ordnance Survey secured its pre-eminence as *the* nationwide mapmaker.

'County maps' were funded by subscriptions, allowing the mapmakers to undertake physical survey on the ground, rather than simply copying earlier maps – and this map is hence specifically 'dedicated to the Nobility & Gentry of the County'. The word 'palatine' is derived from 'palace' and refers to historic royal privileges that were not finally abolished until 1836.

No waggonways or railways are actually shown in the map's key, but double-pecked lines trace the alignment of several rail routes. As shown overleaf, on the north bank of the Wear some miles inland from the coast, three (untitled) separate routes converged on Washington Staithe, where coal was dropped into keel boats which sailed down to Sunderland, with the coal then being loaded into larger sailing ships (colliers) for the journey to the London market.

The waggonway from the colliery at Bunker Hill or Philadelphia, near West Herrington, to Sunderland (marked as 'Railway' below) is the Newbottle Waggonway, which was a major iron-railed route opened in 1812. For 10 years it was the only line direct to the collier staithes at Sunderland, so avoiding the expensive use of keel boats and the transhipment they involved.

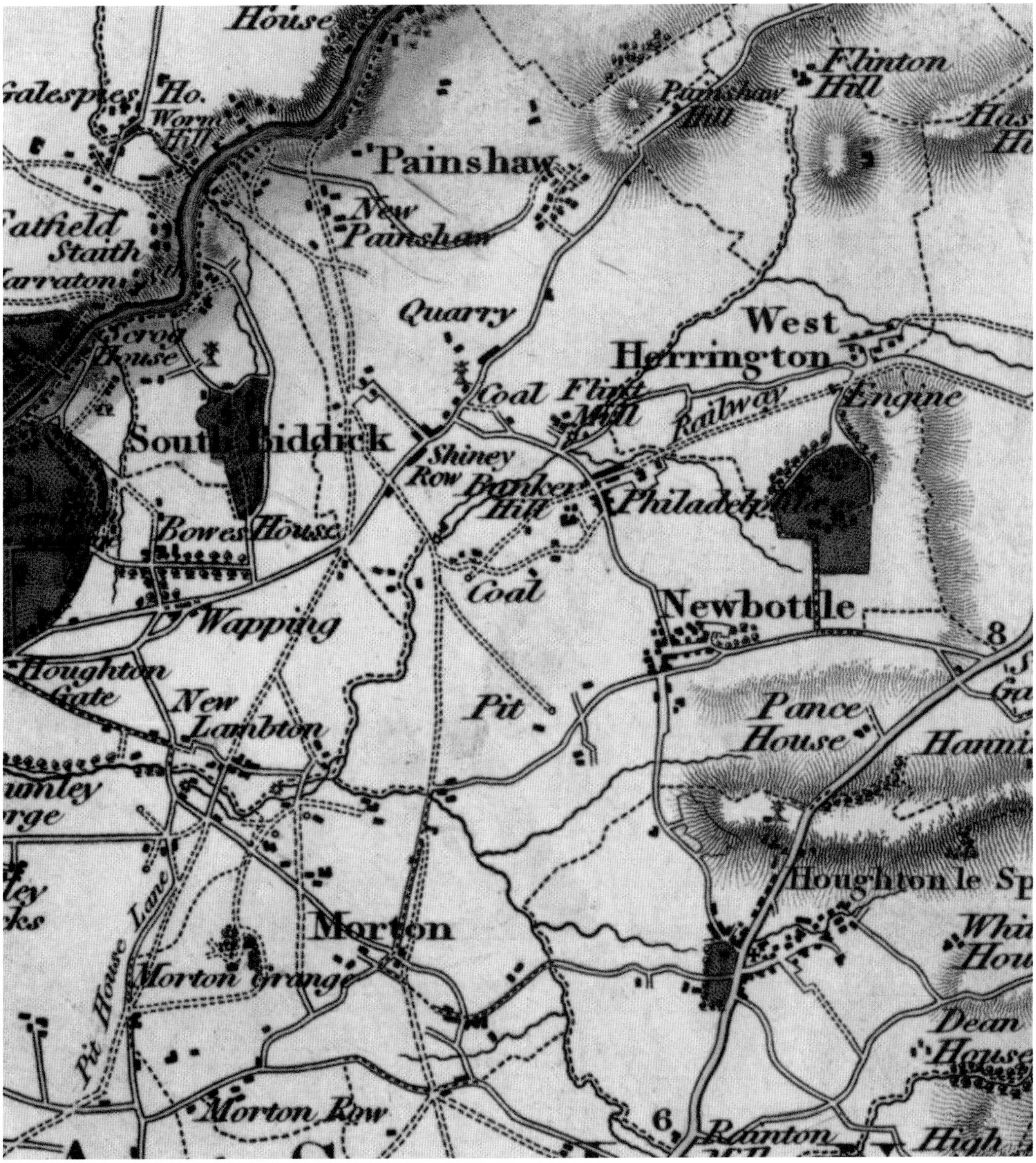

NEWCASTLE upon TYNE
RIVER TYNE
Bladon Staiths
Bladon
Path-head
Derwent Haugh
Winlaton
Swalwell
Park Gate
Whickham
Dunston Hill
Dunston Bank
Low Forge
Gateshead
Park House
T.B.
Deckhams Hall
Felling
Nether Heworth
Winlaton Mill
Norwood
Windy Nook
Over Heworth
Grindstone Quarry
Burnt Houses
Windy Hill
Old Acres Hill
High Hollingside
Riding Burns
Sunnyside
Fellside
Water Gate
Bog le Hole
Farn Acres
Derwent Crook
Chowdean Foot
Chowdean Hall
White House
The Leam
Harlow Green
High Ayton
Old Engine
Ayton Banks
Great Usworth
Lamesley
Ravensworth
Tinkler Row
Marley Hill
Fen House
Low Dunkirk
Low Ayton
Moor Mill
Kibblesworth
Blue House
Havanah
Birtley
Long Acre
T.B.
West House
Hedley
Common
Barcus Close
Tanfield
Causeway Arch
Cooper House
New York
Birtley
Riding
E
S
T
E
R
Causeway Hall
Sledge House
Porto Bello
Birtley Hall
Salt Pans
Tantoby
Leath Hall
Blue House
Oil Mill
Sister Farm
Sparks Ho.
Urpeth
White House
Old Pit
Ewe Hill
Vigo
Shield Row
High Urpeth
Nova Scotia
Harperley Mill
Hill Top
Peels House
Kyo
Stanley Hall
Brackenbeds
The Pea
Stanley
Highhouses
Pelton
Riden Hill
Burnip Field
Hillhead
High Flatts
Little Stanley
Pelton Grange
Jingling Gate
Burnt House
Howlett Hall
Twizdale Hall
Chester Grange
Shaw House
Chester le Street
West House
Craghead
Newfield
Tribley
Wagtail Hall
Whitehill
Cottage
Garden House
Rose Hall
Moor House
New Acres
Holmside
Black House
Broomy Holme
New Houses
Waldridge
Hermitage
T.B.
Burnett Flatt
Old Holmside Hall
Wheatley Green
White Cottage
Waldridge Common
Mirrow Flatt
New Holmside Hall
West Edmonsley
Holm Hill
Great Lumley
In Chester
R
Brecken Hill
D
Highhouse
Edmon House
White Side
East Edmonsley
Nettlesworth
Marlyburn
Black Dene
E
Sacriston Heugh
Plawsworth
Fellside
Whitehouse
Broom House
Acorn
Burnhope Side
Hobbletrott
T.B.
Harbour House
Plawsworth Mill

Jarrow
Harton
Jarrow Red House
Low Simonside
Marsdon Hill
Lodge
Simonside Hall
North Pastures
Monkton
Biddick Hall
High Simonside
Sunnyside
Holden House
Hope House
Brockley Whins
Hedworth
Field House
Newton Garths
Cleadon
Coal Farm
Fell Gate
Harden
Tile Sheds
South Wardley
Cold Side
West House
Whitburn
Pike House
T.B.
Scotch House
West Boldon
East Boldon
Mount Pleasant
Belle Vue
North Follingsby
Strother House
Fulwell
Hilton Bridge
Usworth Red House
North Moor Farm
Hilton Castle
Hill Top
West Moor Farm
Cowsland
Red House
Usworth Place
Southwick
Little Usworth
Dean Haus
Monk Wearmouth
South Moor House
Park House
Pallion
Hilton Place
Low Ford Pottery
Hylton Hall
SUNDERLAND
Barmston
Ford
Ford
Bishp. Wearmouth
Washington
Building Hill
Barmston Forge
High Ford
Railway
North Hall
Eden
Thorn Hill
Blue House
Wood House
Offerton
Broad Meadows
Bainbridge Holme
Washington Staith
Low Barns
Grindon
Hathey House
Flinton Hill
Painshaw
Hasting Hill
Engine
Spring Well House
Field House
Painshaw
New Painshaw
Thorney Close
Maiden's Paps
Quarry
Nor Moor Bank
Tunstall
West Herrington
Middle Herrington
Silksworth
Ryhope
Philadelphia
Engine
East Herrington
Tunstall Lodge
Coal
Cherry Know Hill
Newbottle
Stoney Gate
Burdon
Pance House
Thistley House
Burn Hall
Seaham Grange
Hanning
Old Burdon
Field House
Houghton le Spring
Warden Law
Seaham
Sharpley House
Seaton cum Slingley Hill
Morton
White House
High Sharpley
Dean House
Sharpley Hall
Morton Row
High Downs
Dean House
Rainton Mills
Great Eppleton
Dalden Hall
East Rainton
Dalton le Dale
Hetton
S
N
G
T
O
N

MILEPOST
2

The horse-drawn Surrey Iron Railway was the world's first publicly subscribed railway. It opened for goods traffic between Croydon and Wandsworth in 1803.

Prior to George Stephenson's dramatic impact on the Railway Age, the world's first publicly subscribed railway had already opened in 1803. The nine-mile Surrey Iron Railway was a horse-drawn 4 ft 2 in. gauge cast-iron goods plateway that ran between Croydon and Wandsworth. Built instead of a planned canal, the railway charged tolls to private users in much the same way as our modern railways are run.

The world's first passenger-carrying public railway was also a horse-drawn affair. Opened in 1806 as a 4 ft gauge horsedrawn line, the Swansea & Mumbles Railway started carrying passengers a year later. Slightly ahead of its time, passenger services ceased when a new turnpike road was opened in the 1820s! The railway did however re-open as a standard gauge (4 ft 8½ in.) horse-drawn line in 1855, subsequently converted to steam then electric haulage, prior to closure in 1960.

Cheered on by a large crowd, the opening ceremony of the Stockton & Darlington Railway on 27 September 1825 is depicted in this modern painting by Terence Cuneo.

Back in Newcastle the up-and-coming engineer George Stephenson had been appointed engineer of the Killingworth colliery railway in 1813. His first steam locomotive, the 'Blücher', ran a year later – in most respects it was similar to Blenkinsop's machines on the Middleton Railway but, importantly, it ran on normal rails without the use of rack-and-pinion. Stephenson further developed his steam locomotives but their use on passenger-carrying public railways had to wait until he was appointed Chief Engineer of the Stockton & Darlington Railway. Appointed in 1821, Stephenson planned a 4 ft 8 in. (later widened by ½ inch to reduce friction) gauge railway to run from inland collieries at Witton Park to the River Tees at Stockton via Darlington. While the western end of the line was operated by inclines with stationary steam engines, the 20 miles from Shildon to Stockton Quay were to be operated by steam locomotives. The 1821 Act of Parliament authorizing the building of the railway also included provision for the use of 'moveable engines' and the carrying of passengers – the latter a unique aspect for a steam railway of that period.

The opening day of the Stockton & Darlington Railway on 27 September 1825 was as important a date in world history as 20 July 1969, when man first walked on the Moon. On this day and watched by thousands of people, George Stephenson's 'Locomotion' hauled a passenger-carrying steam train on a public railway for the first time in the world. Around 600 passengers on this first train were carried in the railway's only passenger coach ('The Experiment') and in numerous converted coal wagons. 'Locomotion' was built at Stephenson's new locomotive works in Newcastle and featured, for the first time, driving wheels connected by coupling rods instead of gears. T-section wrought iron rails, connected by 'fishplates' at 15 ft intervals, were laid on heavy stone blocks designed to take the weight of heavily loaded coal trains.

The world's first steam-hauled passenger train arrived at Stockton Quay to be greeted by a crowd of 40,000 and a 21 gun salute! The Railway Age had arrived.

GEORGE STEPHENSON

George Stephenson was born of illiterate parents in Wylam, Northumberland in 1781. His father worked as a fireman for the Wylam colliery stationary pumping engine. At a young age George went to work as a brakesman, controlling the steam-powered winding gear at several collieries in the area. At the same time he attended night classes to learn to read and write and to study mathematics. After marrying in 1802, his son, Robert (who became an important railway engineer in his own right), was born a year later. Stephenson's knowledge of steam engines was greatly enhanced when he was appointed engineer for Killingworth collieries in 1813. His first steam locomotive, the 'Blücher' (named after the Prussian general at the battle of Waterloo) was built in a colliery workshop behind his house in 1814. In 1820 he built the Hetton Colliery Railway, a combination of rope-hauled inclines and loco-hauled sections, the first railway to be built not to use horse power.

Appointed Chief Engineer of the Stockton & Darlington Railway in 1821 and opening his locomotive works in Newcastle in 1823, Stephenson went on to engineer the world's first inter-city railway, the Liverpool & Manchester Railway, which opened in 1830. His locomotive, 'Rocket', won the Rainhill Trial in the previous year, setting a world speed record for steam haulage of 29.1 m.p.h.

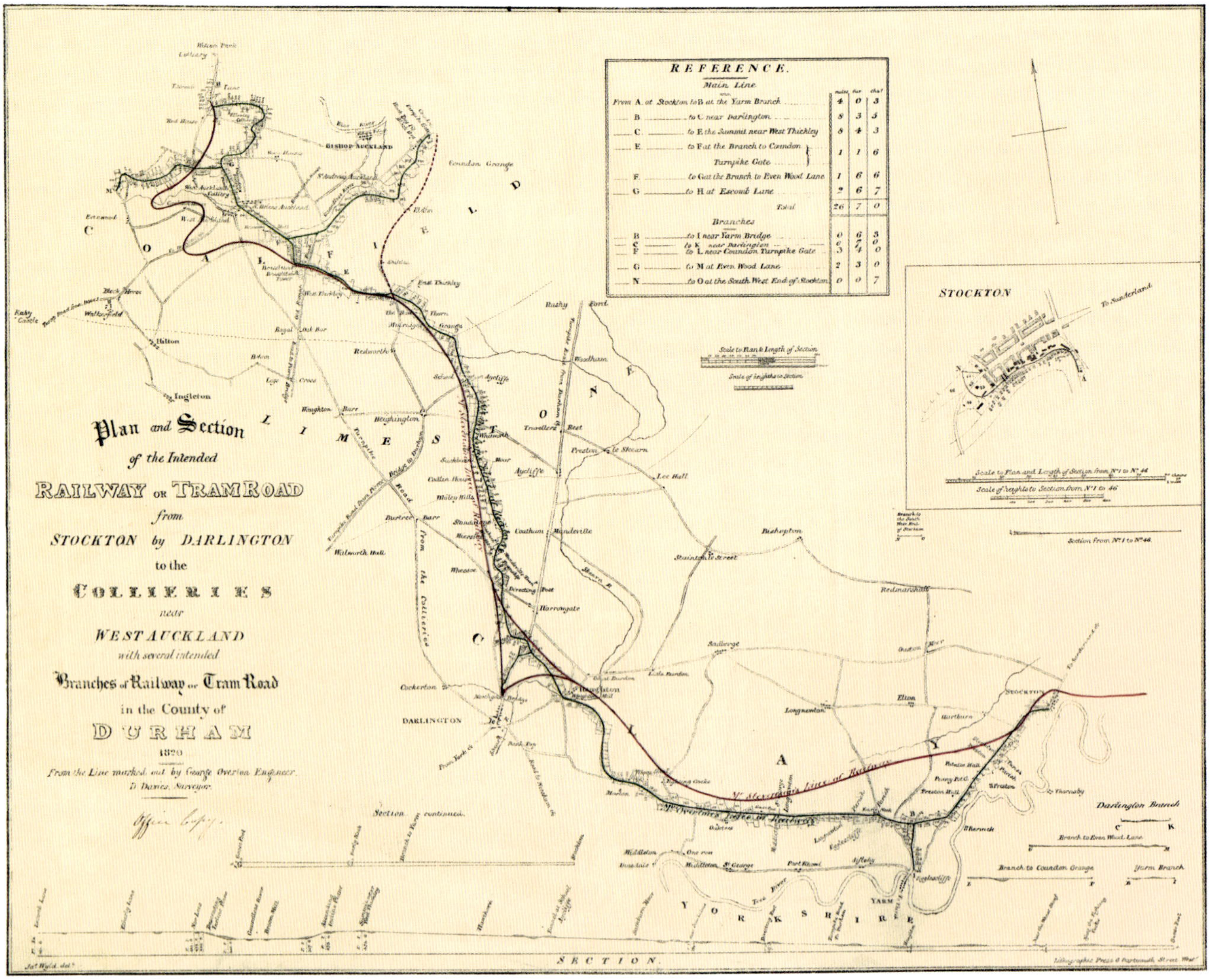

Title	Plan and Section of the Intended Railway or Tram Road from Stockton by Darlington to the Collieries near West Auckland
Engineer	George Overton
Surveyor	D Davies
Date	1820

From 1803 railways of all types required Parliamentary approval, and the provision of maps and plans of the surveyed route was a critical component of the legislative process. Inspired by wealthy local woollen manufacturer Edward Pease, this railway was originally intended to be a conventional horse-drawn waggonway or 'plateway' for the movement of coal. However, George Stephenson had been developing his steam engines at Killingworth for some years, and persuaded Pease that he should resurvey the route and work it by steam.

It seems likely this 1 inch to the mile map represents the first Stockton & Darlington Railway scheme, which was defeated in Parliament. A further Bill was placed before Parliament and received the Royal Assent in 1823, incorporating changes recommended by George Stephenson, consequent on his survey carried out between July 1821 and January 1822 – these included alterations to the route, and a clause added to permit the use of 'loco-motive or movable engines', setting the scene for the opening of the world's first steam-powered passenger-carrying public railway.

The history of the design of this railway is complex, not least because both George Stephenson and Robert Stevenson (no relation) were involved. As shown overleaf, the key features of this map (dated 1820) are the two different alignments shown – 'Mr Overton's Line of Railway' and 'Mr Stevenson's Line of Railway', the former designed in 1818 for horse haulage and the latter designed in 1819 for steam haulage.

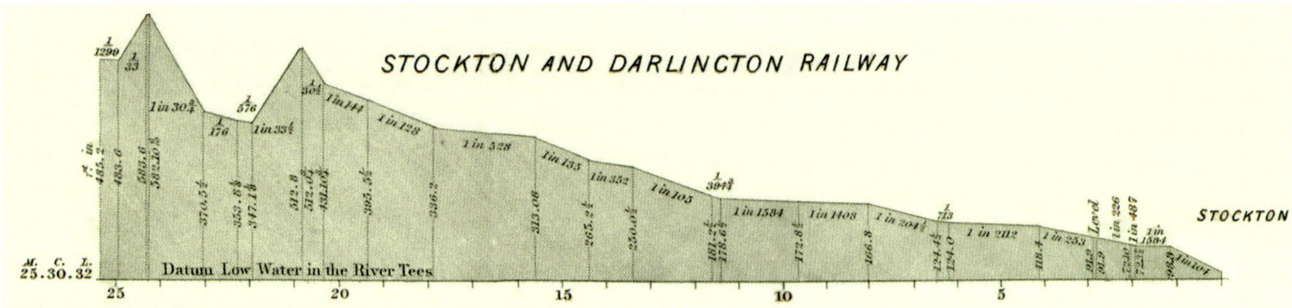

Taken from Bradshaw's map of Great Britain (pages 36–39), this gradient profile illustrates the virtually uninterrupted downhill gradient from the West Auckland collieries to Stockton.

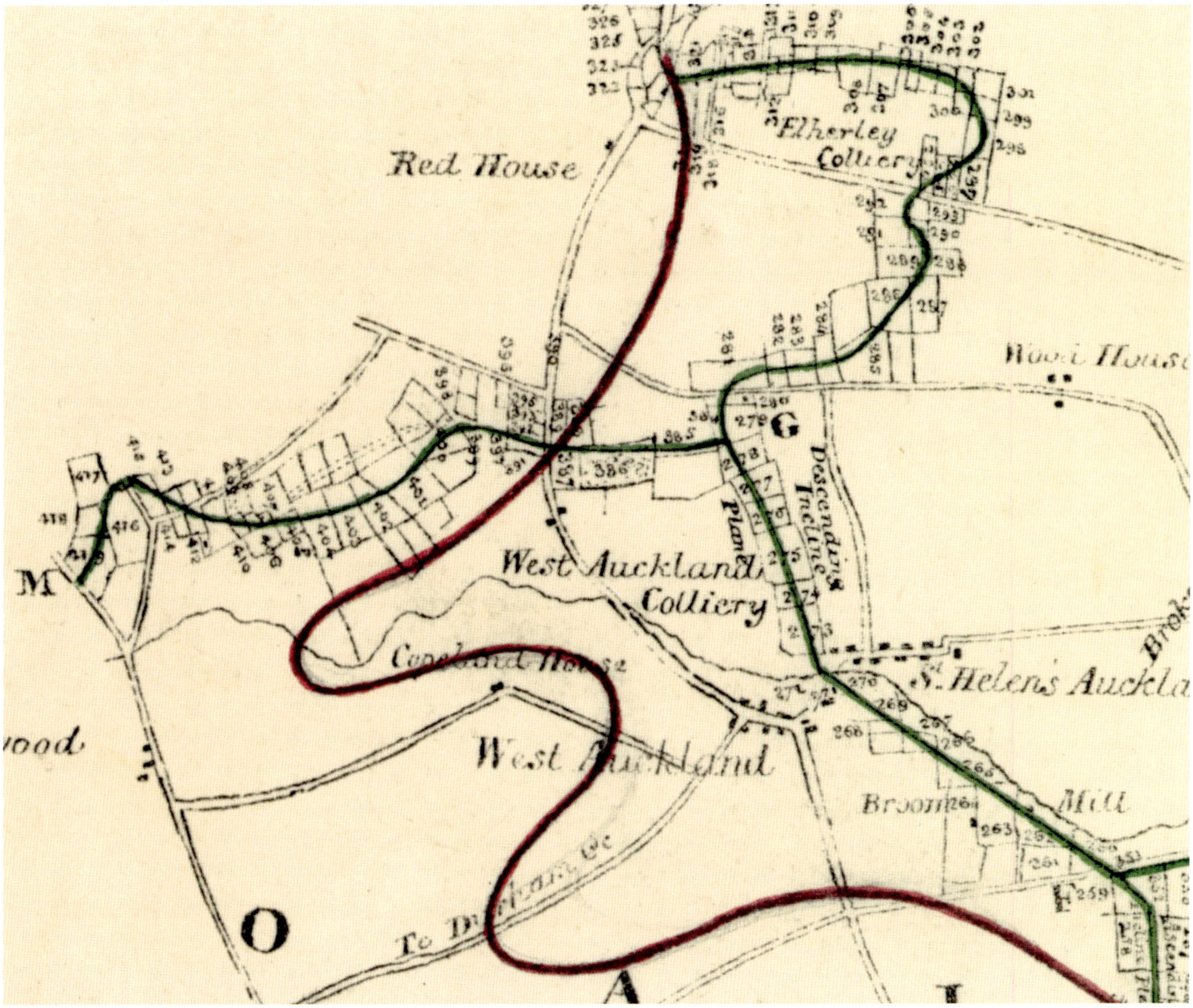

Overton's alignment shows sharper curvature, presumably following the lie of the land (and certainly running immediately adjacent to public roads in places), while Stevenson's is characterized by more sweeping curves.

Numbered plots of land for purchase are marked throughout the length of the proposed waggonway, but not in the case of the railway – suggesting that the development of the railway project was less advanced when the map was completed.

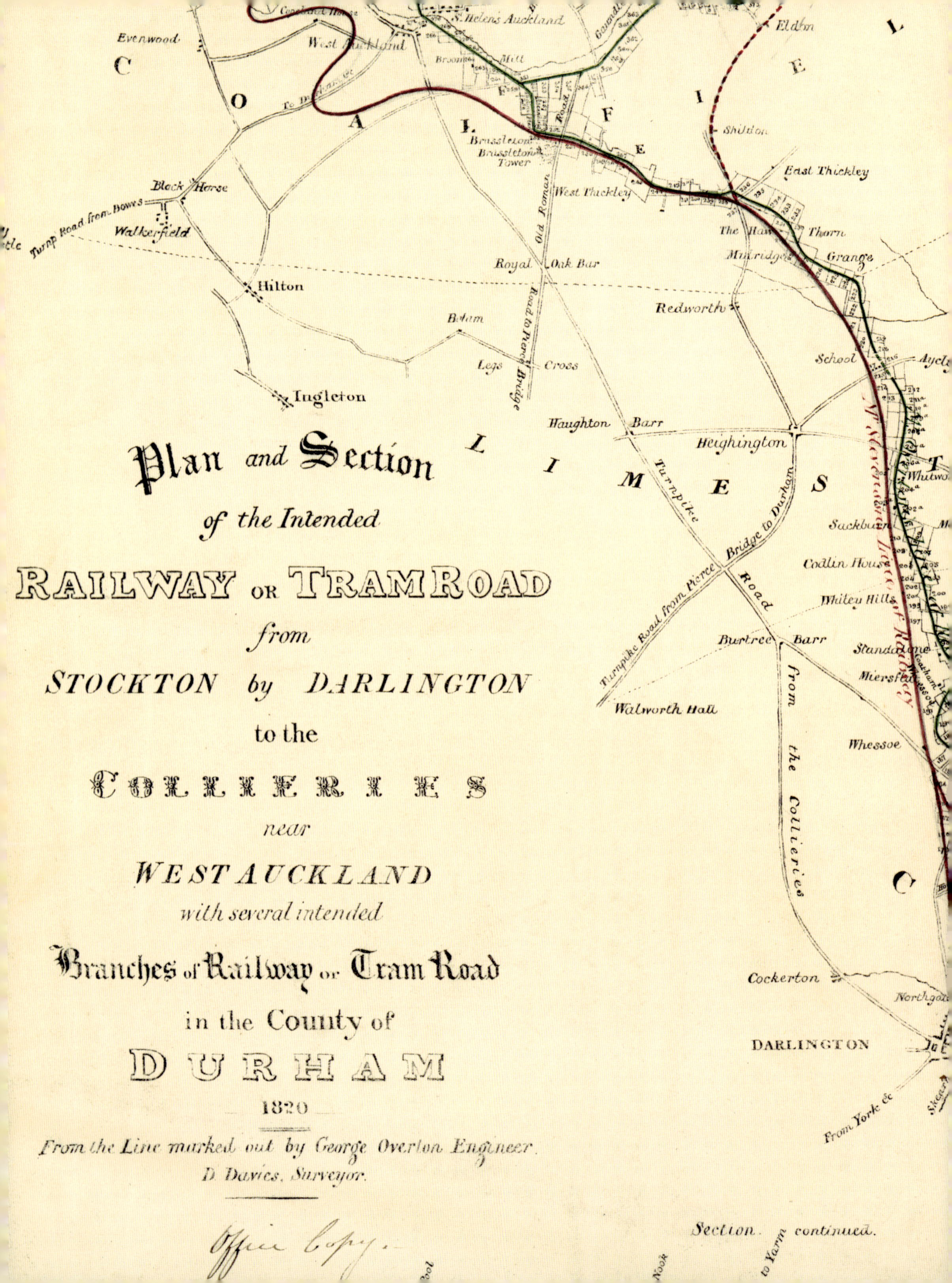
Plan and Section
of the Intended
RAILWAY OR TRAM ROAD
from
STOCKTON by DARLINGTON
to the
COLLIERIES
near
WEST AUCKLAND
with several intended
Branches of Railway or Tram Road
in the County of
DURHAM
1820
From the Line marked out by George Overton Engineer.
D. Davies, Surveyor.
Office Copy.
Section continued.
Evenwood
West Auckland
St. Helens Auckland
Brusselton
Brusselton Tower
Mill
Eldon
Shildon
East Thickley
West Thickley
Black Horse
Walkerfield
Turnp Road from Bowes
Hilton
Royal Oak Bar
Old Roman Road to Pierce Bridge
Redworth
The Haw
Thorn
Grange
Bolam
Legs Cross
Ingleton
School
Waughton Barr
Heighington
Turnpike Road
Bridge to Durham
Turnpike Road from Pierce Bridge to Durham
Sackburn
Codlin House
Whiley Hills
Burtree Barr
Walworth Hall
from the Collieries
Whessoe
Cockerton
DARLINGTON
From York &c
C O A L F I E L
L I M E S T

Total 26 7 0
Branches
B to I near Yarm Bridge 0 6 3
C to K near Darlington 0 7 0
F to L near Coundon Turnpike Gate 3 4 0
G to M at Even Wood Lane 2 3 0
N to O at the South West End of Stockton 0 0 7
STOCKTO
Scale to Plan & Length of Section
Scale of heights to Section
Rushy Ford
Turnpike Road from Durham &c
Woodham
N
E
O
Travellers Rest
Preston le Skearn
Lee Hall
Mandeville
Skearn R
Post
Harrowgate
Bishopton
Stainton le Street
Redmarshall
Branch to the South West End of Stockton
N
O
Sadberge
Ouston Moor
Great Burdon
Little Burdon
Haughton
Haughton Mill
Elton
Longnewton
Hartburn
L
A
Y
Road to Neesham &c
Wheat Sheaf
Fighting Cocks
Morton
Mr. Stevenson's Line of Railway
Mr. Overton's Line of Railway
Oaktree
Goose Pool
Parish
Early Nook
B
Potatoe Hall
Penny Pot G.
Preston Hall
Berwick
Stockton
Preston

George and Robert Stephenson's 'Rocket' won the Rainhill Trials on the Liverpool & Manchester Railway in 1829.

Even before the Stockton & Darlington Railway had opened, a group of wealthy Lancashire businessmen had founded the Liverpool & Manchester Railway Company (LMR). Designed to provide cheap transport for raw materials and finished goods between the port of Liverpool and the textile mills of east Lancashire, the railway was surveyed by George Stephenson in 1824. As his son, Robert, was in South America, George relied too much on his subordinates to check his calculations and the survey's inaccuracies led to the first Bill being rejected by Parliament in 1825. Stephenson was sacked and replaced by George and John Rennie as engineers and Charles Vignoles as surveyor. This time they got it right and the second Bill received Royal Assent in 1826.

When the Rennies proved too expensive, George Stephenson was reappointed as engineer with Joseph Locke as his second-in-command. Built to a gauge of 4 ft 8½ in., the 35-mile line was heavily engineered with 64 bridges and viaducts, a 1 mile 490 yd tunnel under Liverpool, the two-mile Olive Mount Cutting and the crossing of Chat Moss on a raft of wood and heather hurdles. It was a triumph of engineering.

While construction of the line was proceeding, Stephenson was also developing his steam locomotive, the 'Rocket', in preparation for a competition that was to be held at Rainhill in October 1829. Competing against three other steam locomotives and a horse-powered version, Stephenson's 'Rocket' won hands down, scooping the £500 prize and setting a speed record of 29.1 m.p.h. Stephenson was also given the contract to produce locomotives for the railway.

The LMR opened to great celebrations on 15 September 1830. Attended by the Duke of Wellington, then Prime Minister, and other political and business celebrities, the festivities were somewhat spoilt when a local MP, William Huskisson, was killed when the 'Rocket' ran over him. However, the LMR, the world's first inter-city railway, was an instant success, particularly with the travelling public.

The success of the LMR soon saw proposals for a host of new railways throughout the land. Of these, the 82-mile Grand Junction Railway (GJR), from Birmingham to a junction with the LMR at Newton, and the 112-mile London & Birmingham Railway were by far the most important. The former, engineered by George Stephenson and Robert Locke, was authorized by Parliament in 1833 and opened on 4 July 1837. The latter, engineered by Robert Stephenson, was also authorized in 1833 but its opening on 28 June 1838 had been delayed by extensive engineering works such as at Tring Cutting (through the Chilterns) and Kilsby Tunnel (southeast of Rugby). At the London end trains were cable-hauled between Euston and Camden until 1844.

The LMR merged with the Grand Junction in 1845. A year later the GJR merged with the London & Birmingham to form the London & North Western Railway – the West Coast Main Line, as we know it today, was already taking shape.

North of the border the Scots were not far behind – the Kilmarnock & Troon Railway, a coal-carrying line, had already opened in 1808 and in 1817 became the first railway in Scotland to use steam locomotive haulage. The first inter-city line in Scotland was the Edinburgh & Glasgow Railway which was authorized in 1838 and opened on 21 February 1842. The original Edinburgh terminus at Haymarket remains in use today.

The Liverpool & Manchester Railway became the world's first inter-city railway in 1830. Here a train is seen passing through the formidable Olive Mount Cutting near Liverpool – blasted out of sandstone rock the cutting is two miles long and 80 ft deep.

Engineered by Robert Stephenson, the London & Birmingham Railway opened in 1838. Here a train is seen emerging from Primrose Hill Tunnel in North London.

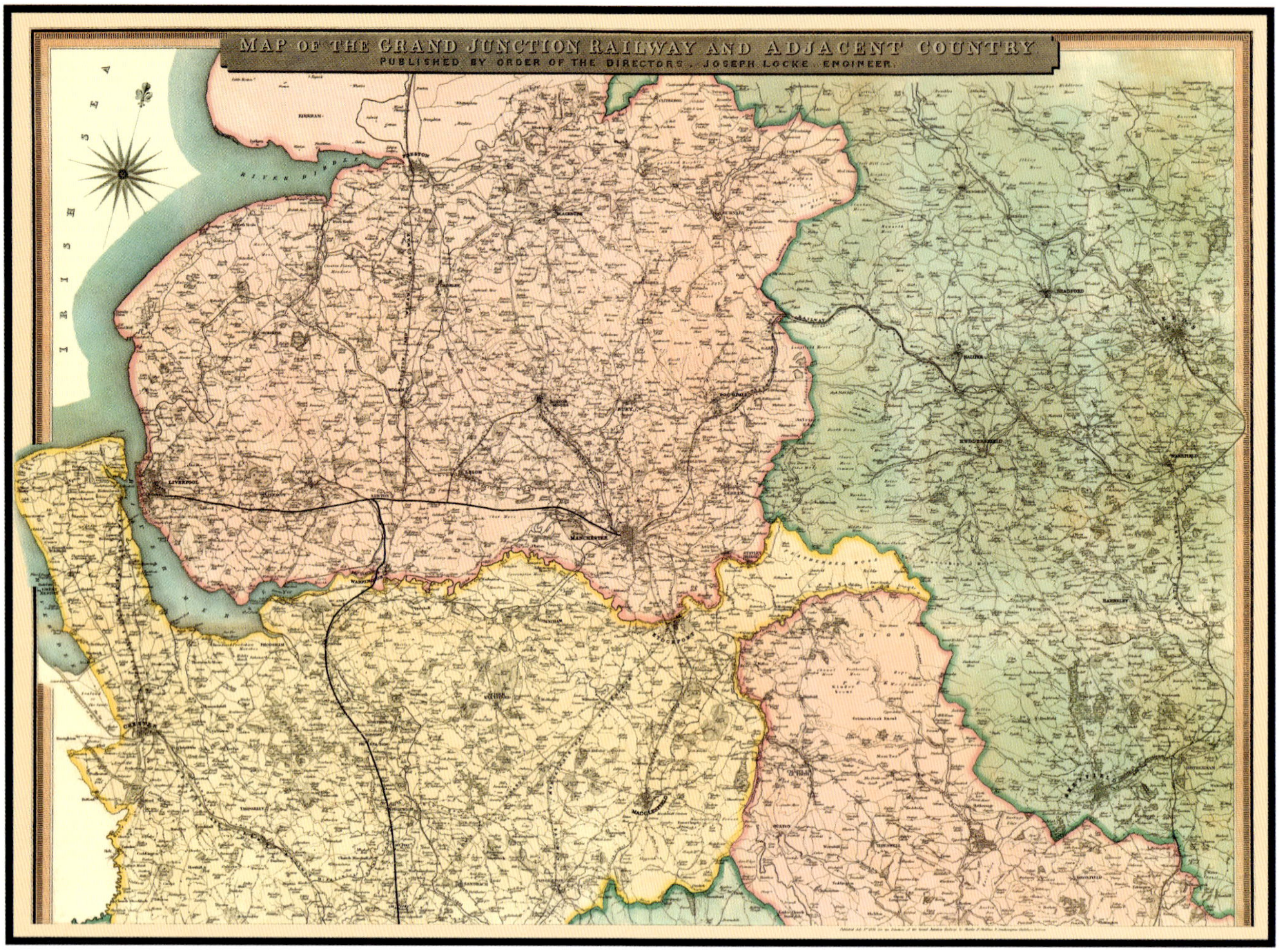

Title	Map of the Grand Junction Railway and Adjacent Country
Publisher	Charles F Cheffins
Date	1836

Following the success of the Liverpool & Manchester Railway (LMR), which opened on 15 September 1830, many schemes were proposed for the building of further railways. Birmingham businessmen were planning a link to London, and the group of financiers involved with the LMR saw Birmingham as an ideal goal for expansion. In 1831 the Warrington & Newton Railway opened; it ran about five miles southwards from Newton Junction (Earlestown) at the centre of the LMR, to Warrington and the River Mersey. This was considered to be an ideal starting point for a route to Birmingham, and after much surveying a route was found that was practical and avoided conflicts with landowners.

The eventual triangular junction at Earlestown was to play an important part in the history of the London & North Western Railway (LNWR), because it was on these very tight curves that the engines then in use frequently broke their axles. This led to the design of the outside-cylinder 'Crewe Type' engines with no crank axles to break, and eventually to the LNWR building its engines in-house. These tight curves remain in use to this day.

The map extract opposite shows the gap in what would become the West Coast Main Line, between Warrington and the railway heading north towards Wigan from Parkside Junction, east of Newton Junction on the LMR. In due course this missing link was completed by crossing under the LMR west of Parkside Junction. Also shown overleaf are a number of routes in planning, including what would become the main line from Crewe Junction via Sandbach to Manchester.

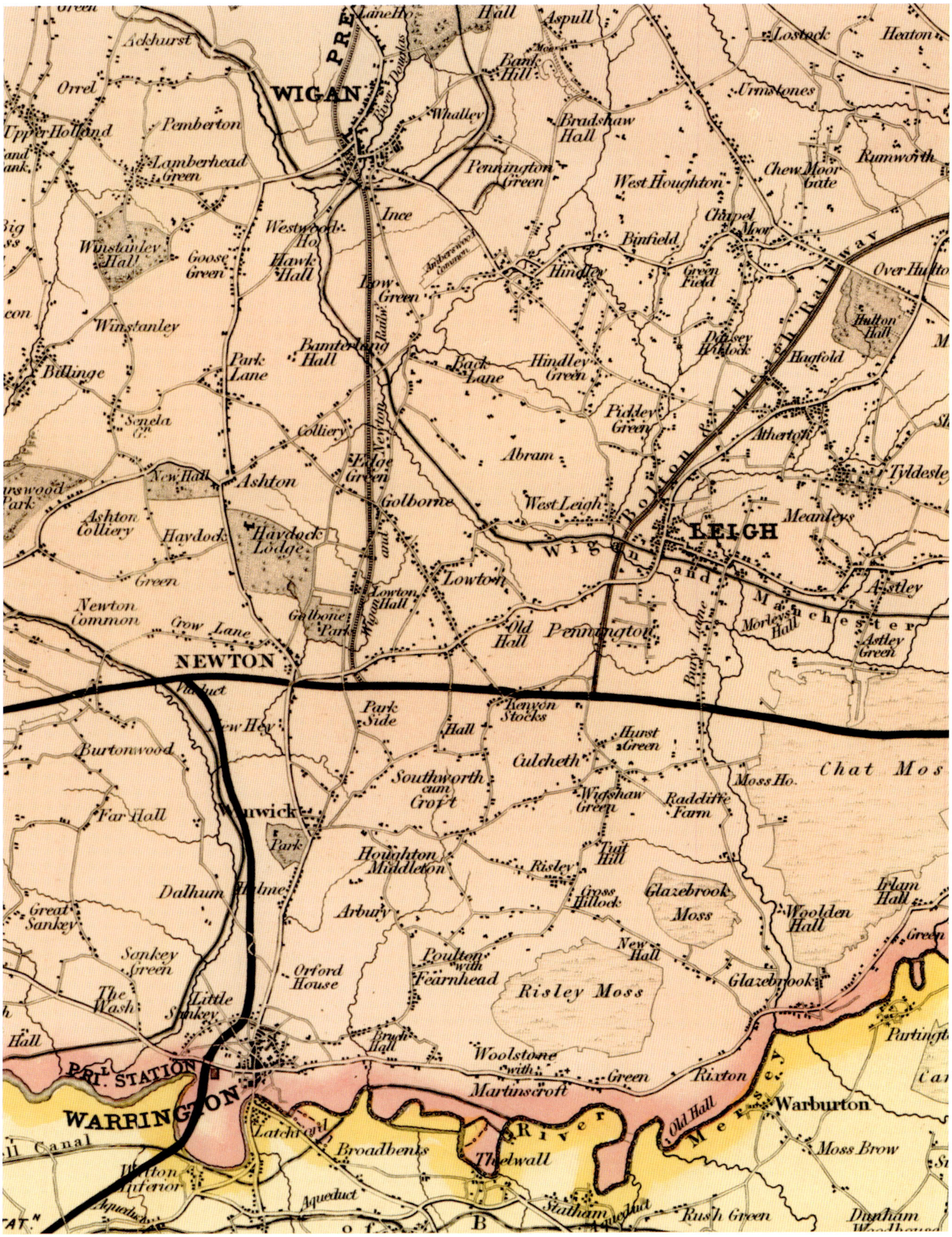
WIGAN
Ackhurst
Orrel
Upper Holland
Pemberton
Lamberhead Green
Winstanley Hall
Goose Green
Winstanley
Billinge
Aspull
Bank Hill
Whalley
Bradshaw Hall
Pennington Green
West Houghton
Lostock
Heaton
Urmstones
Rumworth
Chew Moor Gate
Chapel Moor
Binfield
Ince
Westwood Ho.
Hawk Hall
Low Green
Hindley
Green Field
Over Hulton
Hulton Hall
Hagfold
Daisey Hillock
Bamfurlong Hall
Park Lane
Back Lane
Hindley Green
Fiddler Green
Atherton
Tyldesley
Colliery
Ashton
New Hall
Abram
Edge Green
Golborne
West Leigh
LEIGH
Meanleys
Ashton Colliery
Haydock
Haydock Lodge
Green
Lowton
Lowton Hall
Astley
Newton Common
Crow Lane
Golbone Park
Old Hall
Pennington
Morleys Hall
Astley Green
NEWTON
Viaduct
Park Side
Kenyon Stocks
Hall
Hurst Green
Burtonwood
Culcheth
Southworth cum Croft
Wigshaw Green
Radcliffe Farm
Moss Ho.
Chat Mos
Far Hall
Park
Houghton Middleton
Risley
Tuft Hill
Dalhum
Holme
Cross Hillock
Glazebrook Moss
Irlam Hall
Great Sankey
Arbury
Woolden Hall
New Hall
Sankey Green
Orford House
Poulton with Fearnhead
Risley Moss
Glazebrook
The Wash
Little Sankey
Partington
Hall
Woolstone with Martinscroft
Green
Rixton
PRI.L STATION
WARRINGTON
Warburton
Mersey
Old Hall
River
Latchford
Broadbents
Thelwall
Moss Brow
Canal
Walton Inferior
Aqueduct
Statham
Rush Green
Dunham

Little Woolton
Much Woolton
Allerton
Woolton Hall
Rainhill Stoops
Henshaws
Cronton
Bold Heath
Great Sankey
Dalham
Houghton Middleton
Arbury
Risley
Tarbock
Hough Green
Mossbrook
Farnworth
Sankey Green
Orford House
Poulton with Fearnhead
Risley Moss
Garston
Ditton
Penketh Lodge
Little Sankey
Speke
Woodend Hill
Old Hut
Halewood
Ditton Grove
Widnes
Guerdley
Fidlers Ferry
Hall
The Wash
Woolstone with Martinscroft
Speke Hall
Dungeon Salt Works
Widness & Ditton Marsh
New Cut
Salt Marsh
PRI. STATION
WARRINGTON
Oglet
Hale Hall
Hale
Hale Gate
West Bank
Runcorn Gap
Mersey & Irwell Canal
Latchford
Broadheath
Thelwall
River
Statham
Duke of Bridgewater's Canal Basin
Lower Runcorn
INT. STAT.
Moore
Walton Inferior
Walton Superior
Aqueduct
Grappenhall
West Point
Higher Runcorn
Norton Priory
Halton
The Elms
Daresbury
Appleton Cross
Grappenhall Heys
Houghshots
Point
Stanlow
Weston
Norton
Hatton
Stretton
Bradley Hall
Clifton or Rocksavage
Stockham
Newton by Daresbury
Appleton
Swineyard Lane
Ince Pier
Ferry House
Preston Brook
Preston on the Hill
Holme House
Ince
Ince & Frodsham Marshes
FRODSHAM
Sutton
Whitley Reed
Crowley
Ruley Moors
Higher Whitley
Northwood
Elton
Helsby Marsh
Netherton
Overton
Aston Hall
Middleton Grange
Dutton
Thornton le Moors
Helsby
Five Crosses
Bradley
Lower Whitley
Antrobus
Hall
Green
Hapsford
Seven Oaks
Grandsires Green
Arley Park
Woodhouse
Wimbolds Trafford
Dunham on the Hill
Newton
Kingsley
The Pole
Belmont
Heath
Alvanley
Newton Gutter
Crow Wood Common
Cogshall Hall
Great Budworth
Aston
Bridge Trafford
Manley Hall
Kingsley Common
Little Leigh
Budworth Mere
Marbury
Marston
Pickmere
Mill Green
Plemondstall
Kingswoods
Norley
Crowton
Smithy Bank
Acton Mill
Barnton
Anderton
Salt Works
Mickle Trafford
Mouldsworth
Onston
Weaverham
Shawls Green
Little Barrow
Wimmington Hall
NORTHWICH
Guilden Sutton
Peel Hall
Cuddington
Delamere House
Wallerscoat
Witton cum Twambrook
Gr. Barrow
Ashton
Ashton Heys
Gorstage
Sandiway Vale
Castle Northwich
HARFORD
Lostock Graham
Sandiway
Shurlach cum Bradford
Rudheath
Kelsall
Eddisbury
Grab Tree Green
PRI. STATION
Leftwich
Tarvin
Holme Street
Birches
Plough Lane
Delamere
New Park
Davenham
Shipbrook
Pryor Heys
Oakmere
Eaton Hall
Hockenhall
Lach Dennis
Cotton Abbots
Bruen Stapleford
Tarley Gate
Newchurch Common
Marton
Newhall
Vale Royal
Newchurch or Whitegate
Foulk Stapleford
Willington Hall
Race Course
Whatcroft Hall
Stublach
Quarry Bank
Budworth Common
Moulton
Bostock
Waverton
Duddon Green
Marton Hall
Gravel Green
Moulton Hall
Chester Lane
Bradford Wood
Bostock Hall
Byley cum Yatehouse
Burton
Utkinton Lodge
White Hall
The Oaklands
Wharton
Utkinton
Little Budworth
Over
Hargrave
Iddinshall
Oulton Hall
Winsford Br.
Croxton
Ravenscroft
Huxley Hall
Hoofield
Huxley
Rushton
Eaton
Stanthorn Br.
INT. STAT.
Spurston
Golborn Bellow
TARPORLEY
Eaton Bank
Darley Hall
Weaver
MIDDLEWICH
Newton
Brassey Green
Oulton Low
Manor Hall
Kinderton cum Hulme
Tattenhall
Canal
Darnhall
Lea Hall
Sutton
Tilstone Lodge
Wettenhall Hall
Newton
Beeston Castle
Tilstone Fearnall
Ashbrook
Twelve Acre
Occleston
Peor Ho.
Tattenhall Woods
Beeston
Alpraham
Wettenhall
Lea Green
Stocks
The Old Hough
Tetton
Chowley
Common
Minshall Hall
INT. STAT.
Burwardsley
Barretts Green
Calveley Hall
Church Minshull
Park Hall
Sparrow Grove
Bolesworth Castle
Heath
Bunbury
Moat Fm.
Rectory
Moston
Harthill
Park
Peckforton
Spurstow
Cholmondeston
Hall
Warmingham
Wardle Bank
Bickerton Hill
Sparrow Ho.
Middlewich
Bradfield Green
Springfield
Minshull Moss
Stonyshaw Fm.
Broxton
Ridley Hall
Haughton
Wardle
Aston juxta Mondrum
Elton
Abbeyfield
Bickerton
Bulkeley
Radmore Green
Worleston
Trent

MANCHESTER
Flixton
Urmstone
Stretford
Hulme
Moss Side
Bradford
Openshaw
Gorton
Droylesden
ASHTON under Line
Partington
Carrington
Carrington Moss
Ashton upon Mersey
Warburton
Sale
Chorlton
Levenshulme
Denton
Hyde
Moss Brow
Sinderland
Dunham Woodhouse
Broad Heath
Timperley
Baguley
Northerden
Didsbury
Withington
Heaton Norris
Reddish
ALTRINCHAM
Bowdon
Hale
Cheadle
STOCKPORT
Bredbury
Offerton
Marple
Millington
Rosthern
Tatton Hall
Ashley Hall
Ringway
Bramhall
Poynton
NETHER KNUTSFORD
Mobberley
Wilmslow
Lindow Common
Woodford
Tabley Hall
Over Knutsford
Marthall
Chorley
Great Warford
Lyme Hall
Nether Alderley
Over Alderley
Prestbury
Bollington
Rainow
Capesthorne Hall
Allostock
Goostrey
Siddington
MACCLESFIELD
Macclesfield Common
Macclesfield Forest
Forest Chapel
Sutton
Langley
Gawsworth Hall
Marton
Swettenham
Brereton cum Smethwick
North Rode
Eaton
CONGLETON
Dane Bank
Bug Lawton
Allgreave
Dunkirk
Wincle
Sandbach
Dubthorne
MANCHESTER AND BIRMINGHAM RAILWAY
MACCLESFIELD BRANCH
BRANCH
Macclesfield Canal

Title	Map of the Newcastle & Carlisle Railway, Shewing its Junction with the Carlisle and Maryport Railway, and thus Uniting the German Ocean with the Irish Sea.
From	*Scott's Railway Companion*
Publisher	H Scott
Date	1837

As Victorians increasingly took to exploring hitherto unknown countryside opened up by the rail network, so the rail companies sought to further expand this market with maps and guides. The Newcastle–Carlisle railway crosses attractive landscape, running close to Hadrian's Wall over its western half – and the railway company was clearly well aware of tourism opportunities, as this Railway Companion was published a year before the line was fully opened.

The first suggestion for a new link between the two towns, made in 1794, had been for a broad canal to Hexham on the south bank of the Tyne. Parliamentary opposition and the Napoleonic Wars delayed progress, and it was not until 1825 that a prospectus for the Newcastle & Carlisle Railway (NCR) was issued. The NCR was, however, a true pioneer, being incorporated in May 1829, five months before the famous Rainhill Trials for the Liverpool & Manchester Railway. The Act authorizing the railway had one serious flaw, namely that it forbade the use of locomotives, and an objecting landowner brought a temporary halt to proceedings – but this was soon overcome by an amendment to the original Act.

The map is part of a 105-page booklet which gives not just a route description and short guides to Carlisle and Newcastle,

but also fares, mileages and even details of inns near the intermediate stations – and informs the reader that the estimated total cost of the line was £496,731. Unusually, the geology of the area is identified on the map – notably coal, limestone and red sandstone.

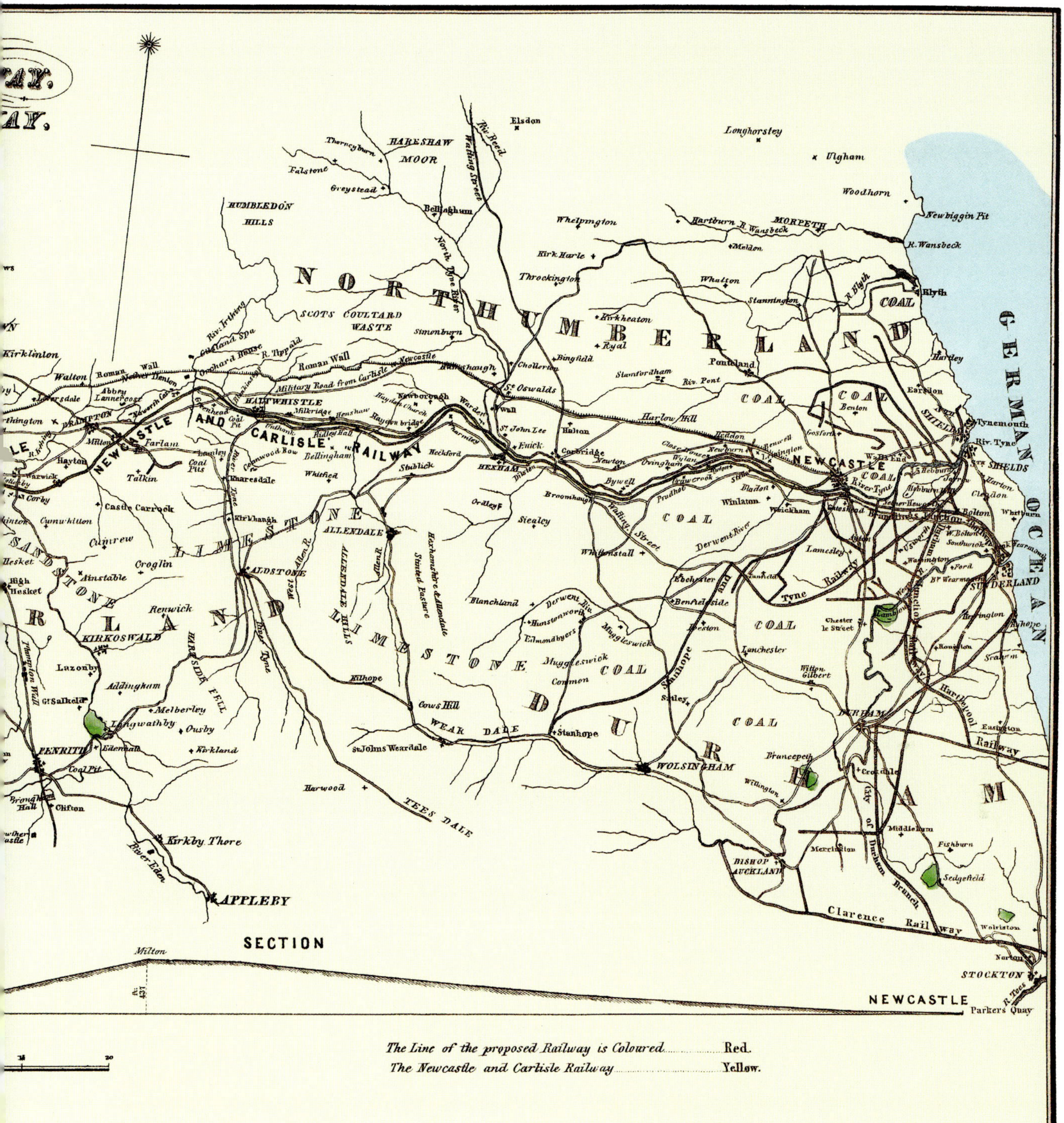

Demonstrating that the Victorians had pride in all kinds of development beyond the stereotypical tourist interest, the lineside description starting out from Carlisle comments: 'The first object which will be presented to the notice of the traveller as the trains begin to start, is the cotton factory of Messrs Rothwell & Co. ...' A passenger train operating company today would be most unlikely to draw attention in its 'windowgazer' brochure to, say, an aluminium smelter or paper mill, even if they were still major customers of the railway.

In partnership with his London agent, William Adams, Lancashire cartographer and Quaker George Bradshaw published the world's first railway timetable in 1839. Sold with a cloth cover, *Bradshaw's Railway Time Tables and Assistant to Railway Travelling* (the title was shortened to *Bradshaw's Railway Companion* in 1840) was initially a fairly small publication of eight pages, costing 6d, but as new railways quickly opened it grew in size until it had reached a staggering 1,000 pages by the end of the century. From 1841 it was published monthly under the new title *Bradshaw's Monthly Railway Guide*, compiled from information supplied by the plethora of railway companies, along with a map of Britain's rail system. Each line on the map was cross-referenced with the number of the page on which its timetable appeared. It was soon putting some order into the chaotic railway system, and started to include information on places to visit, hotels and shipping services. It was not an easy guide to use – there were so many companies that it was difficult to standardize the information, and the multiplicity of footnotes and miniscule type that accompanied each timetable added more confusion. However, despite being the butt of Victorian cartoonists and music hall jokes, it became a unique national institution and was often referred to in famous works of fiction such as *The Adventures of Sherlock Holmes* and *Around the World in Eighty Days*. For over a hundred years it was an invaluable tool for the travelling public.

Bradshaw – his name already synonymous with railway timetables – quickly expanded his railway publishing empire by introducing the *Continental Railway Guide* in 1847 (discontinued in 1939) and, later, the *Railway Manual* and the *Railway Shareholders' Guide*, both of which stayed in print until the early 1920s. Way ahead of its time, *Bradshaw's Air Guide* was also published in 1934–9.

Bradshaw did not face any real competition until the 1930s, when the Big Four railway companies started issuing their own timetables – apart from the Great Western Railway's, even these were produced by the publisher of *Bradshaw*

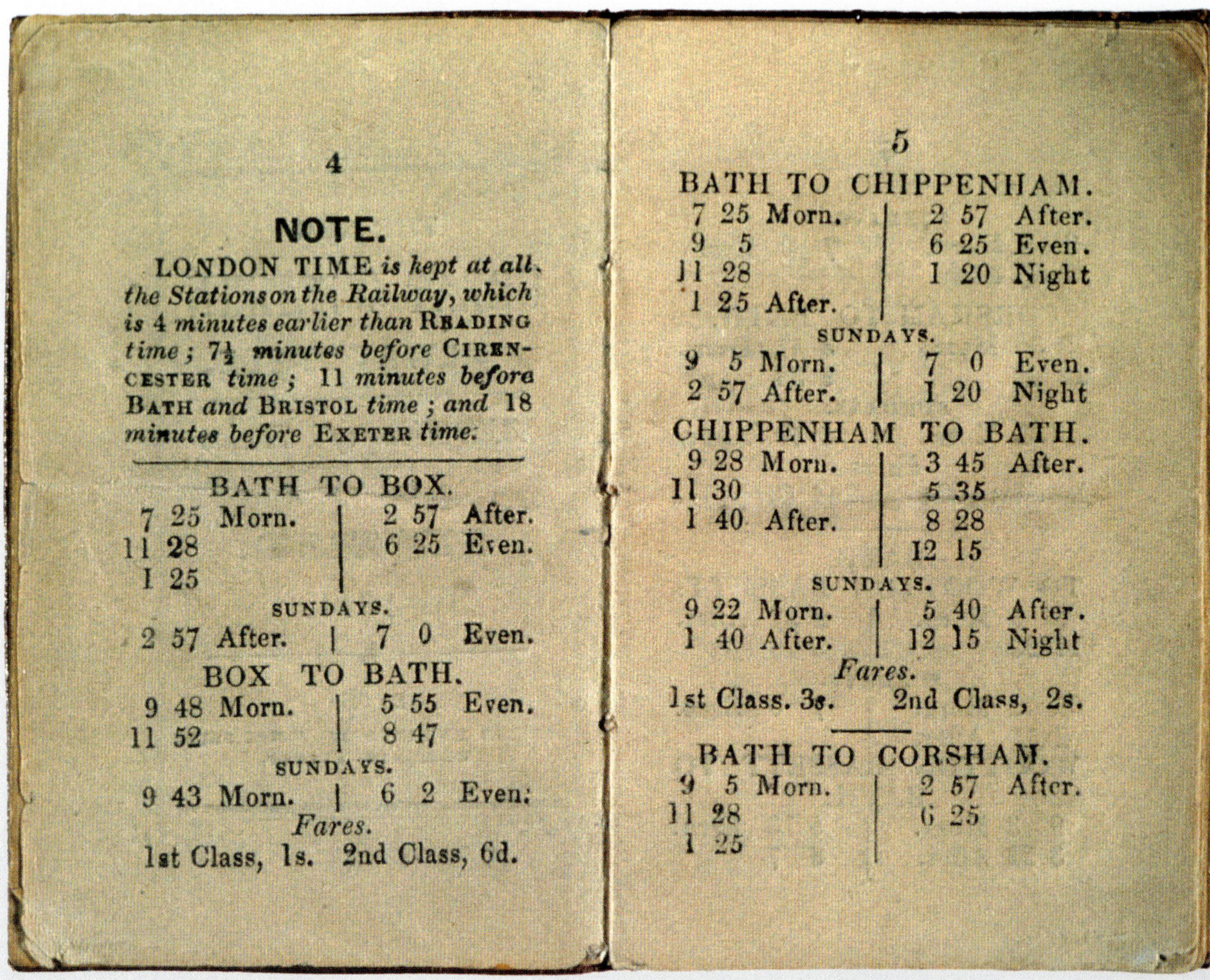

4

NOTE.

LONDON TIME *is kept at all the Stations on the Railway, which is 4 minutes earlier than* READING *time; 7½ minutes before* CIRENCESTER *time; 11 minutes before* BATH *and* BRISTOL *time; and 18 minutes before* EXETER *time.*

BATH TO BOX.

7 25 Morn.	2 57 After.
11 28	6 25 Even.
1 25	

SUNDAYS.

2 57 After.	7 0 Even.

BOX TO BATH.

9 48 Morn.	5 55 Even.
11 52	8 47

SUNDAYS.

9 43 Morn.	6 2 Even.

Fares.

1st Class, 1s. 2nd Class, 6d.

5

BATH TO CHIPPENHAM.

7 25 Morn.	2 57 After.
9 5	6 25 Even.
11 28	1 20 Night
1 25 After.	

SUNDAYS.

9 5 Morn.	7 0 Even.
2 57 After.	1 20 Night

CHIPPENHAM TO BATH.

9 28 Morn.	3 45 After.
11 30	5 35
1 40 After.	8 28
	12 15

SUNDAYS.

9 22 Morn.	5 40 After.
1 40 After.	12 15 Night

Fares.

1st Class. 3s. 2nd Class, 2s.

BATH TO CORSHAM.

9 5 Morn.	2 57 After.
11 28	6 25
1 25	

The problem in running trains to a standardized time is noted in this early timetable for Brunel's Great Western Railway.

until 1939. Despite this competition, and then the availability of regional timetables issued by the nationalized British Railways after 1948, *Bradshaw* was the only all-line timetable and continued to be available until 1961 when the final edition, No. 1521, was published.

It took until 1974 before British Railways first published their own all-line timetable and this was last published (by Network Rail) in 2007. Rail users now access timetables online, or can collect printed leaflets on a company-by-company or route-by-route basis – almost taking us back to the pre-Bradshaw situation.

Complete with map, the first edition of Bradshaw's *Railway Companion* was published in 1840.

GEORGE BRADSHAW

Born in Pendleton, near Salford, in Lancashire, on 29 July 1801, George Bradshaw began his career working for J Beale, a Manchester engraver. After a short spell in Belfast, Bradshaw had become an accomplished cartographer and his first work, a map of his native county, was published in 1827. In 1830 he published a series of map of the canals of Yorkshire and Lancashire known as *Bradshaw's Maps of Inland Navigation.*
His *Bradshaw's Monthly Railway Guide* was published in 1841 and he soon became a household name throughout the land. His printing house, Bradshaw & Blacklock, became internationally renowned for its maps, guides, books and the *Manchester Journal.*

In his private life Bradshaw was a Quaker, a fervent peace activist and philanthropist in his native Manchester. While on a visit to Norway in 1853 he contracted cholera and died. He is buried in the grounds of Oslo cathedral.

MAP & SECTIONS
of the
Railways
OF
GREAT BRITAIN.
Dedicated by Permission
TO JAMES WALKER, F.R.S. L & E.
President of the Institution of Civil Engineers
BY GEORGE BRADSHAW.
January 1839
SECTIONS.

Title	Map & Sections of the Railways of Great Britain
Publisher	George Bradshaw
Date	1839

This classic Bradshaw map was produced in very large format (62 in. by 40 in.) to enable the whole country to be shown, as well as considerable detail of settlements, roads and the relatively few railways which had been built or were under construction by 1839. Even proposed railways were incorporated in the map, including the yellow-marked line from Cambridge north to Lincoln (shown overleaf), which was never built on this alignment. Across the northern half of the map an unknown scribe has roughly sketched by hand (in red) possible additional railways – some of which were built and some which were not. The map is edged by numerous 'sections' or gradient profiles of the principal railways – a form of information which was later deemed not to be of interest to the average passenger, as opposed to railway professionals and enthusiasts.

The Kent, Surrey and Sussex extract below illustrates the very early stages of the development of the dense rail network across what was to become archetypal commuter territory. Just two principal lines south of the Thames are shown – from London to Brighton, and from a junction at Redhill (between London and Brighton) to Folkestone.

By the 1830s Brighton was the most popular seaside resort in Britain, with over 2,000 people a week visiting the town, and the London & Brighton Railway Company was set up to tap this market, with Robert Stephenson advising on the best possible route. The ultimate choice was a direct alignment which involved building four long tunnels and a viaduct across the Ouse valley. In July 1837, Parliament gave permission for the proposed railway; construction started in July 1838, and the first train arrived at Brighton Station on 21 September 1841.

The first railway to the ports of Folkestone and Dover from London was the circuitous route via Redhill (on the London–Brighton line), Tonbridge and Ashford, with Dover finally being reached in 1844. Subsequently, a more direct route for the 'South East Main Line' was built between London and Tonbridge via Sevenoaks, leaving the Tonbridge–Redhill section as a minor but useful cross-country line, which survives to this day.

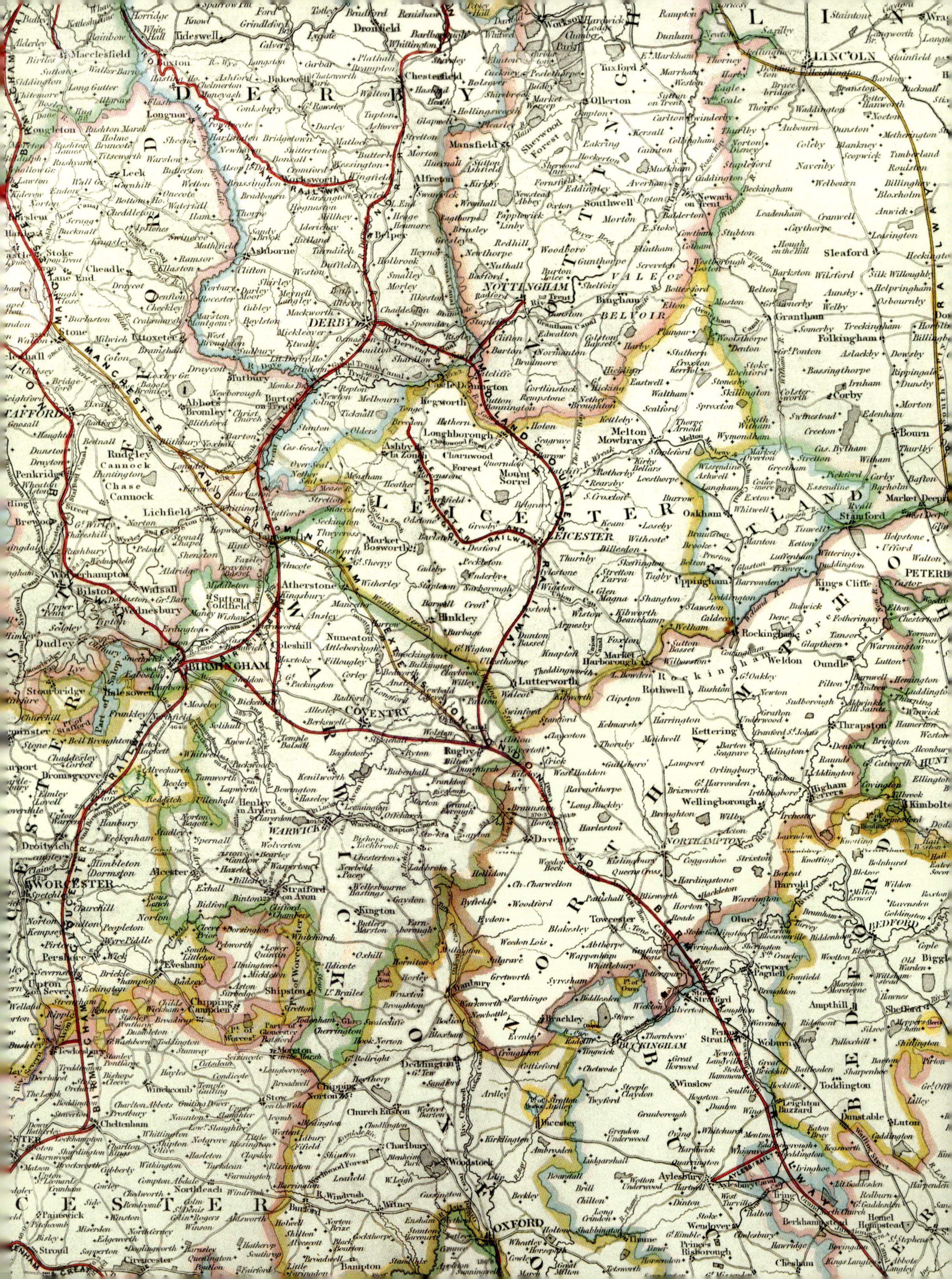

DERBY
NOTTINGHAM
LEICESTER
LINCOLN
BIRMINGHAM
COVENTRY
WARWICK
NORTHAMPTON
OXFORD
WORCESTER
BEDFORD
BUCKINGHAM
RUTLAND
VALE OF BELVOIR
Chesterfield
Mansfield
Newark on Trent
Sleaford
Grantham
Stamford
Oakham
Melton Mowbray
Loughborough
Ashby de la Zouch
Hinkley
Lutterworth
Market Harborough
Rockingham
Kettering
Wellingborough
Daventry
Towcester
Banbury
Stratford on Avon
Evesham
Chipping Campden
Tewkesbury
Cheltenham
Cirencester
Woodstock
Bicester
Aylesbury
Leighton Buzzard
Dunstable
Luton
Wolverhampton
Walsall
Lichfield
Tamworth
Stafford
Stone
Cheadle
Ashborne
Bakewell
Matlock
Belper
Burton on Trent
Uttoxeter
Kenilworth
Rugby
Oundle
Thrapston
Newport Pagnell
Woburn
Olney
Stourbridge
Dudley
Bromsgrove
Droitwich
Pershore
Alcester
Shipston
Chipping Norton
Deddington
Brackley
Winslow
Tring
Wendover
Thame
Witney
Macclesfield
Leek
Congleton
Southwell
Alfreton
Bingham
Atherstone
Nuneaton

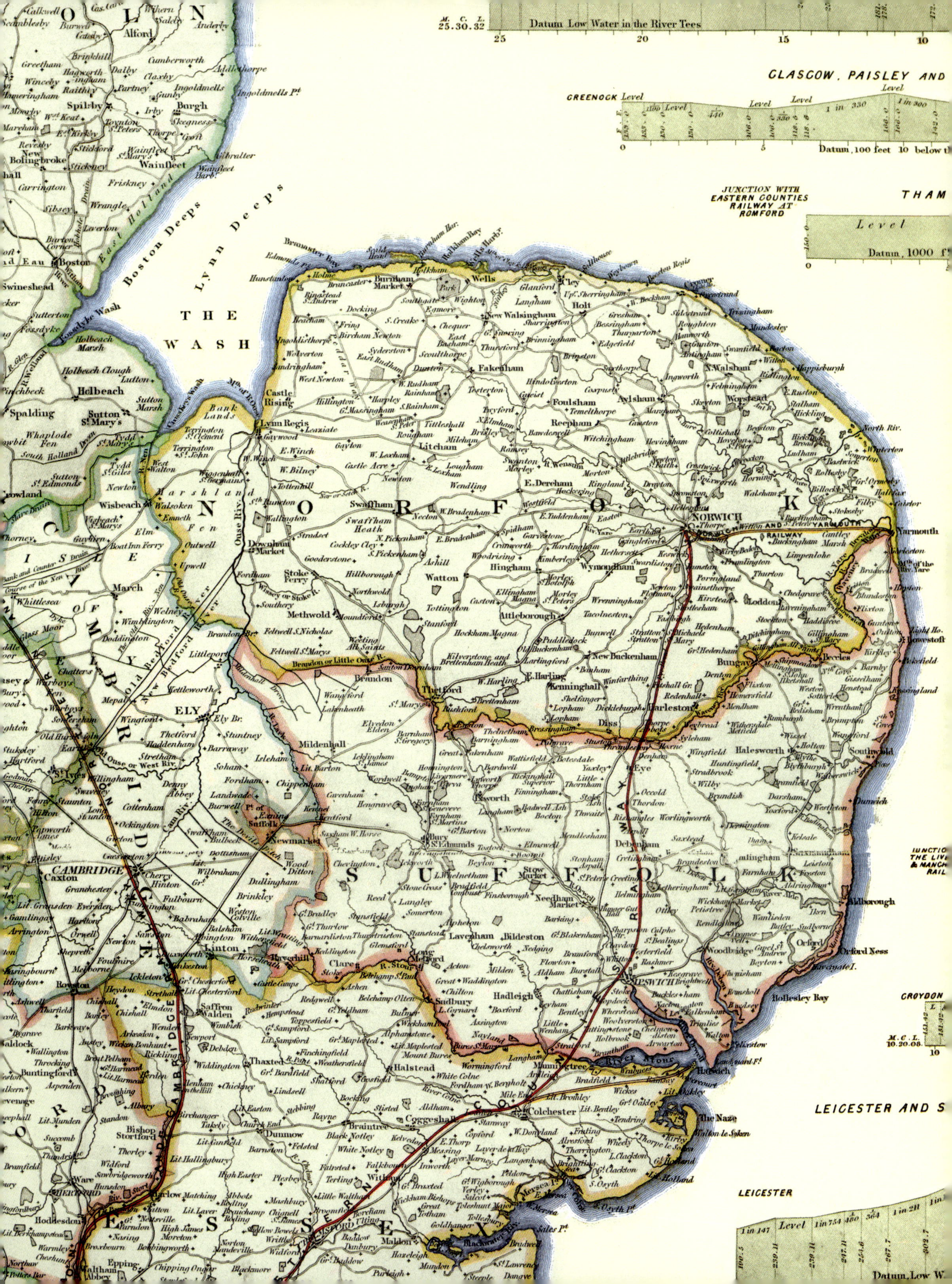

Datum Low Water in the River Tees
M. C. L. 25.30.32
25
20
15
10
GLASGOW, PAISLEY AND
GREENOCK Level
Level
1 in 330
Datum, 100 feet 10 below
JUNCTION WITH EASTERN COUNTIES RAILWAY AT ROMFORD
THAM
Level
Datum, 1000 f
THE WASH
Boston Deeps
Lynn Deeps
NORFOLK
SUFFOLK
CAMBRIDGE
ELY
NORWICH
Yarmouth
Lowestoft
Thetford
Swaffham
Fakenham
East Dereham
Attleborough
Wymondham
Aylsham
North Walsham
Lynn Regis
Downham Market
Wisbeach
Spalding
Holbeach
Boston
Spilsby
Alford
Wainfleet
Newmarket
Bury St. Edmunds
Stowmarket
IPSWICH
Woodbridge
Sudbury
Hadleigh
Haverhill
Colchester
Harwich
Halstead
Coggeshall
Braintree
Witham
Maldon
Chelmsford
Saffron Walden
Royston
Bishop Stortford
Dunmow
Hertford
Ware
Waltham Abbey
Epping
Beccles
Bungay
Halesworth
Southwold
Aldborough
Orford Ness
Hollesley Bay
The Naze
Framlingham
Eye
Diss
Harleston
Mildenhall
Brandon
Soham
Fordham
March
Whittlesea
Caxton
Linton
Thaxted
Loddon
CROYDON
LEICESTER AND S
LEICESTER
Datum Low W.

NEWCASTLE upon TYNE
Gateshead
Swalwell
Whickham
Axwell Park
Nether Heworth
SOUTH SHIELDS
Monkton
Jarrow
Brandling Junction
Ravensworth
Lamesley
Kibblesworth
Hedley
Birtley
Washington
Barmston
CHESTER LE STREET
UNION
North Biddick
Ouston
Urpeth
Pelton
Harraton
Stanley
Lambton Park
Lumley Castle
Great Lumley
Chester le Street
Waldridge
Plawsworth
Kimblesworth
Holmside
Newbottle
Painshaw
Houghton
East Rainton
West Rainton
Great Usworth
Little Usworth
Beamish Hall
South Biddick

Title	Map of the County Palatine of Durham
Publisher	William Colling Hobson
Date	1840

After the year of this map's production, 1840, there is no record of any new county survey maps, only revisions of earlier maps. With public funding and mass production, the Ordnance Survey had become pre-eminent, so Hobson's map of County Durham represents one of the last of the era. The publisher may have seen the writing on the wall, as the language of the full title is profusely ingratiating to his funding base:

> *This Map of the County Palatine of Durham is most respectfully dedicated to the Nobility, Clergy, Gentry, &c. &c. By their most obliged and very humble Servant, the Proprietor, William Colling Hobson*

Unlike Greenwood's 1820 map of County Durham (see pages 16-19), rail routes are marked distinctly, with conventional railways in red and waggonways in ticked grey lines. The Washington and Newbottle waggonways shown in 1820 are identified, together with others not previously evident, linking collieries with the rivers Tyne and Wear and direct to the North Sea coast at Seaham Harbour.

The burgeoning railway network has spread in a different pattern from the waggonways – new economic forces were encouraging links between cities, towns and ports, although coal was still very much a driving force.

This satirical sketch attacks the speculative investment fever of the 'Railway Mania' that reached its peak in 1846.

Following the financial ruin that had befallen speculating investors in the South Sea Company, otherwise known as the 'South Sea Bubble', the British Government introduced the Royal Exchange and London Assurance Corporation Act (commonly known as the 'Bubble Act') in 1719. The Act was designed to prevent similar speculation and fraudulent trading in worthless business ventures, limiting their formation and requiring all new joint-stock companies to be incorporated by Act of Parliament. In its wisdom Parliament repealed the 'Bubble Act' in 1825.

The opening of the Liverpool & Manchester Railway in 1830 and its subsequent success in transporting goods and people between the two cities soon led to proposals for other such railways around Britain. To start with, progress was slow. A combination of banks not lending money due to political and social unrest, coupled with a sluggish British economy,

Opened in 1839, York's first railway station was deliberately sited within the city walls by George Hudson, the chief promoter of the York & North Midland Railway (see page 52).

where the high interest rates offered by Government bonds sucked in the majority of investments, put the brakes on railway development.

By the early 1840s the British economy was bouncing back, interest rates were cut and manufacturing industry was experiencing boom times – the period of the Victorian entrepreneur had dawned. The consequent success of the few railways that had already been built, such as the Grand Junction, the Great Western and the London & Birmingham, soon attracted investors who had previously put their money into Government bonds.

Rapidly rising railway share prices soon saw a speculative frenzy in a plethora of overambitious and sometimes fraudulent proposals for new railways across Britain. Known as 'Railway Mania' this period reached its peak in 1846 when 272 Acts of Parliament were passed, authorizing the building of 9,500 miles of new railway. This level of authorization was hardly surprising as many MPs, such as the 'Railway King' George Hudson, were also directors of these fledgling railways. Seeing the chance of a quick profit, the general public were also sucked in, purchasing shares for only a 10 per cent deposit, with the railway company having the right to call in the remainder when they pleased. Sadly, many of these prospective schemes soon collapsed, leaving the investors having to stump up the balance for an already worthless investment – the bubble had burst and, not for the last time, the British middle classes had become the victims. By the late 1840s and early 1850s new railway construction was mainly limited to the established railway companies who had profited by buying failed companies for a fraction of their original value.

GEORGE HUDSON, THE 'RAILWAY KING'

The son of a Yorkshire farmer George Hudson was born in 1800 and by the age of 27 had become a director in a firm of drapers in York. He had the good luck to inherit a fortune from his great uncle in 1827 and went on to establish the York Union Banking Company. Hudson was elected to York City Council in 1835, becoming Lord Mayor two years later.

Forming a business relationship with George Stephenson, Hudson promoted the York & North Midland Railway which was authorized in 1837. He was also a major shareholder in three other railways – the Midland Counties Railway, the North Midland Railway and the Birmingham & Derby Junction Railway – that were to form the mighty Midland Railway, also led by Hudson, in 1844. To the north of York, Hudson's York & Newcastle Railway was also nearing completion and by 1846, the peak of 'Railway Mania', he controlled a 1,000-mile railway empire and became known throughout the land as the 'Railway King'. By now a personal friend of the Duke of Wellington and not content with his railway empire-building, Hudson was also elected MP for Sunderland in 1845, a position he held, despite his eventual fall from power, until 1859.

However, behind all of this success was a world of fraud and corruption on a big scale. Hudson's world started to fall apart after the bursting of the 'Railway Mania' bubble at the end of 1847 when he was forced to retire as chairman of the railway companies he owned. A committee of investigation then found that the unscrupulous Hudson had bribed MPs to vote for his railway projects, manipulated share prices by using insider information, sold land he did not own and paid dividends out of capital. Admitting his guilt Hudson was eventually imprisoned for debt in York Castle in 1865 but was soon released after friends raised the necessary funds. He died a broken man in 1871.

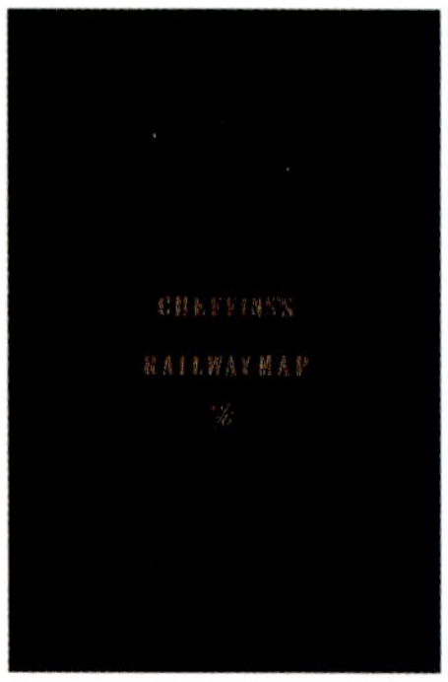

Title	Cheffins's Map of the English & Scotch Railways
Publisher	Charles F Cheffins
Date	1848

The geographical extent of the railway mania is graphically illustrated in Cheffins's 14 miles to 1 in. map dating from 1848. Other than south of the Great Western Railway main line to Bristol, and in Devon and Cornwall, there were as many lines planned (in red) as there were actually in operation or under construction (in black). The linkage between Victorian railways, coal and industry is demonstrated by the dense network of lines all the way from the South Midlands to Lancashire and Yorkshire – in contrast, the swathe of southeast England, long regarded as classic rail commuter territory, had a relatively sparse network at that time. A notable feature, shown overleaf, is the distinctively marked Great Western broad gauge network radiating from London (Paddington) to Basingstoke, Hungerford, Oxford, Cirencester, Gloucester, Westbury, Bath, Bristol and Plymouth.

This map extract of North Yorkshire particularly illustrates the fact that many lines which secured Parliamentary approval, in this case during 1846 or 1847, were never actually built. Thirsk on the Great North of England Railway – presumably an inspiration for the GNER of the 1990s' privatized railway – would have become a more important junction on the East Coast Main Line had the lines to Leyburn and Malton ever been constructed. Equally, Richmond to Barnard Castle, York to Wetherby, and Lockington to Hornsea were all railways which never saw the light of day. The small coastal town of Whitby, however, was eventually to be served by three lines, although at this stage only the Whitby & Pickering Railway (whose construction had begun from the coastal terminus, in complete isolation from the rest of the network) provided a through connection to the rest of the country. Much of this line is preserved to this day as the steam and heritage North Yorkshire Moors Railway.

NOTE.
Lines completed & open for Traffic
Lines in course of Construction
Lines for which Acts were obtained in 1846
D^o. D^o. D^o. in 1847
Broad Guage Lines
SHREWSBURY
Wellington
Newport
Madeley Market
Broseley
Much Wenlock
Easthope
Bridgenorth
Wolverhampton
Walsall
Wednesbury
Bilston
Dudley
Stourbridge
Halesowen
Kidderminster
Bewdley
Stourport
LICHFIELD
BIRMINGHAM
Solihull
Bromsgrove
Droitwich
Beoley
Henley in Arden
Alcester
Bradley
WORCESTER
Upton Snodbury
Norton
Evesham
Pershore
Powick
Upton
Bredon
Tewkesbury
Chipping Campden
Moreton
Winchcombe
CHELTENHAM
Stow in the Wold
GLOCESTER
Painswick
Northleach
Burford
Bibury
Cirencester
Fairford
Lechlade
Cricklade
Highworth
Swindon
Malmesbury
Tetbury
Dursley
Berkeley
Wotton under Edge
Chipping Sodbury
Badminton
Marshfield
Chippenham
Calne
Marlborough
Hungerford
BATH
BRISTOL
Keynsham
Brislington
Clevedon
Portishead
Devizes
Trowbridge
Westbury
Warminster
Frome
WELLS
Shepton Mallet
Glastonbury
Bridgewater
Taunton
Wellington
Milverton
Wiveliscombe
Dulverton
Bampton
Tiverton
Collumpton
Crediton
EXETER
Honiton
Axminster
Lyme Regis
Bridport
Chard
Ilminster
Crewkerne
Yeovil
Sherborne
Ilchester
Langport
Somerton
Castle Cary
Wincanton
Bruton
Mere
Shaftesbury
Hindon
Heytesbury
Wilton
SALISBURY
Amesbury
Romsey
SOUTHAMPTON
Ringwood
Wimborne Minster
Blandford
Cerne Abbas
Beaminster
Stalbridge
Cranborne
Lydiard
Ludgershall
Everley
Tilshead
Chitern All Saints
Burbage
Hereford
HEREFORD
Ledbury
Newent
Ross
Monmouth
Abergavenny
Crickhowell
Brecon
Builth
Llandovery
Hay
Kington
Leominster
Ludlow
Tenbury
Bromyard
Knighton
Presteign
N. Radnor
Rhayader
Llanidloes
Machynlleth
Welch Pool
Llanfair
Bishops Castle
Church Stretton
Munslow
Aston
Cleobury Mortimer
Bromfield
Wigmore
Pembridge
Kimbolton
Stretton
Madley
Longtown
Goodrich
Coleford
Blakeney
Newnham
Mitchell Dean
Chepstow
Usk
Pontypool
Newport
Caerleon
Risca
Caerphilly
Merthyr Tydvil
Aberdare
Neath
Swansea
Swansea Bay
Llandilo Vawr
Llangadock
Trecastle
Llandaff
Cardiff
Cowbridge
Llantrissant
Bridgend
Penarth
Barry Is.
St. Nicholas
Michaelstone
Llanelly Rail.
Llandebie
Dunster
Watchet
Nether Stowey
Minehead
Bridgewater Bay
Yannery
Martinhoe
BRISTOL CHANNEL
Bristol 270 Miles
Cardiff to Bristol
GREAT WESTERN RAILWAY
BRISTOL & EXETER RAILWAY
BRISTOL & GLOUCESTER RAILWAY
CHELTENHAM & GREAT WESTERN UNION RAILY
BIRMINGHAM & GLOUCESTER RAILY
TAFF VALE RAILWAY
Severn R.
Wye R.
Usk R.
Avon R.
Teme R.

Loughborough
LEICESTER
Melton Mowbray
Oakham
Stamford
PETERBOROUGH
Spalding
Boston
Holbeach
Wisbeach
LYNN REGIS
Hunstanton
Burnham Westgate
New Walsingham
Snettisham
Downham Market
Swaffham
East Dereham
Market Harborough
Rugby
Daventry
NORTHAMPTON
Kettering
Wellingborough
Thrapston
Oundle
Huntingdon
St Ives
March
Ely
Thetford
Mildenhall
Newmarket
CAMBRIDGE
Bury St
Stow Market
Needham
Bedford
Olney
Newport Pagnell
Banbury
Buckingham
Stony Stratford
Wolverton
Leighton Buzzard
Dunstable
Luton
Hitchin
Stevenage
Biggleswade
Royston
Saffron Walden
Newport
Haverhill
Lavenham
Sudbury
Hadleigh
Halstead
Braintree
Coggeshall
COLCHESTER
Manningtree
Bicester
OXFORD
Aylesbury
Tring
Hemel Hempstead
St Albans
Hatfield
Hertford
Ware
Bishop Stortford
Dunmow
Witham
Maldon
Chelmsford
Abingdon
Wallingford
High Wycombe
Great Marlow
Rickmansworth
WATFORD
Barnet
Epping
Chipping Ongar
Romford
Brentwood
Billericay
Rochford
Southend
Reading
Windsor
Henley on Thames
Maidenhead
Uxbridge
Hampstead
LONDON
Greenwich
Woolwich
Gravesend
ROCHESTER
Chatham
Sheerness
Sheppy
Wokingham
Egham
Twickenham
Kingston
Sutton
Epsom
Croydon
Dartford
Bromley
Basingstoke
Farnham
Guildford
Dorking
Reigate
Godstone
Westerham
Seven Oaks
Maidstone
Sittingbourne
Faversham
CANTERBURY
Tunbridge
Tunbridge Wells
Ashford
Folkstone
Hythe
New Romney
Rye
Tenterden
Horsham
Crawley
East Grinstead
Cuckfield
Petersfield
Midhurst
Havant
Steyning
Alton
Alresford
Winchfield
Godalming
Haslemere
THE WASH
RIVER THAMES
SOUTH EASTERN RAILWAY
LONDON & BRIGHTON RAILWAY
SOUTH WESTERN RAILWAY
EASTERN COUNTIES RAILWAY
LONDON & NORTH WESTERN RAILWAY
MIDLAND
GREAT WESTERN RAILWAY

ENTERED AT STATIONERS HALL
6d PLAIN
REID'S
1s COLD
"RAILWAY RIDE"
FROM
LONDON to EDINBURGH
(On the London & North Western, Sn Midland, Nn Midland, York & Nn Midland, York & Newcastle, Ncastle & Berwick & Nn British Railways)
SHEWING THE TOWNS, VILLAGES, ROADS & DISTANCE OF EACH STATION FROM LONDON.
Engraved, Printed & Published by
ANDREW REID, 117, PILGRIM STREET, NEWCASTLE UPON TYNE.
TOTAL LENGTH 430 MILES

196
Jackling
NORMANTON STATION
WAKEFIELD
OKENSHAW STATION
192½ From Manchester
Grofton
Griggleston
Walton Park
ROYSTON & NOTTON STATION
188¾ Wolley
Felkirk
Royston
Carlton
Darton
Monk Burton
185½
BARNESLEY STATION
BARNESLEY
Ardsley
Billingley
181¼
DARFIELD STATION
Bolton upon Dearne
DONCASTER
Holyland
179¼ Tankersley
WATH STATION
Wentworth
Swinton
Canisborough
Hooton Roberts
From Manchester
Ecclesfield
Thribergh
MASBRO STATION
Brightside
ROTHERHAM
Wickersley
Griemsthorpe
Darnall
Whiston
SHEFFIELD
Treeton
Handsworth
Aughton
167¾
WOODHOUSE MILL STATION
Aston Netherthorpe
Hackenthorpe
Birley
Waterthorpe
Killamarsh
Eckington
West Thorp
162¾
ECKINGTON STATION
Barlborough Hall
Southgate Ho
160
STAVELEY STATION
Barlborough
Whittington
Staveley
Newbold
Brimmington
156½
CHESTERFIELD
MIDLAND RAILWAY

Title	Reid's 'Railway Ride' from London to Edinburgh
Publisher	Andrew Reid
Date	*c.*1848–9

As the longer-distance main lines spread across the country, Victorian mapmakers introduced 'strip maps' which transformed the representation of rail routes into straighter lines – facilitating reproduction in booklet form for the rail traveller interested in observing and understanding better the scenes through the carriage window.

When this map was produced in 1848–9, there was as yet no direct route between London and York, so an 'East Coast' journey from London to Scotland involved a departure from the London & North Western Railway's Euston terminus to Rugby, then by Leicester, Derby, Chesterfield, Rotherham, Swinton and Normanton to York, where the present East Coast Main Line was joined. While vastly quicker than its predecessor, the stagecoach, the relatively circuitous route followed by the train (430 miles) meant journeys of 16 to 18 hours between the two capitals.

The excerpt opposite from Chesterfield to Normanton is particularly interesting in that it features parts of the route which can no longer be followed in the 21st century. The first section, to Rotherham, which bypassed Sheffield, was superseded in 1870 by a line directly serving the city. The original line remained open as an important freight artery, a function it still fulfils today (see page 265), and as a diversionary route for passenger trains when engineering works are being undertaken on the main line. The section from Swinton to Normanton has now been lost completely as a through route, although short lengths remain open for freight.

This strip map features one of the first known representations of railway stations by solid red circles – a practice that continues to this day, notably in the Ordnance Survey 1:50,000 'Landranger' series.

ENTERED AT STATIONERS HALL

6

6^D PLAIN | 1^S COL^D

REID'S "RAILWAY RIDE" FROM LONDON to EDINBURGH

(On the London & North Western, S^o Midland N^o Midland, York & N^o Midland, York & Newcastle, N'castle & Berwick, & N^o British Railways.)

SHEWING THE TOWNS, VILLAGES, ROADS & DISTANCE OF EACH STATION FROM LONDON.

Engraved, Printed & Published by
ANDREW REID, 117, PILGRIM STREET, NEWCASTLE UPON TYNE.

Continued at the bottom of next Column

TOTAL LENGTH 430 MILES.
Fare at 1^d per Mile £1.15.10.

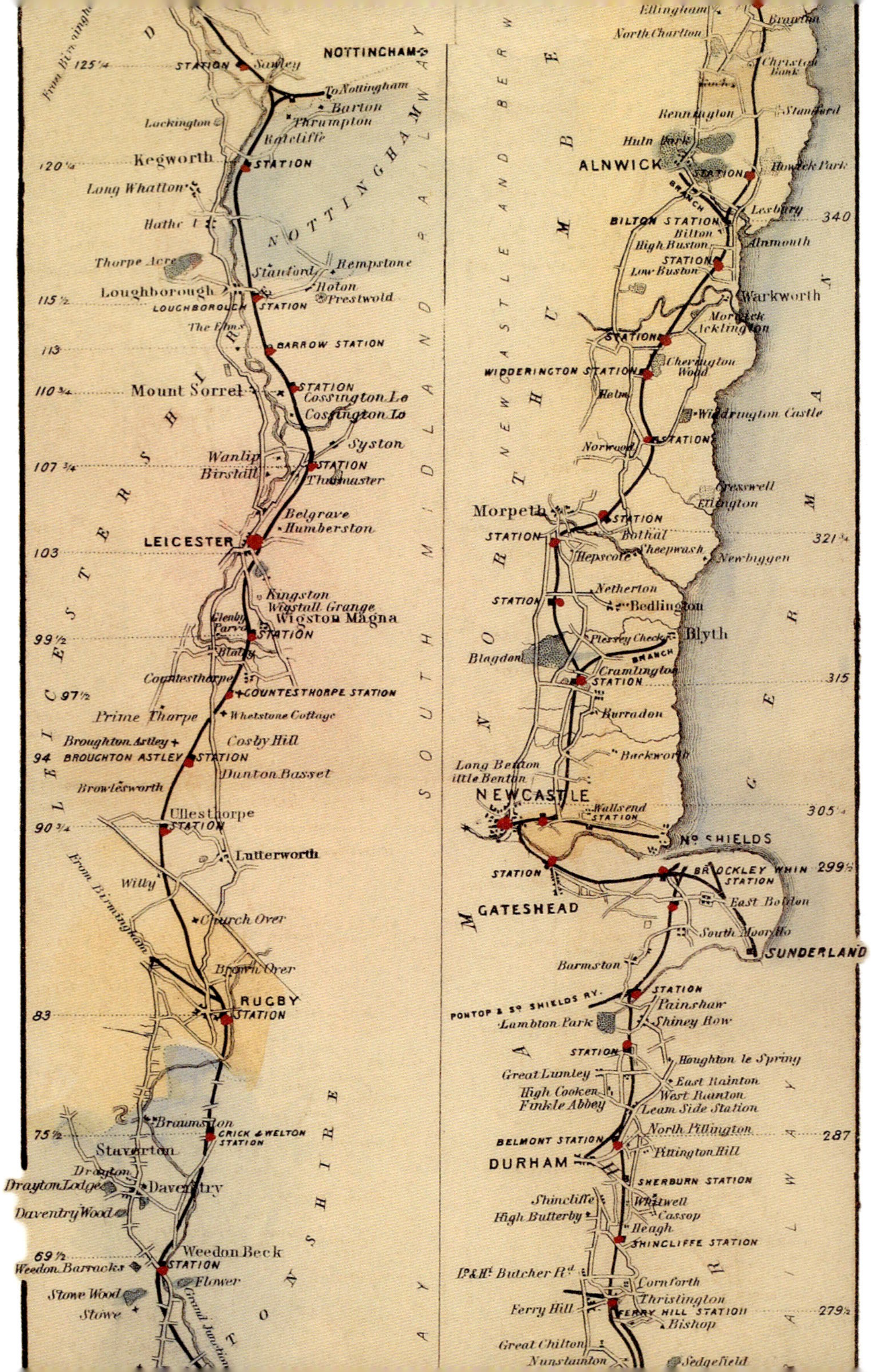

WAKEFIELD
NORMANTON STATION
BARNESLEY
BARNESLEY STATION
ROYSTON & NOTTON STATION
DARFIELD STATION
WATH STATION
DONCASTER
ROTHERHAM
MASBRO STATION
SHEFFIELD
WOODHOUSE MILL STATION
ECKINGTON STATION
STAVELEY STATION
CHESTERFIELD
CLAY CROSS STATION
STRETTON STATION
WINGFIELD STATION
Alfreton
Belper
DUFFIELD STATION
DERBY
SPONDON STATION
BOROWASH STATION
NOTTINGHAM
NORTH MIDLAND RAILWAY
HADDINGTON
LONG KIDDRY STATION
BALLENCRIEF STATION
LINTON STATION
DUNBAR
Cockburnspath
COCKBURNSPATH STATION
GRANTS HOUSE STATION
RESTON STATION
AYTON STATION
BERWICK STATION
TWEEDMOUTH STATION
KELSO BRANCH
BEAL STATION
BELFORD STATION
Belford
NEWHAM STATION
NORTH BRITISH RAILWAY
HADDINGTON
BERWICK
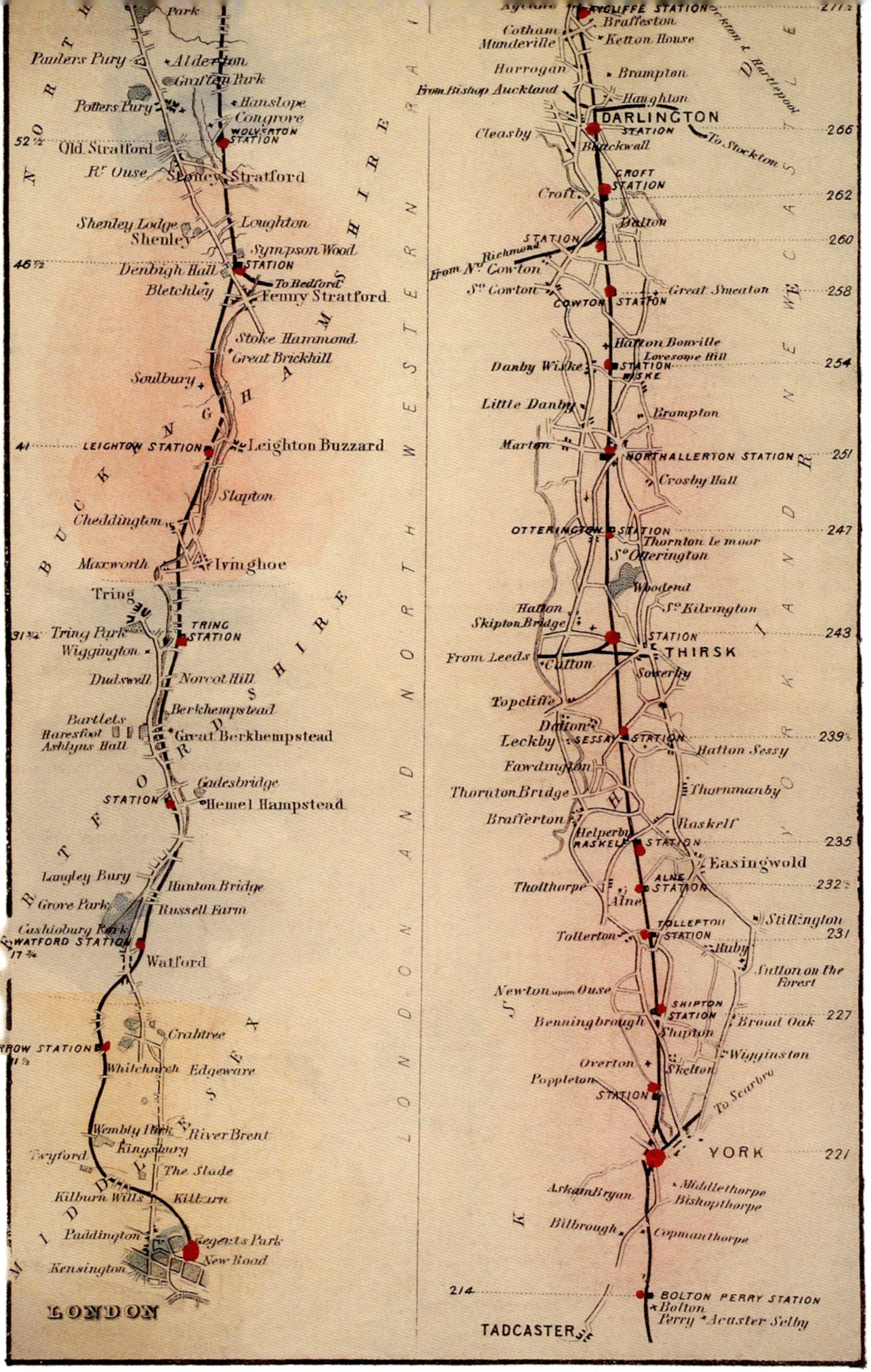
Old Stratford
Stoney Stratford
WOLVERTON STATION
Fenny Stratford
Leighton Buzzard
LEIGHTON STATION
TRING STATION
Tring
Great Berkhampstead
Hemel Hampstead
Watford
WATFORD STATION
HARROW STATION
LONDON
LONDON AND NORTH WESTERN RAILWAY
BUCKINGHAMSHIRE
HERTFORDSHIRE
MIDDLESEX
DARLINGTON STATION
CROFT STATION
COWTON STATION
NORTHALLERTON STATION
OTTERINGTON STATION
THIRSK
SESSAY STATION
RASKELF STATION
Easingwold
ALNE STATION
TOLLERTON STATION
SHIPTON STATION
YORK
BOLTON PERRY STATION
TADCASTER

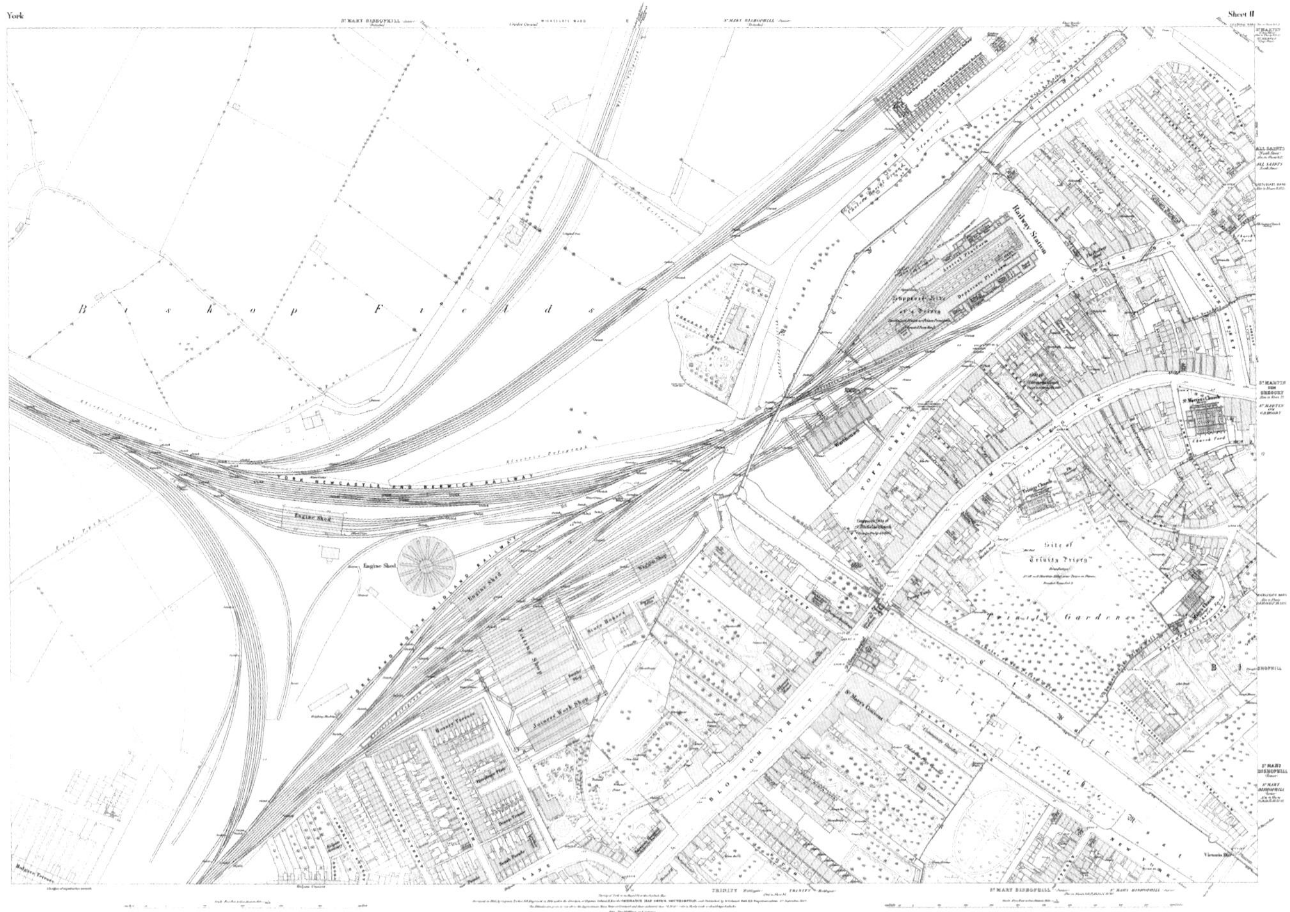

Series	Ordnance Survey 60 in. to 1 Mile
Sheet	York Sheet II
Date	1851

York – which was to become one of Britain's great railway centres – first joined the railway network with the opening of the York & North Midland Railway (YNM) in 1839, the first of the many companies to be chaired by George Hudson, the 'Railway King.' Initially trains terminated at a temporary wooden structure in Queen Street, just outside York's medieval city wall, but the latter was soon breached by a Gothic arch which took the railway into a new permanent station, opened in 1841, on a site previously occupied by the city's House of Correction. Traffic growth led to the enlargement of the station in 1845–6 and the construction of a second arch through the walls.

This early Ordnance Survey map of 60 in. to 1 mile scale shows in fascinating detail the layout of the station and associated railway infrastructure – note in particular the separate Arrival and Departure platforms, and the small turntables used to turn goods wagons through 90 degrees to be positioned in short unloading sidings straddled by warehouses southwest of the passenger station. As shown above, two coal depots stand on the site of the future permanent station just outside the city wall.

The combination of operational inconvenience – trains from London to Newcastle had to reverse out of the station in order to continue their journeys – and restricted space for expansion, led to the construction of the present York station outside the wall, opened in 1877. Railway offices remained within the old station, and carriages were stored on its tracks until 1965 when the station was largely demolished to make way for a new office block for British Rail. The old station frontage can, however, still be seen from Tanner Row, and some ironwork and platform remains have survived to the rear of the building.

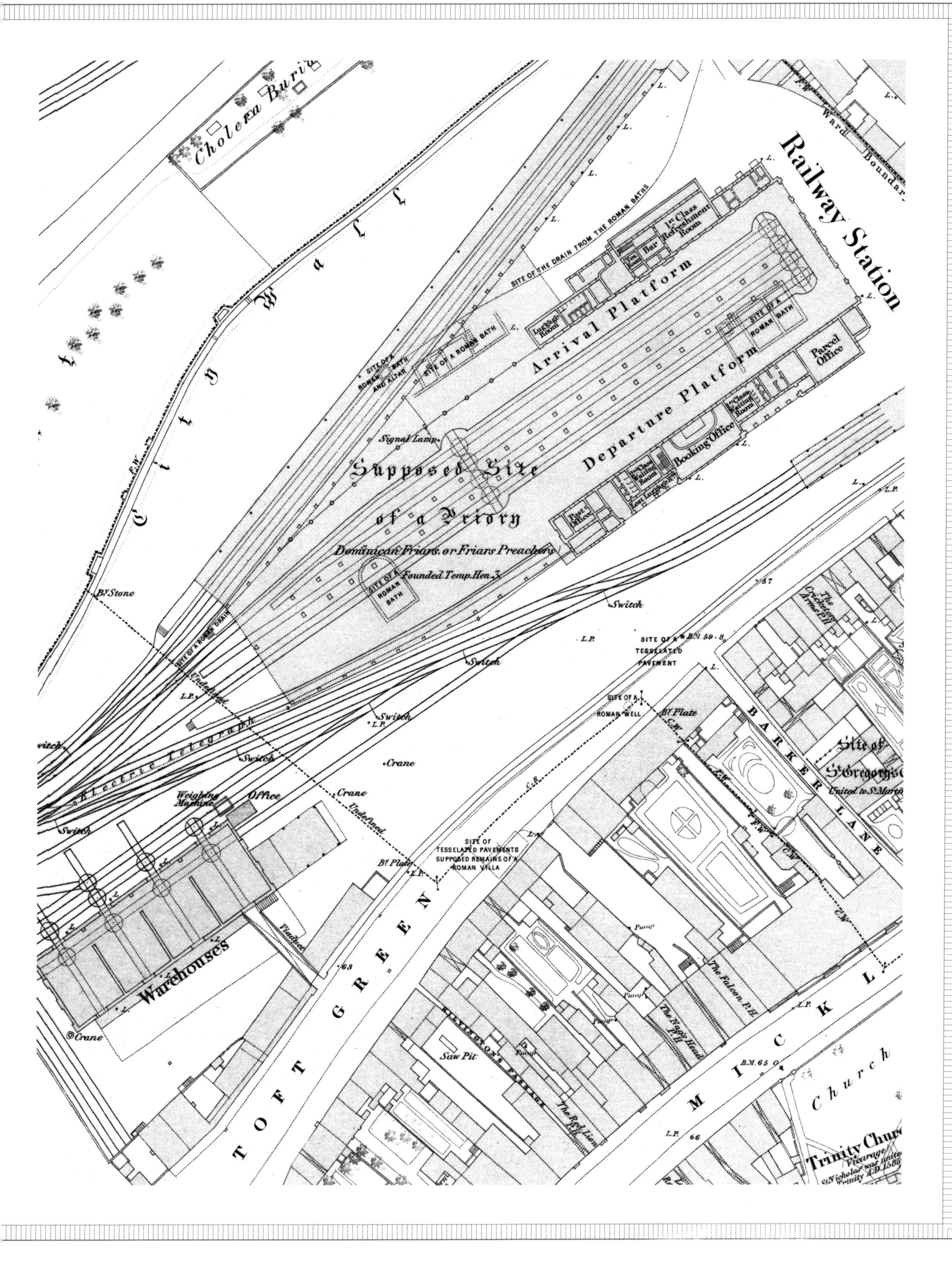
Cholera Burial
City Wall
Railway Station
Ward
Boundary
SITE OF THE DRAIN FROM THE ROMAN BATHS
1st Class Refreshment Room
Bar
Luggage Room
Arrival Platform
SITE OF A ROMAN BATH
SITE OF A ROMAN BATH AND ALTAR
Parcel Office
Departure Platform
1st Class Waiting Room
Booking Office
Signal Lamp
Supposed Site of a Priory
Dominican Friars, or Friars Preachers
Founded Temp. Hen. 3.
Post Office
Lost Luggage
Bry Stone
SITE OF A ROMAN DRAIN
Switch
SITE OF A TESSELATED PAVEMENT
SITE OF A ROMAN WELL
Bry Plate
Electric Telegraph
Undefined
Crane
Weighing Machine
Office
SITE OF TESSELATED PAVEMENTS SUPPOSED REMAINS OF A ROMAN VILLA
Warehouses
Viaduct
TOFT GREEN
BARKER LANE
Site of St. Gregory's Church
United to St. Martin
The Cricketers Arms P.H.
The Falcon P.H.
The Navy Head P.H.
The Red Lion P.H.
Saw Pit
Pump
MICKL
Church
Trinity Church
Vicarage
St. Nicholas was united to Trinity A.D. 1585

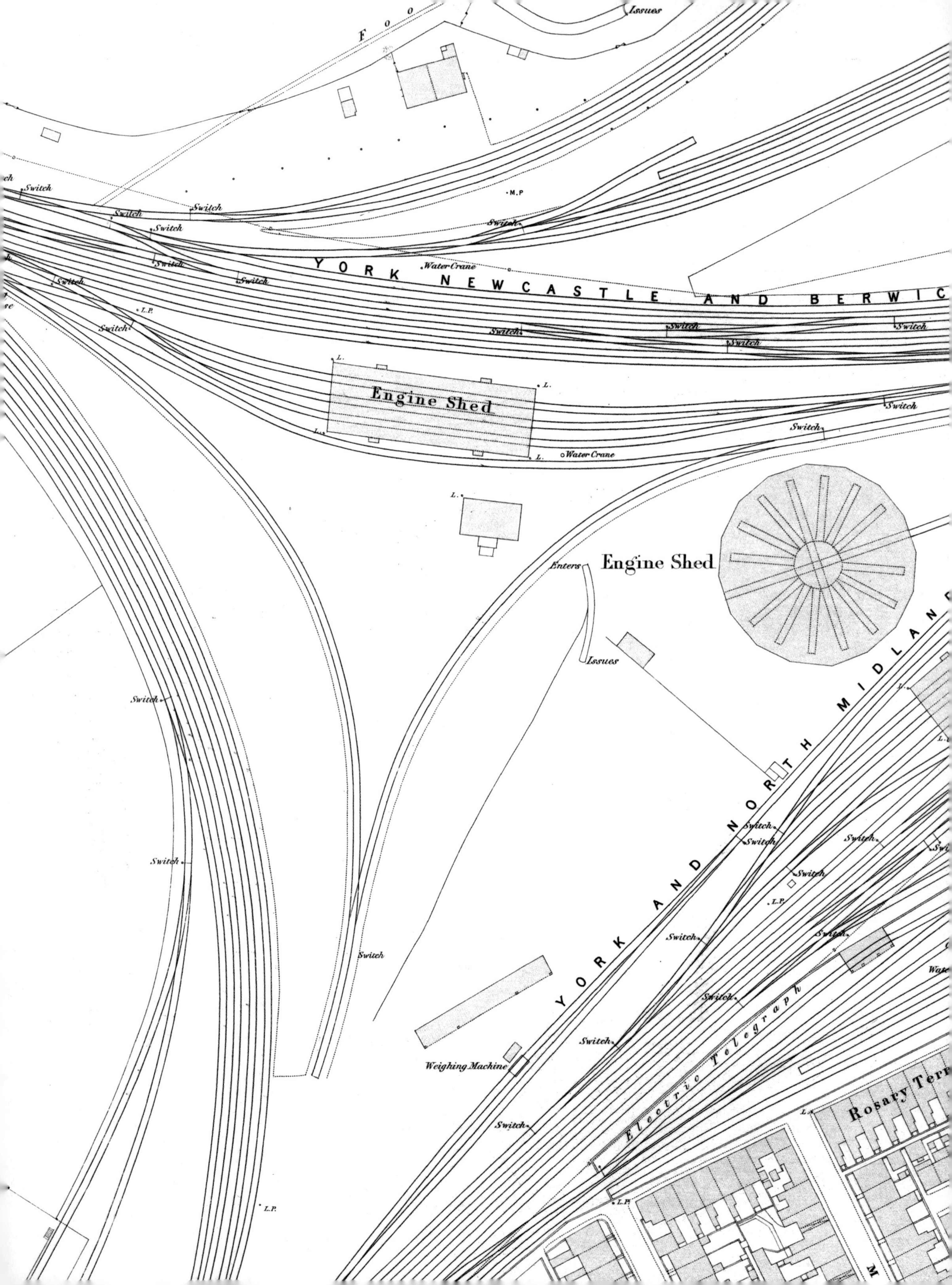
YORK NEWCASTLE AND BERWIC
Engine Shed
Engine Shed
YORK AND NORTH MIDLAND
Electric Telegraph
Rosary Terr
Weighing Machine
Water Crane
Water Crane
Switch
Enters
Issues
Issues
M.P
L.P.
L.

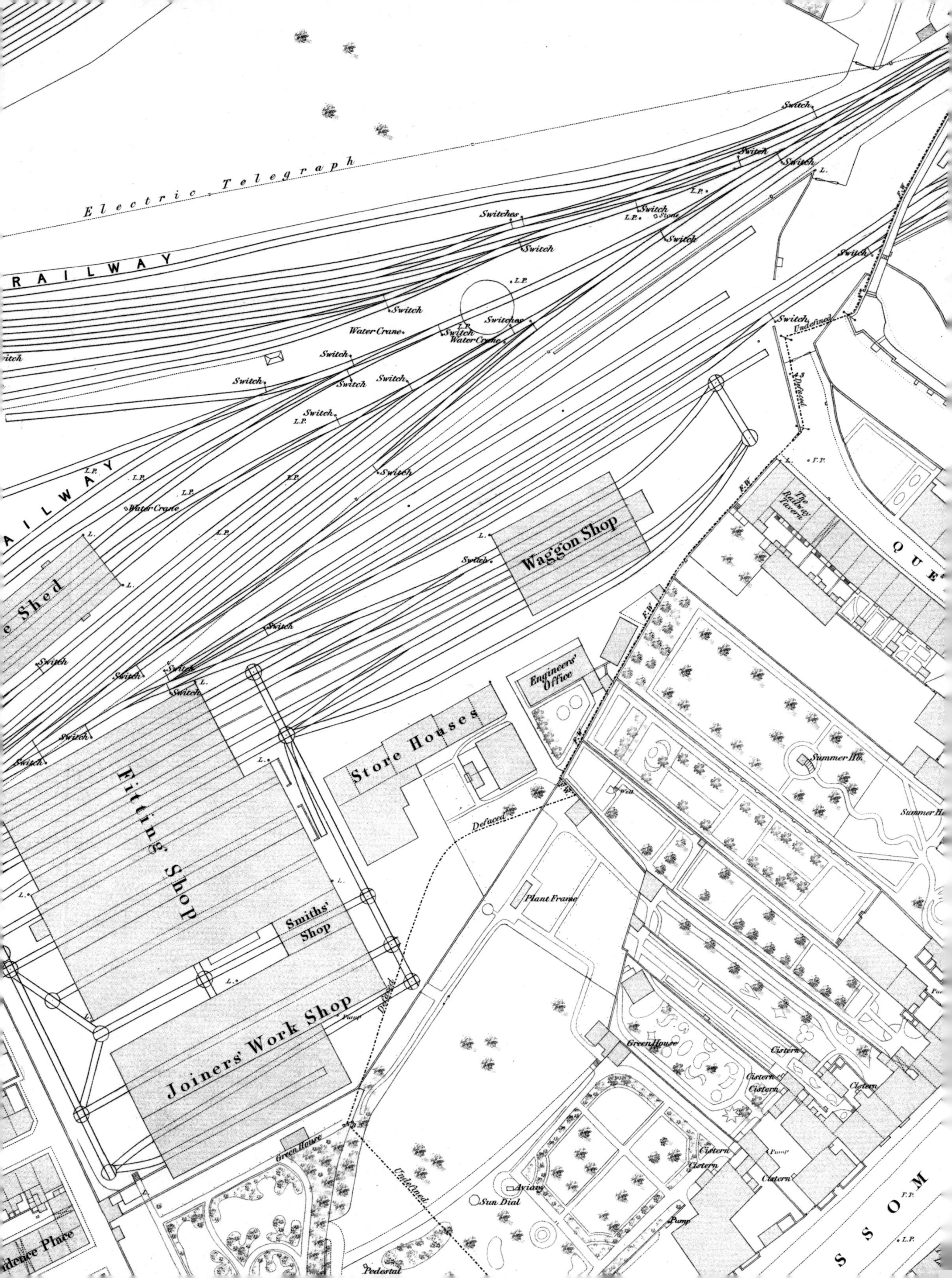

Electric Telegraph
RAILWAY
RAILWAY
Switch
Switches
Water Crane
L.P.
Waggon Shop
e Shed
Fitting Shop
Smiths' Shop
Joiners' Work Shop
Store Houses
Engineers' Office
The Railway Tavern
QUE
Summer Ho.
Summer Ho
Defaced
Undefined
Well
Plant Frame
Green House
Cistern
Pump
Aviary
Sun Dial
Pedestal
dence Place
SSOM

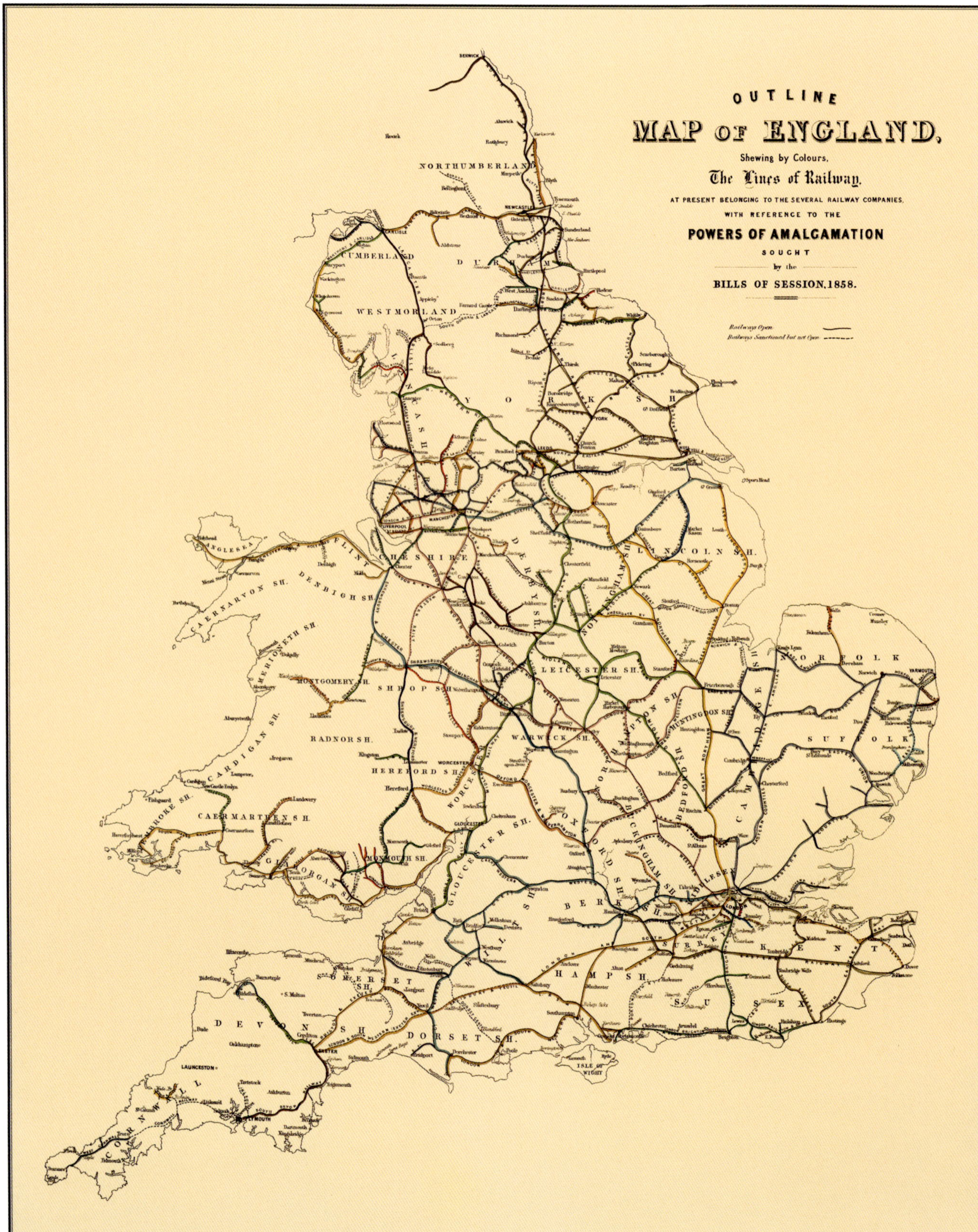
OUTLINE
MAP OF ENGLAND,
Shewing by Colours,
The Lines of Railway,
AT PRESENT BELONGING TO THE SEVERAL RAILWAY COMPANIES,
WITH REFERENCE TO THE
POWERS OF AMALGAMATION
SOUGHT
by the
BILLS OF SESSION, 1858.
Railways Open
Railways Sanctioned but not Open

Title	Outline Map of England, Shewing by Colours, The Lines of Railway, at present belonging to the several railway companies, with reference to the Powers of Amalgamation sought by the Bills of Session, 1858
Printer	Henry Hansard
Date	1858

The British railway system developed over the second quarter of the 19th century in a piecemeal (if not chaotic) manner, comprising railways with a strategic intent and those which were much more local in their scope. This plethora of different companies may have been in tune with the Victorian preference for competition, but it became apparent that it was not the most effective way to run what was by then a near-comprehensive national rail system.

The way forward would be further consolidation, but this was not something that could be done at the mere whim of the different railway companies, with Parliament requiring that a Bill be placed before them for authorization before any mergers could take place. This map produced for the House of Commons to assist in its deliberations is not, however, specific on what exactly was being proposed in 1858 and it is not clear whether all the Bills covered even made it to Parliament.

The map extract below features the northern end of one particular line, the Oxford, Worcester & Wolverhampton Railway (OWWR), which certainly did not form part of any amalgamation in 1858. However, the inevitable rationalization could not be put off for long, and it would disappear under an 1860 Act of Parliament into the West Midland Railway (WMR), formed by absorption of the Newport, Abergavenny & Hereford and Worcester & Hereford railways. Even this would turn out to be a short-lived halfway house, because the WMR was leased by the Great Western Railway in 1861, and taken over by it just two years after that. Having earned for itself the less than flattering soubriquet of the 'Old Worse and Worse', it can be safely assumed that the passing of the OWWR was not mourned by any of its erstwhile users.

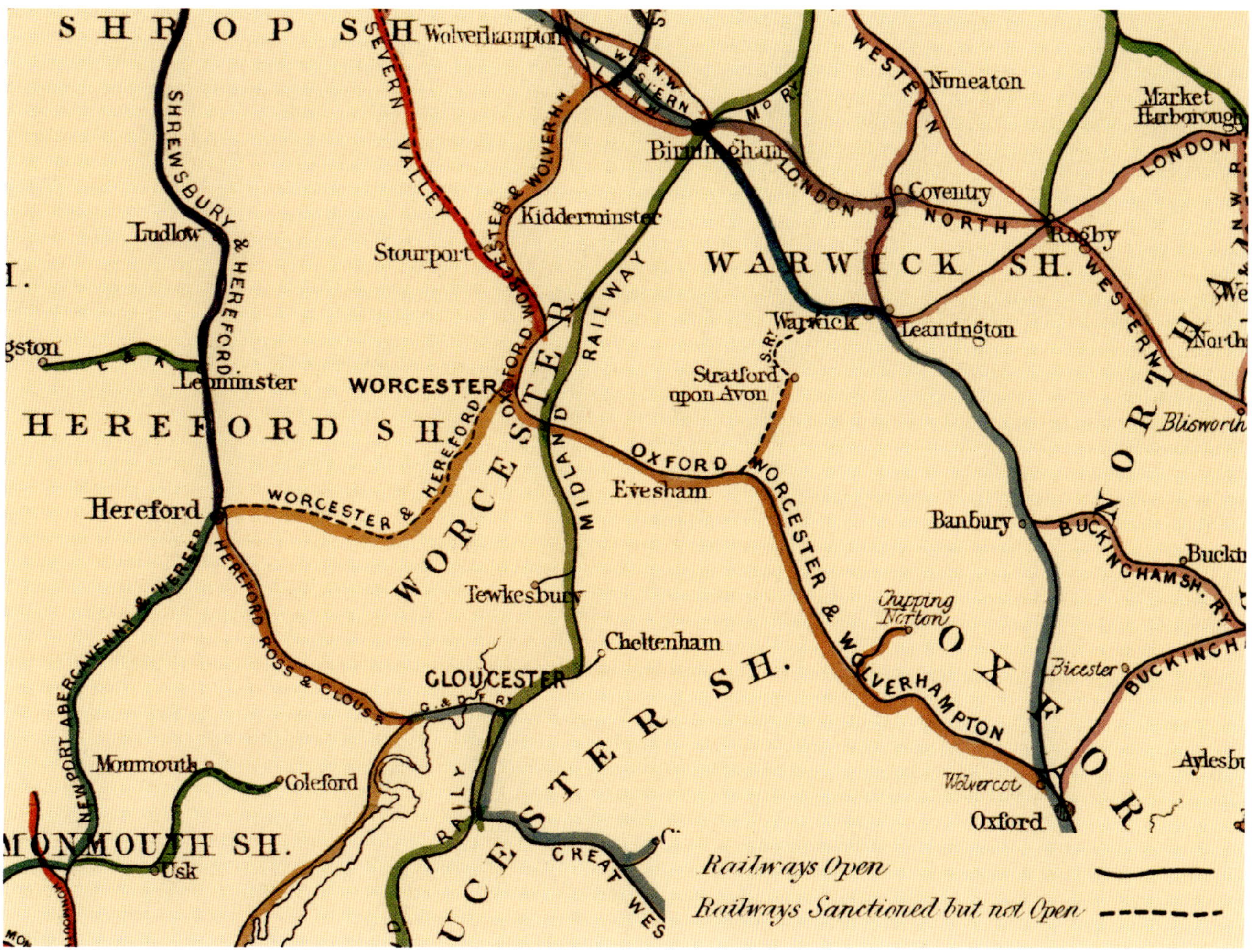

BERWICK
NORTH EASTERN
Alnwick
Hawick
Rothbury
Warkworth
NORTHUMBERLAND
Morpeth
BORDER COUNTIES RY
Bellingham
BLYTH & TYNE
Blyth
Tynemouth
N. Shields
S. Shields
NEWCASTLE
Haltwistle
Hexham
NEWCASTLE & CARLISLE
Gateshead
Sunderland
New Seaham
Port Carlisle
Silloth
CARLISLE
CALEDONIAN RY
MARYPORT & CARLISLE
Wigton
Aldstone
Medomsley
CUMBERLAND
Maryport
Workington
Whitehaven
Egremont
DURHAM
Durham
Stanhope
HARTLEPOOL
Hartlepool
Penrith
LANCASTER
West Auckland
Stockton
Appleby
Barnard Castle
DARLINGTON & BARD CASTLE
Darlington
WESTMORLAND
Orton
SOUTH DURHAM & LANCASHIRE UN.
STOCKTON & DARLINGTON
Stokesley
N. YORKSHIRE
WHITEHAVEN & FURNESS
Ravenglass
Coniston
KENDEL & WINDEMERE
Sedberg
Richmond
N. E. RY.
N. Allerton
Broughton
CONISTON
BEDALE RY
Bedale
Thirsk
Kirby Lonsdale
ULVERSTONE & LANCASTER
Rampside
Arnside
Ingleton
FURNESS
Ripon
Boroughbridge
Poulton
Lancaster
N. WESTERN RY
YORK
Knaresborough
Harrogate
N. EASTERN
Skipton
Fleetwood
LANCASr & PRESTON JN
LANCASHIRE
MIDLd RY
Clitheroe
Colne
Church Fenton
PRESTON & WYRE
Blackpool
Preston
Burnley
Bradford
LEEDS
NORTH MIDLAND
Blackburn
E. LANCASHIRE
Accrington
Bacup
Halifax
Knottingley
Ribble R.
Hesketh
LANCASHIRE & YORKSHIRE
Huddersfield
Southport
N. UNION
Bolton
Bury
L & N. W. R
S. Y. RAILWAY
Ormskirk
Holmforth
Penistone
Barnsley
Wigan
Leigh
Stalybridge
SHEFFIELD
Elsecar

OUTLINE

MAP OF ENGLAND,

Shewing by Colours,

The Lines of Railway,

AT PRESENT BELONGING TO THE SEVERAL RAILWAY COMPANIES,

WITH REFERENCE TO THE

POWERS OF AMALGAMATION

SOUGHT

by the

BILLS OF SESSION, 1858.

Railways Open

Railways Sanctioned but not Open

Series	Ordnance Survey of Scotland 1 in. Map
Sheet	32
Date	Originally published 1858. Printed from an Electrotype taken in 1877.

Before the Forth Bridge was opened in 1890, both sides of the estuary were only served by relatively minor railway lines. On the south side, a branch left the Edinburgh–Glasgow main line of the North British Railway (NBR) at Ratho, running via Kirkliston and Dalmeny to South Queensferry pier, where a passenger ferry operated across the Forth to North Queensferry. Passengers could then continue their journey by NBR train on the branch line to Inverkeithing and onwards through Fife to Kirkcaldy and beyond.

At that time, express trains from both Edinburgh and Glasgow to Perth, Inverness and Aberdeen ran inland via Stirling, and the rail network on the north shores of the Firth comprised modestly engineered routes serving an essentially local market. The legacy of these humble origins is felt to this day, as London–Aberdeen trains capable of running at 125 m.p.h. are forced to drop their speed to 50 m.p.h. or less through tight curves at Aberdour, Burntisland and Kinghorn. The North Queensferry branch closed to passengers when the Forth Bridge opened, but the South Queensferry line carried passenger traffic until 1929, and survived for freight into the 1960s (its mainstay loco latterly being J36 No. 65243 'Maude', now preserved at the nearby Bo'ness & Kinneil Railway).

As shown overleaf, by 1858 Edinburgh's suburban rail network was substantially complete, the main exception being the South Suburban Line loop through Morningside. The lack of built-up development in northwest Edinburgh – despite the presence of a number of railways – is striking. This points towards the importance of goods traffic; and fast passenger trains to the city centre were soon to stimulate widespread housing development in these new suburbs.

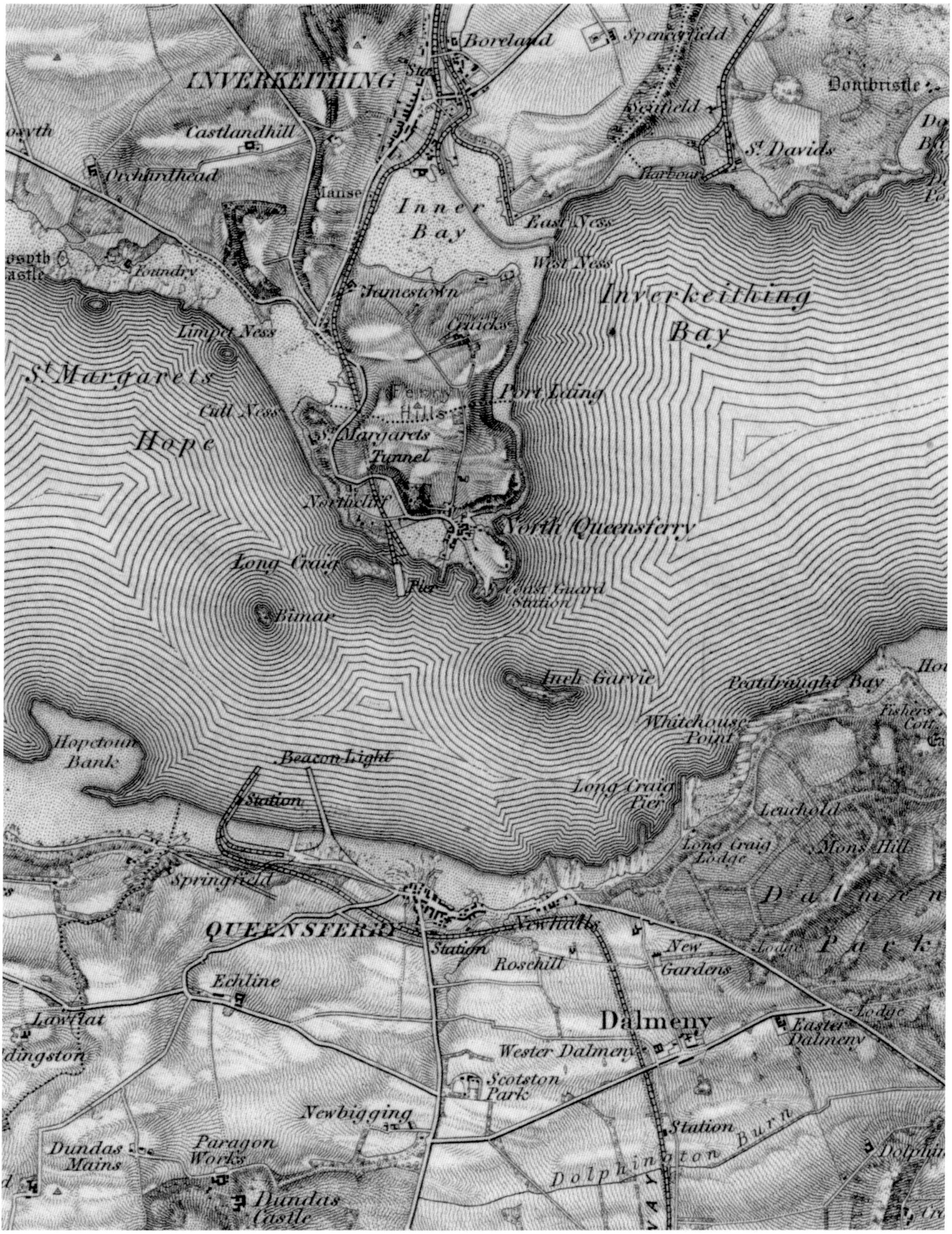
Boreland
Spencerfield
Donibristle
INVERKEITHING
Castlandhill
Orchardhead
St. Davids
Harbour
Manse
Inner Bay
East Ness
West Ness
Foundry
Jamestown
Inverkeithing Bay
Limpet Ness
St. Margarets Hope
Cult Ness
Port Laing
Ferry Hills
St. Margarets Tunnel
Northcliff
North Queensferry
Long Craig
Pier
Coast Guard Station
Bimar
Inch Garvie
Peatdraught Bay
Whitehouse Point
Hopetoun Bank
Beacon Light
Station
Long Craig Pier
Leuchold
Long Craig Lodge
Mons Hill
Springfield
QUEENSFERRY
Newhalls
Station
Rosehill
New Gardens
Echline
Lawflat
Dalmeny
Easter Dalmeny
Wester Dalmeny
Scotston Park
Newbigging
Station
Dolphington Burn
Dundas Mains
Paragon Works
Dundas Castle

North Queensferry
Inchmickery
Inch Garvie
Hound Point
The Buchans
Drum Sands
Cramond Island
The Pinnacle
Queensferry
Dalmeny
Dundas Castle
Kirkliston
River Almond
Cramond Bridge
Davidson's Mains
Corstorphine
Ratho
Currie
Kirknewton
Caledonian Railway
Edinburgh and Glasgow Railway
Newliston
Gogar
Saughton
E

Dolphin Head
Black
Middle Craigs
Rocks
Eastern Craigs
Outer Light House
Inner Light House
Light Ho.
Station
Wardie Bush
Newhaven
Granton
Trinity Grove
Leith
LEITH
Sands
Leith Links
Hillside Ho.
Chemical Works
Rl. Botanic Garden
N. Inverleith Mains
Inverleith Ho.
Hawkhill
Lochend Ho.
Restalrig
Southside Bank
Westbank Brick Works
PORTOBELLO
Seaview Ho.
Hamilton Lodge
Joppa
Fisherrow Sands
Musselburgh Sands
Stockbridge
Piershill Barracks
Northfield
Meadowfield
Southfield
Wester Duddington
Duddington Mains
Duddington Cottage
Easter Duddington
Glen Nurseries
Chemical Works
Fisherrow
Gas Works
Musselburgh Links
Newbigging
Inveresk
Pinkieburn
Brunstane
New Hailes
Stonyhill
Stoneybank
Wanton Walls
New Craighall
The Meadows
Newington
Salisbury Green
Prestonfield
Duddington Ho.
Niddry Mill
Niddry Ho.
Niddry Mains
Cairntows
NORTH BRITISH RAILWAY
Whitehill
Monktonhall
Midfield
Cowpits
Sweethope
Whitecraigs
Halfway
Morningside
Merchiston Castle
Royal Edinburgh Asylum
Powburn
Cameron Ho.
Bridgend
Craigmillar
Egypt
Cloverfoot
Cauldcoats
Backdean
Carlehall
Cairnie
Old Craighall
Shawfair
Hilltown
Monkton Ho.
Carberry
Little France
Edmonstone
Woolmet Ho.
Cocklerow
Redrow
Barclaw
Newton
Station
Easter Millerhill
Wester Millerhill
Edmonston Colliery
Castle Steads
Smeaton
Colinton Mains
Liberton
Liberton Ho.
Fernieside Lodge
Danderhall
Todhills
Sheriffhall Mains
Dalkeith Park
Gilmerton
Campend
Summerside
Dalkeith Ho.
Woodhead
Morton Hall
Burndale
Melville Grange
Five Houses
Broomhill
Bowbridge
Morton Mains
Burdiehouse
Swanston
Straiton
Wester Melville
Melville Mains
DALKEITH
Cowdenfoot
West Cowden
Lowholm
Peaselat
Caplelaw
Langlaw
Lasswade
Newbattle
Newbattle Abbey
Eldindean
Bonnyrigg
Pentland
New Pentland
Pentland Mains
Loanhead
Polton Farm
Polton Hall
Dalhousie Mains
Newton Grange
Newbattle Colliery
Boghall
Honey Hole
Midfield Ho.
Polton South Mains
Burnbrae
Dalhousie Mains
Mayfield
Easthouses

THE RAILWAYS PRE-EMINENT

The railways became by far the most important mode of transport, and the network filled out in every part of the country. In an era of fierce inter-company rivalry, the Great Central became the last main line to reach London, and the Midland pushed across the Pennines from Settle to Carlisle.

MILEPOST 6	Battle of the Gauges – Brunel's Great Western Railway	66
1863	Bills of Session (England & Wales)	68
1863	Cruchley's Travellers' Companion through England & Wales	72
1872	London Metropolitan Railways	76
MILEPOST 7	The Railway Clearing House	80
1877	Airey's England & Wales	82
c.1880	Plan of Proposed Cleator & Workington Junction Line	86
1887	Airey's Staffordshire	90
1887	W H Smith's British Isles	94
1887	London Goods Stations of the Great Eastern Railway	98

1860–1900

MILEPOST 8 The London 'Overground' 100

1887 London & North Western Railway Timetable Map 102

1887 London & North Western Railway Hotel Poster 104

1890 Proposed Channel Tunnel to Northern Ireland 108

c.1890–96 Glasgow 'Citizen' Map 110

1895 Mountain, Moor and Loch on the West Highland Railway 114

1896 Meyers Lexikon's Edinburgh 118

1896 North British Railway Map Poster 120

MILEPOST 9 Wonders of the Victorian World 124

1899 Central Scotland after the Forth Bridge 126

MILEPOST

6

From the days of John Cabot's expedition to the New World in the 15th century, the port of Bristol grew to become one of the most important in Britain, the city prospering from its links with the Caribbean and with the slave trade. However, communication with London was slow and even by the late 18th century, when the first mail coach was introduced, the uncomfortable stagecoach journey along turnpike roads could take at least a day.

By 1833 several schemes to link Bristol and London by railway had already fallen by the wayside. In March of that year, a group of wealthy Bristol businessmen appointed a young engineer by the name of Isambard Kingdom Brunel to survey a route to the capital. Less than six months later a near-level route via Bath and the Thames valley through Reading and Maidenhead had been surveyed, a board of directors had been appointed at each end of the proposed railway, and the Great Western Railway (GWR) had been founded.

Although the GWR received Parliamentary authorization on 31 August 1835, there was surprisingly no stipulation about the gauge of the line (the width between the rails) – nearly all British railways were by then adopting the standard gauge of 4 ft 8½ in. Seizing upon this glaring omission, Brunel, who had been appointed chief engineer of the GWR, successfully advocated a broad gauge of 7 ft 0¼ in. Offering journeys between the two cities at unheard-of speeds and in sumptuous comfort, Brunel's broad gauge looked like a world-beater and, for a while, it was.

Built by thousands of navvies using only basic implements, construction of the level, high-speed route was fairly rapid with the first section from a temporary station at Paddington to Maidenhead opening on 4 June 1838. Completion of the entire route was held up by the construction of the 1 mile 1,452 yd Box Tunnel near Bath, then the longest railway tunnel in the world. Opening day came on 30 June 1841, the same date as the broad gauge Bristol & Exeter Railway, soon to be part of the GWR's route to the southwest, opened its northern section.

In its early years the broad gauge GWR was a great success and heralded a new age of high-speed inter-city rail travel. Indeed, by 1867 the GWR's tentacles, with the help of its neighbouring broad-gauge supporters, had reached Penzance with unheard of journey times of nine hours from London – today it takes just over five hours. Unfortunately the rest of the country's railways were being built to the narrower, standard gauge of 4 ft 8½ in. and where these railways came into contact with the GWR, the scenes of confusion were recorded for posterity by Victorian cartoonists.

Following Brunel's death on 15 September 1859, the individualistic GWR soon came to the inevitable conclusion that the broad gauge dream was incompatible with the rest of the nation's railways. Starting in 1864, broad gauge track was converted to standard gauge and in places mixed gauge track was common for some years. By 1892 all that remained of the broad gauge was the main line from Paddington to Penzance and this was converted by an army of workmen over one weekend in May of that year.

Plenty of physical evidence of broad gauge remains to this day – Brunel's original timber-roofed Temple Meads station at Bristol has been preserved and is now in use as a car park for rail users. Tunnels, such as Box Tunnel, have generous broad gauge proportions, while Sonning Cutting, east of Reading – which was excavated by navvies using just picks, shovels and wheelbarrows – is over a mile long and up to 60 ft deep.

The last Great Western Railway broad gauge train leaves Paddington for the West Country on 20 May 1892.

This 1846 illustration depicts the chaos caused by passengers and luggage being transferred from broad gauge to standard gauge carriages at Gloucester station while on their way to Birmingham.

MAP OF ENGLAND
Shewing
THE LINES OF RAILWAY CONSTRUCTED AND IN
COURSE OF CONSTRUCTION AS WELL AS THOSE
Proposed by the
BILLS OF SESSION, 1863.
The authorized lines which belong to or leased to each Company are distinguished by a Particular Colour.
REFERENCE
Railways Open
Narrow Gauge
Broad Gauge
Mixed Gauge
D^o. Authorized but not Open
PROPOSED RAILWAYS (Session 1863)
NORTHUMBERLAND
CUMBERLAND
WESTMORLAND
YORKSH
LINCOLN SH
NORFOLK
SUFFOLK
SOMERSET SH.
DEVON SH.
DORSET SH.
HAMPSH.
KENT
LONDON
YORK
PLYMOUTH
Day & Son, Lith^rs to the Queen
London. Published 21st Feby 1863. by Day & Son, Lith^rs to the Queen.

Title	Map of England shewing the lines of railway constructed and in course of construction as well as those proposed by the Bills of Session, 1863
By	Day & Son, London
Date	1863

Comparing this map produced for Parliament in 1863 with the equivalent for 1858 (see pages 56–59) one can see countrywide evidence of significant extension to the rail network. The southeast of England had begun to catch up with the early pace of development in the Midlands and North of England, with the shape of the future commuter network becoming more recognizable.

The map extract overleaf reveals a second route to Dover, the first railway on the Isle of Wight, the Great Western Railway extended westwards beyond Hungerford, and significant infill in counties such as Kent, Sussex, Hampshire and north Essex. Unlike the 1858 map, the 1863 version distinguishes between 'narrow gauge' (standard gauge), 'broad gauge' and 'mixed gauge', which was the technical solution permitting trains of both gauges to use the same route corridor. 1860 (the year after Brunel's death) was the last year in which any broad gauge was laid; the broad gauge between London and Oxford was converted to mixed in 1861, thereby allowing through running of standard gauge trains between London and Birkenhead by the time this map was produced.

Broad gauge was progressively converted to standard gauge from 1864, the Hungerford parish magazine recording in 1874 that:

> *Hundreds of labourers have been engaged for the last fortnight in effecting the change from broad to narrow gauge on the Hungerford Branch of the Great Western Railway, the Berks and Hants, and Wilts and Somerset Railways. For five days the traffic was entirely suspended between Hungerford and Marlborough and Devizes; and on Sunday, June 28, the line was entirely closed. It is hoped that the ordinary service of trains will be resumed on Saturday, July 4.*

The writer's expectations were met, as the line did indeed reopen on time.

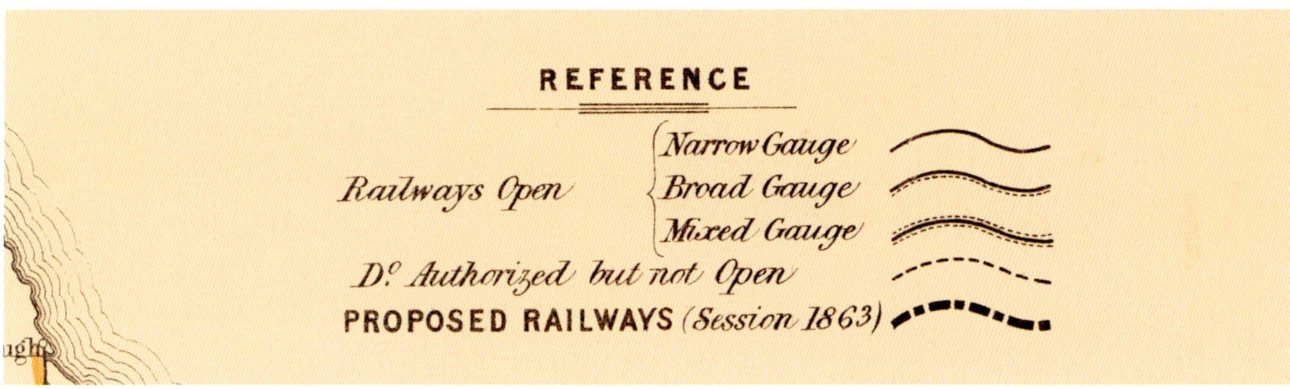

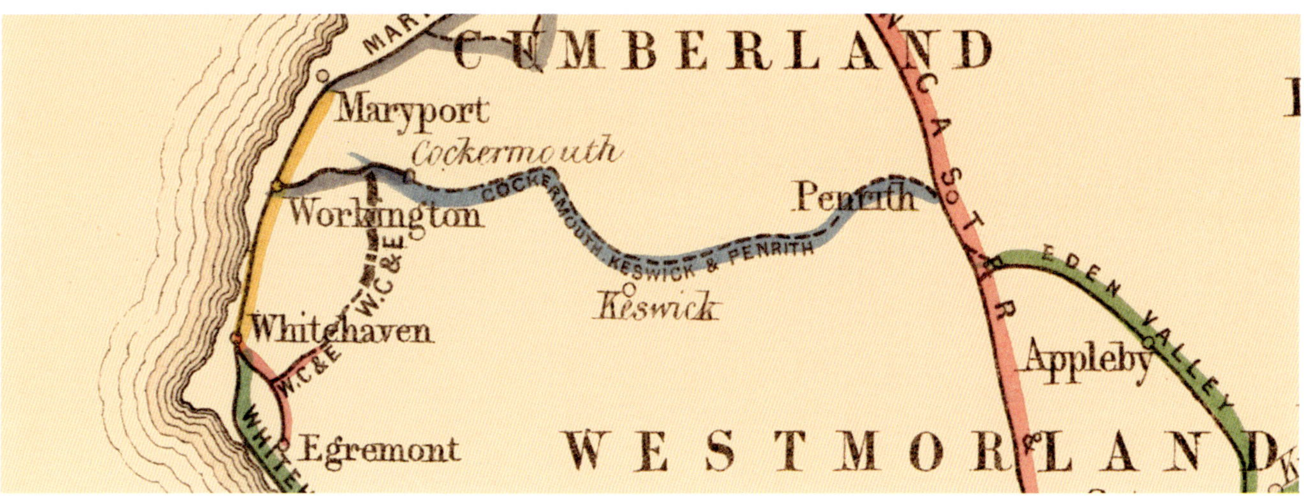

MONTGOMERY SH.
SHROP SH.
Minstrelsey
Newtown
Montgomery
Bishopscastle
Llanidloes
RADNOR SH.
Knighton
Ludlow
Wenlock
Wolverhampton
Bridgenorth
Dudley
Bushbury
Tamworth
Sutton Coldfield
Nuneaton
BIRMINGHAM
Coventry
Halesowen
Kidderminster
Stourport
Bewdley
Tenbury
WARWICK
Redditch
Henly
Warwick
Leamington
Stratford upon Avon
Kingston
Leominster
WORCESTER
HEREFORD SH.
Hay
Hereford
Evesham
Banbury
Llandovery
Talgath
Brecon
Ledbury
Tewkesbury
Newent
Chipping Norton
Llandillofawr
GLOUCESTER
Cheltenham
Bourton
Witney
Abergavenny
Monmouth
Coleford
MONMOUTH SH.
Usk
Merthyr
Aberdare
Neath
GLAMORGAN SH.
Newport
Caerphilly
Porth Cawl
Cardiff
Cowbridge
Barry
Stonehouse
Cirencester
Nailsworth
Dursley
Malmesbury
Faringdon
Abingdon
Swindon
Bristol
Clevedon
Yatton
Bath
Calne
Melksham
Marlborough
Hungerford
Devizes
Bradford
Weston
Radstock
Axbridge
Burnham
Highbridge
Wells
Frome
Westbury
Warminster
Lynmouth
Minehead
Watchet
Bridgewater
Glastonbury
Bruton
Andover
SOMERSET SH.
Langport
Wincanton
Salisbury
S. Molton
Taunton
Yeovil
Shaftesbury
Redbridge
Tiverton
Stalybridge
Southampton
Chard
Blandford
Ringwood
Crediton
DORSET SH.
Colyton
Axmouth
Lyme Regis
Poole
Lymington
Christchurch
EXETER
Sidmouth
Topsham
Exmouth
Bridport
Dorchester
Wareham
Swanage
M. Hampstead
Seaton
Weymouth
Teignmouth
Ashburton
GREAT WESTERN
MIDLAND
LONDON & NORTH WESTERN
LONDON & SOUTH WESTERN
BRISTOL & EXETER
SOMERSET CENTL
SOMERSET & DORSET
SOUTH WALES RAILWAY
CENTRAL WALES
MID WALES
SALISBURY & YEOVIL

Leicester
Peterborough
March
Market Harborough
Thrapstone
Kettering
Huntingdon
Ramsey
Chatteris
St. Ives
Ely
Brandon
Thetford
Diss
Bungay
Beccles
Harleston
Halesworth
Reeham
SUFFOLK
Framlington
Wellingborough
Northampton
Weedon
Cambridge
Bury St. Edmunds
Woodbridge
Ipswich
Bedford
Potton
Sandy
Chesterford
Saffron Walden
Haverhill
Long Melford
Sudbury
Hadleigh
N. Pagnell
Royston
Buckingham
Halstead
Mistley
Harwich
Colchester
Walton
Brightlingsea
Buntingford
Bishop Stortford
Braintree
Hitchen
Dunstable
Luton
Hertford
Ware
Aylesbury
St. Albans
Hatfield
Maldon
Thame
Wendover
Chesham
Amersham
Epping
Chipping Ongar
Loughton
Enfield
Wycombe
Wallingford
Rickmansworth
Edgware
Uxbridge
Leigh
Southend
Henley
Windsor
Staines
Reading
Richmond
Kingston
LONDON
Woolwich
Dartford
Gravesend
Sheerness
Whitstable
Herne Bay
Bromley
Strood
Rochester
Faversham
Canterbury
Croydon
Epsom
Farningham
Farnborough
Seven Oaks
Maidstone
Wokingham
Chertsey
Woking
Leatherhead
Caterham
Deal
Basingstoke
Ash
Guildford
Dorking
Reigate
Tonbridge
KENT
Ashford
Alton
Godalming
E. Grinstead
Tonbridge Wells
Dover
Folkstone
Alresford
Haslemere
Horsham
Hartfield
Petersfield
Petworth
Uckfield
Midhurst
Pulborough
Bishop Waltham
Hailsham
Fareham
Havant
Chichester
Arundel
Shoreham
Lewes
Brighton
Hastings
Bognor
E. Bourn
Portsmouth
Newhaven
Ventnor
SOUTH EASTERN RAILWAY
EASTERN COUNTIES RY
LONDON BRIGHTON & SOUTH COAST

CRUCHLEY'S
TRAVELLERS COMPANION
through
ENGLAND AND WALES,
shewing the
Railways and Stations
to the present time
MAPS OF SCOTLAND & IRELAND TO CORRESPOND.
LONDON, G.F. CRUCHLEY, MAP PUBLISHER & GLOBE MANUFACTURER, 81, FLEET ST
British Statute Miles
IRISH SEA
GERMAN
OCEAN
ST. GEORGE'S CHANNEL
CARDIGAN BAY
BRISTOL CHANNEL
ENGLISH CHANNEL
IRELAND
FRANCE
LONDON, PUBLISHED BY G. F. CRUCHLEY, MAP-SELLER 81, FLEET STREET

Title	Cruchley's Travellers Companion through England and Wales shewing the Railways and Stations to the present time
Publisher	G F Cruchley
Date	1863

The very substantial expansion of the rail network between the late 1840s (see pages 44–47) and 1863 is well illustrated by this map produced by the renowned map publisher George Cruchley – although it is not immediately obvious how convenient a 'companion' this sheet, measuring 50 in. by 40 in., would have been for the average rail traveller!

Most main lines had been built by 1863, notable exceptions being the Great Western between Pewsey and Taunton, the Great Central Railway into London, the Hull & Barnsley, and the Settle & Carlisle line across the Pennines. The latter decades of the 19th century also saw further infill of cross-country and branch lines nationwide, not least in 'peripheral' North Devon and North Cornwall. The map extracts below and overleaf show the gaps yet to be filled before the tentacles of the network finally reached five coastal towns which were to play a key part in the boom in train travel to holiday destinations – St Ives, Newquay, Padstow, lfracombe and Minehead.

The very large size of the map allowed considerable detail to be shown of settlements and roads, as well as county boundaries – but a word of caution should be given. In *Maps for Local History* (1988), Brian Hindle notes that:

> *Publishers like Cruchley . . . simply added railways to existing plates or, worse, just overprinted the lines onto existing maps. Cruchley, for example, was still using Cary's county maps of 1801 as base maps in the 1870s!*

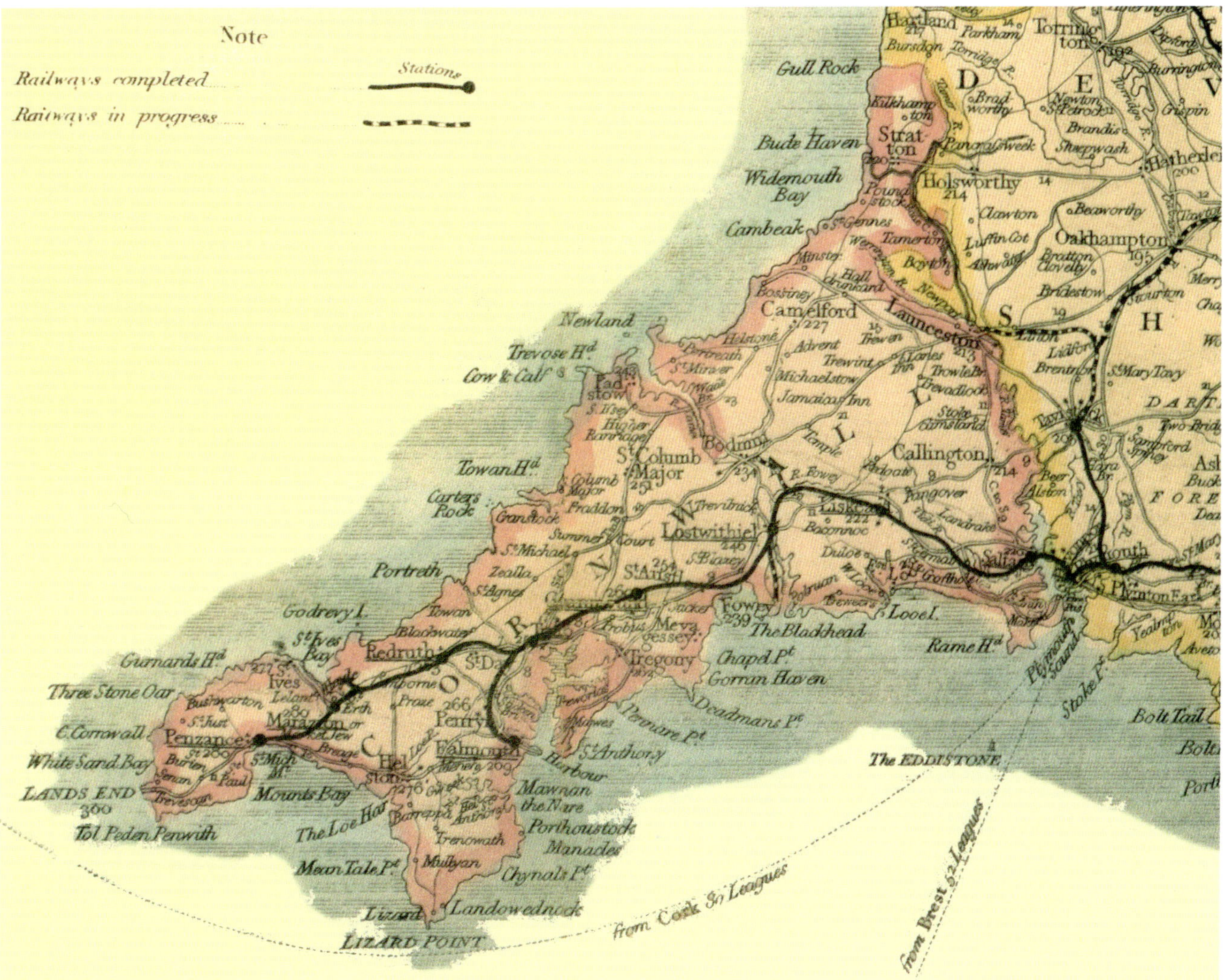

BRISTOL CHANNEL
SWANSEA BAY
Mouth of the Severn
Bridgewater Bay
Barnstaple or Biddeford Bay
TOR BAY
START BAY
PORTLAND RACE
The EDDISTONE
Caermarthen
Llandilo Vawr
Llandovery
Brecon
Crickhowel
Abergavenny
Monmouth
Usk
Merthyr Tydvil
Neath
Swansea
Llantrissant
Cowbridge
Caerphilly
Cardiff
Newport
Chepstow
Cardigan
Lampeter
Builth
Hay
Kington
Leominster
Bromyard
Ledbury
Newent
Ross
Gloucester
Colford
Thornbury
Tetbury
Malmsbury
Chippenham
Bristol
Bath
Calne
Wells
Frome
Shepton Mallet
Glastonbury
Bridgewater
Taunton
Wiveliscombe
Watchet
Dunster
Minehead
Porlock
Exmoor Forest
Barnstaple
Biddeford
Torrington
South Molton
Chumleigh
Tiverton
Bampton
Collumpton
Wellington
Honiton
Ilminster
Chard
Crewkerne
Yeovil
Sherborne
Shaftsbury
Sturminster
Blandford
Dorchester
Bridport
Lyme Regis
Beaminster
Weymouth
Portland Bill
Wareham
Corfe Castle
Axminster
Colyton
Ottery St Mary
Sidmouth
Exeter
Topsham
Crediton
Oakhampton
Hatherleigh
Holsworthy
Stratton
Launceston
Callington
Tavistock
Dartmoor Forest
Moreton Hampstead
Chudleigh
Newton Abbot
Ashburton
Teignmouth
Dawlish
Totness
Dartmouth
Kingsbridge
Modbury
Plymouth
Plympton
Saltash
Looe I.
Rame Hd.
Bolt Tail
Bolt Hd.
Start Pt.
Berry Hd.
Hopes Nose
Stoke Pt.
Portland
Worms Hd.
Mumbles
Nash Pt.
Breaksea Pt.
Barry I.
Sully I.
Steep Holm
Flat Holm
Bull Hd.
Wollacomb Bay
Horsedown Pt.
Countis bury
CAERMARTHEN SHIRE
GLAMORGAN SHIRE
MONMOUTH SHIRE
BRECKNOCK SHIRE
HEREFORD SHIRE
GLOCESTER SHIRE
SOMERSET SHIRE
DEVON SHIRE
DORSET SHIRE

Northampton
Cambridge
Bedford
Buckingham
Aylesbury
Oxford
Hitchin
Hertford
Saffron Walden
Braintree
Chelmsford
Epping
Billericay
Rochford
Maldon
Reading
Henley
Wendover
Chesham
Uxbridge
Dartford
Gravesend
Rochester
Chatham
Maidstone
Croydon
Guildford
Dorking
Reigate
Horsham
Cuckfield
Lewes
Brighton
Worthing
Chichester
Arundel
Petworth
Midhurst
Haslemere
Godalming
Farnham
Alton
Odiham
Basingstoke
Whitchurch
Andover
Stockbridge
Salisbury
Winchester
Southampton
Lymington
Christchurch
Portsmouth
Fareham
Hailsham
Battel
Hastings
Winchelsea
Tunbridge
Cranbrook
Tenterden
Hungerford
Newbury
Swindon
Highworth
Faringdon
Witney
Burford
Woodstock
Banbury
Deddington
Shipston
Kineton
Towcester
Winslow
Tring
Luton
Stevenage
Baldock
Royston
Haverhill
Halstead
Coggeshall
Sudbury
Newport Pagnell
Woburn
Ampthill
Abingdon
Wallingford
Wantage
Thame
Marlow
Maidenhead
Windsor
Wokingham
Chertsey
Ewell
Sevenoaks
Westerham
Steyning
Petersfield
Havant
Gosport
Ringwood
Fordingbridge
Romsey
Ludgershall
Amesbury
Kingsclere
Whitchurch
Epsom
Crawley
East Grinstead
Pevensey
Beachy Hd.
Eastbourne
Newhaven Harbour
Selsea Bill
Bognor Rocks
Spithead
The Solent
St. Helens Rd.
Culver Cliffs
Sandown Bay
Dunnose
Ventnor
St. Lawrence
Chale Bay
Niton
Freshwater Bay
the Needles
Hurst Cas.
Yarmouth
New Forest
Sheerness
Milton
Southend
Mouth of the Thames
H C H A N N E L
London to Guernsey 378 Miles
To Jersey 400 Miles
25 Leagues
Submarine Telegraph
London to Havre 213 Miles
Southampton to Havre de Grace 34 Le.
B E R K S H I R E
H A M P S H I R E
S U R R E Y
S U S S E X
K E N T
E S S E X
H E R T F O R D
B E D F O R D
O X F O R D S H I R E
B U C K I N G H A M S H I R E
I S L E O F W I G H T

TUNNEL
TUNNEL
MIDLAND
FINCHLEY STATION
FARM
CHURCH
SCHOOL FOR THE BLIND
FELLOWS ROAD
ADELAIDE ROAD
CRICKET GROUND
PRIMROSE HILL & PARK
GYMNASIUM
ZOOLOGICAL GARDENS
REGENTS PARK
BRIDGE
ROYAL BOTANIC GARDENS
LORDS CRICKET GROUND
SUTHERLAND CRES
84
64
96
87

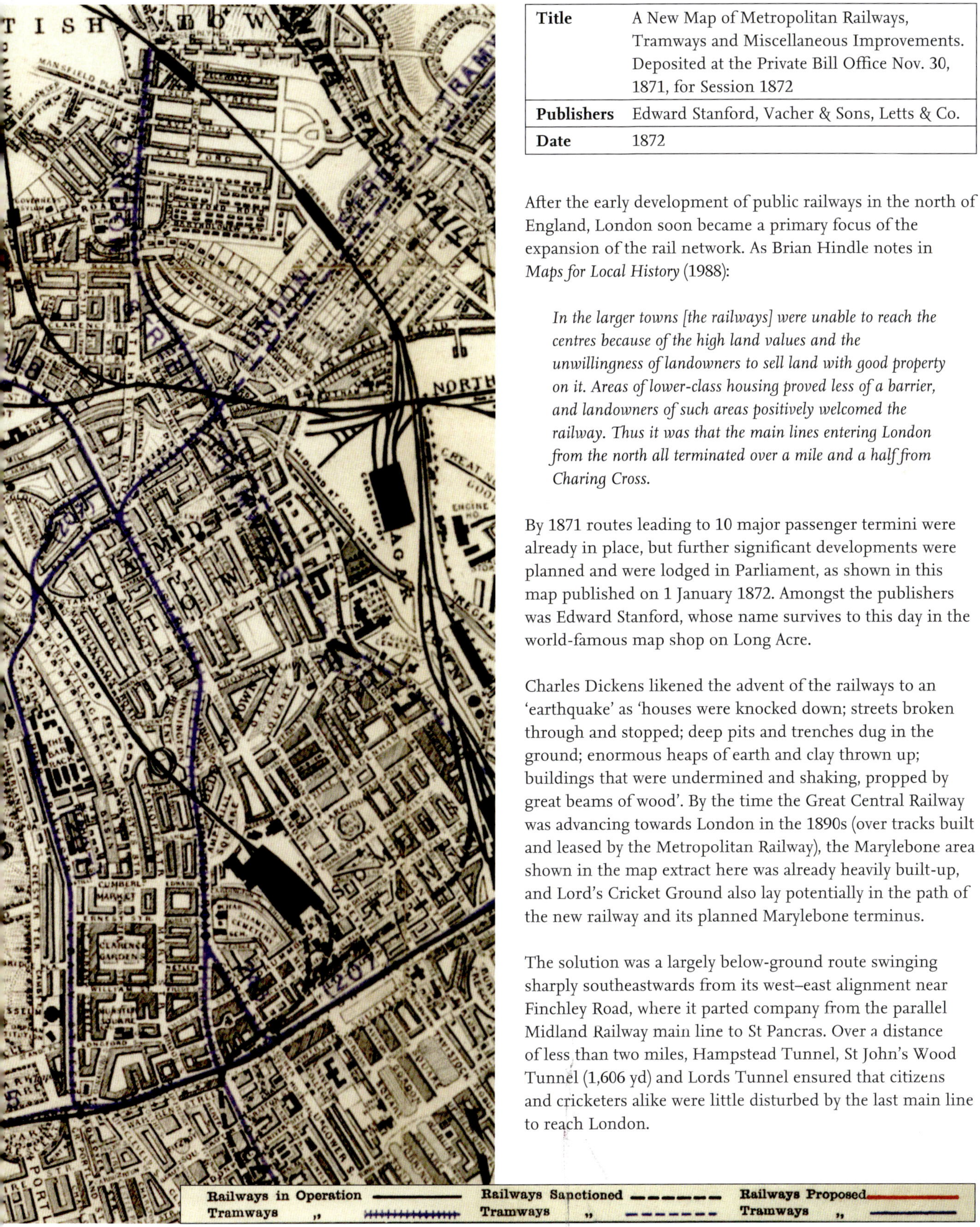

Title	A New Map of Metropolitan Railways, Tramways and Miscellaneous Improvements. Deposited at the Private Bill Office Nov. 30, 1871, for Session 1872
Publishers	Edward Stanford, Vacher & Sons, Letts & Co.
Date	1872

After the early development of public railways in the north of England, London soon became a primary focus of the expansion of the rail network. As Brian Hindle notes in *Maps for Local History* (1988):

> *In the larger towns [the railways] were unable to reach the centres because of the high land values and the unwillingness of landowners to sell land with good property on it. Areas of lower-class housing proved less of a barrier, and landowners of such areas positively welcomed the railway. Thus it was that the main lines entering London from the north all terminated over a mile and a half from Charing Cross.*

By 1871 routes leading to 10 major passenger termini were already in place, but further significant developments were planned and were lodged in Parliament, as shown in this map published on 1 January 1872. Amongst the publishers was Edward Stanford, whose name survives to this day in the world-famous map shop on Long Acre.

Charles Dickens likened the advent of the railways to an 'earthquake' as 'houses were knocked down; streets broken through and stopped; deep pits and trenches dug in the ground; enormous heaps of earth and clay thrown up; buildings that were undermined and shaking, propped by great beams of wood'. By the time the Great Central Railway was advancing towards London in the 1890s (over tracks built and leased by the Metropolitan Railway), the Marylebone area shown in the map extract here was already heavily built-up, and Lord's Cricket Ground also lay potentially in the path of the new railway and its planned Marylebone terminus.

The solution was a largely below-ground route swinging sharply southeastwards from its west–east alignment near Finchley Road, where it parted company from the parallel Midland Railway main line to St Pancras. Over a distance of less than two miles, Hampstead Tunnel, St John's Wood Tunnel (1,606 yd) and Lords Tunnel ensured that citizens and cricketers alike were little disturbed by the last main line to reach London.

REGENT PARK
ROYAL BOTANIC GARDENS
GREEN PARK
ST JAMES PARK
RIVER
NEW PARLIAMENT HOUSES
BOROUGH ROAD
CHELSEA EMBANKMENT 1868
CHELSEA REACH
BATTERSEA PARK
SOUTH LAMBETH
NORTH BRIXTON
MARKET GARDENS

TOWER HAMLETS
TOWER HAMLETS CEMETERY
EASTERN
1866
BROMLEY MARSH
EAST INDIA IMPORT DOCK
LONDON DOCK
THAMES
WAPPING
THE POOL
SHADWELL
204
WEST INDIA IMPORT DOCK
WEST INDIA SOUTH DOCK
TIMBER DOCK
MILLWALL DOCKS
LIMEHOUSE REACH
BLACKWALL REACH
LONDON BLACKWALL & MILLWALL 1865
ISLE OF DOGS
ROYAL DOCK YARD
GREENWICH REACH
BRICKLAYER ARMS
124
MARKET GARDENS
GRAND SURREY CANAL
PECKHAM NEW TOWN
DOVER ROAD

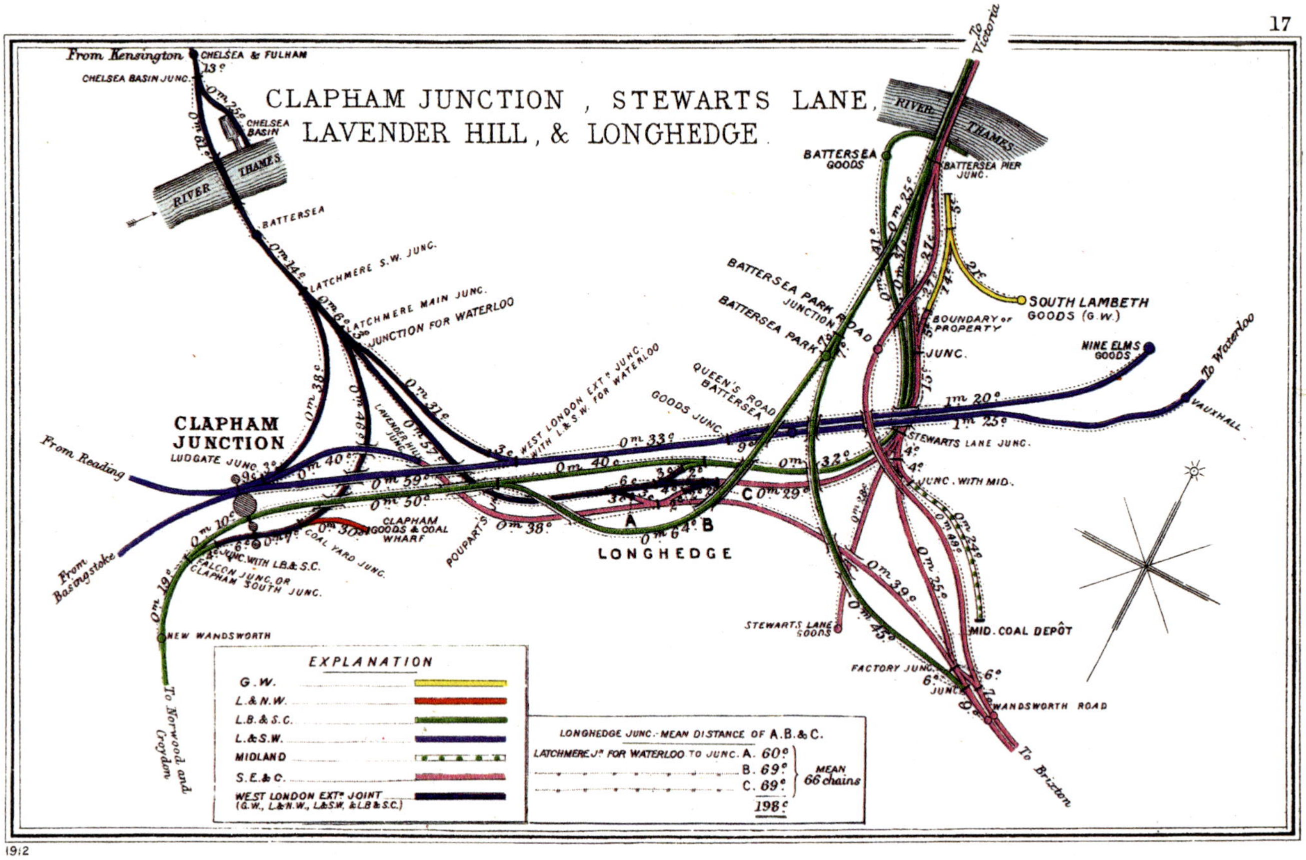

John Airey's Junction Diagram depicting the network of lines around Clapham Junction was published by the Railway Clearing House in 1912.

By the early 1840s railways were spreading like tentacles across the map of Britain and at first sight it appeared that they were all linked, offering the opportunity of through passenger and goods traffic for the first time. This was far from the truth, however, as there was virtually no commonality between them. Despite most companies opting for the standard gauge of 4 ft 8½ in. – the Great Western Railway and its later constituent companies were different as they chose Brunel's 7 ft 0¼ in. broad gauge – that is where the commonality usually ended.

A selection of confusing early railway signals, 1837–46.

In many cases through carriages from one company to another were not allowed and some companies refused to carry third-class passengers from other lines. There being no standardization of tickets, the through booking of a journey that involved using several railways was not possible either. There was no standardization of goods vehicles, no common classification of goods and there were constant disputes over the allocation of receipts from goods traffic as there were no agreed distances between stations for calculation of chargeable mileage rates. Even station clocks and consequently timekeeping were not synchronized to a common time.

On the safety side, signalling was a very hit and miss affair with some companies using red discs to indicate 'clear road ahead' while others used red discs to indicate the opposite. Goods wagons, whether privately owned or railway owned, had different braking systems and even buffers that did not align.

All of the above caused enormous confusion and was certainly a stumbling block to the continuing successful development of the railways. Painfully aware that some form of organization and standardization needed to be introduced, the Chairman of the London & Birmingham Railway invited representatives of nine other railway companies to a meeting at Euston in 1842. Thus was the Railway Clearing House (RCH) formed, although some companies, notably the Great Western Railway, took another 20 years before they joined.

The agreed objectives of the RCH were to organize the through booking of passengers, carriages and horses between different companies and to divide receipts based on mileage. For goods traffic the RCH also organized a classification of goods, and tried to encourage a rate for these on a per mile basis. Of course they could not set these rates as each company charged differently, but it was a start. Member companies paid the RCH a fixed rate based on their number of stations and also a small percentage of their receipts.

Before long the RCH had standardized railway timekeeping using Greenwich Mean Time, published distance tables, encouraged the use of Edmondson ticket-printing and date-stamping machines and had produced a general classification of goods. Inter-company debts were also settled at the RCH.

The RCH also published many railway maps, the first one being produced by Zachary Macauley in 1851. His work was continued by John Airey who first published his Junction Diagrams in 1867. By the end of the 19th century his hand-coloured maps of the entire rail system, with distances shown between stations and junctions in miles and chains, were an important tool for every railway company when calculating rates.

By the outbreak of the First World War the RCH had a staff of over 3,000, many of them employed in the painstaking task of number-taking to check the numbers and contents of goods wagons travelling over the system. In fact, by the time of the 'Grouping' of the railways into just four companies in 1923, there were around 4,000 private owners using 650,000 goods wagons, and a sheet of paper had to be produced recording each journey made by each wagon – an enormous task.

Despite the Grouping, the RCH remained an important statistic-gathering organization until nationalization in 1948. However, the writing was now on the wall for the RCH, although its much-reduced staff were still called upon by British Railways to provide statistics until 1963, when it finally closed.

The journeys of all private-owner wagons, such as these seen at Birkenhead docks in 1924, were painstakingly recorded by the Railway Clearing House.

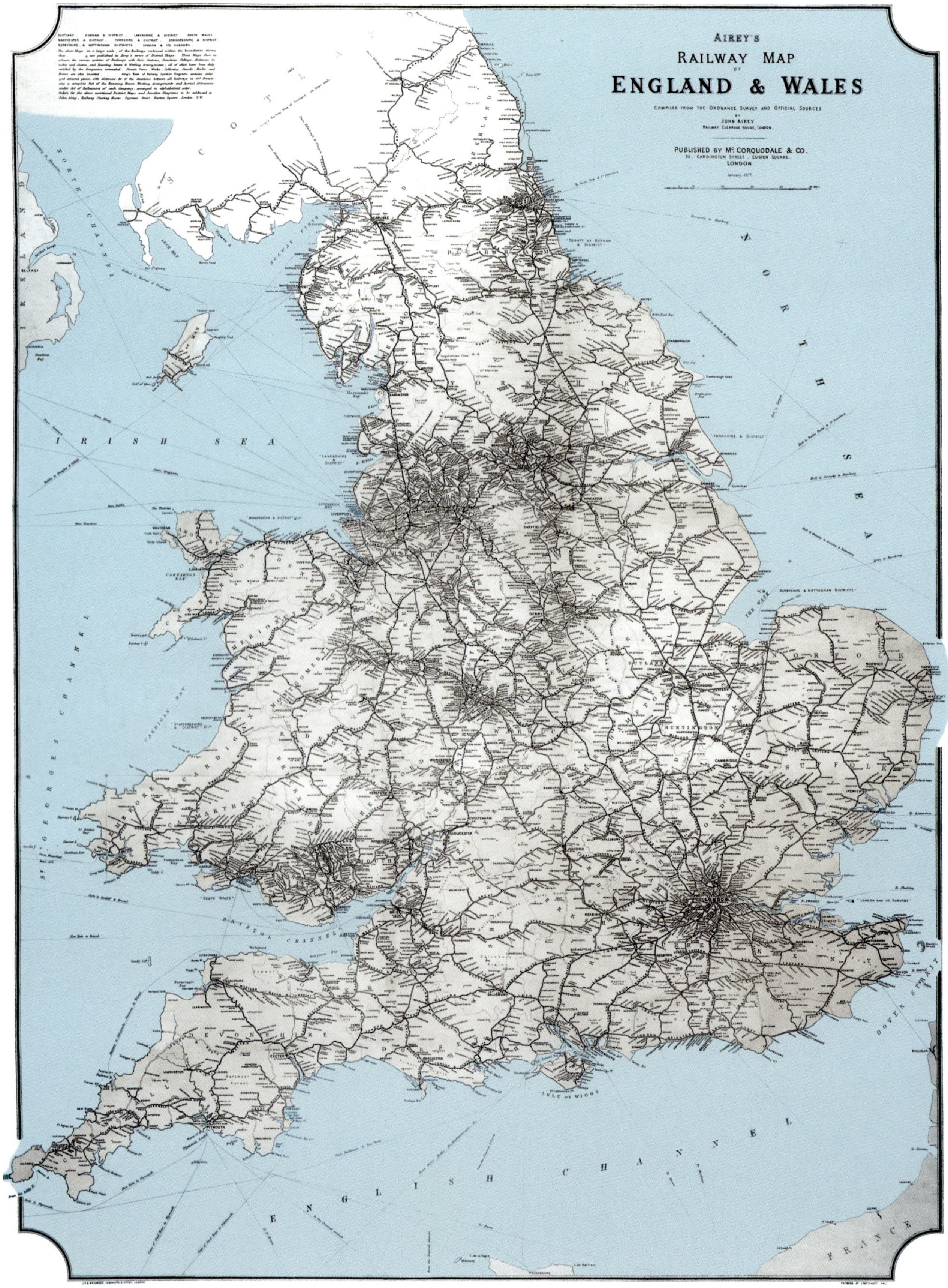
AIREY'S
RAILWAY MAP
OF
ENGLAND & WALES
COMPILED FROM THE ORDNANCE SURVEY AND OFFICIAL SOURCES
BY
JOHN AIREY
RAILWAY CLEARING HOUSE, LONDON.
PUBLISHED BY Mc CORQUODALE & CO.
55, CARDINGTON STREET, EUSTON SQUARE,
LONDON
January, 1877
IRISH SEA
NORTH SEA
ENGLISH CHANNEL
BRISTOL CHANNEL
ST GEORGES CHANNEL
DOVER STRAIT
FRANCE
ISLE OF WIGHT
LIVERPOOL
NORFOLK

Title	Airey's Railway Map of England & Wales compiled from the Ordnance Survey and Official Sources by John Airey
By	Railway Clearing House
Date	1877

Maps were a key tool of the Railway Clearing House set up in 1842 to handle inter-company dealings across the multiple ownerships of the rail network. A Clearing House employee, John Airey, prepared what were essentially working maps for railwaymen, comprising regional sheets showing ownership, junctions and distances, as well as more detailed 'junction diagrams' dealing with the key interfaces between companies.

The Clearing House maps increasingly came into the public domain, and this large (47 in. by 35 in.) 1877 sheet by John Airey clearly had a wider purpose, covering as it did all of England and Wales plus the Borders and southwest Scotland.

The map extract below illustrates the intricate network of railways that developed in County Durham, Tyneside and Teesside, building on the inheritance of nearly two centuries of waggonways linking collieries to the coast. Branch lines penetrated westwards far into the Pennines to tap mineral resources for the burgeoning industries along the coast from Blyth in the north through Tynemouth, Sunderland and Hartlepool to Middlesbrough, while around Newcastle a dense network of commuter lines served and stimulated the massive housing development of the late 19th century.

In response to competition from new electric street trams, much of the Tyneside passenger network was electrified by the North Eastern Railway in 1904, using a 'third rail' system similar to that so familiar in commuter country south of London. In response to employment and population changes, the system was de-electrified by British Railways between 1963 and 1967 – and replaced by diesel trains – but the key routes were re-electrified (with overhead wires) and new links built, as the Tyne and Wear Metro, the first section of which opened in 1980.

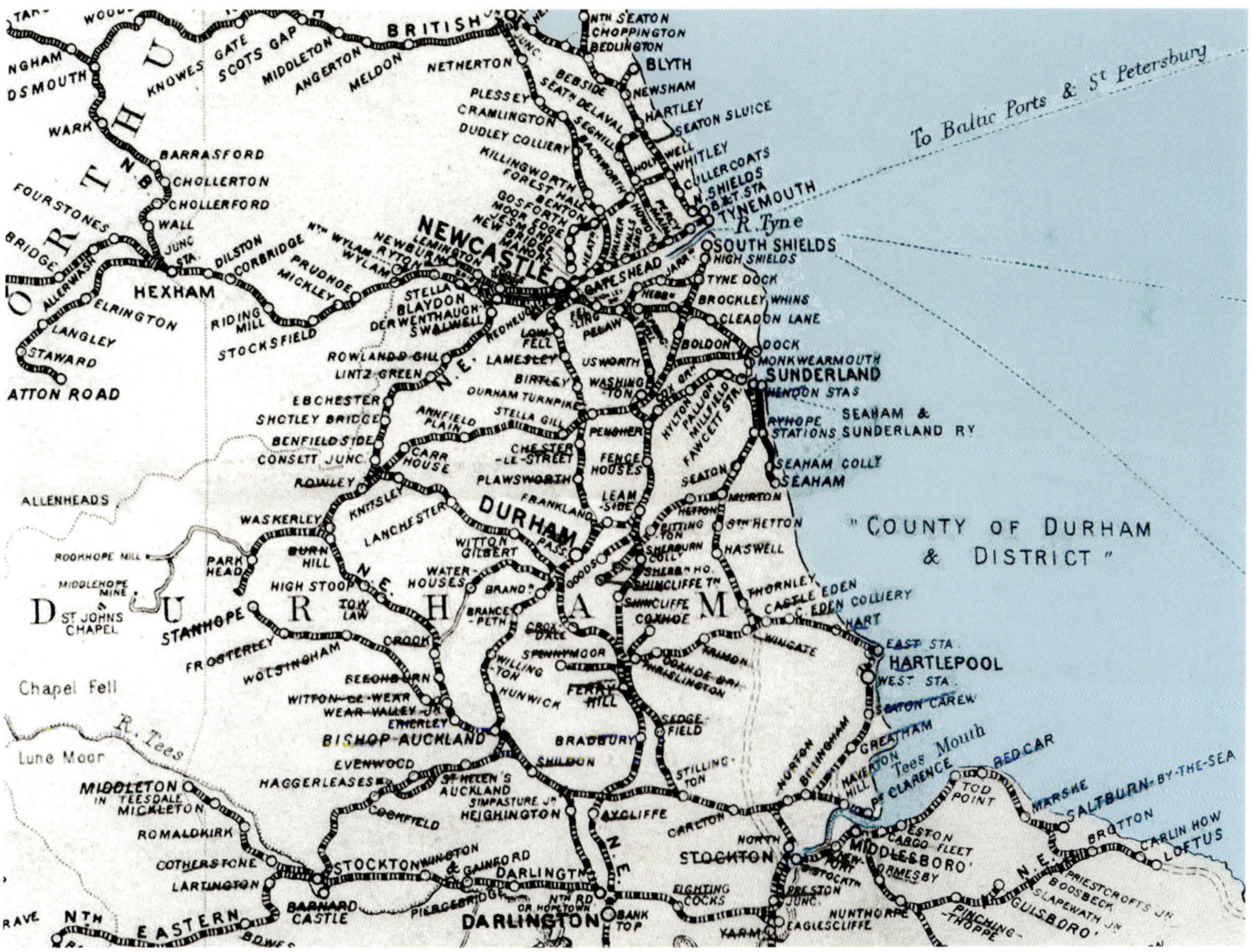

MALVERN LINK
GRT MALVERN
MALVERN WELLS
MORETON ON LUGG
WITHINGTON
STOKE EDITH
COLWALL
SHIPERTON
BARTON JN
BARRS COURT
ROTHERWASS JN
LEDBURY
HOLM LACY
FAWLEY
ROSS
NEWENT
TEWKESBURY
ASHCHURCH
CLEEVE
WINCHCOMB
CHELTENHAM
HIGH STR.
LANSDOWN
G.W.
CHURCHDOWN JUNC.
GLOUCESTER
MITCHELDEAN ROAD
LONGHOPE
KERNE BRIDGE
LYDBROOK JN
SYMOND'S YAT
GRANGE COURT
OAKLE STR.
NEWNHAM
BULLO PILL DOCKS
COLEFORD
MILKWALL
PARKEND
WHITECROFT
LYDNEY
S.&W.
WOOL ASTON
TIDENHAM
STOW
SHARPNESS
BERKELEY
BERKELEY RD.
BASIN
R. Severn
SEVERN TUNNEL
NEW PASSAGE
PILNING
THORNBURY
TYTHERINGTON
IRON ACTON
PATCHWAY
FRAMPTON COTTERELL
YATE
WESTERLEIGH JN
SHIREHAMPTON
SEA MILLS
FILTON
CLIFTON DOWN
MONTPELLIER
STAPLETON RD.
FISH PONDS
MANGOTSFIELD
WARMLEY
BRISTOL
CLIFTON BRIDGE
LAWRENCE HILL
PILL
BEDMINSTER
BRISLINGTON
BITTON
KELSTON
WESTON
BATH
BATHAMPTON
BOX
CORSHAM
MARSHFIELD
SALTFORD
TWERTON
PENSFORD
CLUTTON
HALLATROW
MIDSOMER NORTON
CHILCOMPTON
BINEGAR
MASBURY
WELTON
RADSTOCK
WELLOW
MIDFORD
MELLS & BABINGTON
VOBSTER
FROME
SHEPTON MALLET
DOULTING
CRANMORE
WANSTROW
WITHAM
EVERCREECH
EVERCREECH JN
CASTLE CARY
PYLLE
PENNARD
WELLS
DORSET
COLE
BRUTON
WINCANTON
TEMPLECOMBE
HENSTRIDGE
STALBRIDGE
MILBORNE PORT
SHERBORNE
YEOVIL
YEOVIL JUNC.
SUTTON BINGHAM
YETMINSTER
EVERSHOT
CERNE ABBAS
MAIDEN
SPARKFORD
MARSTON
MARTOCK
STURMINSTER NEWTON
SHILLINGSTONE
BLANDFORD
SPETISBURY
BAILEY GATE
WEST MOORS
CRANBORNE
MIDLAND JOINT
EVESHAM
BENGEWORTH
HINTON
ASHTON UNDER HILL
BECKFORD
BREDON
FLADBURY
DEFFORD
ECKINGTON
UPTON
RIPPLE
HONEYBOURNE
CAMPDEN
BLOCKLEY
SHIPSTON
MORETON IN THE MARSH
CHIPPING NORTON
ADDLESTROP
STOW ON THE WOLD
BOURTON ON THE WATER
CHIPPG NORTH JUNC.
SHIPTON
ASCOT
CHARLBURY
NORTHLEACH
BURFORD
BANBURY
CROPREDY
KINGS SUTTON
AYNHO
DEDDINGTON
SOMERTON
HEYFORD
KIRKLINGTON
WOODSTOCK ROAD
HANDBORO
YARNTON
OXFORD
OXFORD RD. JUNC.
ISLIP
LONDON
WITNEY
SOUTH LEIGH
EYNSHAM
BAMPTON
ALVESCOT
LECHLADE
FAIRFORD
R. Thames
KENNINGTON JUNC.
LITTLEMORE
RADLEY
ABINGDON
CULHAM
STROUD
STONEHOUSE
STANDISH JUNC.
HARESFIELD
BRIMSCOMB
CIRENCESTER
MINCHINHAMPTON
NAILSWORTH
WOODCHESTER
DUDBRIDGE
RYEFORD
COALEY JUNC.
CAM
DURSLEY
TETBURY
TETBURY RD.
KEMBLE JUNC.
WOOTTON
CHARFIELD
WICKWAR
MIDLAND
Cotswold Hills
MALMESBURY
MINETY
CRICKLADE
PURTON
HIGHWORTH
SWINDON
FARINGDON
STEVENTON
DIDCOT
CHALLOW
UFFINGTON
WANTAGE ROAD
WANTAGE
SHRIVENHAM
Ilsley Downs
EAST ILSLEY
LAMBOURN
MOULSFORD
GORING
PANGBOURNE
WOOTTON BASSETT
DAUNTSEY
CALNE
CHIPPENHAM
THINGLEY
R. Avon
Marlborough Downs
MARLBOROUGH
LIMPLEY STOKE
FRESHFORD
BRADFORD
MELKSHAM
SEMINGTON
HOLT
SEEND
TROWBRIDGE
DEVIZES
WOODBOROUGH
PEWSEY
SAVERNAKE
BURBAGE
BEDWIN
HUNGERFORD
KINTBURY
NEWBURY
THATCHAM
Inkpen Beacon
KINGS CLERE
WESTBURY
WARMINSTER
HEYTESBURY
CODFORD
WILEY
LANGFORD
WISHFORD
WILTON
Salisbury Plain
Stonehenge
AMESBURY
LUDGERSHALL
BASINGSTOKE
OAKLEY
WHITCHURCH
OVERTON
ANDOVER TOWN
ANDOVER JUNC.
CLATFORD
FULLERTON
GRATELEY
PORTON
MICHELDEVER
ITCHEN ABBAS
STOCKBRIDGE
HORSEBRIDGE
WINCHESTER
MOTTISFONT
KIMBRIDGE JUNC.
DUNBRIDGE
ROMSEY
BISHOPSTOKE
CHANDLERS FORD
BISHOPS WALTHAM
SALISBURY
DINTON
TISBURY
SEMLEY
GILLINGHAM
SHAFTESBURY
MERE
HINDON
ALDERBURY JUNCTION
DEAN
DOWNTON
BREAMORE
FORDINGBRIDGE
DAGGON'S ROAD
VERWOOD
RINGWOOD
New Forest
LYNDHURST ROAD
TOTTON
REDBRIDGE
MILLBROOK
SOUTHAMPTON
BITTERNE ROAD
WOOLSTON
SHOLING
NETLEY
BOTLEY
BROCKHURST
GOSPORT
HOLMSLEY
BROKENHURST JUNCTION STA.
GREAT WESTERN
LONDON & SOUTH WESTERN
L. & S. W.
WORCESTER
GLOUCESTER
OXFORD
BERKS
WILTSHIRE
SOMERSET
HAMPSHIRE

LONDON
BUCKINGHAM
BEDFORD
HERTFORD
MIDDLESEX
ESSEX
SURREY
KENT
SUSSEX
BERKSHIRE
WOLVERTON
NEWPORT PAGNELL
BLETCHLEY
LEIGHTON
AYLESBURY
TRING
BERKHAMSTEAD
WATFORD
RICKMANSWORTH
ST ALBANS
LUTON
DUNSTABLE
HARPENDEN
HITCHIN
STEVENAGE
BIGGLESWADE
ROYSTON
BALDOCK
HATFIELD
HERTFORD
WARE
BUNTINGFORD
BISHOPS STORTFORD
SAFFRON WALDEN
AUDLEY END
NEWPORT
ELSENHAM
STANSTEAD
DUNMOW
BRAINTREE
HARLOW
ROYDON
BROXBOURNE
CHESHUNT
WALTHAM
ENFIELD
EPPING
ONGAR
CHELMSFORD
INGATESTONE
BRENTWOOD
BILLERICAY
ROMFORD
HAROLD WOOD
RAINHAM
PURFLEET
GRAYS
TILBURY
SOUTHEND
GRAVESEND
ROCHESTER
HIGHAM
STRATFORD
EUSTON
PADDINGTON
VICTORIA
WATERLOO
CLAPHAM
BRENTFORD
HOUNSLOW
RICHMOND
KINGSTON
WIMBLEDON
CROYDON
SURBITON
EPSOM
SUTTON
CATERHAM
REDHILL
REIGATE
DORKING
GUILDFORD
WOKING
WEYBRIDGE
CHERTSEY
STAINES
WINDSOR
SLOUGH
MAIDENHEAD
HENLEY
TWYFORD
MARLOW
HIGH WYCOMBE
PRINCES RISBORO'
THAME
UXBRIDGE
HARROW
ALDERSHOT
FARNBOROUGH
FARNHAM
GODALMING
HASLEMERE
ALTON
PETERSFIELD
MIDHURST
CHICHESTER
HAVANT
EMSWORTH
ARUNDEL
HORSHAM
CRAWLEY
THREE BRIDGES
EAST GRINSTEAD
HAYWARDS HEATH
LEWES
BRIGHTON
HOVE
SHOREHAM
STEYNING
HAILSHAM
POLEGATE
TUNBRIDGE
TUNBRIDGE WELLS
SEVENOAKS
WESTERHAM
DARTFORD
BEXLEY
CHISLEHURST
GREAT WESTERN
LONDON & NORTH WESTERN
MIDLAND
GREAT NORTHERN
GREAT EASTERN
SOUTH EASTERN
LONDON & SOUTH WESTERN
L.B.& S.C.
LONDON BRIGHTON & SOUTH COAST
Woolmer Forest
Tilgate Forest
South Downs
The Weald

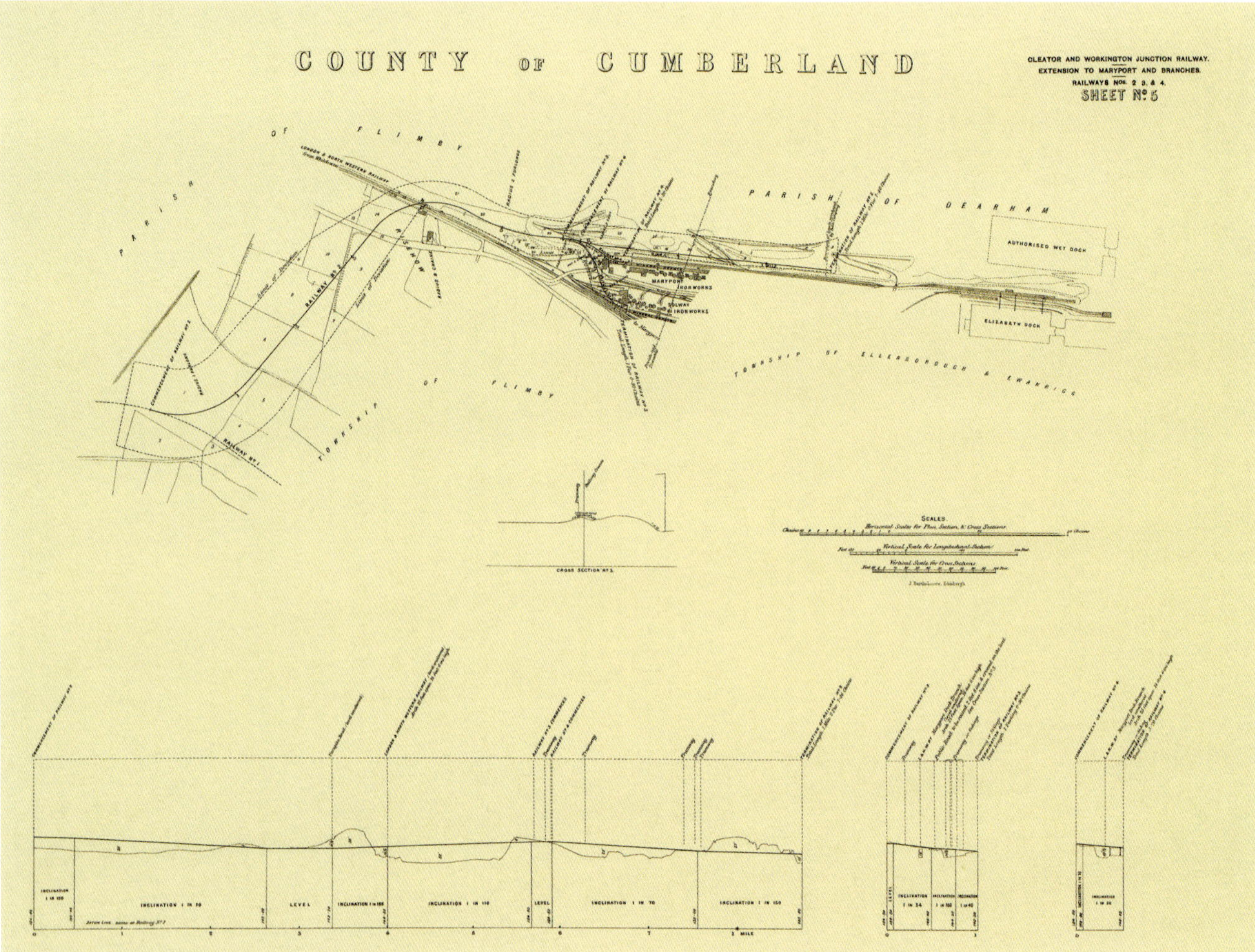

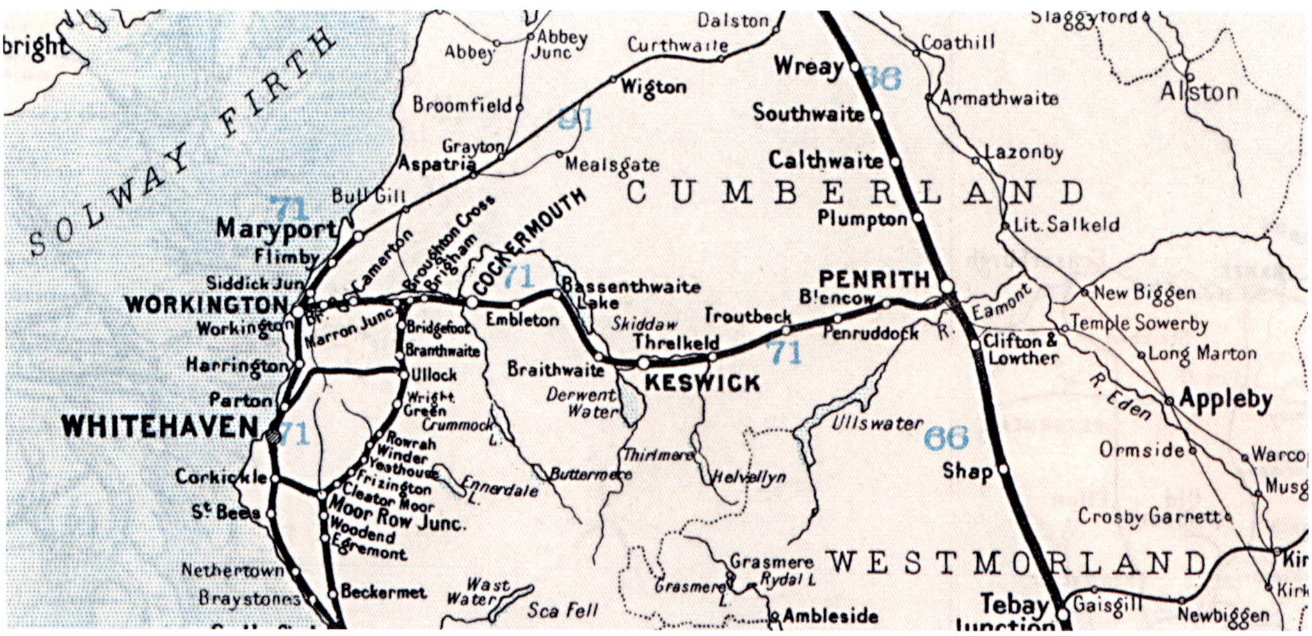

An extract from an LNWR timetable map of 1887, showing Maryport's position in the wider rail network.

Title	Sheet No. 5 – Cleator & Workington Junction Railway. Extension to Maryport and branches. Plans and sections. Session 1880–81
Engineers	John Wood, A H Strongitharm
Publisher	J Bartholomew, Edinburgh
Date	c.1880

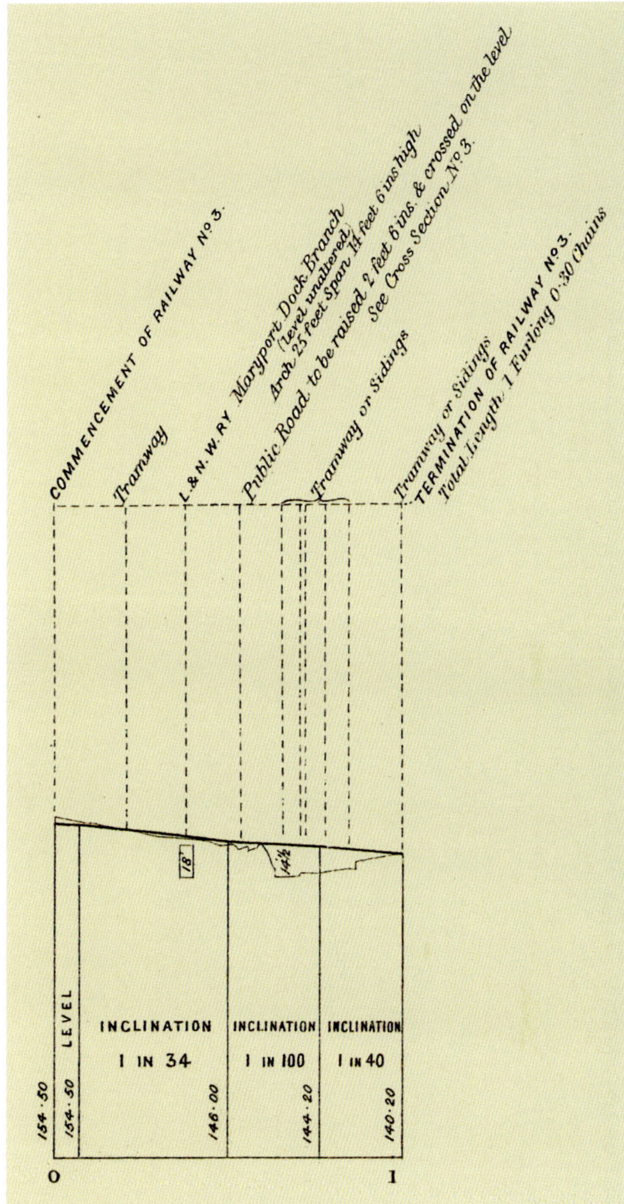

The map opposite and overleaf, originally printed at a scale of 1 in. to 5 chains (110 yd), illustrates the level of detail which had to be submitted to Parliament when railway companies proposed to build new lines. The wider network context is shown by the extract opposite from a London and North Western Railway (LNWR) timetable map.

The Cleator & Workington Junction Railway (CWJR) was an independent concern promoted by iron and coal interests in west Cumberland. The original goal had been to completely avoid the coastal route and high charges of the LNWR by constructing an inland line from Cleator Moor to Maryport, where connection would have been made with the Maryport & Carlisle Railway (MCR) for onward access to Scotland and the Newcastle area.

Initially, only the section from Cleator Moor to Siddick, on the north side of Workington, was built, opening in 1879. At Siddick connection was made with the LNWR and it would be some years before an independent extension to join up with the MCR could be realised.

The proposed junction just south of Maryport shown on this map was actually part of the second attempt at effecting this extension – the original intention had been to join the MCR just east of Maryport – but this would also be dropped in its turn in favour of a more ambitious scheme to make a direct connection with the Solway Junction Railway (and hence into Scotland) at Brayton Junction between Aspatria and Wigton. Foreseeing a loss of income, the MCR objected to this proposal.

The CWJR eventually got its independent northern outlet in 1887 by making a junction with the MCR's Bullgill to Brigham branch at Linefoot some three miles southeast of Maryport. So the time and money invested in surveying the route shown on the map, and in producing the map and associated documentation, were all wasted – except in terms of the interest for subsequent map readers.

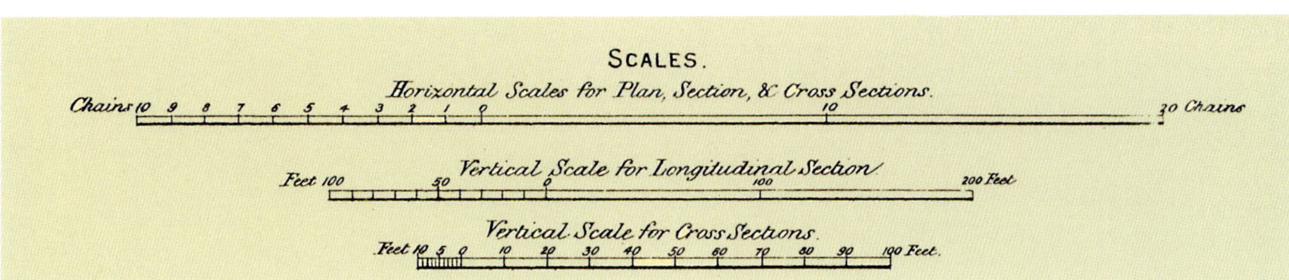

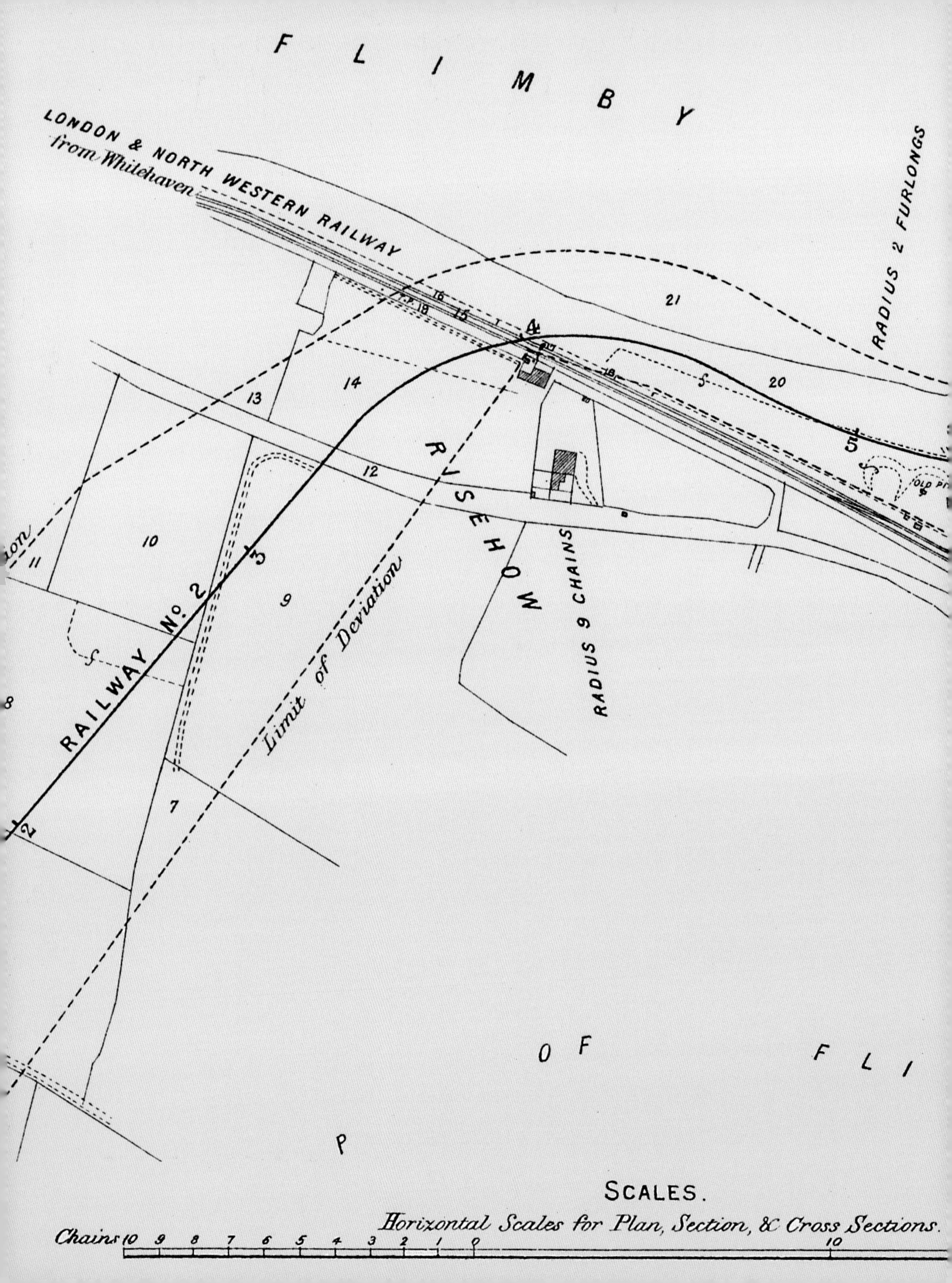

FLIMBY
LONDON & NORTH WESTERN RAILWAY
from Whitehaven
RADIUS 2 FURLONGS
RYSEHOW
RADIUS 9 CHAINS
RAILWAY No. 2
Limit of Deviation
OF FLI
P
SCALES.
Horizontal Scales for Plan, Section, & Cross Sections.
Chains 10 9 8 7 6 5 4 3 2 1 0 10

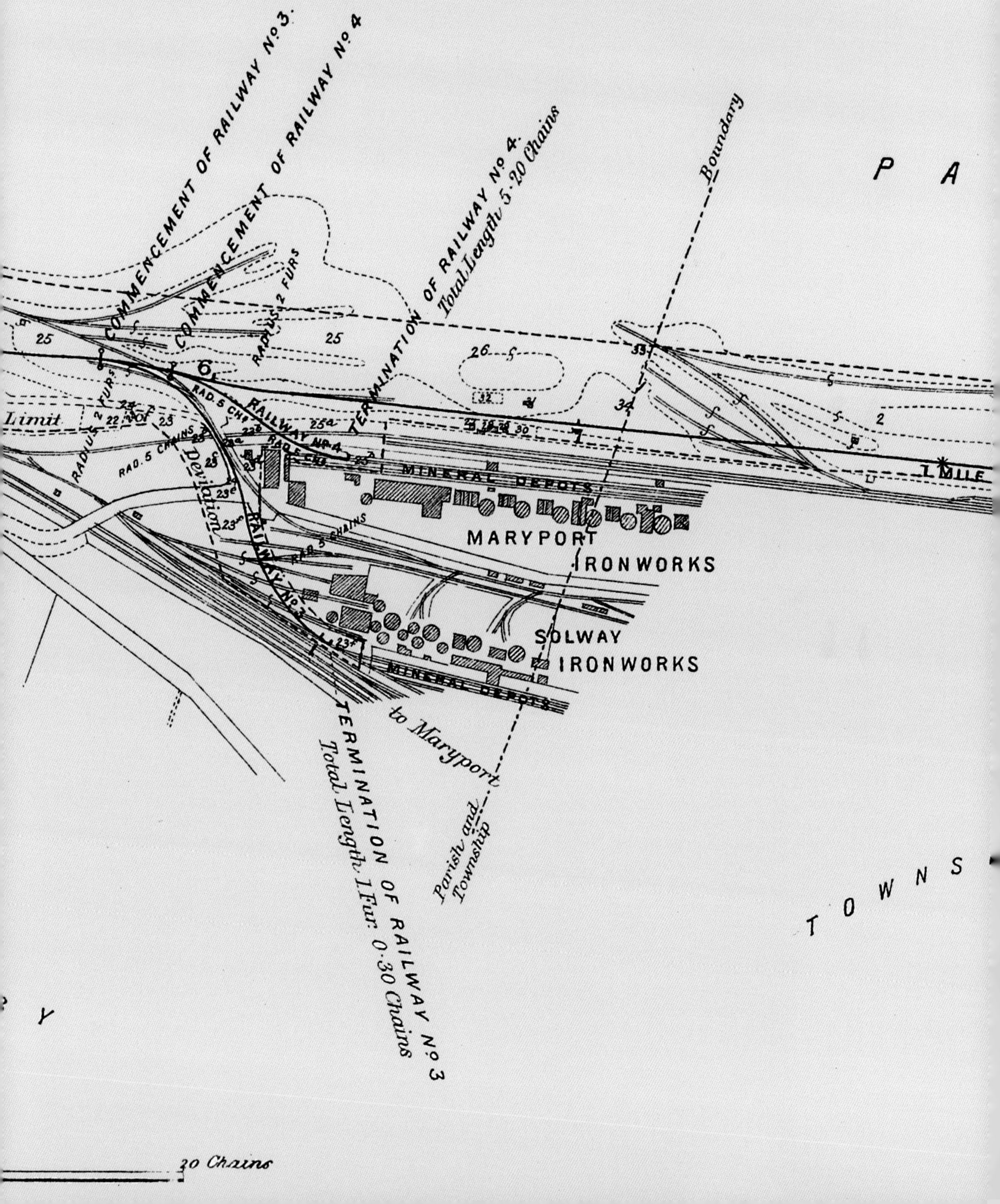
COMMENCEMENT OF RAILWAY No. 3.
COMMENCEMENT OF RAILWAY No. 4
TERMINATION OF RAILWAY No. 4.
Total Length 5·20 Chains
Boundary
P A
RAILWAY No. 4
RAILWAY No. 3
Deviation
Limit
MINERAL DEPOTS
MARYPORT
IRONWORKS
SOLWAY
IRONWORKS
MINERAL DEPOTS
to Maryport
TERMINATION OF RAILWAY No. 3
Total Length 1 Fur. 0·30 Chains
Parish and Township
T O W N S
Y
1 MILE
30 Chains

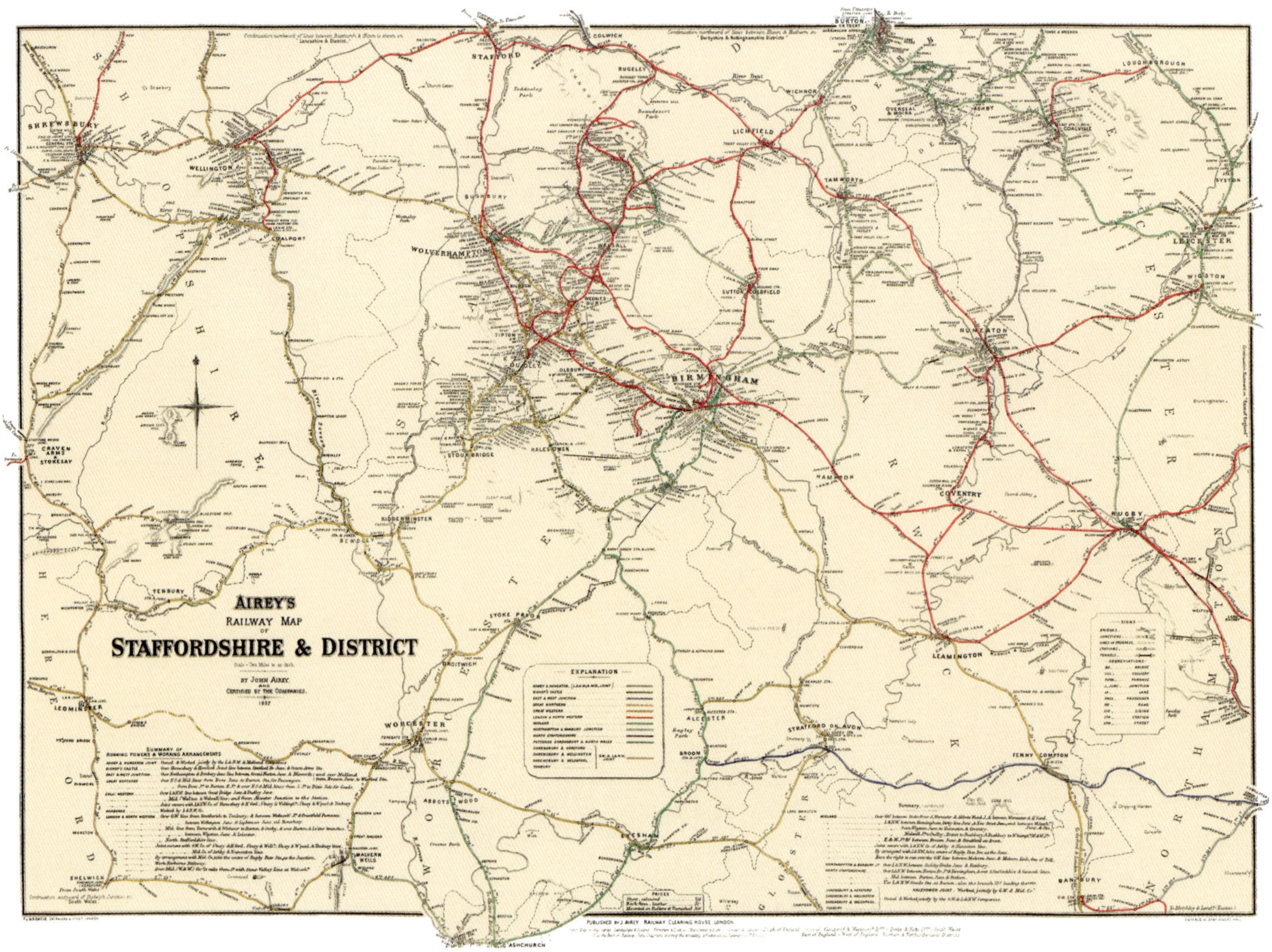

Title	Airey's Railway Map of Staffordshire & District
By	Railway Clearing House
Date	1887

This sheet, at a scale of 2 miles to 1 in., was one of a series of 13 regional maps available at the time from the Clearing House. In addition to railways, county boundaries are marked, together with some topographical detail of hills, rivers and canals. The map key contains not just the usual features such as bridges, roads and streets, but also 'FURN' for furnace, illustrating a widespread manifestation of the Industrial Revolution during that period.

To the west of the map and in the extract opposite is the Severn Valley Railway, whose north–south alignment and lack of direct connection to Birmingham made it an inevitable casualty in the modern era. It was already going through the closure procedure when the Beeching Report was published in 1963, but fortunately was later revived, as far north as Bridgnorth, as one of the longest preserved steam railways in Britain. Nearby Wellington was a major junction – partly through inter-company competition – where routes converged from six different directions, but is nowadays just an intermediate stop on the surviving Wolverhampton–Shrewsbury line.

This map provides some evidence for David Smith's comment in *Maps and Plans for the Local Historian and Collector* (1988) that, 'Despite Airey's high reputation, many of his maps are difficult to date and some were certainly prepared by careless and inexpert staff.' A Bridgnorth to Wolverhampton line is shown as 'in progress', but in fact it was never built.

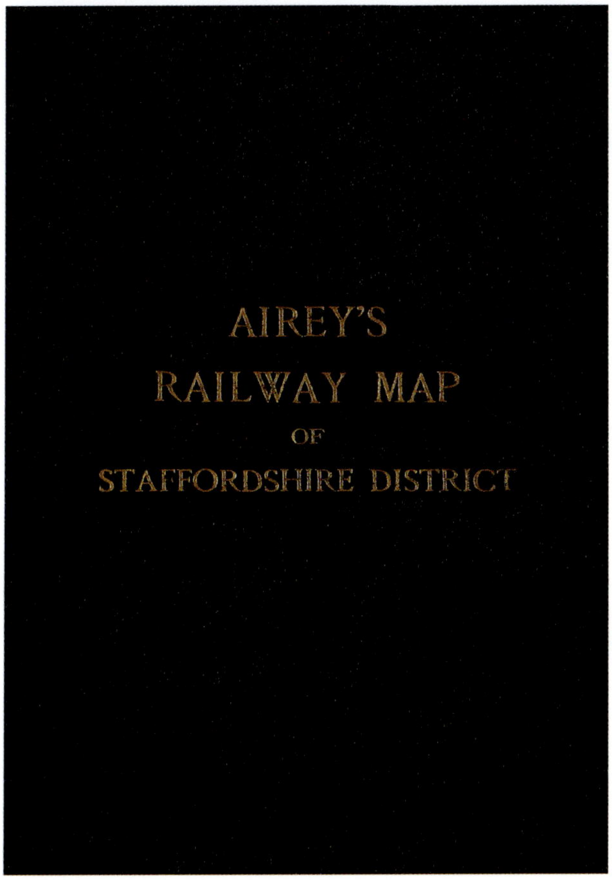
AIREY'S
RAILWAY MAP
OF
STAFFORDSHIRE DISTRICT

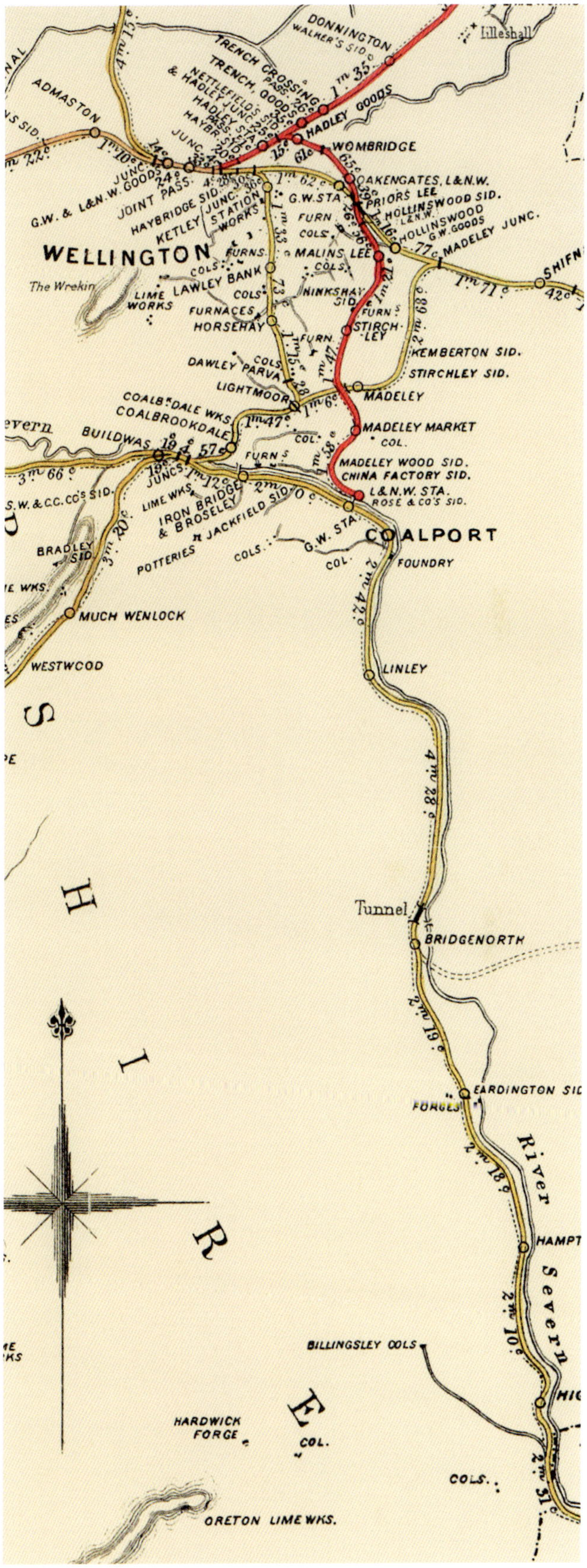
WELLINGTON
DONNINGTON
WALKER'S SID.
TRENCH CROSSING
TRENCH, GOODS
HADLEY GOODS
WOMBRIDGE
ADMASTON
G.W. & L.&N.W. GOODS
JOINT PASS.
HAYBRIDGE SID.
KETLEY JUNC.
OAKENGATES, L.&N.W.
G.W. STA.
PRIORS LEE
HOLLINSWOOD SID.
HOLLINSWOOD
G.W. GOODS
MADELEY JUNC.
SHIFN
MALINS LEE
The Wrekin
LIME WORKS
LAWLEY BANK
FURNACES
HORSEHAY
NINKSHAY SID.
STIRCH-LEY
KEMBERTON SID.
STIRCHLEY SID.
DAWLEY PARVA
LIGHTMOOR
MADELEY
COALB^{K}DALE WKS.
COALBROOKDALE
BUILDWAS
MADELEY MARKET
MADELEY WOOD SID.
CHINA FACTORY SID.
L.&N.W. STA.
ROSE & CO'S SID.
S.W. & C.C. CO'S SID.
LIME WKS.
IRON BRIDGE & BROSELEY
JACKFIELD SID.
BRADLEY SID.
POTTERIES
G.W. STA.
COALPORT
FOUNDRY
MUCH WENLOCK
WESTWOOD
LINLEY
Tunnel
BRIDGENORTH
EARDINGTON SID.
FORGES
River Severn
HAMPT
BILLINGSLEY COLS.
HARDWICK FORGE
COL.
COLS.
ORETON LIME WKS.
S H I R E

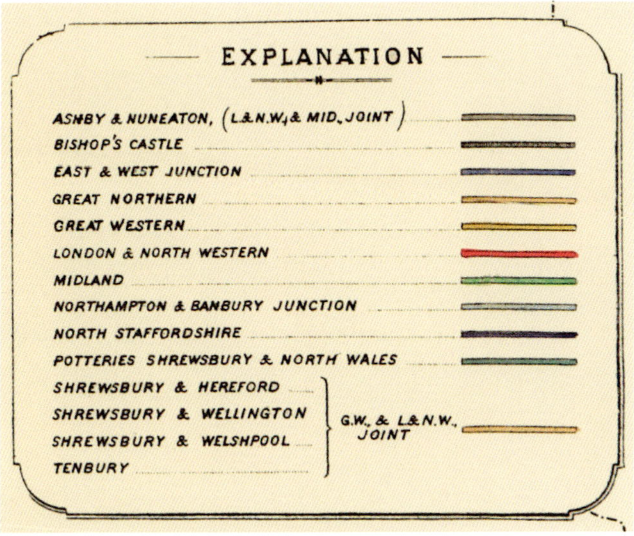
EXPLANATION
ASHBY & NUNEATON, (L.&N.W. & MID. JOINT)
BISHOP'S CASTLE
EAST & WEST JUNCTION
GREAT NORTHERN
GREAT WESTERN
LONDON & NORTH WESTERN
MIDLAND
NORTHAMPTON & BANBURY JUNCTION
NORTH STAFFORDSHIRE
POTTERIES SHREWSBURY & NORTH WALES
SHREWSBURY & HEREFORD
SHREWSBURY & WELLINGTON
SHREWSBURY & WELSHPOOL
TENBURY
G.W. & L.&N.W. JOINT

STAFFORD
RUGELEY
CANNOCK CHASE
Teddesley Park
Church Eaton
Wheaton Aston
PENKRIDGE
GAILEY
BREWOOD
FOUR ASHES
Chillington Hall
Boscobel Oak
White Ladies
SHIFNAL
ALBRIGHTON
CODSALL
BUSHBURY
Wrottesley Park
WOLVERHAMPTON
WILLENHALL
WALSALL
BILSTON
WEDNESBURY
Sedgley
Wombourne
TIPTON
DUDLEY
OLDBURY
STOURBRIDGE
HALESOWEN
KIDDERMINSTER
HAGLEY
CLENT HILLS
HAMPTON LOADE
HIGHLEY
ARLEY
River Severn
Kinver
HARBORNE
SELLY OAK
NORTHFIELD
BIRMINGHAM & LIVERPOOL JUNCTION CANAL
STAFFORD & WORCESTER CANAL
Lappal Tunnel
BROMSGROVE

River Trent
WICHNOR
LICHFIELD
TRENT VALLEY STA.
TRENT & MERSEY CANAL
ARMITAGE
ALREWAS
CROXALL
BARTON & WALTON
GRAND TRUNK CANAL
HASELOUR & ELFORD
HAMMERWICH
CANNOCK CHASE
SHENSTONE
BLAKE STREET
FOUR OAKS
STREETLY
SUTTON COLDFIELD
MIDLAND STA.
WYLDE GREEN
CHESTER ROAD
ERDINGTON
GRAVELLY HILL
PENNS
TAMWORTH
AMINGTON COL. SID.
TAMWORTH COL. SID.
POOLEY HALL COL. SID.
POLESWORTH
KETTLEBROOK
WILNECOTE COL.
WILNECOTE & FAZELEY
TAME VALLEY COL. JN
HOCKLEY HALL SID.
DOSTHILL COL. SID.
WHATELEY COL. SID.
R. Tame
KINGSBURY WOOD COL. SID.
BAXTERLEY PARK OR BADDESLEY COL.
KINGSBURY
WHITACRE, GOODS
SHUSTOKE
COVENTRY CANAL
BIRMINGHAM & FAZELEY CANAL
ATHERSTONE
BADDESLEY COL. CO'S SID.
HARTSHILL SID.
MANGANESE PITS
ANSLEY COLS.
NUNEATON
WEDDINGTON JUNC.
ABBEY STA. MID.
STOCKINGFORD STA.
BRIDGE STA.
LEICESTER LINE JUNC.
STANLEY SID.
CHILVERS COTON
GRIFF COL. SID.
SEAL'S SID.
ASHBY CANAL
CHARITY COL. SID.
BEDWORTH
BULKINGTON
LIME WORKS
HAWKESBURY COL. SID.
HAWKESBURY LANE
LONGFORD & EXHALL
WYKEN COL.
FOLESHILL
COUNDEN ROAD STA.
COVENTRY
Coomb Abbey
WHITLEY COAL WHARF
ARLEY & FILLONGLEY
COLESHILL
WATER ORTON STA.
CASTLE BROMWICH
BROMYARD FORGE
BIRMINGHAM
STECHFORD FOR YARDLEY STA.
ADDERLEY PARK
SAINT ANDREWS JUNC.
BORDESLEY JUNC.
SMALLHEATH & SPARKBROOK
ACOCK'S GREEN & S.TH YARDLEY
CAMP HILL & BALSALL HEATH
BRIGHTON ROAD
MOSELEY
KING'S HEATH
OLTON
SOLIHULL
KNOWLE
MARSTON GREEN
MIDLAND STA.
HAMPTON
L.& N.W. STA.
BERKSWELL STA.
TILE HILL
Tunnel
SWADLINCOTE
WOODVILLE
GRESLEY
OVERSEAL & MOIRA
MOIRA
ASHBY
DONISTHORPE STATN
MEASHAM
SNARESTONE
SHACKERSTONE
HEATHER STA.
Twycross
MARKET BOSWORTH
SHENTON
STOKE GOLDING
HIGHAM ON THE HILL
DERBY
W
A
R

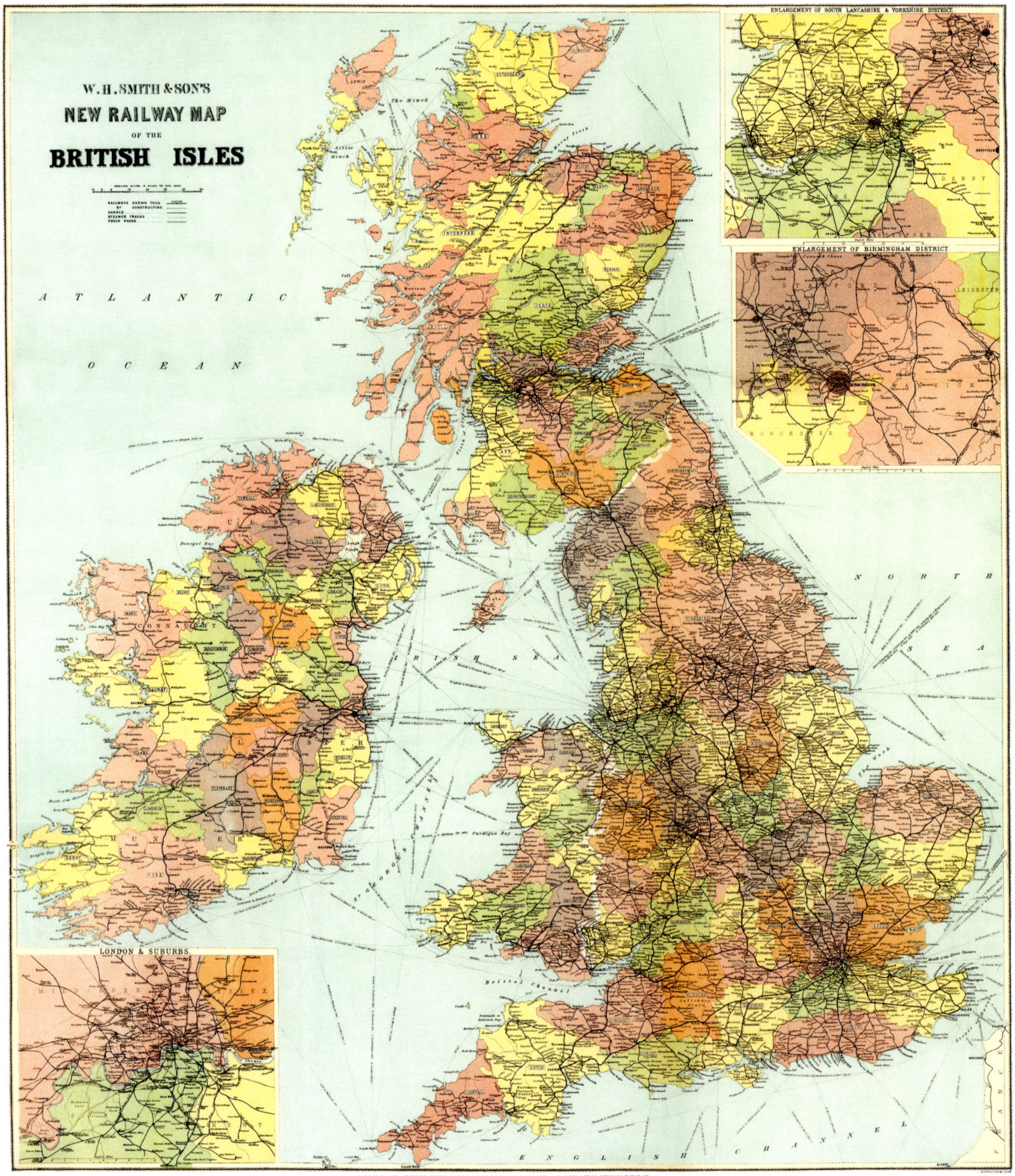

Title	W H Smith & Son's New Railway Map of the British Isles
By	J Bartholomew, Edinburgh
Date	1887

The long-established High Street name of W H Smith (whose origins lie in 18th-century London) took advantage of the Victorian railway boom by developing a network of news-stands at stations throughout England and Wales (in Scotland, John Menzies dominated the market until their takeover by Smith's in the late 20th century).

It was only natural that Smith's would want to commission rail network maps, in this case from the famous Edinburgh mapmaker, John Bartholomew. The original is a fold-up map, allowing different sections of the country to be viewed without spreading the entire large sheet out. As the map and extracts show, a very large proportion of the rail network was complete by 1887.

The extract below focuses on the counties of Devon, Dorset, Somerset and Wiltshire, within which by far the most notable absence is the completed main lines of the Great Western Railway (GWR). A further two decades were to elapse before the GWR had opened all its important 'cut-offs' which shortened the distance between Paddington station in London and Temple Meads in Bristol (by means of the Swindon–Badminton–Bristol section in 1903), and between Paddington and Exeter (Stert to Westbury in 1900, and Castle Cary to Taunton in 1906).

Amongst the relatively few other lines still to be built were the Blagdon branch in Somerset, the Lyme Regis branch on the Devon/Dorset border, the meandering cross-country route from Halwill Junction in north Devon via Launceston to Padstow in north Cornwall, and one of the very last railways to be constructed in the pre-modern era – the Halwill Junction to Torrington link which was opened as a standard gauge light railway as late as 1925. Across the Bristol Channel in south Wales, the Vale of Glamorgan Railway from Cardiff to Bridgend via Barry had also yet to open.

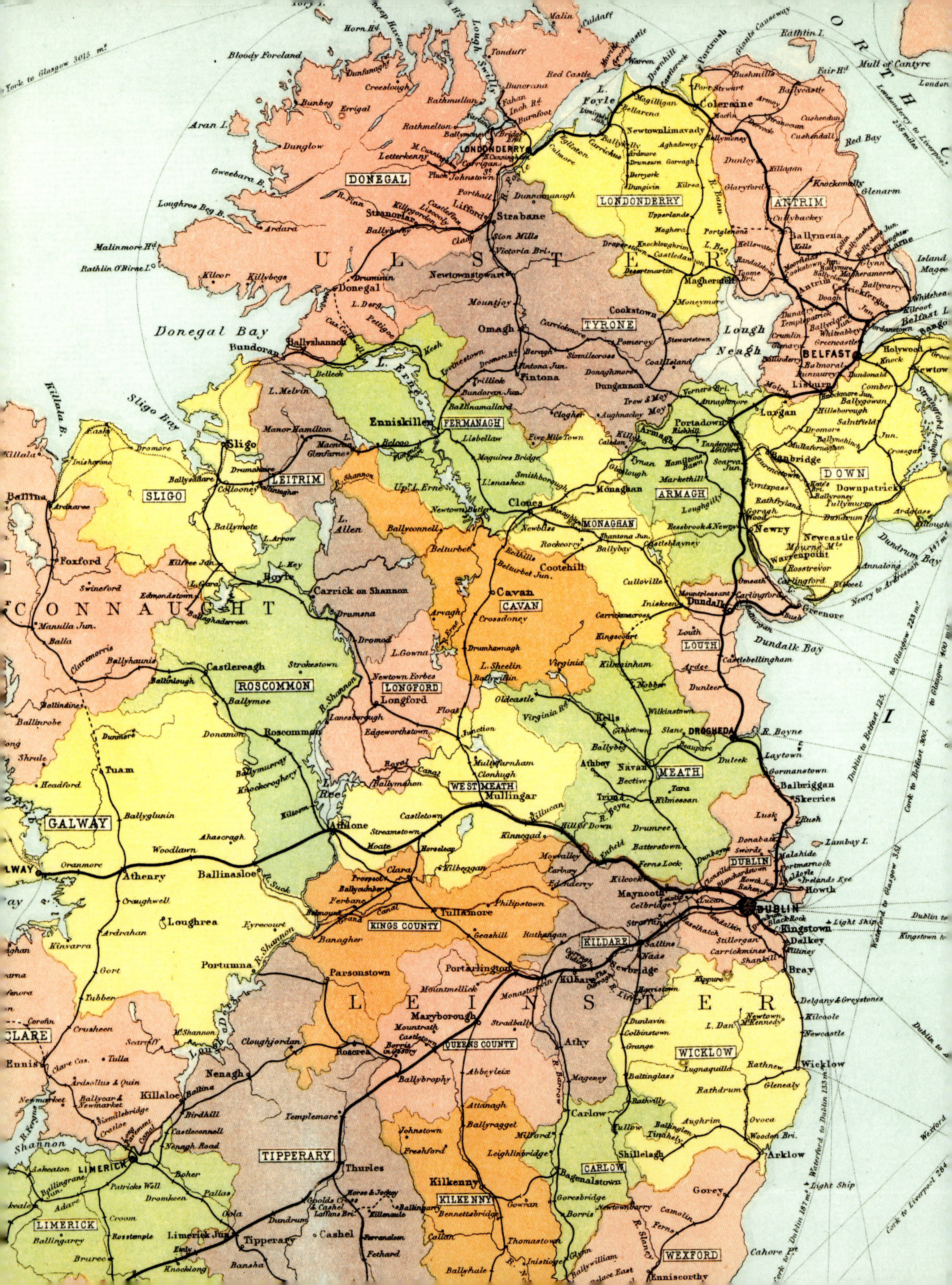

ULSTER
CONNAUGHT
LEINSTER
DONEGAL
LONDONDERRY
ANTRIM
TYRONE
FERMANAGH
DOWN
ARMAGH
MONAGHAN
CAVAN
LOUTH
SLIGO
LEITRIM
ROSCOMMON
LONGFORD
MEATH
WEST MEATH
DUBLIN
GALWAY
KINGS COUNTY
KILDARE
QUEENS COUNTY
WICKLOW
CARLOW
KILKENNY
TIPPERARY
LIMERICK
CLARE
WEXFORD
LONDONDERRY
BELFAST
DUBLIN
LIMERICK
DROGHEDA
Donegal Bay
Sligo Bay
Dundalk Bay
Dundrum Bay
Lough Neagh
Bloody Foreland
Aran I.
Malin
Rathlin I.
Mull of Cantyre
Lambay I.
Coleraine
Portrush
Ballymena
Larne
Lisburn
Portadown
Newry
Dundalk
Armagh
Enniskillen
Omagh
Strabane
Letterkenny
Sligo
Ballina
Foxford
Boyle
Carrick on Shannon
Longford
Mullingar
Athlone
Athenry
Ballinasloe
Tuam
Loughrea
Tullamore
Maryborough
Parsonstown
Roscrea
Nenagh
Thurles
Tipperary
Cashel
Kilkenny
Carlow
Athy
Kildare
Maynooth
Kingstown
Bray
Wicklow
Arklow
Gorey
Enniscorthy
Navan
Kells
Cavan
Clones
Monaghan
Ennis
Killaloe
Howth
Warrenpoint
Newcastle
Downpatrick
Greenore

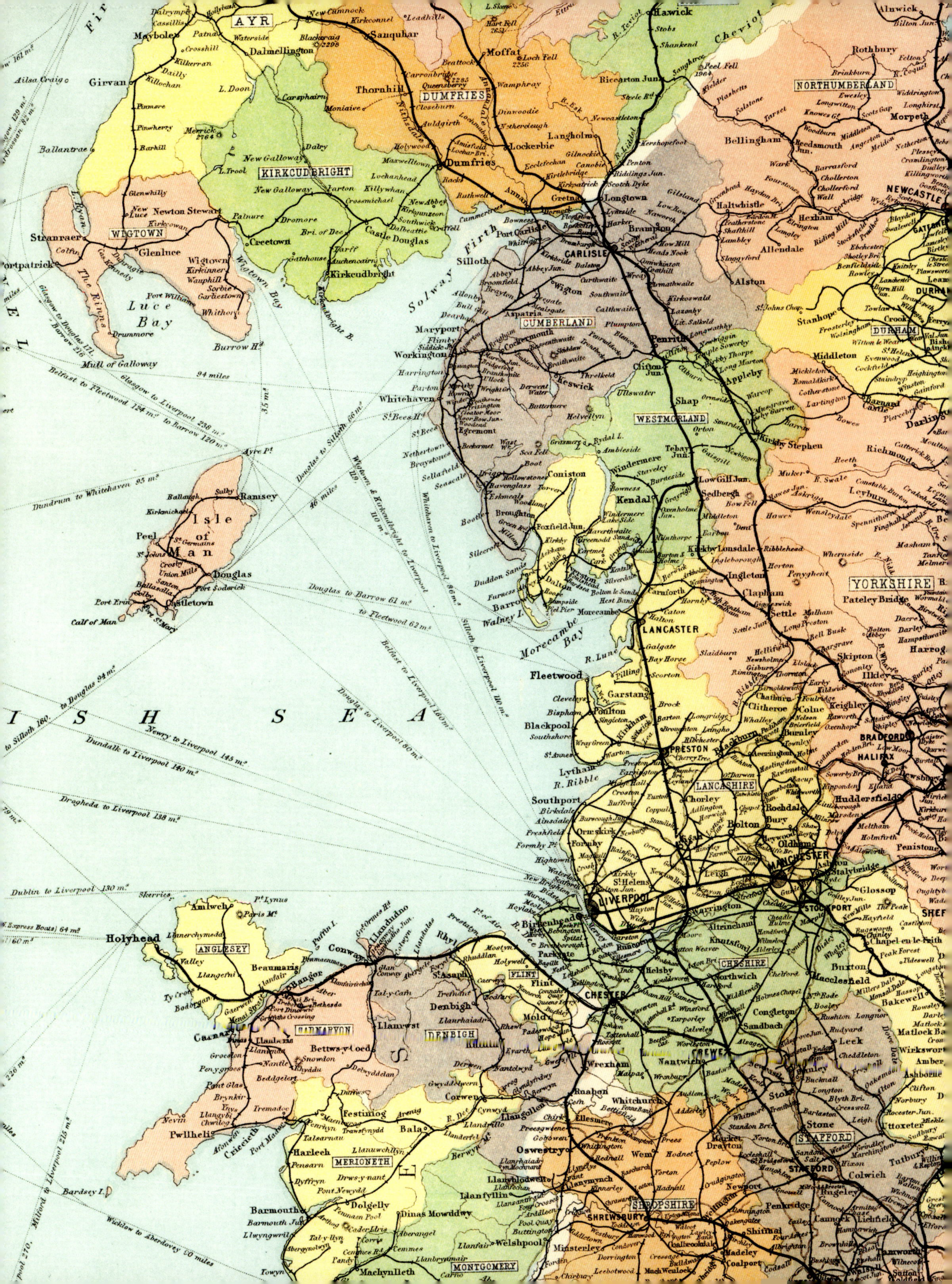

AYR
DUMFRIES
KIRKCUDBRIGHT
WIGTOWN
NORTHUMBERLAND
CUMBERLAND
DURHAM
WESTMORLAND
YORKSHIRE
LANCASHIRE
ANGLESEY
CARNARVON
DENBIGH
FLINT
CHESHIRE
MERIONETH
SHROPSHIRE
STAFFORD
MONTGOMERY
Isle of Man
Luce Bay
Solway Firth
Wigtown Bay
Morecambe Bay
I S H S E A
Mull of Galloway
Calf of Man
Carlisle
Dumfries
Newcastle
Lancaster
Preston
Liverpool
Manchester
Chester
Stafford
Shrewsbury
Holyhead
Stranraer
Douglas
Kendal
Blackpool
Bradford
Halifax
Stockport
Crewe
Wrexham
Penrith
Whitehaven
Workington
Maryport
Fleetwood
Southport
Bolton
Oldham
Rochdale
Belfast to Fleetwood 124 m.s
Glasgow to Liverpool 239 m.s
Douglas to Barrow 61 m.s
Dublin to Liverpool 130 m.s
Dundalk to Liverpool 140 m.s
Drogheda to Liverpool 138 m.s
Newry to Liverpool 145 m.s
Douglas to Liverpool 80 m.s

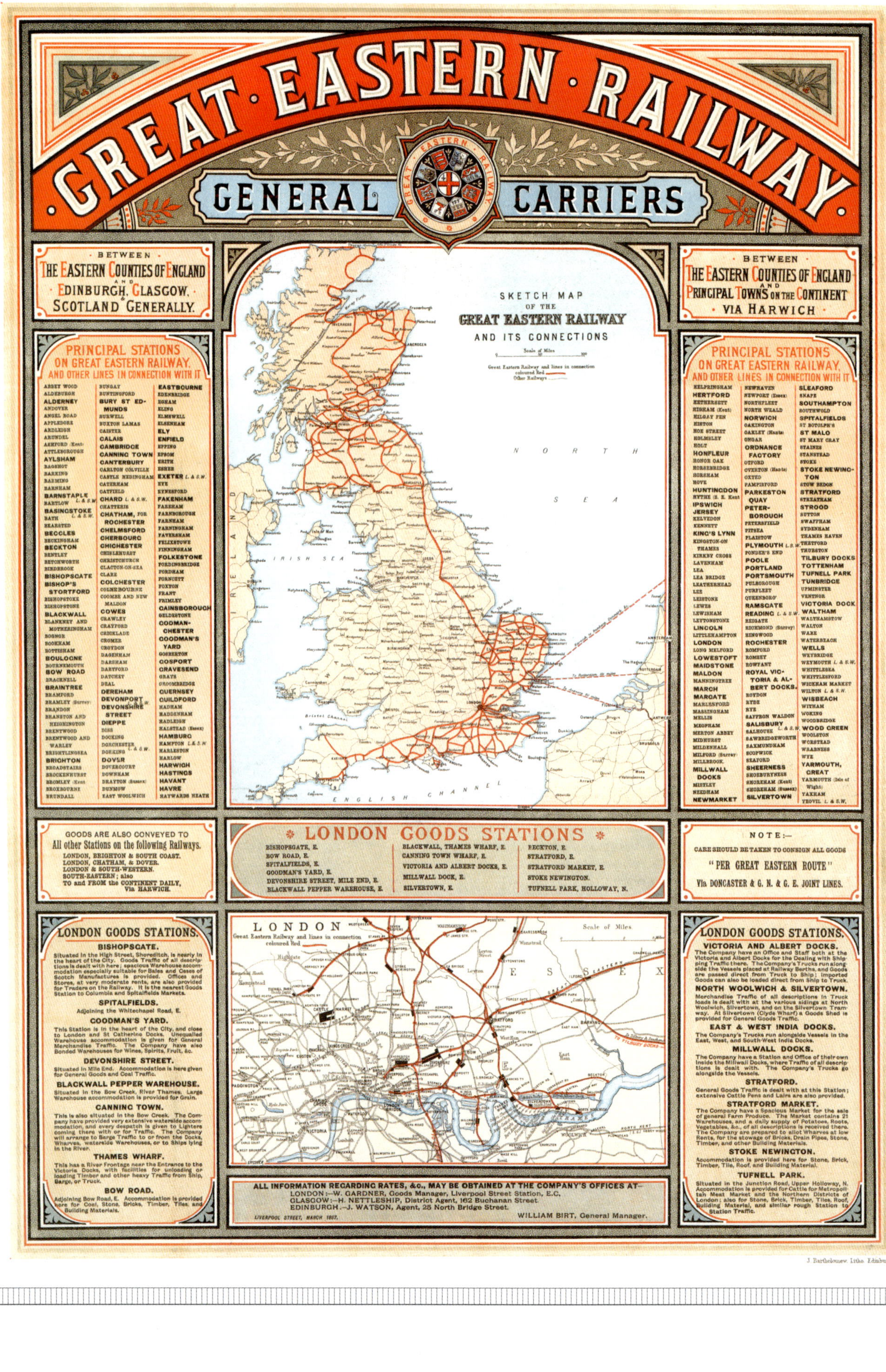

J. Bartholomew Litho. Edinburgh

Title	Great Eastern Railway. General carriers
By	J Bartholomew, Edinburgh
Date	1887

This map poster illustrates the major role of the railways in serving the merchandise needs of the nation in the late 19th century. Interestingly, in the main map, the Great Eastern Railway (GER) gives the impression of serving two quite distinct and discrete territories – southeast England and Scotland – with just one railway, the East Coast Main Line, linking the two. In practice the GER would offer through goods services to locations the length and breadth of the country, with the Railway Clearing House sorting out the complex business of financial allocations between those companies over whose rails the goods wagons travelled.

The detailed map of London identifies no less than 16 goods stations operated by the GER in its main zone of influence in the centre and east end of the city. Just about anything would be carried on the railways at that time, and the depot descriptions demonstrate the sheer variety of commodities handled by thousands of railway staff on a daily basis. Within the sprawling expanse of the Bishopsgate depot, which survived until the 1970s:

> *Goods traffic of all descriptions is dealt with here; spacious warehouse accommodation especially suitable for Bales and Cases of Scotch Manufactures is provided.*

A couple of miles east at the Bow Road depot, 'Accommodation is provided here for Coal, Stone, Bricks, Timber, Tiles and Building Materials'. Today, there is still a rail freight depot in Bow, albeit relocated to the main line, less than a mile away from the original site. Here, stone from the Mendips and East Midlands, and concrete blocks from Yorkshire, arrive by rail for distribution round London by road; while East London's waste is sent by train to landfill in Oxfordshire.

At Stratford Market, yet further east, the poster announces that:

> *The Company have a Spacious Market for the sale of general Farm Produce. The Market contains 21 warehouses, and a daily supply of Potatoes, Roots, Vegetable, &c., of all descriptions is received there.*

The railways also worked very closely with shipping services at London's many docks on the River Thames. At the Victoria & Albert Docks:

> *The Company's Trucks run alongside the vessels placed at Railway Berths, and Goods are passed direct from Truck to Ship.*

MILEPOST
8

Demolished in 1962, Philip Hardwick's Doric Arch was built at the entrance to London & Birmingham Railway's Euston station in 1836.

During the 19th century the map of London changed beyond all recognition. In less than a century the villages and green fields that once surrounded the city were replaced by the sprawl of Victorian development, all made possible by the coming of the railways and the consequent birth of a modern, mobile society.

By the 1840s the railways were approaching London from all points of the compass – the first true inter-city line to reach the capital was the London & Birmingham Railway (LBR) which opened to Euston in 1838. It took another 11 years before Philip Hardwick's famous Doric Arch and his son's Great Hall at Euston were completed. Sadly these were both demolished by British Railways in the early 1960s and replaced by the current bland modern building.

Although Euston was the earliest terminus of an inter-city railway in London, the first terminus to be built in the capital was at London Bridge which was opened by the London & Greenwich Railway in 1836. This was later developed as two stations, becoming the terminus of the London, Brighton & South Coast Railway and a through station for South Eastern & Chatham Railway trains to Charing Cross and Cannon Street.

Next on the scene was the Eastern Counties Railway (ECR) which opened its Bishopsgate terminus in 1840. By numerous takeovers of other railways the ECR became the main constituent company of the newly formed Great Eastern Railway (GER) in 1862. Outgrowing its cramped Bishopsgate terminus the GER built a grand new station on the site of Bethlem Hospital – Liverpool Street, as it was known, was opened in 1874.

By the time that the LBR had reached Euston in 1838, Brunel's broad gauge Great Western Railway was nearing completion. With its terminus at Paddington, the line to Bristol opened throughout on 30 June 1841 and it soon became a byword for speed and comfort, although the Paddington station we know today was not completed until 1854.

Many other railways were soon reaching London, all of them desperate to take a share in the burgeoning business of rail travel. The London & South Western Railway opened its terminus at Waterloo in 1848 – it was subsequently enlarged on several occasions and is now the busiest station in Britain; the Great Northern Railway opened King's Cross in 1852; the London, Brighton & South Coast Railway opened the west side of Victoria station in 1860 and the London, Chatham & Dover Railway opened the east side in 1862. Built on the site of Hungerford Market, Charing Cross station was opened by the South Eastern Railway in 1864; and the North London Railway's suburban line terminus at Broad Street opened in 1865 – it closed in 1986 and was subsequently demolished. Cannon Street was opened by the South Eastern Railway in 1866; built on the site of slums, a church and a graveyard, George Gilbert Scott's splendid gothic-revival

St Pancras was opened by the Midland Railway in 1868 - at the time its train shed had the largest single-span roof in the world. Escaping demolition in the 1960s, it has since become the terminus of High Speed trains to the Continent via the Channel Tunnel; Holborn Viaduct was opened by the London, Chatham & Dover Railway in 1874, and it closed in 1990.

Last, but by no means least, the Great Central Railway opened its London Extension to Marylebone station, one of the smallest and most charming of the London termini, in 1899. Although it lost its main line services to Nottingham in 1966, Marylebone has since seen major growth in traffic on the Chiltern Lines to Aylesbury, Bicester and Birmingham, with its four platforms increased to six in recent years.

A busy scene on Platform 1 at Paddington station in 1931.

Painted by Fred Taylor, this Midland Railway poster shows the interior of St Pancras station in 1910.

L. & N. W. R. TIME TABLE MAPS.—SOUTHERN SECTION. *For continuation Northwards*

The Blue Numbers—thus, 36—refer to the pages in the Co.'s Time Book on which the Train Service can be found. *For continuation South*

Title	L. & N.W.R. Time Table Maps – Southern Section
By	London & North Western Railway
Date	1887

The railway companies issued detailed maps with their public timetables, and the maps would typically, as in this example from the London & North Western Railway (LNWR), show the appropriate train service table number against each line on the map. Companies would often downplay the importance of their rivals, and the LNWR was no exception – relegating Great Western Railway (GWR) routes to the narrowest of lines on the map, yet elevating northwest Wales narrow-gauge lines which connected with the LNWR to the same stature as the latter.

By 1887 there were just a few rural rail routes yet to be built in Wales – including the line to the small town of Aberaeron (on the coast south of Aberystwyth), whose station did not open until 1911, and closed to passengers just 40 years later.

The LNWR managed to penetrate well beyond its core territory – deep into the South Wales Valleys at Merthyr, Ebbw Vale, Tredegar and Nine Mile Point, via the Shrewsbury–Abergavenny line, and far into the Pembrokeshire countryside via the largely single-track Central Wales Line which terminated at Swansea Victoria station – a miraculous survivor, as far south as Llanelli, to this day. This penetration was achieved by joint ownership (with the GWR) in the case of the sections of line between Llandeilo and Llandovery, and Shrewsbury and Hereford, and by 'running powers' over the GWR between Hereford and Abergavenny.

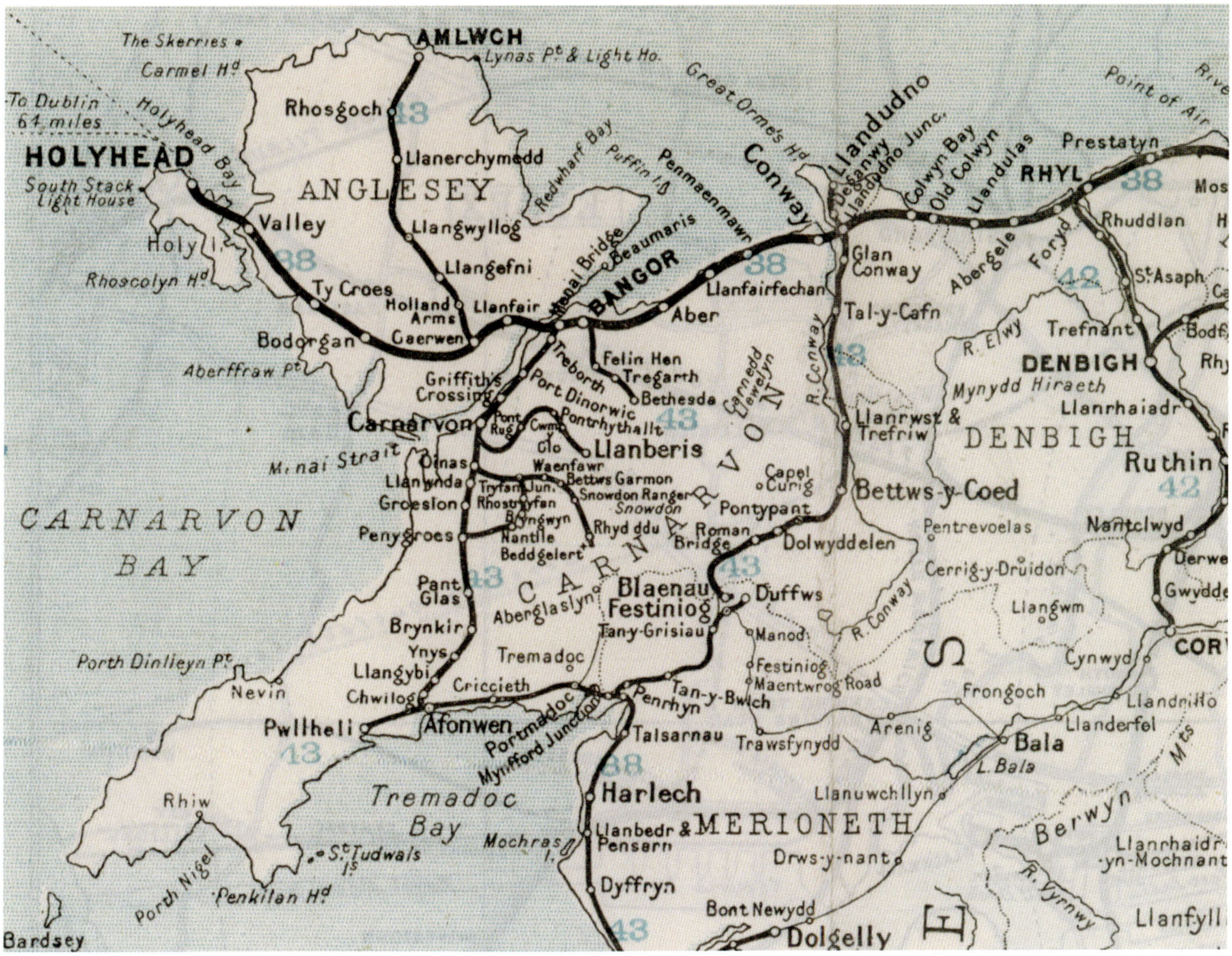

Title	General Sketch Map of the L & N.W.R. and its communications
Publisher	London & North Western Railway
Mapmaker	J Bartholomew, Edinburgh
Date	1887

Railway companies played a key part in developing the concept of the Victorian 'grand hotel', opening large and prestigious establishments the length and breadth of the country, from the 1830s through to the 1930s. The London & North Western Railway (LNWR) and its predecessor company played a pioneering role, as noted by the *Oxford Companion to British Railway History* (1997):

> *The first hotel for rail travellers was built alongside Crewe station by Lord Crewe, opened in 1837, leased by the London & North Western Railway in 1864, and bought outright in 1877. The first to be built by a railway were the twin hotels at Euston, opened in 1839 by the London & Birmingham Railway.*

A note handwritten in the margin of this LNWR 'hotel show bill' (or small poster – possibly also for use inside railway carriages) indicates that 10,025 copies were produced. It features 10 grand hotels served by the LNWR or connecting rail and shipping services, representing the apotheosis of comfort and style in the Victorian era. Fashions changed in due course, and the railway-owned hotels had varying fortunes. In June 1963 *Modern Railways* magazine recorded that:

> *The latest casualty amongst the historic buildings at the old station is Euston Hotel, which shut down on May 13 in readiness for demolition. It was the world's first railway hotel. Designed by P. Hardwick and opened about the year 1840, it was originally called 'Victoria Hotel and Dormitories' and consisted of two separate buildings with some 40 bedrooms, a large coffee room and a lounge. In 1881 the two wings were connected by a large central block, bringing the total number of bedrooms to 141, and the whole building was renamed Euston Hotel, with separate entrances to the east and west wings.*

The remaining hotels in British Rail ownership were privatized in 1981, and only a few of those shown in this brochure still exist – notably the Central Hotel in Glasgow which has recently been restored to its former grandeur.

GENERAL
SKETCH MAP
OF THE
L.&N.W.R.
AND ITS
COMMUNICATIONS
ROYAL MAIL
WEST COAST ROUTE
BETWEEN
ENGLAND
AND
SCOTLAND
Scale of English Miles
J. BARTHOLOMEW, EDINR
Steamers to Quebec and Montreal
Steamers to New York
Newcastle to Hamburg 480 miles
Hull to Hamburg 430 miles
GLASGOW
EDINBURGH
LIVERPOOL
MANCHESTER
BIRMINGHAM
LONDON
DUBLIN
GREENORE

LONDON & NORTH WESTERN
LONDON & NORTH WESTERN RAILWAY COMPANY
RAILWAY COMPANY'S HOTELS
GLASGOW, CAL RY.
PRESTON
EUSTON
EUSTON HOTEL LONDON
GLASGOW
DUBLIN
LONDON
BLAENAU FESTINIOG
NORTH WESTERN HOTEL
LONDON & NORTH WESTERN RAILWAY Co.
DUBLIN
SAUCHIEHALL ST.
BATH STREET
REGENT ST.
QUEEN ST. STATION
GEORGE SQUARE
ST VINCENT STREET
GORDON
HOTEL
CENTRAL STATION
ARGYLE STREET
TRONGATE
ST ENOCH STATION
BROOMIELAW
RIVER CLYDE
GREAT CLYDE STREET
BRIDGE ST. STA.
RIVER LIFFEY
HOTEL
GRAND CANAL
EUSTON HOTEL
GOWER STR
PORTLAND ROAD
BAKER STR.
OMNIBUS ROUTE
FARRINGDON STREET
ALDERSGATE STR
BROAD STREET
MOORGATE STR
CITY OF LONDON
HOLBORN VIADUCT
LUDGATE HILL
TEMPLE
BLACKFRIARS
MANSION HO.
CANNON STR.
MAYFAIR
CHARING CROSS
BLACKFRIARS BRI.
WATERLOO JUN
WATERLOO (South Western)
BOROUGH ROAD
WESTMINSTER BRIDGE
ST JAMES PARK

LONDON & NORTH WESTERN
RAILWAY COMPANY'S HOTELS
LONDON & NORTH WESTERN RAILWAY COMPANY
GREENORE
BIRMINGHAM
LIVERPOOL
LONDON & N.W. HOTEL, LIVERPOOL
BIRMINGHAM
HOLYHEAD
ENVIRONS OF LIVERPOOL
CREWE

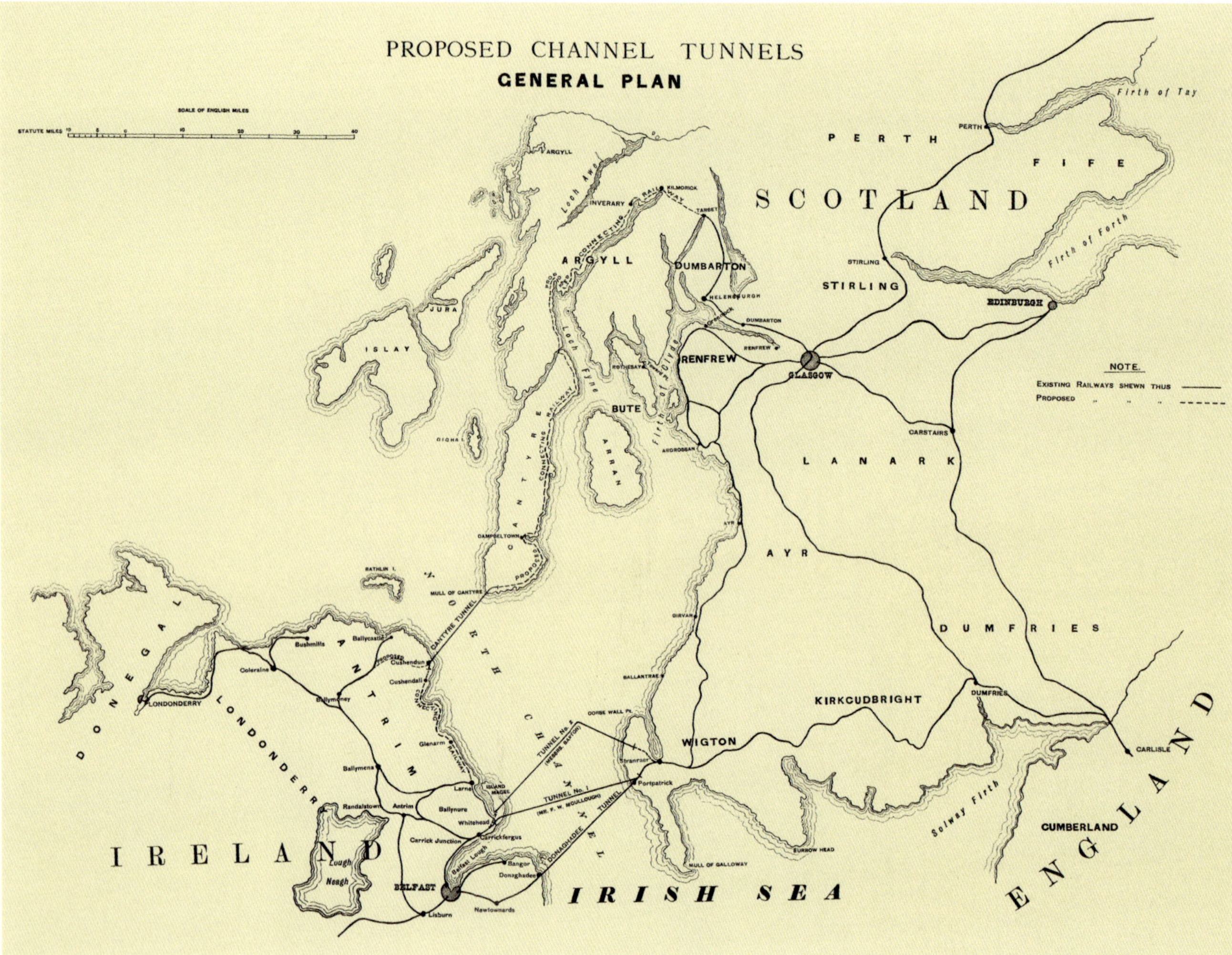

Title	The Proposed Channel Tunnel to Northern Ireland and through trains between Edinburgh, Glasgow, Manchester, Belfast, and Londonderry
By	L Livingston Macassey
Date	1890

This map comes from a fascinating prospectus setting out a variety of options for a little-known Victorian proposal to take trains under or across the North Channel of the Irish Sea between Scotland and Northern Ireland. Increased demand for travel at faster speeds is cited as a key justification for the scheme, but there were also more prosaic reasons:

> *There is, however, one thing in which time has made no change in the public mind, and that is the dread of sea sickness . . . . They would undergo the fatigue of a hundred miles trip by rail rather than risk the horrors of twenty miles in a rough sea.*

No less than seven options are reviewed, perhaps the most fantastic being a submerged tubular bridge, comprising a steel and cement outer shell and an 'inner tube' containing the railway, with the whole structure 'to be kept in position by means of chains and anchors'. Prospective passengers would perhaps not have been wholly reassured about their safety in transit:

> *At intervals of 500 feet were water-tight doors so arranged as to close in cases of emergency. The rere [sic] of trains to be so designed as to act as a piston in order that in case of an inrush of water the train might be forced out of the tube.*

At an estimated cost of £5.25 million, this would have been a bargain compared to the projected £70 million for a 'solid causeway' between the Mull of Cantyre [sic] and Northern Ireland. A Channel Bridge at £30 million is also rejected, and the map focuses on four tunnel options ranging in cost from £7.6 million to £16 million. The overall impact

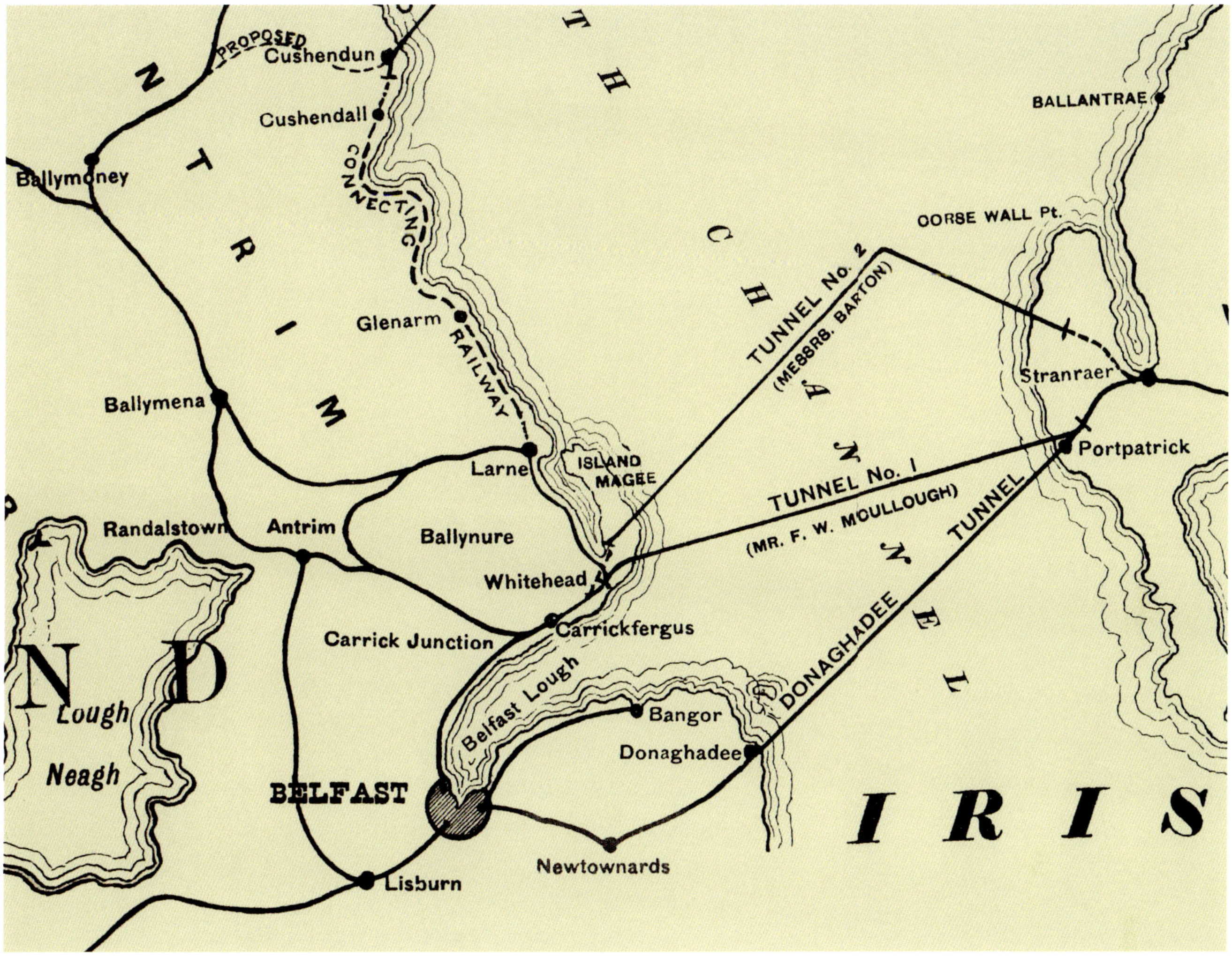

of the prospectus is, however, undermined by its concluding comments to the effect that:

> *. . . no amount of traffic likely to arise would make the tunnel a dividend-paying concern . . . the channel tunnel must be constructed at the expense of the state. No railway company or body of speculators would ever venture upon an undertaking of so doubtful a character.*

Unsurprisingly, no tunnel, bridge or causeway was ever built. If it had been, there would have been the additional difficulty of Britain's 4 ft 8½ in. track gauge trains meeting Ireland's distinctive 5 ft 3 in. gauge – however, to this day there are plenty of places on mainland Europe where broad gauge meets standard gauge, notably at the Spanish–French border, and suitable technical solutions have been developed. Possibly the Victorians would have provided 'mixed gauge' track from the tunnel mouth to Belfast, but we shall never know.

THE PROPOSED CHANNEL TUNNEL
AND THROUGH TRAINS BETWEEN
EDINBURGH, GLASGOW, MANCHESTER, BELFAST, AND LONDONDERRY.
BY
L. LIVINGSTON MACASSEY,

GLASGOW:
ROBERT FORRESTER, 1 ROYAL EXCHANGE SQUARE.
BELFAST:
WILLIAM MULLAN & SON, 4 DONEGALL PLACE.
1890.
PRICE SIXPENCE.

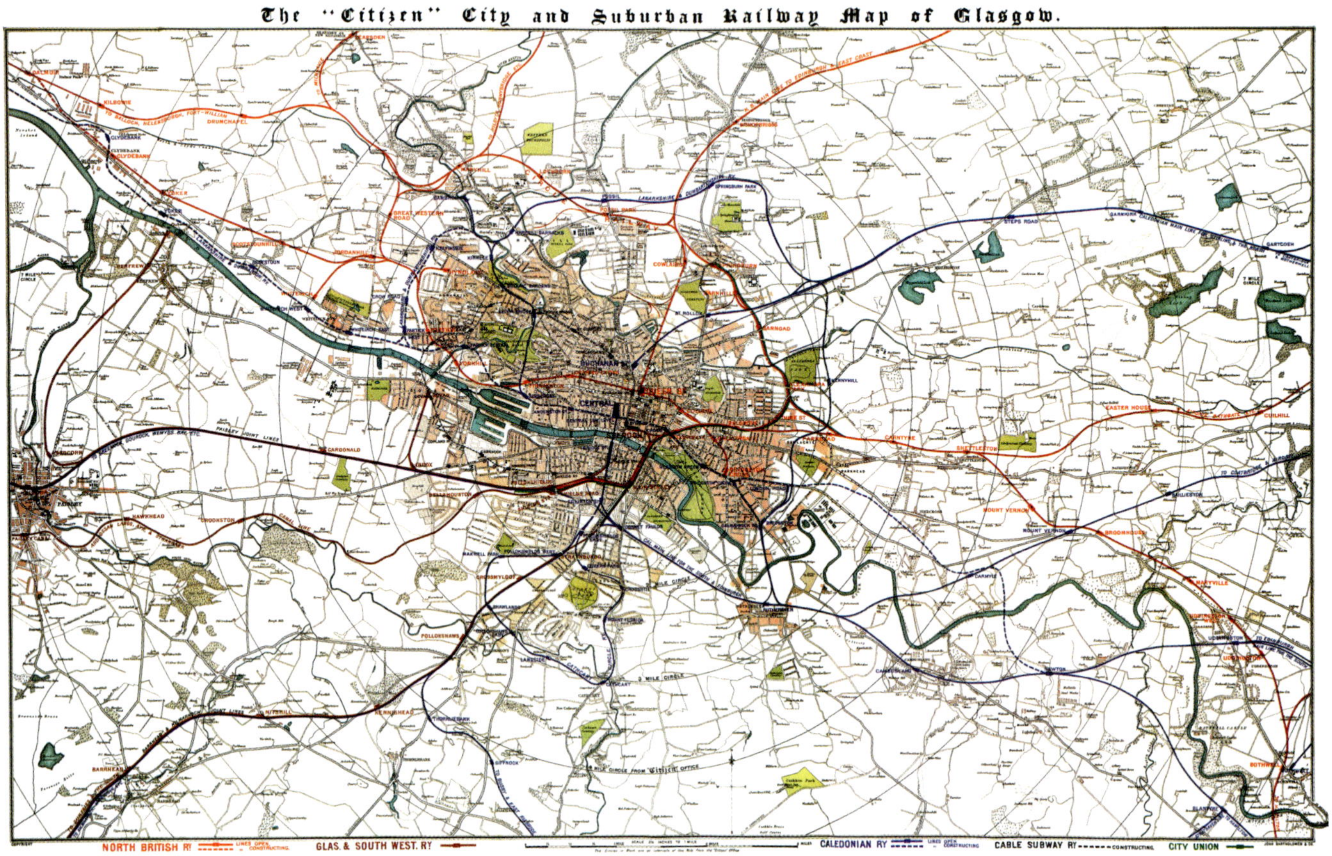

Title	The 'Citizen' City and Suburban Railway Map of Glasgow
By	J Bartholomew, Edinburgh
Date	*c.*1890–96

The famous Edinburgh mapmaking company of Bartholomew produced this clear and colourful map of Glasgow's developing rail network for the *Evening Citizen*, one of three evening newspapers in the then 'Second City' of the British Empire. The map's date is not shown, but reference to the 'Cable Subway Ry' (the Glasgow Underground) being under construction narrows the period down to 1890–96.

Bartholomew's map illustrates strikingly the competition between the two large Scottish railway companies battling for traffic in the more developed area of the city north of the River Clyde. This was the North British Railway's heartland in the city, with routes fanning out below ground and above ground from the jewel in the crown, Queen Street station – terminus of Scotland's first inter-city railway, from Edinburgh. The Caledonian Railway worked hard to extend its influence beyond its core territory south of Glasgow Central, with the Lanarkshire & Dumbarton Railway striking west, close to the north bank of the River Clyde and the big markets for freight and passenger traffic in its dense concentration of dockyards and shipyards. Two offshoot lines converged at Maryhill and then meandered around the northern and eastern edges of the city through Springburn Park and Parkhead to Bridgeton. All the latter system has now disappeared, but the 'Argyle Line' linking Bridgeton, Central Low Level, Stobcross (now Exhibition Centre) and the new Partick interchange station, is a key link in the modern Strathclyde rail system.

Suburban housing development stimulated by new rail routes was still in its infancy south of the Clyde at the time of this map's publication, with two key lines yet to be opened from Cathcart – southwest to Neilston and Ayrshire (1903) and east through Kirkhill to the Caledonian main line junction at Newton (1904).

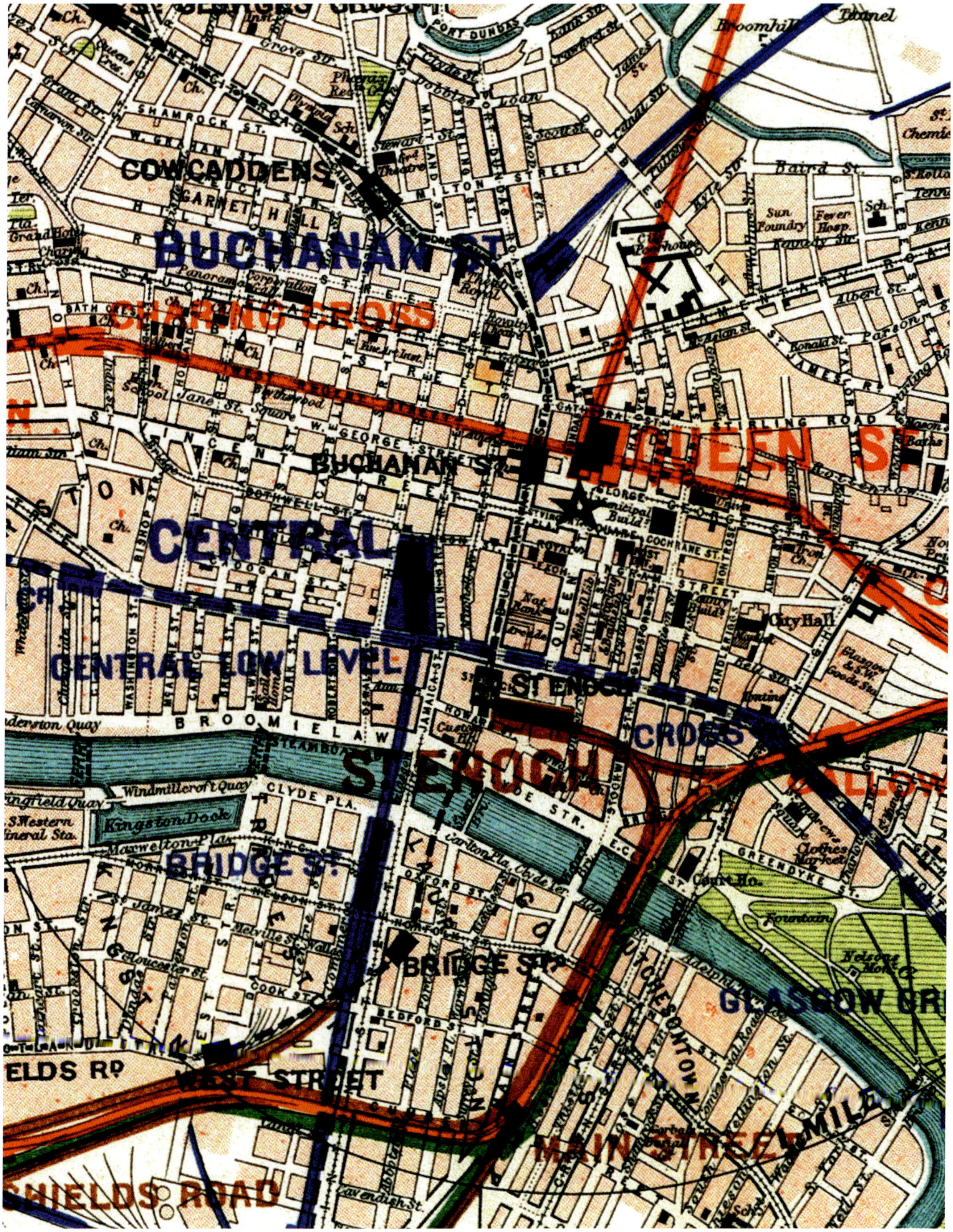
COWCADDENS
BUCHANAN ST.
CHARING CROSS
BUCHANAN ST.
QUEEN ST.
CENTRAL
CENTRAL LOW LEVEL
ST. ENOCH
CROSS
ST. ENOCH
BROOMIELAW
BRIDGE ST.
BRIDGE ST.
GLASGOW GR
WEST STREET
MAIN STREET
SHIELDS ROAD
ELDS RD
City Hall
Kingston Dock
Windmillcroft Quay
Grove St.
Shamrock St.
Garnet Hill
Baird St.
George Square
Cochrane St.
Sauchiehall St.
Bothwell St.
Cadogan St.
Clyde Pla.
Stirling Road
Tunnel
Broomhill
Fountain
Greendyke St.
Court Ho.
Clothes Market
Nelsons Mont.
Oxford St.
Norfolk St.
Bedford St.
Cook St.
Cavendish St.
Hutchesontown

MARYHILL
GREAT WESTERN ROAD
DAWSHOLM
JORDANHILL
KELVINSIDE
KIRKLEE
MARYHILL BARRACKS
HYNDLAND
BOTANIC GARDENS
LANARKSHIRE & DUMBARTONSHIRE RY.
CROW ROAD
HILLHEAD
KELVIN BRIDGE
WHITEINCH
WHITEINCH EAST
PARTICK WEST
PARTICK
PARTICK EAST
PARTICK CENTRAL
YORKHILL
FINNIESTON
STOBCROSS
ANDERSTON
GOVAN
COPLAND RD
CESSNOCK
KINNING PARK
CARDONALD
IBROX
POLLOKSHIELDS
BELLAHOUSTON
SHIELDS STA.
RIVER CLYDE
ELDER PARK
Merryflats Poorhouse
Craigton Cemetery
Cessnock Dock
Queen's Dock

LANARKSHIRE & DUMBARTONSHIRE RY
SPRINGBURN PARK
POSSIL
POSSIL PARK
RAILWAY
COWLAIRS
SPRINGBURN
BARNHILL
ST ROLLOX
GARNGAD
BUCHANAN ST
QUEEN ST
CENTRAL
COLLEGE
DUKE ST
BELGROVE
ALEXANDRA PARK
KENNYHILL
PARKHEAD
GALLOWGATE
ST ENOCH
BRIDGE ST
GLASGOW GREEN
BRIDGETON CROSS
LONDON ROAD
MAIN STREET
BRIDGETON
DALMARNOCK RD
CUSHET FAULDS
ST GEORGES CROSS
COWCADDENS
CHARING CROSS
Sighthill Cemetery
Necropolis
Southern Necropolis
Springburn Park
Barony Poorhouse
Belvidere Hospital
Eastern Necropolis

Title	Mountain, Moor and Loch, Illustrated by Pen and Pencil, on the Route of the West Highland Railway
Publisher	Sir Joseph Causton & Sons, London, for the North British Railway
Date	1895

As the railway historian John Thomas writes in his introduction to the 1972 reprint, the North British Railway (NBR) had several motives for producing *Mountain, Moor and Loch* to celebrate the opening of the West Highland Railway from Glasgow to Fort William in 1894:

> *With its wealth of delightful pen sketches and accurately observed descriptive material it was one of the most distinctive and best-loved railway guides ever produced. The primary function of the book was to publicize the railway, but the North British used it quite ruthlessly as a political weapon to further the promotion of the westward extension of the railway. Less than two weeks before the opening of the railway the West Highland had obtained an Act for the extension of the railway to Mallaig. However, work on the extension could not begin until the company had a second Act authorizing a Government subsidy for the construction of the line . . . . The North British hoped that the opening of the West Highland would focus on attention on the need for railways in those remote parts and help to break down opposition to the public subsidizing of the Mallaig Extension.*

Thomas notes that the 1894 edition of *Mountain, Moor and Loch* ran to 14,000 copies and that a second edition of 10,000 copies was printed in 1895. The overview map and five 'strip maps' of the railway in this book are reprinted from the latter edition. The text of the route guide ('A Bird's Eye View') begins:

> *If you look at the map of Scotland you will see that the country on the West, is simply torn to tatters by lochs, either inland sheets of fresh water or far-reaching arms of the Atlantic . . . . Such a wild and picturesque district is a very Paradise for the tourist, but hitherto this 'Land of Mountain, Moor and Loch' has been remarkably difficult of access . . . . The West Highland Railway breaks fresh ground from start to finish of its hundred-mile run; carrying the traveller through what is, perhaps, the most sublime and characteristic portion of Scotland.*

ITINERARY.

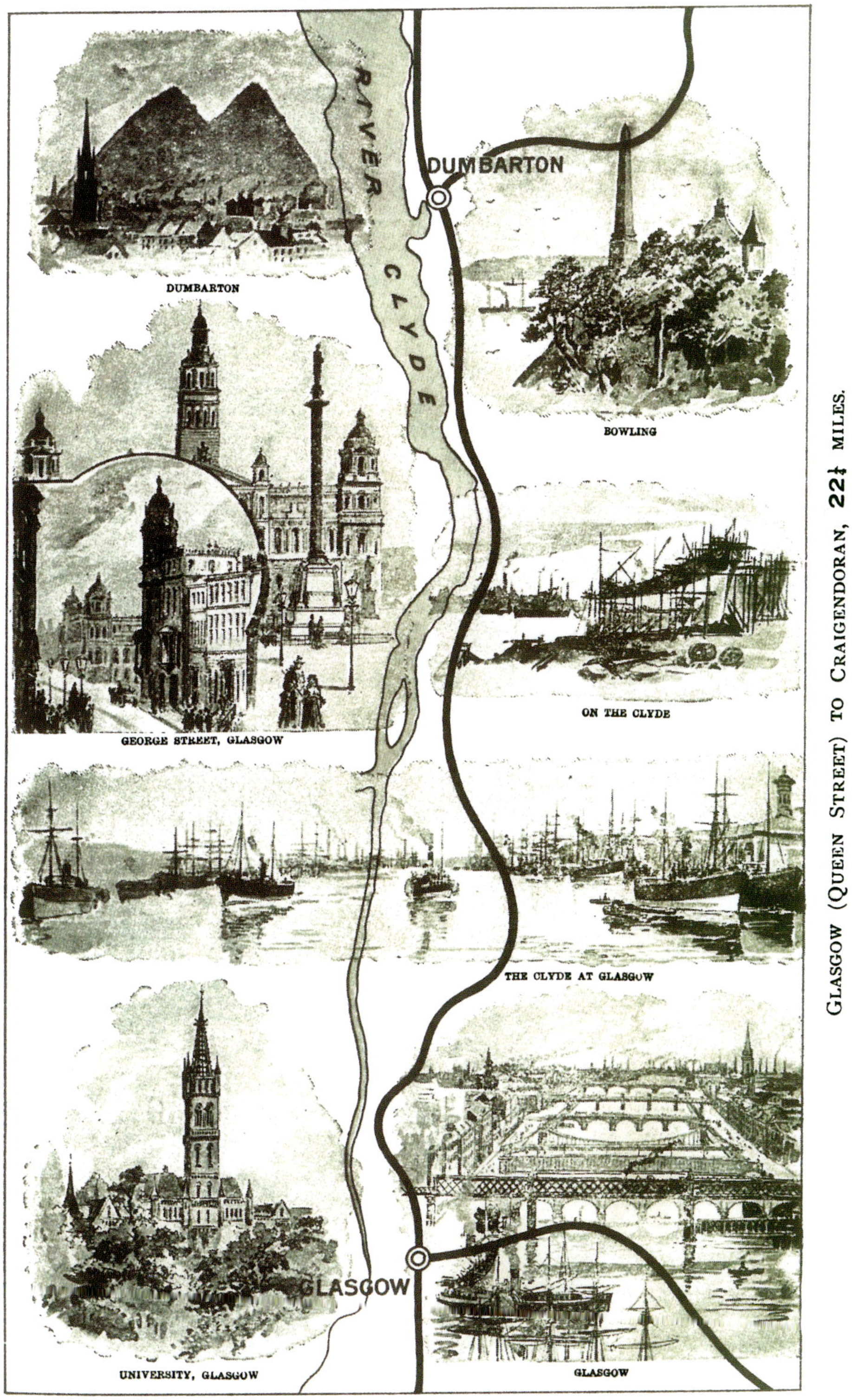

GLASGOW TO CRAIGENDORAN.

SECTION 1.

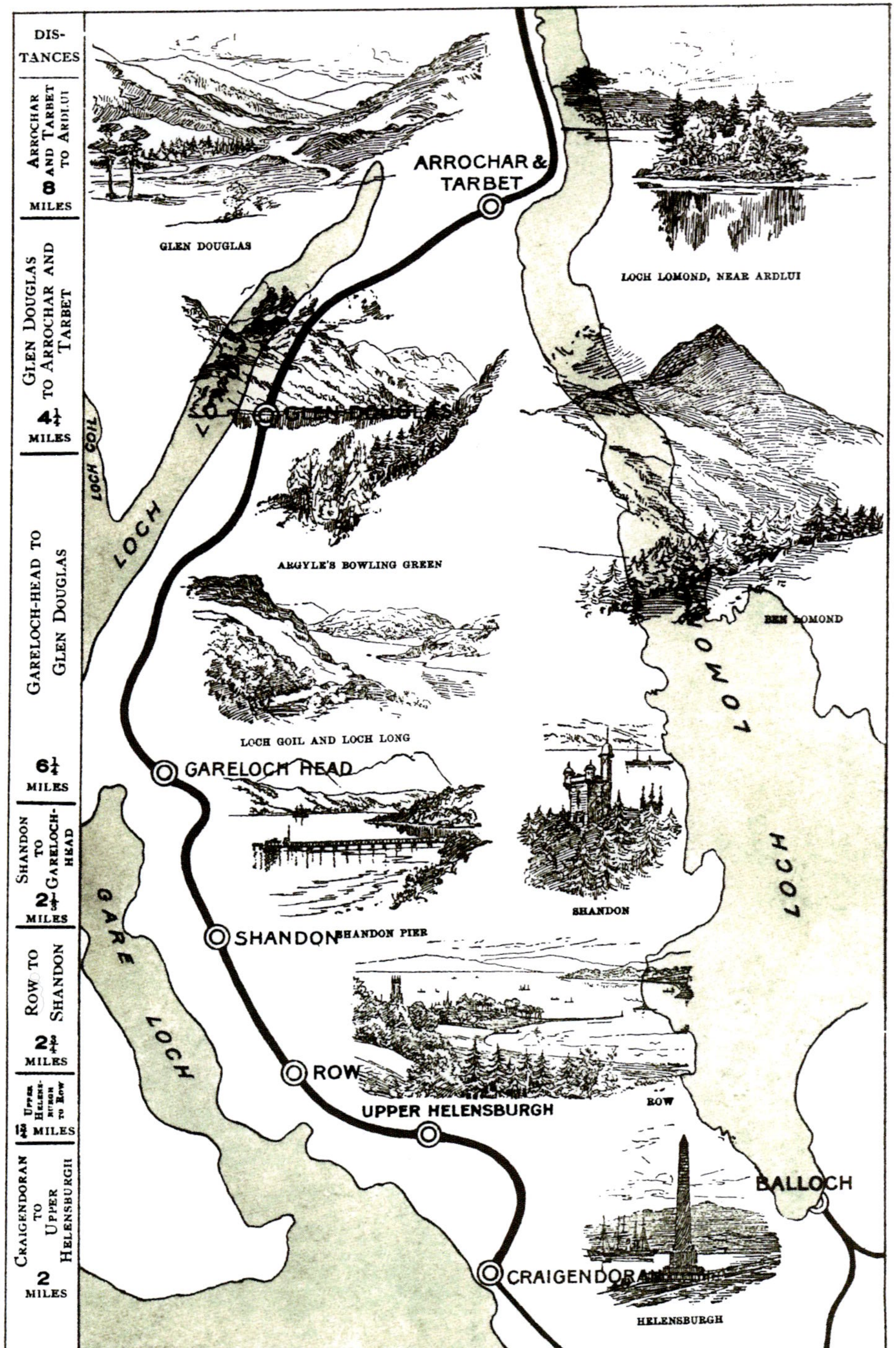

CRAIGENDORAN TO ARROCHAR AND TARBET.
SECTION 2.

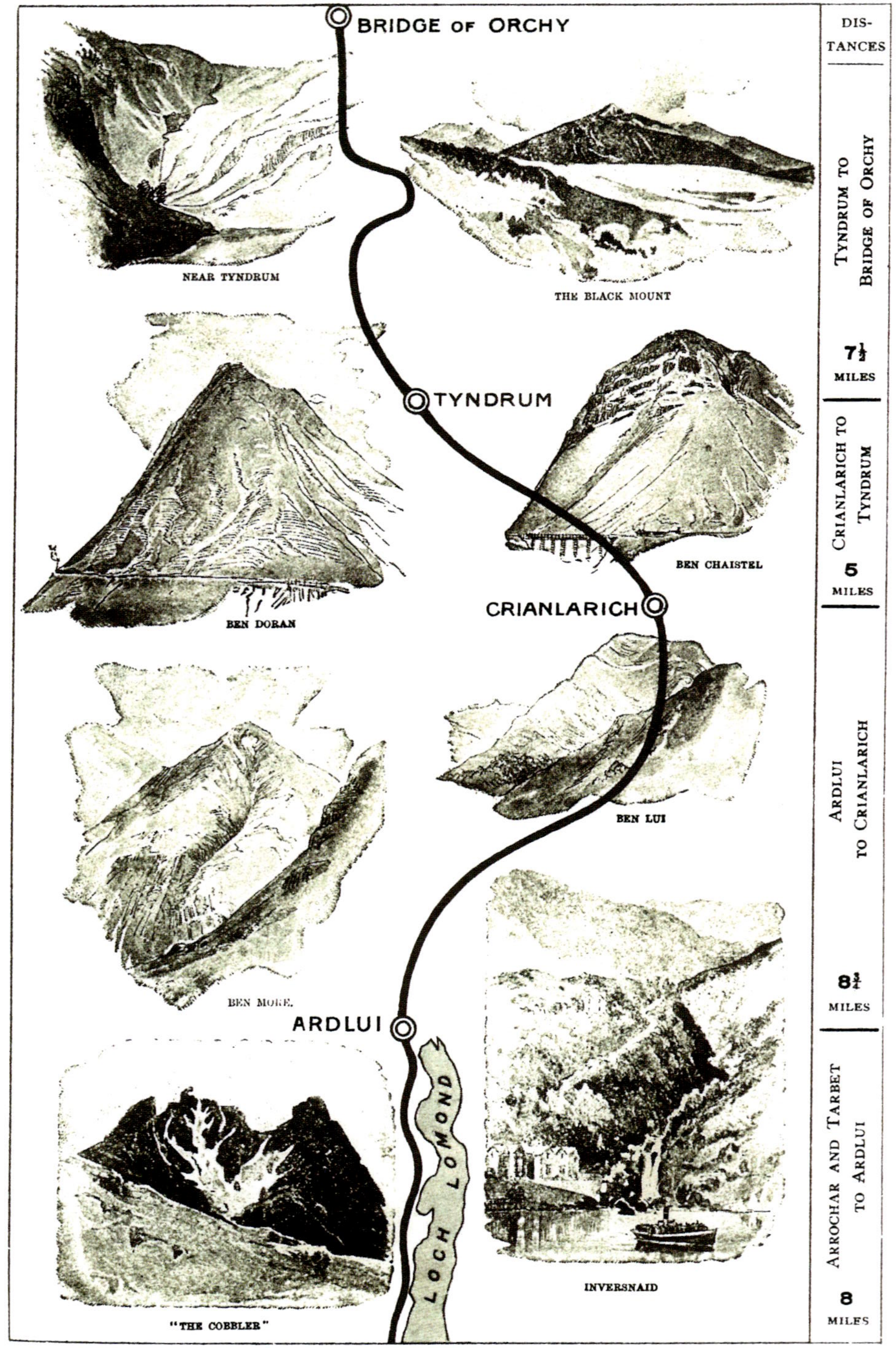

ARROCHAR AND TARBET TO BRIDGE OF ORCHY.
SECTION 3.

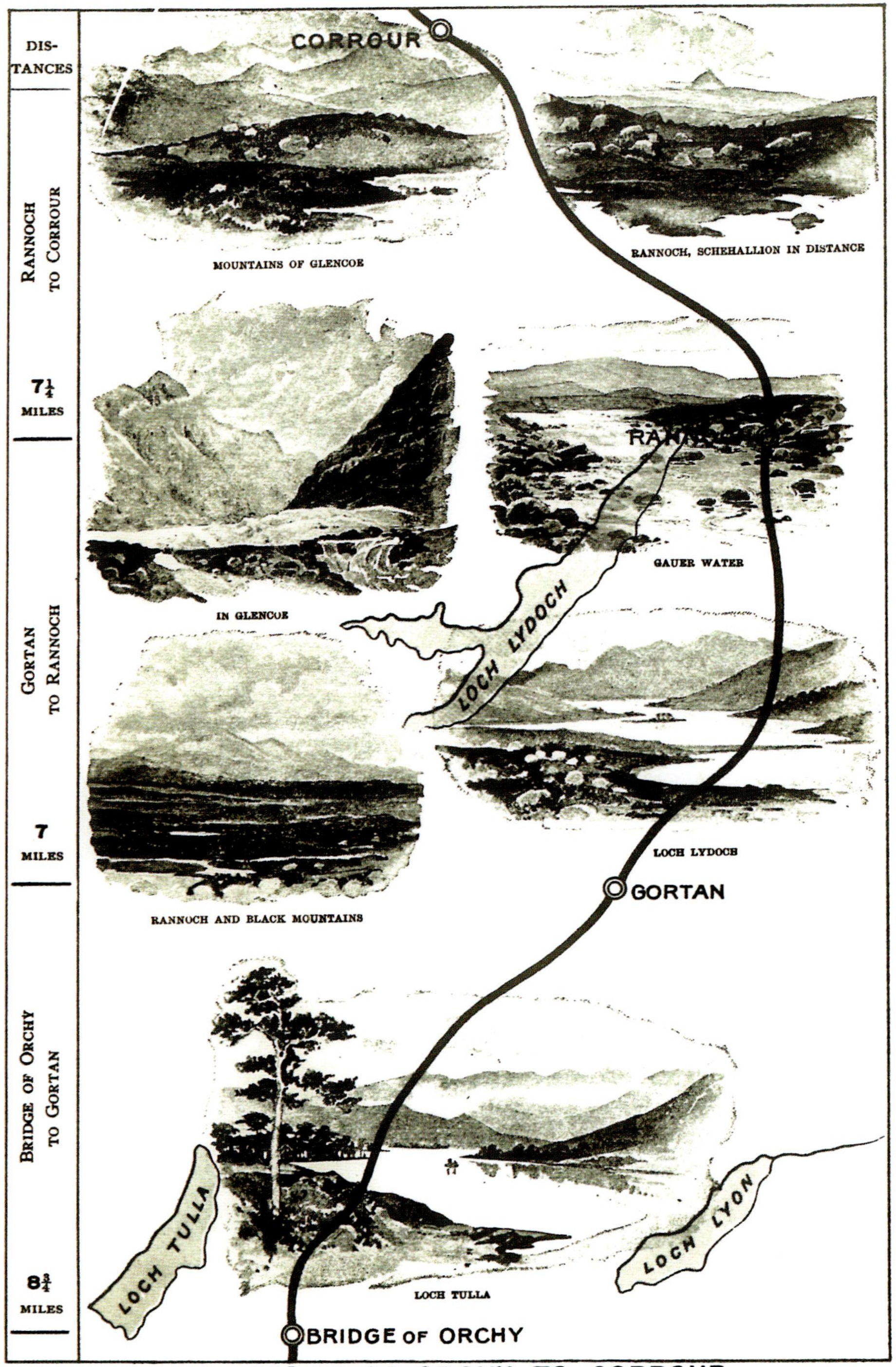

BRIDGE OF ORCHY TO CORROUR.
SECTION 4.

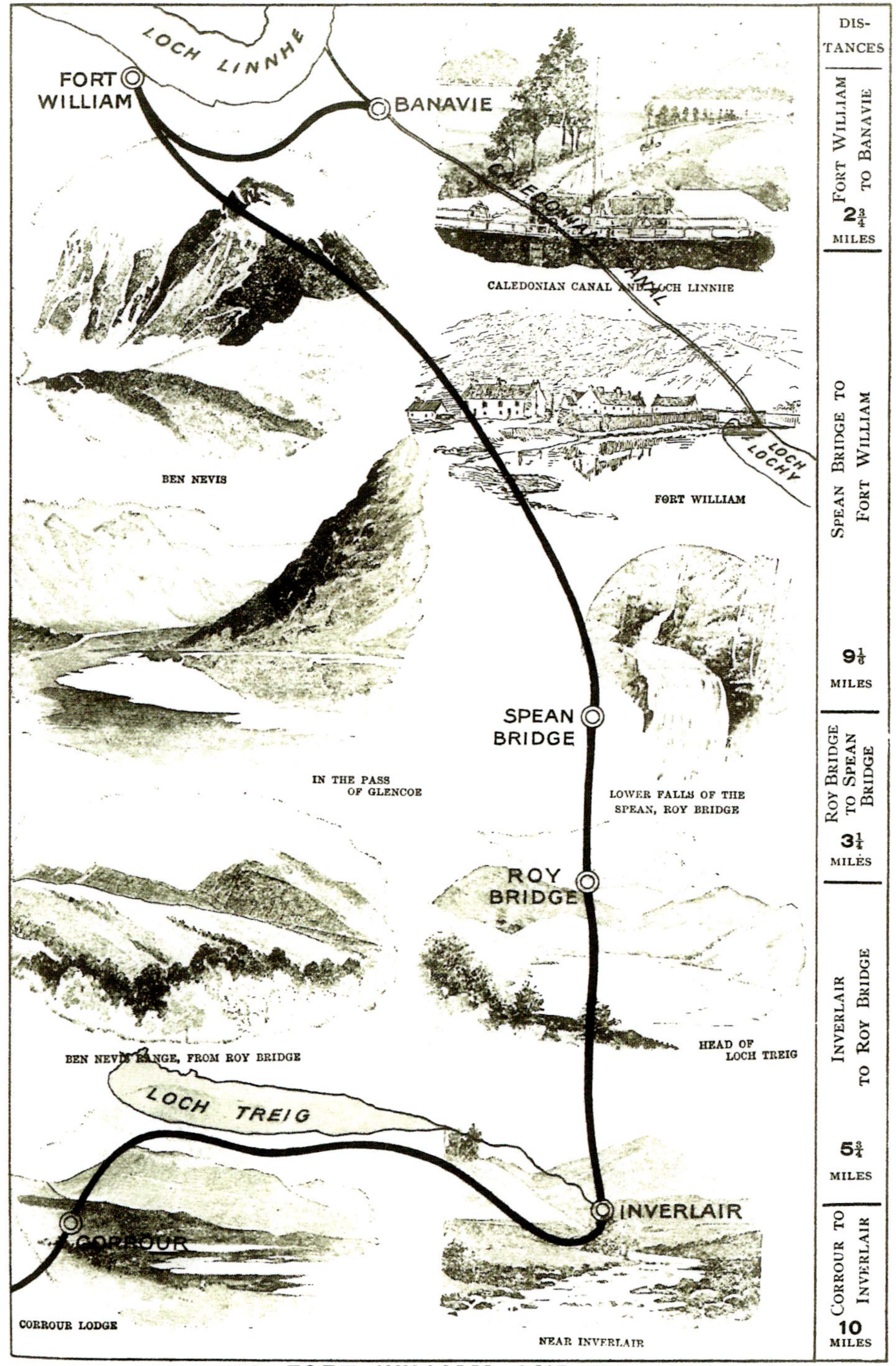

CORROUR TO FORT WILLIAM AND BANAVIE.
SECTION 5.

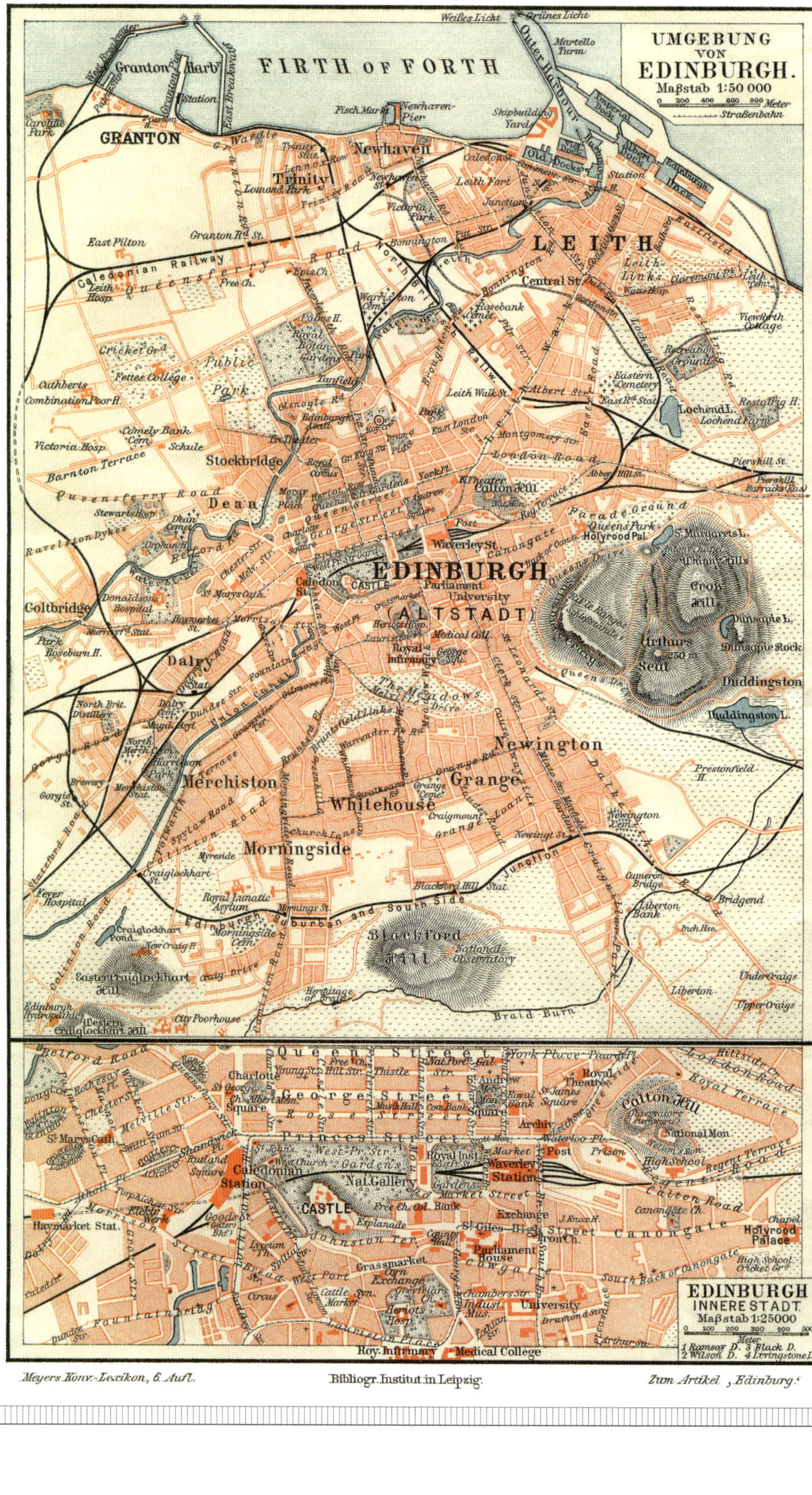
UMGEBUNG VON EDINBURGH.
Maßstab 1:50 000
Meter
Straßenbahn
FIRTH OF FORTH
GRANTON
Newhaven
Trinity
LEITH
Caledonian Railway
Stockbridge
Dean
EDINBURGH
(ALTSTADT)
Waverley St.
Calton Hill
Arthurs Seat
Duddingston
Dalry
Merchiston
Whitehouse
Grange
Newington
Morningside
Blackford Hill
Edinburgh Suburban and South Side Junction
EDINBURGH INNERE STADT
Maßstab 1:25000
Caledonian Station
Waverley Station
CASTLE
Holyrood Palace
Meyers Konv.-Lexikon, 6. Aufl.
Bibliogr. Institut in Leipzig.
Zum Artikel „Edinburg.“

Title	*Umgebung von Edinburgh* (Surroundings of Edinburgh) and *Edinburgh Innere Stadt* (Inner City)
Publisher	Meyers Konversations-Lexikon
Date	1896

Meyers Lexikon was a major German encyclopedia published between 1839 and 1984. This attractive map was probably made in Britain, as the district and street names and topographical features are overwhelmingly in English – although there are some exceptions, such as *Altstadt* (old town), *Güter Bhf.* (goods station) and, intriguingly, *Weisses Licht* (white light) and *Grünes Licht* (green light) at the sea entrance to Leith Docks.

Edinburgh's relatively extensive suburban rail network is clearly delineated in black, while its well-developed tram system (*Strassenbahn*) is shown by hatched lines, including the long stretch linking Edinburgh and Leith (formerly two separate burghs) down Leith Walk, recently scheduled to be re-opened as part of a new Edinburgh tram route.

The routes of trains heading for the train ferries across the Firth of Forth (prior to the opening of the Forth Bridge in 1890) can be readily traced. The original route north from Canal Street station (at right angles to Waverley station) ran underground on a steep gradient down to Scotland Street station (by Royal Crescent) where the line emerged above ground and continued north through Trinity to Granton Harbour. The first section of this route, opened in 1847, had to be operated with cable haulage; the first train ferry in the world – from Granton to Burntisland – began operation in 1850; and by 1868 an eastern diversionary route was opened via Abbeyhill and Leith Walk stations, enabling standard locomotive haulage to replace the cable-operated incline.

The Scotland Street tunnel was then closed to passengers, but was used for some time to store freight wagons, and subsequently for growing mushrooms! Coal traffic, however, continued to arrive by train at Scotland Street goods yard (at the tunnel mouth) from the Granton direction until as late as 1967. The tunnel entrances are now sealed up, but the tunnel itself remains *in situ*, perhaps awaiting a new use in future times.

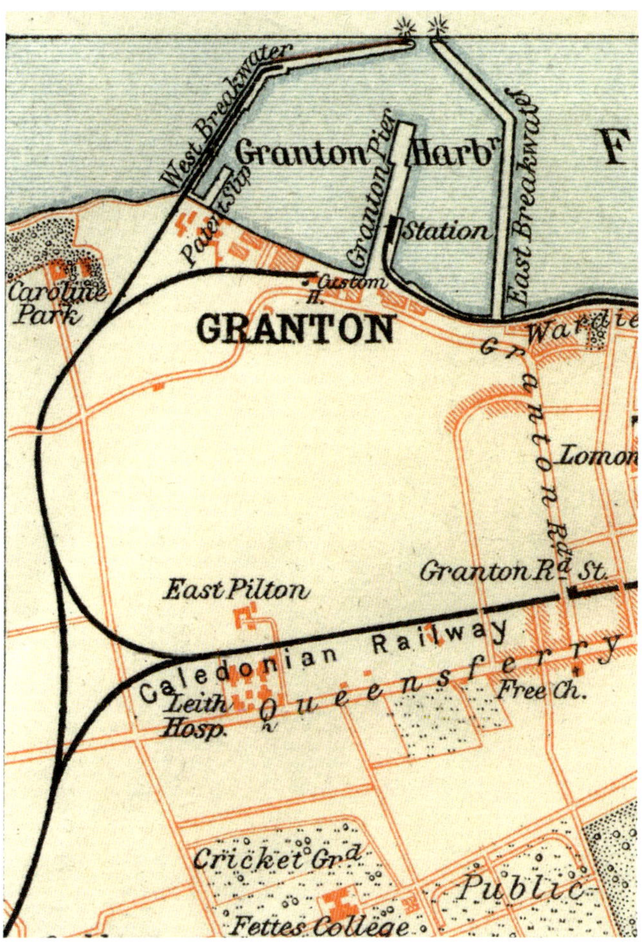

Coal wagons at Scotland Street goods yard, Edinburgh, in 1966, a year before its closure. From 1850 to 1868, passenger trains from Edinburgh Canal Street to the train ferry at Granton passed through the tunnel mouth in the background. A cycle track now follows the old rail route northwards from here to Granton.

NORTH BRITISH RAILWAY
AND ITS COMMUNICATIONS
THURSO
WICK
LYBSTER
HELMSDALE
GOLSPIE
DORNOCH
BONAR BRIDGE
TAIN
STORNOWAY
LEWIS
HARRIS
ULLAPOOL
POOLEWE
INVERGORDON
CROMARTY
DINGWALL
FORTROSE
STRATHPEFFER
BEAULY
INVERNESS
NAIRN
FORRES
ELGIN
HOPEMAN
BURGHEAD
LOSSIEMOUTH
BUCKIE
CULLEN
PORTSOY
BANFF
MACDUFF
FRASERBURGH
PETERHEAD
KEITH
TURRIFF
HUNTLY
DUFFTOWN
ROTHES
OLD MELDRUM
KINTORE
ALFORD
ABERDEEN
BALLATER
BANCHORY
STONEHAVEN
BERVIE
LAURENCEKIRK
MONTROSE
BRECHIN
FORFAR
ARBROATH
CARNOUSTIE
DUNDEE
BROUGHTY FERRY
TAYPORT
PORTREE
SKYE
APPLECROSS
STROME FERRY
KYLE
BROADFORD
GRANTOWN
AVIEMORE
KINGUSSIE
FORT AUGUSTUS
INVERMORISTON
MALLAIG
ARISAIG
BANAVIE
FORT WILLIAM
DALNASPIDAL
BLAIR ATHOLL
PITLOCHRY
BALLINLUIG
DUNKELD
BLAIRGOWRIE
ABERFELDY
KENMORE
TOBERMORY
MULL
OBAN
DALMALLY
TYNDRUM
CRIANLARICH
KILLIN
LOCHEARNHEAD
COMRIE
CRIEFF
PERTH
NEWBURGH
CUPAR
ST ANDREWS
LADYBANK JUNC.
LARGO
ANSTRUTHER
CRAIL
INVERARAY
CALLANDER
DUNBLANE
ALLOA
ABERFOYLE
STIRLING
DUNFERMLINE
KIRKCALDY
BURNTISLAND
THORNTON JUNC.
LEVEN
METHIL
GRANGEMOUTH
LARBERT
FALKIRK
KILSYTH
NORTH BERWICK
DUNBAR
LEITH
HADDINGTON
EDINBURGH
DALKEITH
ESKBANK
LINLITHGOW
BATHGATE
AIRDRIE
COATBRIDGE
GLASGOW
PAISLEY
HELENSBURGH
DUNBARTON
GREENOCK
BALLOCH
ROTHESAY
ARDRISHAIG
TARBERT
JURA
ISLAY
PORT ELLEN
KINTYRE
CAMPBELTOWN
ARRAN
BRODICK
LAMLASH
ARDROSSAN
KILMARNOCK
AYR
TROON
HAMILTON
PEEBLES
GALASHIELS
MELROSE
ST BOSWELLS JUNC.
SELKIRK
JEDBURGH
HAWICK
KELSO
COLDSTREAM
DUNS
EYEMOUTH
BERWICK
ALNWICK
ALNMOUTH
ROTHBURY
MORPETH
NEWCASTLE
TYNEMOUTH
SUNDERLAND
HEXHAM
RICCARTON JUNC.
LANGHOLM
LOCKERBIE
MOFFAT
GIRVAN
DUMFRIES
CARLISLE
SILLOTH
PORT CARLISLE
STRANRAER
PORT PATRICK
NEWTON STEWART
WIGTOWN
WHITHORN
CASTLE DOUGLAS
KIRKCUDBRIGHT
DURHAM
ENGLAND
IRELAND
LONDONDERRY
COLERAINE
PORTRUSH
LARNE
BELFAST
HEAD OFFICES, WAVERLEY STATION, EDINBURGH
JOHN BARTHOLOMEW & CO., THE EDINBURGH GEOGRAPHICAL INSTITUTE.
J. CONACHER, GENERAL MANAGER.

Title	North British Railway and its communications
Mapmaker	J Bartholomew, Edinburgh
Date	1896

The North British Railway (NBR) was one of the two big railway companies which dominated Central Scotland up to the 1923 'Grouping'. As noted in the *Oxford Companion to British Railway History* (1997), competition with the Caledonian Railway (CR) was:

> *a key feature of the railway's existence, the two companies fighting each other over control of Fife and its coal; services from the south to Aberdeen, culminating in the races of 1895; suburban services in the Glasgow area – the CR's home territory; and the provision of great hotels, a contest which the NBR won hands down with its impressive building next to Waverley station, Edinburgh.*

The depth of rivalry between the two companies is aptly illustrated by aspects of this NBR map poster, which uses a simplified geographic style with a distinctly modern feel. The NBR was clearly keen to show all its potential future routes, and those of companies with which it collaborated – the map represents with a pecked line the Highland Railway's direct Aviemore–Carrbridge–Inverness line (not opened until late 1898) and its own planned Mallaig Extension, which did not come into service until 1901.

The map extract of southern Scotland overleaf demonstrates that the lines of companies with whom the NBR had working agreements – in this case the Glasgow & South Western to Stranraer and Dumfries – were shown just as prominently on the map as its own routes. You have to look very carefully, however, to identify the Glasgow/Edinburgh–Carlisle main line of its great rival, arguably the premier Anglo-Scottish railway. Elsewhere on the map, key connecting horse-bus services are shown prominently, in some cases far from NBR territory – such as the link from the Great North of Scotland Railway's Ballater terminus to nearby Braemar.

The extent of railway-owned and rail-connected shipping services on the Clyde Estuary is also well represented – with routes to the islands of Bute and Arran, to the Kintyre peninsula and across the North Channel to Northern Ireland.

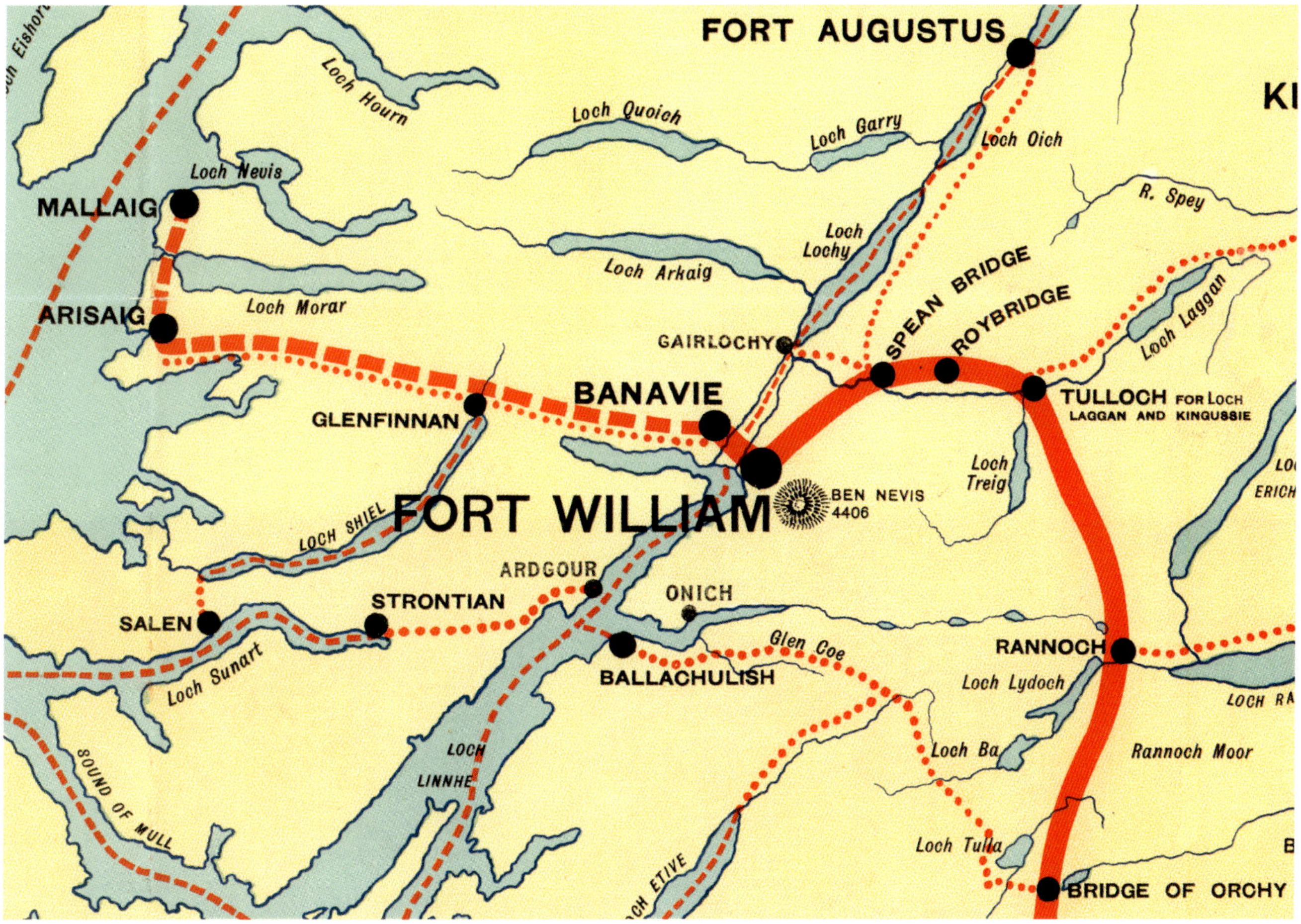

LOCHGOILHEAD
2891
TARBET
Ben Lomond
3192
L. Vennachar
L. of Menteith
FORD
STRACHUR
L. Goil
Loch Long
ROWARDENNAN
ABERFOYLE
LOCH FYNE
LUSS
LOCH LOMOND
BALMAHA
Crinan Canal
ARDRISHAIG
LOCHGILPHEAD
GARELOCHHEAD
Loch Eck
SHANDON
ROW
JURA
HELENSBURGH
BALLOCH
Loch Tarbert
Sound of Jura
DUNOON
CRAIGENDORAN PIER
DUNBARTON
INELLAN
CLYDEBANK
Kyles of Bute
GREENOCK
TARBERT
GLASGOW
QUEEN
FIRTH OF CLYDE
ROTHESAY
BUTE
LARGS
PAISLEY
of Islay
CANTYRE
Sound of Bute
BEITH
MILLPORT
CORRIE
ARDROSSAN
Goat Fell
2966
Kilbrennan Sound
KILMARNOCK
BRODICK
FIRTH OF CLYDE
ARRAN
TROON
LAMLASH
PRESTWICK
MAUCHLINE
AYR
R. Ayr
CAMPBELTON
Pladda I.
Ardrossan to Belfast
MAYBOLE
R. Doon
Mull of Cantyre
Sanda I.
Ailsa Craig
Loch Doon
GIRVAN
R. Stinchar
Merrick
2764
Glasgow to Belfast
Ayr to Belfast
NEW GALLOWAY
R. Cree
Loch Ryan
NEWTON STEWART
D
LARNE
STRANRAER
WIGTOWN
PORT PATRICK
Wigtown Bay
LUCE BAY
WHITHORN

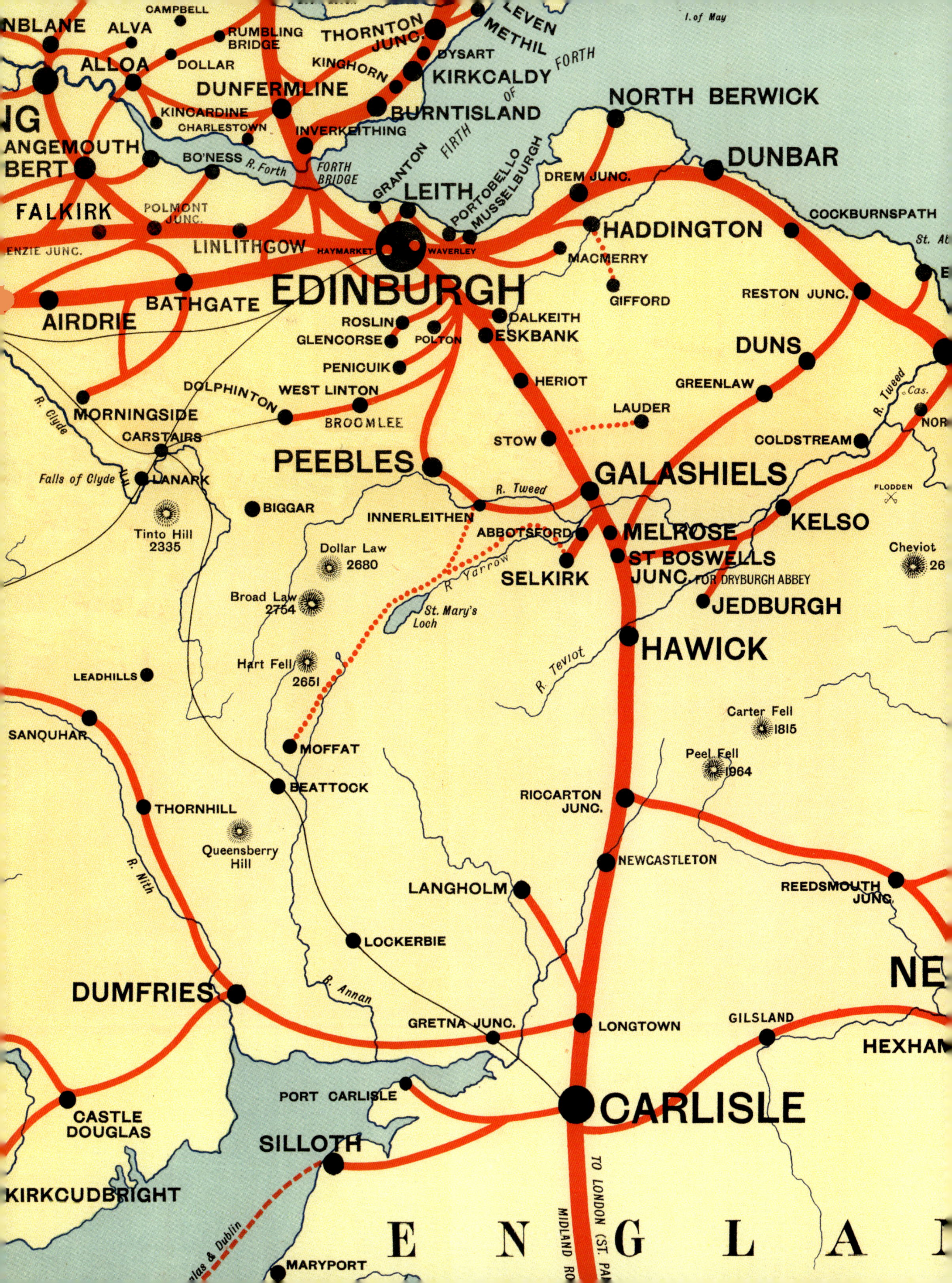
EDINBURGH
HAYMARKET
WAVERLEY
LEITH
GRANTON
PORTOBELLO
MUSSELBURGH
FIRTH OF FORTH
FORTH BRIDGE
R. Forth
NORTH BERWICK
DUNBAR
DREM JUNC.
HADDINGTON
COCKBURNSPATH
MACMERRY
GIFFORD
RESTON JUNC.
DUNS
GREENLAW
DALKEITH
ESKBANK
ROSLIN
GLENCORSE
POLTON
PENICUIK
HERIOT
LAUDER
STOW
GALASHIELS
MELROSE
ST BOSWELLS JUNC.
FOR DRYBURGH ABBEY
JEDBURGH
KELSO
COLDSTREAM
FLODDEN
Cheviot
R. Tweed
HAWICK
R. Teviot
Carter Fell 1815
Peel Fell 1964
RICCARTON JUNC.
NEWCASTLETON
REEDSMOUTH JUNC.
LANGHOLM
LONGTOWN
GILSLAND
HEXHAM
CARLISLE
TO LONDON (ST. PA
MIDLAND RO
ENGLA
GRETNA JUNC.
PORT CARLISLE
SILLOTH
MARYPORT
KIRKCUDBRIGHT
CASTLE DOUGLAS
DUMFRIES
R. Annan
LOCKERBIE
R. Nith
THORNHILL
Queensberry Hill
SANQUHAR
LEADHILLS
BEATTOCK
MOFFAT
Hart Fell 2651
Broad Law 2754
Dollar Law 2680
St. Mary's Loch
R. Yarrow
SELKIRK
ABBOTSFORD
INNERLEITHEN
R. Tweed
PEEBLES
BROOMLEE
WEST LINTON
DOLPHINTON
BIGGAR
Tinto Hill 2335
LANARK
Falls of Clyde
CARSTAIRS
R. Clyde
MORNINGSIDE
AIRDRIE
BATHGATE
LINLITHGOW
POLMONT JUNC.
FALKIRK
BO'NESS
ANGEMOUTH
BERT
CHARLESTOWN
KINCARDINE
ALLOA
DOLLAR
ALVA
CAMPBELL
NBLANE
RUMBLING BRIDGE
DUNFERMLINE
INVERKEITHING
BURNTISLAND
KINGHORN
KIRKCALDY
DYSART
THORNTON JUNC.
LEVEN
METHIL
NE
I. of May

Ribblehead Viaduct is the longest and one of 23 viaducts built on the Settle & Carlisle line in the 1870s.

The Victorian railway builders certainly left their mark on the map of Britain. Standing the test of time, their graceful bridges and viaducts stride majestically across our landscape while their long, dark tunnels burrow deep beneath. The majority are still in use today.

Faced with natural obstacles such as rivers and ranges of hills, the early railway builders' achievements were spectacular. In particular the three trans-Pennine rail routes between Yorkshire and Lancashire merit special mention as they were built not only before the introduction of mechanical excavators but also before the invention of dynamite. The first of these railways was the Manchester & Leeds Railway which opened in 1841 and included the 1 mile 1,109 yd Summit Tunnel, at that time the longest in the world. This was soon followed by the Sheffield, Ashton-under-Lyne & Manchester Railway's 3 mile 13 yd Woodhead Tunnel, which opened in 1845, and the London & North Western Railway's 3 mile 57 yd Standedge Tunnel, which opened in 1848. (To the south the Midland Railway's 3 mile 950 yd Totley Tunnel was not completed until 1893.)

Without doubt the most amazing Victorian overland railway achievement was the building of the Midland Railway's Settle & Carlisle line which opened in 1876. Benefiting from the then recent invention of dynamite, a workforce of 6,000 navvies blasted their way up the bleak Pennine hills, building 13 tunnels and 23 viaducts. Of the former, Blea Moor is the longest at 1 mile 869 yd while of the latter, Ribblehead is the longest at 440 yd.

Down south the Great Western Railway, nicknamed 'Great Way Round' because of its circuitous route from London to South Wales via Gloucester, eventually bit the bullet and built, under the River Severn, what was then the longest underwater tunnel in the world. Beset by problems from flooding during construction, the 4 mile 624 yd Severn Tunnel opened in 1886 and was the longest rail tunnel in Britain until the opening of the High Speed 1 tunnel under London in 2007.

Meanwhile, in Scotland, the rail crossings of the Firth of Tay and Firth of Forth linking Edinburgh with Dundee and Aberdeen had run into problems. Officially opened by Queen Victoria in 1878, Thomas Bouch's 2¼ mile single-track Tay Bridge was then the longest bridge in the world and significantly cut journey times between Edinburgh and Dundee – although passengers still had to cross the Firth of Forth by ferry. However, Bouch's design was flawed and there were weaknesses in the cast ironwork used in its construction – only 18 months after it had been opened, a gale force wind swept away the centre section of its 85 spans while a passenger train was crossing. All 75 passengers along with the crew perished in the icy waters of the Tay. Thomas Bouch was disgraced and died a broken man, his design for a suspension bridge across the Firth of Forth being consigned to the wastepaper bin.

Aided by new developments in the use of steel, John Fowler and Benjamin Baker were able to design a cantilever bridge to carry a double-track railway across the Forth – when completed in 1890 it was the wonder of the Victorian world. Around 65,000 tons of steel were used in the construction of the 1½-mile bridge, then the largest cantilever bridge in the world. The Forth Bridge was built by William Arrol, a prolific contractor at that time as he had also built the replacement double-track Tay Bridge and London's Tower Bridge. These three structures have all stood the test of time and are fitting memorials to Britain's innovative Victorian engineers.

Rescuers search in vain in the windswept waters of the Firth of Tay for survivors from the 1879 Tay Bridge disaster.

Containing around 65,000 tons of steel, the Forth Railway Bridge was the largest cantilever bridge in the world when it opened in 1890.

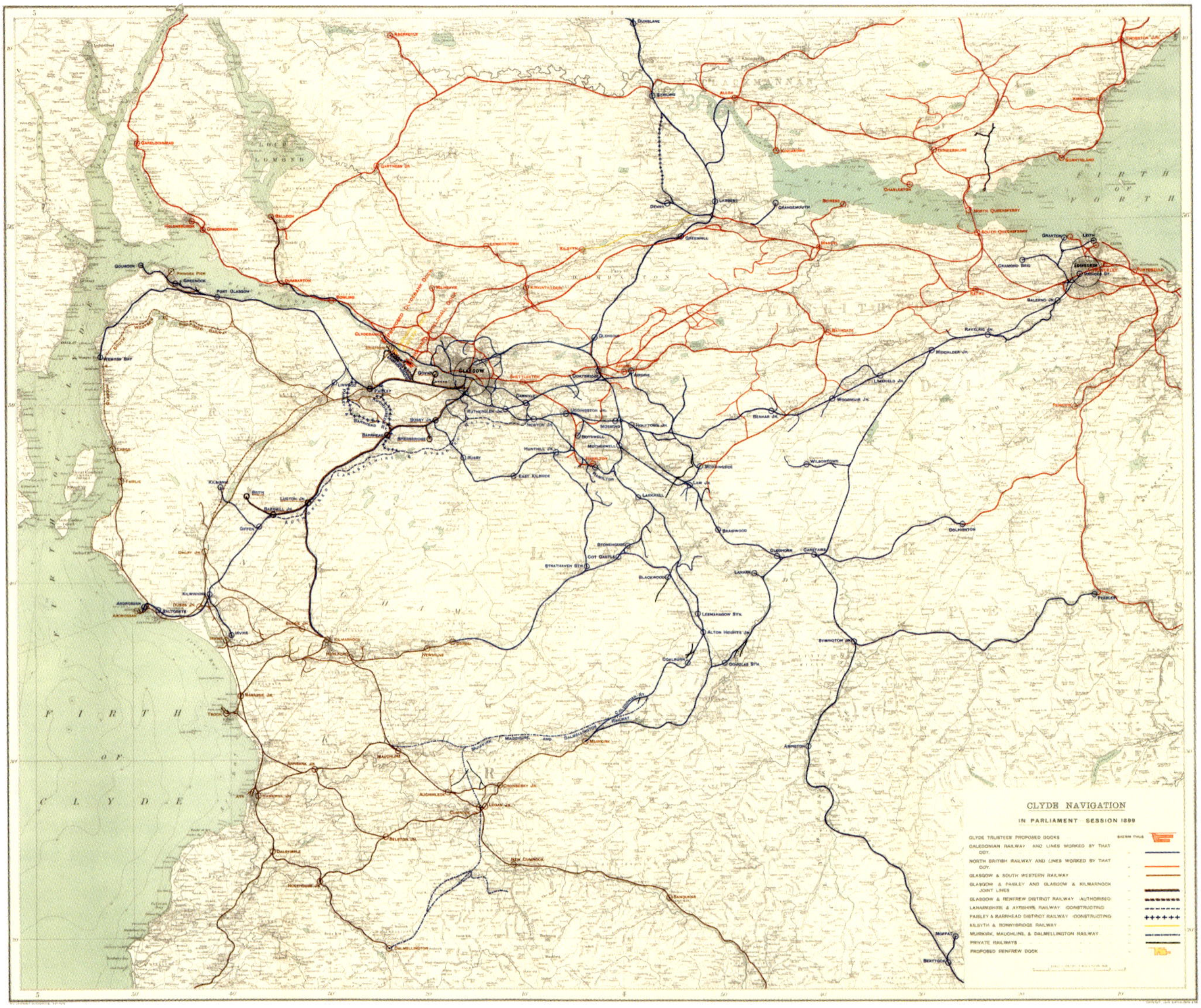

Title	Clyde Navigation In Parliament – Session 1899
Publisher	The Edinburgh Geographical Institute
Mapmaker	J Bartholomew, Edinburgh
Date	1899

Shipping-related dock schemes were presumably the primary motivation for production of this map, but in so doing Bartholomew's also created a clear and colourful snapshot of the largely completed expansion of the three big railway companies serving Central Scotland.

The Caledonian Railway had made a major incursion into Glasgow & South Western Railway (GSW) territory with its long isolated branch from the Glasgow suburbs to the key port of Ardrossan, while on the East Coast the 'Caley' pushed through North British Railway (NBR) country to reach the prized port destinations of Granton and Leith.

The NBR was more modest in its territorial expansion, with only a few notable outreach lines, such as that to Hamilton in Lanarkshire. The smaller GSW remained largely self-contained, but it is interesting to note on the map the proposed railway from Largs to Wemyss Bay and onwards to Kilmacolm and Port Glasgow – this line was never built, and the absence of rails between Largs and Wemyss Bay has always been one of the more striking gaps in the Scottish rail network.

The map extract opposite illustrates how the opening of the Forth Bridge in 1890 had transformed the network immediately south of the Forth Estuary. A well-engineered main line had been built southeast towards Edinburgh,

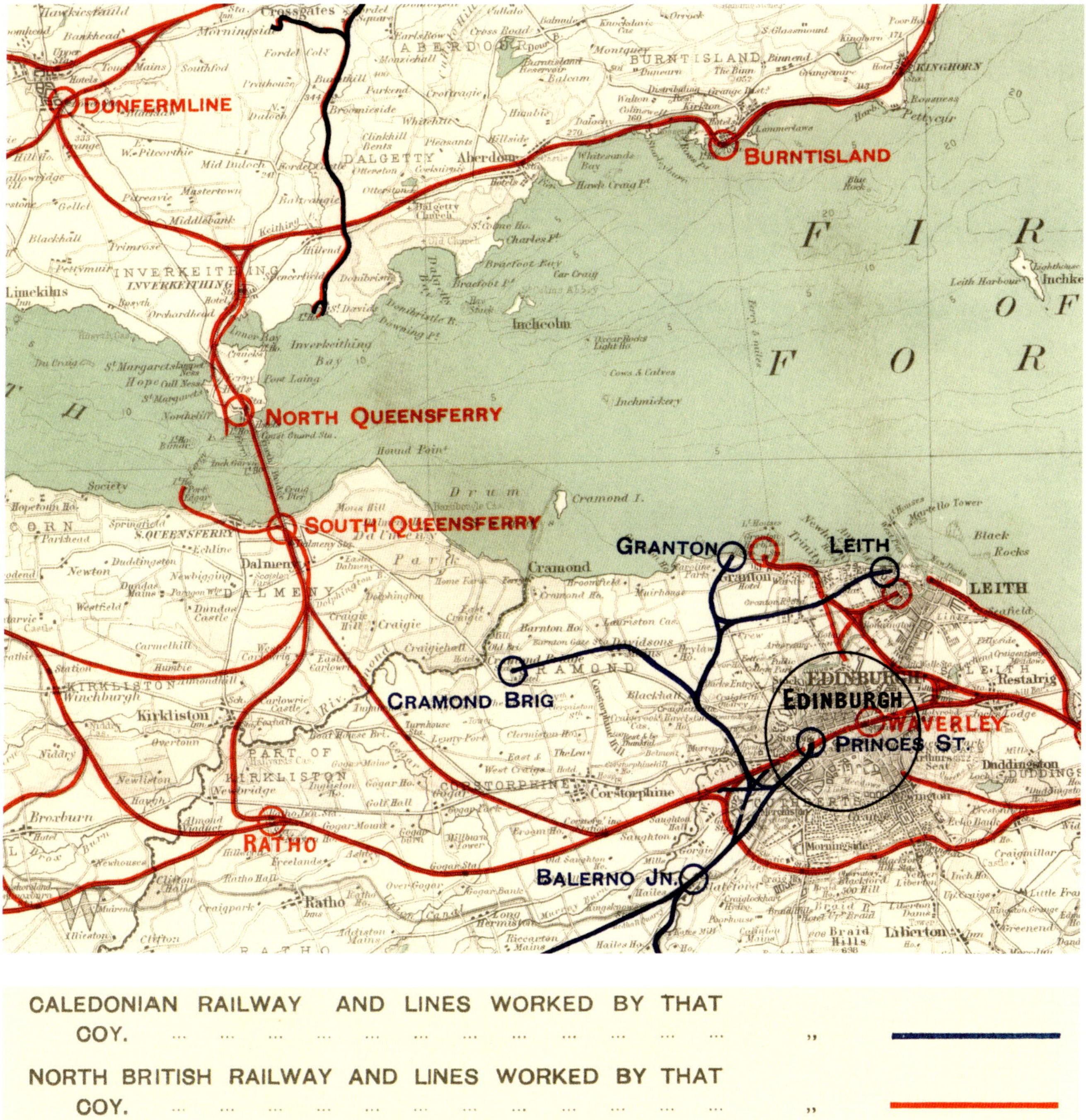

carrying express trains from Aberdeen, Perth and Inverness, while the important chord line southwest to Winchburgh Junction, on the Edinburgh–Glasgow main line, allowed fast direct connection from Fife to Scotland's biggest city.

A number of smaller independent railways are shown, typically built for coal transport, as in the case of the Fordell Railway in Fife, a former waggonway which had been converted to steam haulage and linked by means of interchange sidings to the standard gauge network by 1868. The line – which remained at the unusual 4 ft gauge throughout its life – ran from collieries in the Cowdenbeath and Crossgates areas down to St David's Harbour on Inverkeithing Bay. The southern end of the line, and the harbour, were last used for coal exports in 1946, and the remaining section of the line closed in 1966 when the last of the mines it served also closed.

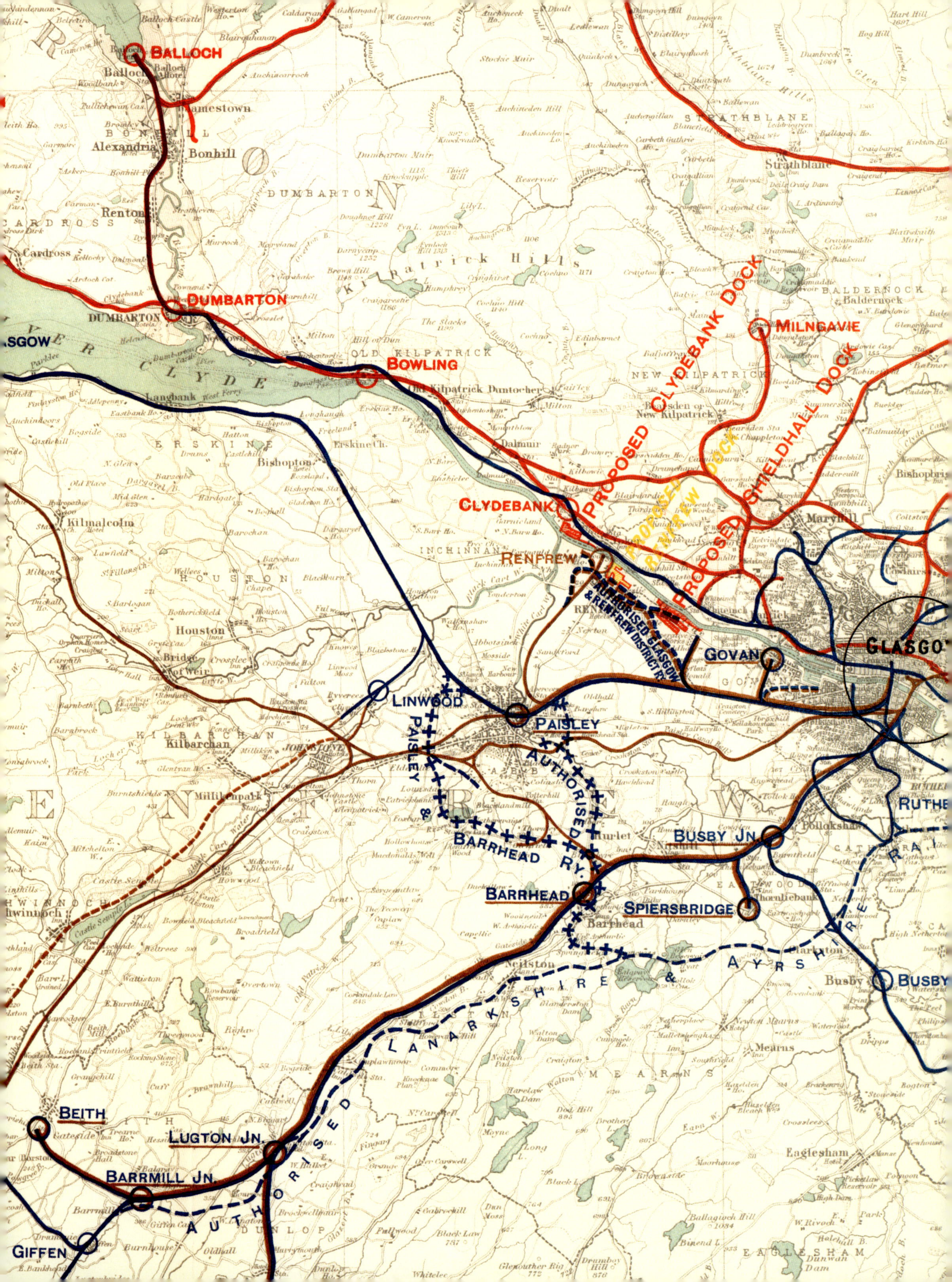

BALLOCH
DUMBARTON
BOWLING
MILNGAVIE
CLYDEBANK
RENFREW
GOVAN
GLASGO
LINWOOD
PAISLEY
BUSBY JN.
BARRHEAD
SPIERSBRIDGE
BUSBY
BEITH
LUGTON JN.
BARRMILL JN.
GIFFEN
PROPOSED CLYDEBANK DOCK
PROPOSED SHIELDHALL DOCK
PROPOSED RENFREW DOCK
AUTHORISED GLASGOW & RENFREW DISTRICT RY.
PAISLEY & BARRHEAD AUTHORISED RY.
LANARKSHIRE & AYRSHIRE RA
AUTHORISED
RIVER CLYDE
KILPATRICK HILLS
DUMBARTON
STRATHBLANE

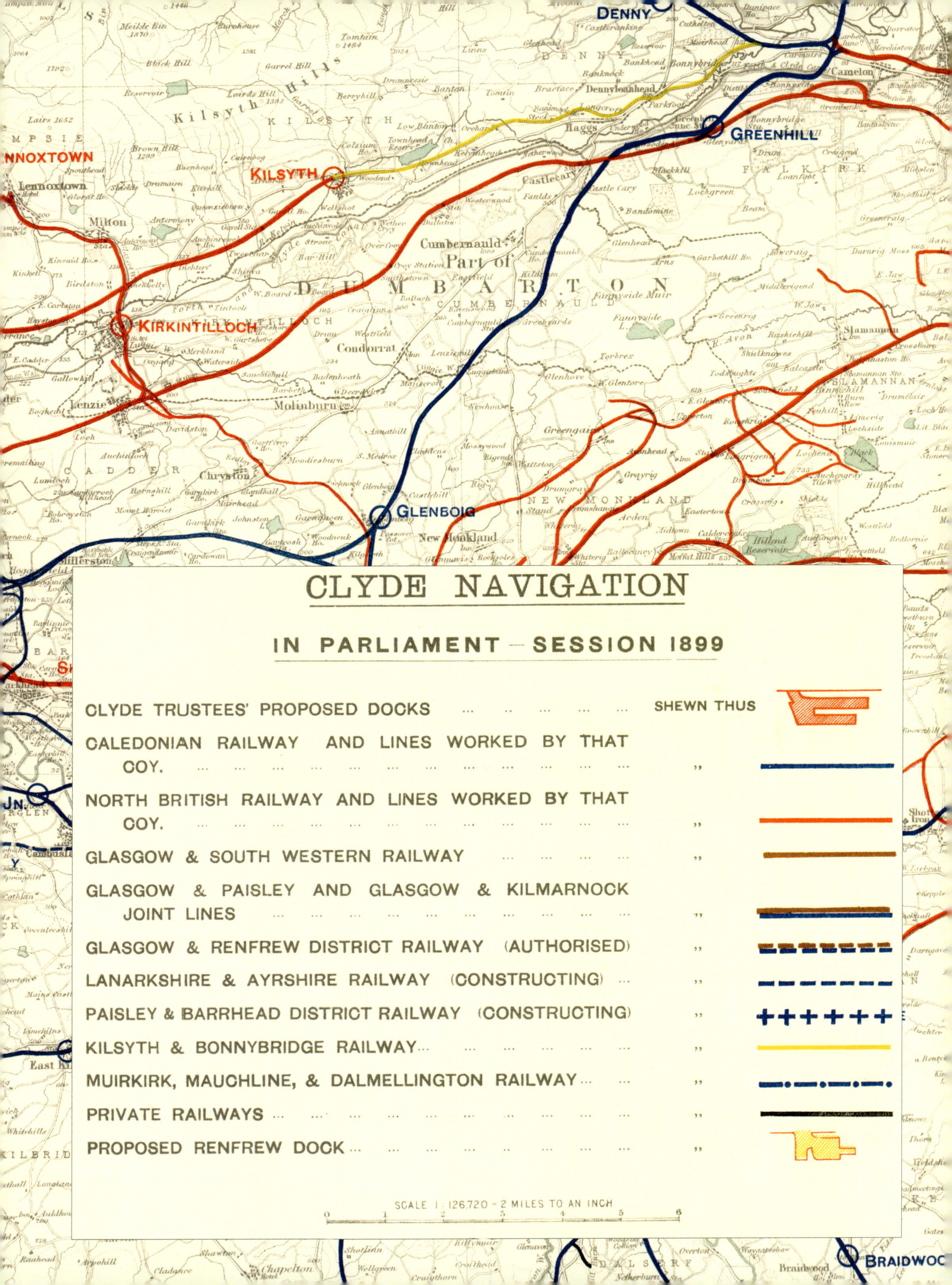
CLYDE NAVIGATION
IN PARLIAMENT — SESSION 1899
CLYDE TRUSTEES' PROPOSED DOCKS ... SHEWN THUS
CALEDONIAN RAILWAY AND LINES WORKED BY THAT COY. ... ,,
NORTH BRITISH RAILWAY AND LINES WORKED BY THAT COY. ... ,,
GLASGOW & SOUTH WESTERN RAILWAY ... ,,
GLASGOW & PAISLEY AND GLASGOW & KILMARNOCK JOINT LINES ... ,,
GLASGOW & RENFREW DISTRICT RAILWAY (AUTHORISED) ,,
LANARKSHIRE & AYRSHIRE RAILWAY (CONSTRUCTING) ... ,,
PAISLEY & BARRHEAD DISTRICT RAILWAY (CONSTRUCTING) ,,
KILSYTH & BONNYBRIDGE RAILWAY ... ,,
MUIRKIRK, MAUCHLINE, & DALMELLINGTON RAILWAY ... ,,
PRIVATE RAILWAYS ... ,,
PROPOSED RENFREW DOCK ... ,,
SCALE 1 : 126,720 = 2 MILES TO AN INCH
0 1 2 3 4 5 6
DENNY
GREENHILL
KILSYTH
KIRKINTILLOCH
GLENBOIG
LENNOXTOWN
BRAIDWOOD
Part of DUMBARTON
Cumbernauld
Lenzie
Chryston
New Monkland

THE ZENITH OF THE NETWORK

The national network reached its maximum extent, with the last new rural branch lines and important extensions of the Great Western main line. However, after the railways had made their immense contribution to the First World War effort, early signs of road competition appeared.

MILEPOST 10	Railway Shipping Services	132
1907	A Cornish China Clay Branch Line	134
1908	Bartholomew's London and Suburbs	136
1908	Bartholomew's Central London	140
1908	The Sportsman's Guide to Scotland	144
c.1900–10	North Eastern Railway Tile Map	146
MILEPOST 11	Railways and Coal	150
1901–22	Jowett's Upper South Wales Valleys (East)	152
1913	Great Central Railway Map Poster	154

1901–1922

MILEPOST 12	The Railways at War	158
1846/1915/ 1990	Jowett's Hull	160
MILEPOST 13	Light Railways	162
1909	The Kent & East Sussex Railway	164
1919	Light Railways of Scotland	166
1850/1920/ 1991	Jowett's Canterbury	168
1921	Bartholomew's Half-Inch of South Wales	170

An aerial view of the Midland Railway's harbour facilities at Heysham, Lancashire, in 1910.

Britain has always depended on seaborne traffic for trade and communications with the rest of the world. By the mid-19th century the burgeoning industrial strength of Britain had become totally dependent on railway operations – imported raw materials and exported finished goods were all carried between docksides and factories by rail. The days of slow canal transport were largely over.

In the very early years of the railways, companies were not allowed to operate their own ships but by 1863 they were given powers to operate their own passenger ships on specific routes. Although cargo shipping was subject to separate arrangements, the passenger shipping provisions stayed unaltered until 1967 when route impositions were finally swept away.

The first railway shipping service came in 1846 when the Great Grimsby & Sheffield Junction Railway started operating a ferry across the Humber Estuary between Hull and New Holland. Two years later the Chester & Holyhead Railway started operating the all-important Holyhead to Kingstown (now Dún Laoghaire) 'packet' service – packet boats being small vessels designed to carry mail, scheduled cargo and passengers. The South Eastern Railway began the first packet service to the Continent in 1853 with their Folkestone to Boulogne service. By the end of the century railway passenger shipping services had become so successful that there were routes across the North Sea to Scandinavia and the Low Countries, across the English Channel to France and the Channel Islands, across the Solent to the Isle of Wight and across the Irish Sea to Ireland and the Isle of Man; over river estuaries such as the Severn and Clyde; along English lakes and Scottish lochs; and over the sea to the Scottish islands. Special boat trains also ran in connection with these services along with ocean liner expresses to and from Plymouth, Southampton, Liverpool and Greenock. Car ferries using cranes were first introduced (between Stranraer and Larne) in 1939, while the first roll on-roll off vehicle ferry (Dover to Boulogne) came into service in 1952.

Freight traffic was even more important and to handle the vast quantities of raw materials and finished goods many of the larger railway companies had built their own dock complexes by the early 20th century. Principal among these

A Great Central Railway poster showing the deep-water dock at Immingham, near Grimsby, in the 1930s.

were the London & South Western Railway's modern docks at Southampton which later proved to be of vital strategic importance during both World Wars, and the Great Central Railway's vast freight handling facilities at Grimsby and Immingham. In the early 20th century Britain was also a major exporter of coal, with companies like the North Eastern Railway leading the way in the northeast and the Barry Railway in South Wales – the former company's docks at Tyne Dock were exporting seven million tons of coal a year while the latter's facility at Barry Docks was exporting nine million tons each year.

Following nationalization of the railways in 1948, the newly formed British Railways took over the running of all railway shipping services. From 1970 until 1984, when they were sold off, these services were known as Sealink, with an interesting offshoot in the Seaspeed hovercraft operation across the English Channel. Nowadays international rail passengers speed through the Channel Tunnel.

The British Railways Sealink ferry *Cambridge* leaves Harwich carrying railway wagons and lorries bound for Zeebrugge in Belgium, 1966.

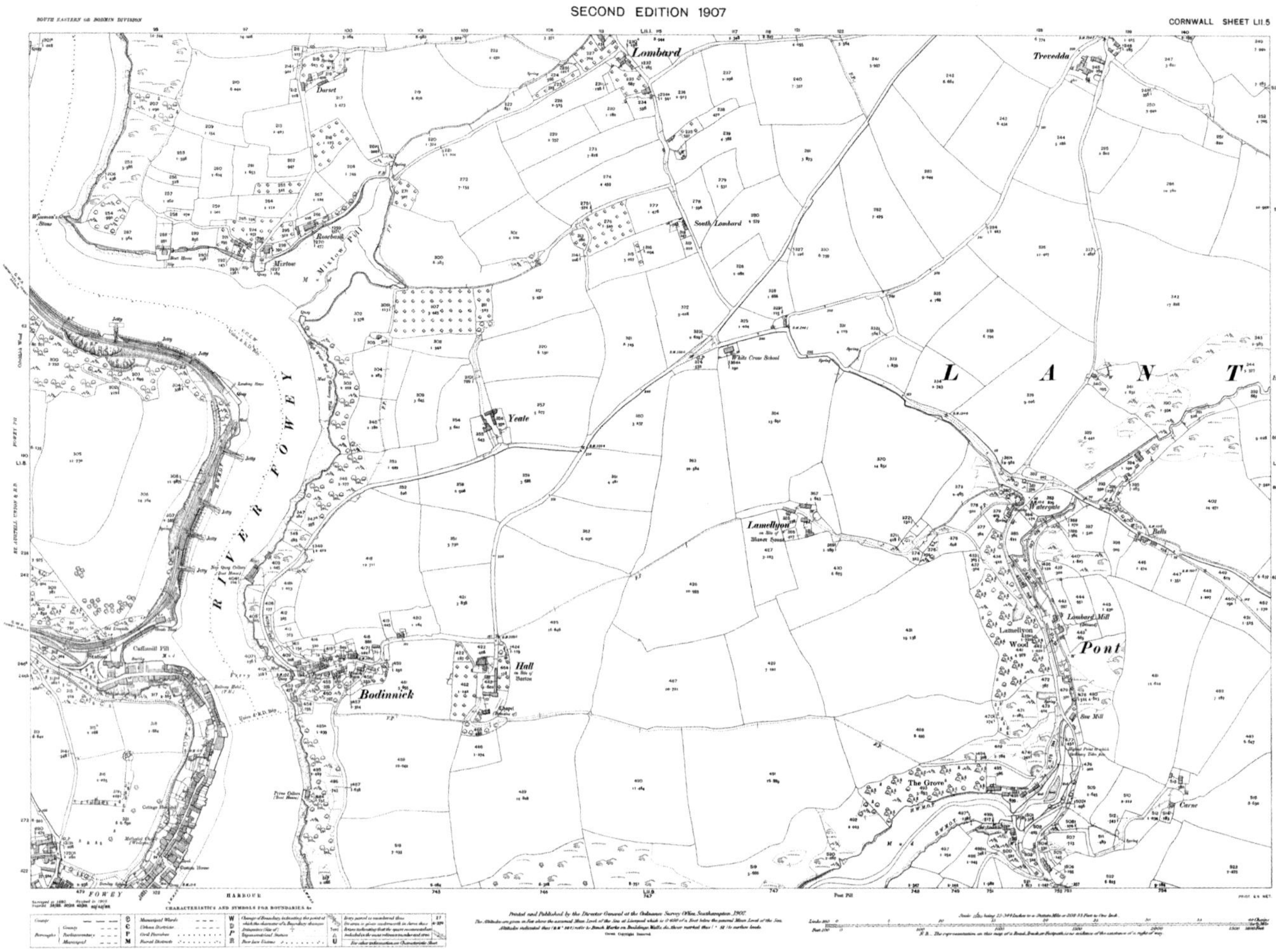

Series	Ordnance Survey Second Edition 1907. 25 in. to 1 mile
Sheet	Cornwall LII.5
Date	Surveyed 1880, revised 1906, published 1907

Although Cornwall had seen the development of an isolated network of tramways and railways serving tin and copper mines since the early 19th century, it took the arrival of the 'main line' railway from the east in 1859 to intensify the exploitation of the county's mineral resources, not least its china clay. The deep water harbour at Fowey on the south coast, and the opening in 1869 of a broad gauge (but independently owned) branch line south from the Cornwall Railway (later Great Western Railway) main line at Lostwithiel, gave the village of Fowey strategic advantage as a point of export for china clay which it enjoys to this day. The rail route ran (and still runs) along the western shore of the River Fowey, hugging steep hills and waterside woodland – a scenic trip normally the preserve of freight train drivers only.

In 1873 the tracks of a rival company, the standard-gauge Cornwall Minerals Railway (CMR), reached jetties around Carne Point north of Fowey, from Newquay and Par to the west. After six years of intense competition between the two railways, the broad gauge line closed in 1879 – but was reopened in 1895 by the CMR as a single-track standard gauge line. The two branches were connected at Fowey, and passenger trains were also introduced.

This Ordnance Survey map shows seven rail-served jetties, while Fowey station – at the edge of the village, due to the steep topography – had its own signal box ('SB' on the map) and goods shed, as well as platforms for passenger trains. The line from Par saw a very early withdrawal of passenger services, in 1934, but clung on for freight traffic until 1968, after which it was converted into a private 'haul road' for lorry movement of china clay from 'the dries' at Par to the docks at Fowey.

Passenger trains from Fowey to Lostwithiel were a Beeching casualty in 1965, and the terminus of the remaining freight branch line was pulled back to the north of Carne Point (west of all the jetties shown), where it now handles trainloads of export china clay several times daily from pits a dozen miles to the west.

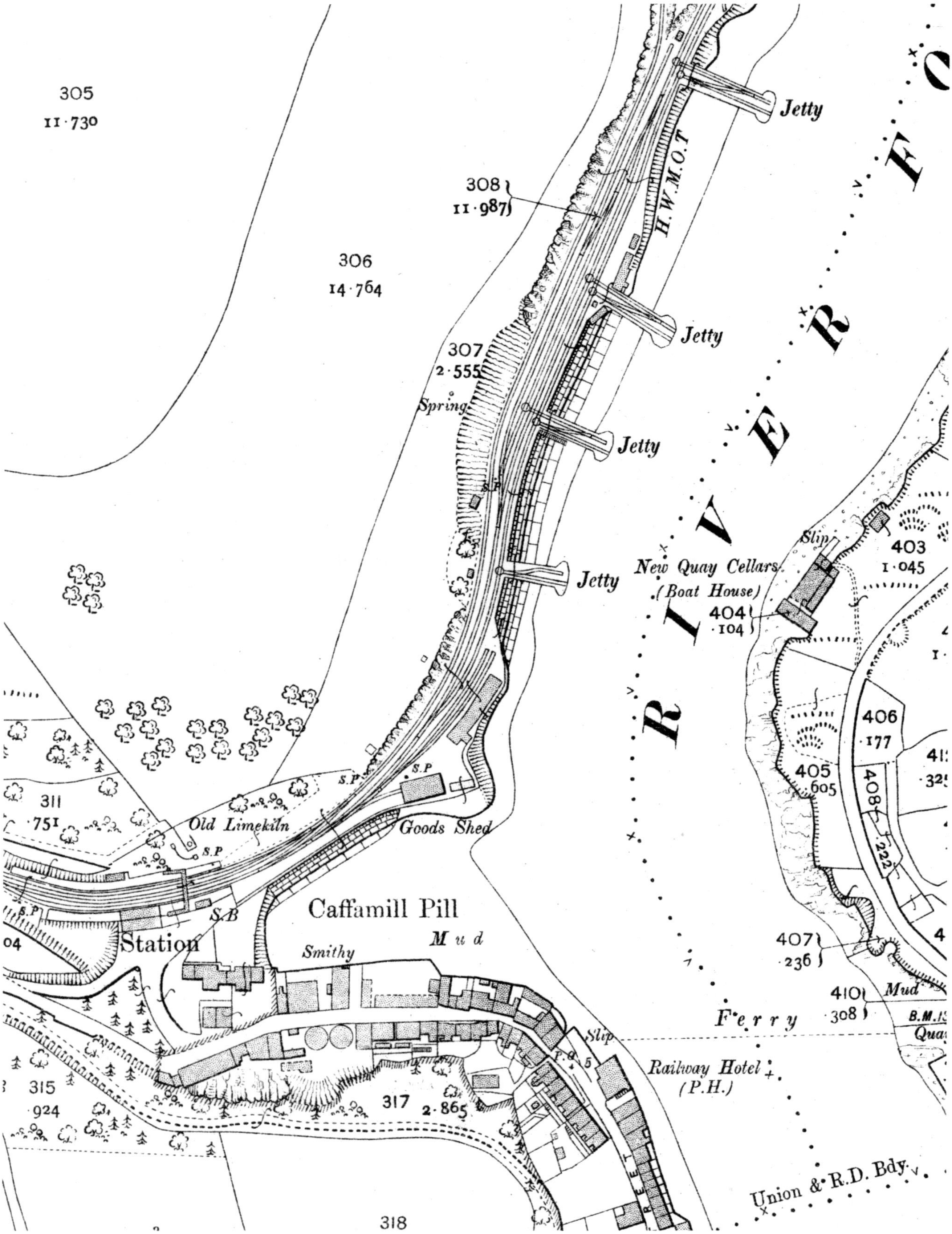
305
11·730
306
14·764
307
2·555
308
11·987
Spring
Jetty
Jetty
Jetty
Jetty
H.W.M.O.T
RIVER FOWEY
New Quay Cellars
(Boat House)
Slip
403
1·045
404
·104
405
·605
406
·177
407
·236
408
·222
410
·308
Mud
B.M.
Ferry
Railway Hotel
(P.H.)
Union & R.D. Bdy.
Old Limekiln
Goods Shed
S.P
S.B
Station
Caffamill Pill
Mud
Smithy
Slip
311
·751
315
·924
317
2·865
318

RAILWAY MAP OF LONDON AND SUBURBS, WITH POSTAL DISTRICTS - 1908

Title	Railway Map of London and Suburbs, with Postal Districts – 1908
Mapmaker	J Bartholomew, Edinburgh
Date	1908

By the early 20th century London's inner suburban rail network was largely complete, both in terms of main lines and local routes with a primarily commuter role. The railways were at the core of the daily life of the metropolis, as demonstrated by this map and the following map, both from the *Pocket Guide and Atlas of London* (1908) produced by the Edinburgh publisher and mapmaker, J Bartholomew.

This is a particularly clear and attractive map, which avoids the temptation to overload the sheet with details of built-up areas, roads and other topographical information not essential for a pocket rail guide. Railways are represented in black – which had become a virtually standard treatment in general mapping – and station names are produced in clear black fonts. Wider topographical context – and some colour variety – is provided by representation of the bigger parks (in green), the River Thames (in blue) and postal district boundaries (in red).

The crucial role of mass commuting by rail into and around the capital has ensured that the London rail network has survived more unscathed than in any other British city, and the pattern of routes on this map is still overwhelmingly recognizable today. However, there have been some changes. A number of freight branches have disappeared, notably as a result of closure of most of the commercial docks on the upper stretch of the Thames, while local factors led to the demise by 1963 of a few passenger branch lines – such as those to Alexandra Palace, Crystal Palace (High Level), Greenwich Park and Palace Gates (Wood Green).

London's passenger rail network survived the Beeching threat almost untouched, with reprieves for the Richmond–Camden arm of the North London Line, the Gospel Oak–Barking route, and the Clapham Junction–Kensington Olympia link – all of which now play a key part in the London commuter system. The only significant loss has been the 1986 closure of the line from Dalston Junction to the Broad Street terminus, just west of Liverpool Street station, with North London Line trains then diverted to run east to North Woolwich.

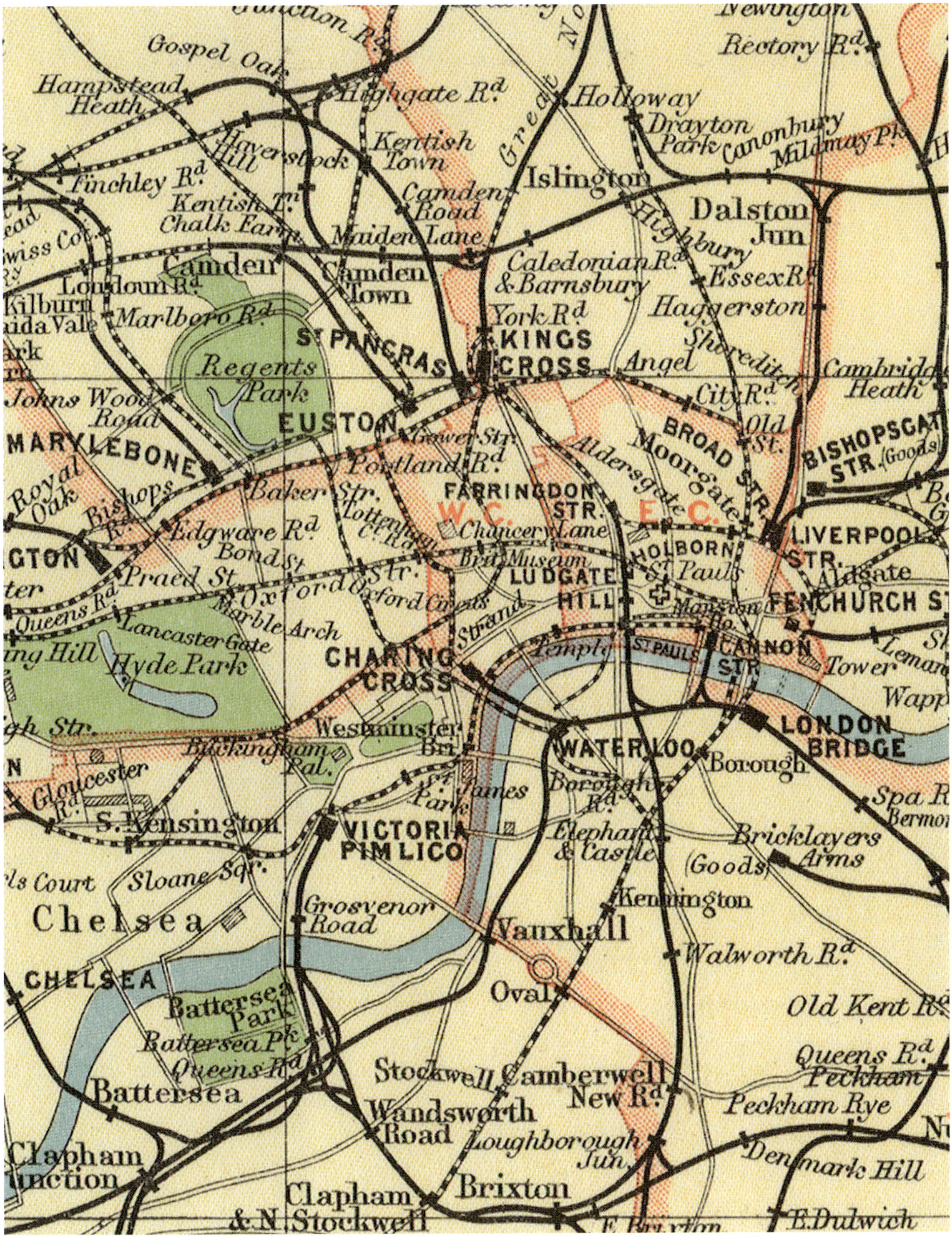
Gospel Oak
Hampstead Heath
Highgate Rd.
Holloway
Newington
Rectory Rd.
Drayton Park
Canonbury
Mildmay Pk.
Haverstock Hill
Kentish Town
Great
Finchley Rd.
Kentish Tn.
Chalk Farm
Camden Road
Islington
Dalston Jun.
Maiden Lane
Highbury
Camden
Camden Town
Caledonian Rd. & Barnsbury
Essex Rd.
Loudoun Rd.
Kilburn
Marlboro Rd.
York Rd.
Haggerston
St. PANCRAS
KINGS CROSS
Shoreditch
Regents Park
Angel
Cambridge Heath
City Rd.
Johns Wood Road
EUSTON
Gower Str.
Old St.
MARYLEBONE
Portland Rd.
Aldersgate
Moorgate
BROAD STR.
BISHOPSGATE STR. (Goods)
Royal Oak
Baker Str.
FARRINGDON STR.
Bishops Rd.
W.C.
E.C.
Tottenham Ct. Rd.
Chancery Lane
Edgware Rd.
HOLBORN
LIVERPOOL STR.
Bond St.
Brit. Museum
LUDGATE HILL
St. Pauls
Aldgate
Praed St.
Oxford Str.
Oxford Circus
Mansion Ho.
FENCHURCH ST.
Queens Rd.
Marble Arch
Strand
Lancaster Gate
Temple
ST. PAULS
CANNON STR.
Leman
Hyde Park
CHARING CROSS
Tower
Wapp
Westminster Br.
LONDON BRIDGE
WATERLOO
Borough
Buckingham Pal.
Gloucester Rd.
Borough Rd.
St. James Park
Spa R
Bermo
S. Kensington
VICTORIA
PIMLICO
Elephant & Castle
Bricklayers Arms
(Goods)
Sloane Sqr.
Kennington
Chelsea
Grosvenor Road
Vauxhall
Walworth Rd.
CHELSEA
Oval
Old Kent Rd.
Battersea Park
Battersea Pk.
Queens Rd.
Queens Rd. Peckham
Stockwell
Camberwell New Rd.
Peckham Rye
Battersea
Wandsworth Road
Loughborough Jun.
Clapham Junction
Denmark Hill
Brixton
Clapham & N. Stockwell
E. Dulwich

RAILWAY MAP OF LONDON AND SU

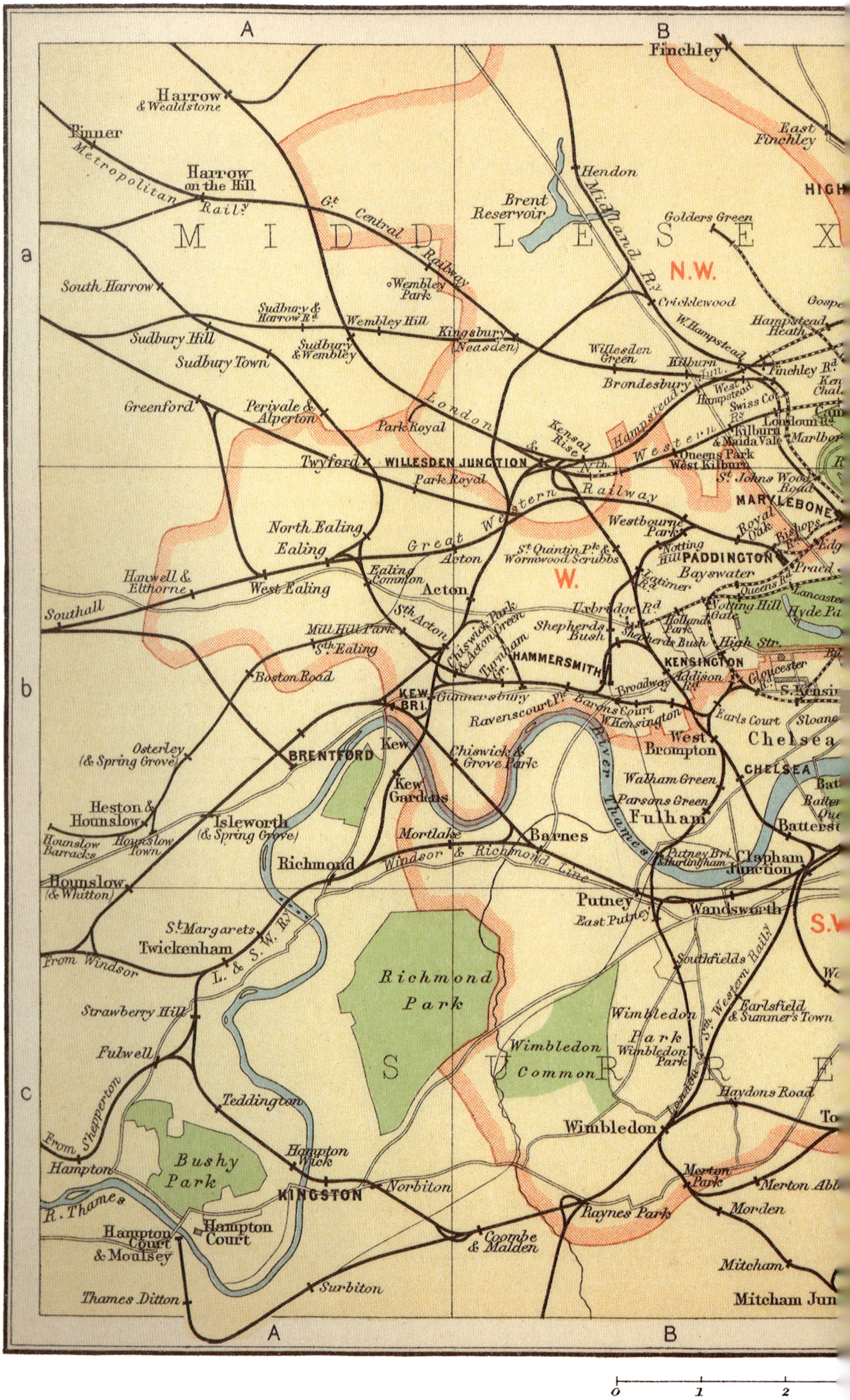

BS, WITH POSTAL DISTRICTS - 1908

RAILWAY MAP OF CENTRAL LONDON - 1908

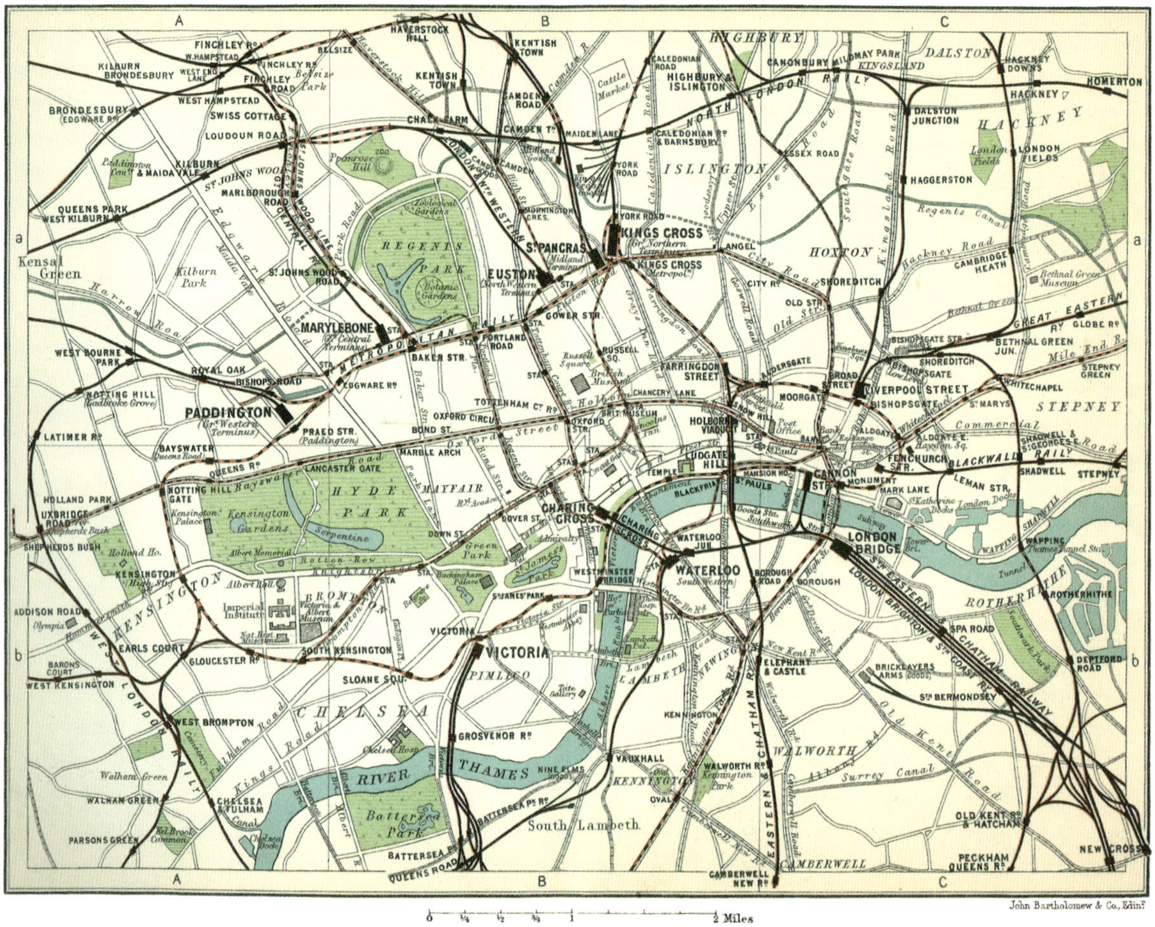

Title	Railway Map of Central London – 1908
Mapmaker	J Bartholomew, Edinburgh
Date	1908

By the time of publication of this map, all of London's main line rail termini had been completed, and mass commuting over an extensive inner and outer suburban rail network was an established part of daily life – the *Oxford Companion to British Railway History* (1997) recording that 'in 1904 passenger arrivals at London main-line terminals before 10.30 a.m. were 318,000', and also noting that:

> *The period from 1900 to 1914 was crucial for London railway and traffic development. Electrification of the Metropolitan and the District, its beginning on the LBSCR [London, Brighton & South Coast Railway] South London line, and the building of the new tubes, together with a growing interest in London suburban possibilities shown by the LNWR [London & North Western Railway] and the GWR [Great Western Railway] . . . led to a surge of local traffic that the railways both encouraged and found hard to cope with.*

Given the larger scale of the map (just over 2 in. to 1 mile) compared to the previous map, much more detail has been shown – but again carefully selected to avoid visual clutter. Principal roads and street names are included – as are the new electric tram routes which were to play a largely complementary role to the railways rather than competitive, except in parts of the inner suburbs.

The tube and above-ground lines of the Underground are shown in full (in hatched black and red) despite being below-ground in most of Central London. This allows us to see the extent to which key routes followed the alignment of the main roads, helping to keep construction costs down.

Major public buildings are also shown, as are additional topographical details within the principal public parks. The map may have been produced with the passenger railway in mind, but the big goods stations, which played a crucial role in the inner London economy, are also clearly marked – note in particular Bricklayers Arms and Nine Elms south of the Thames, and Bishopsgate, Camden and King's Cross north of the river.

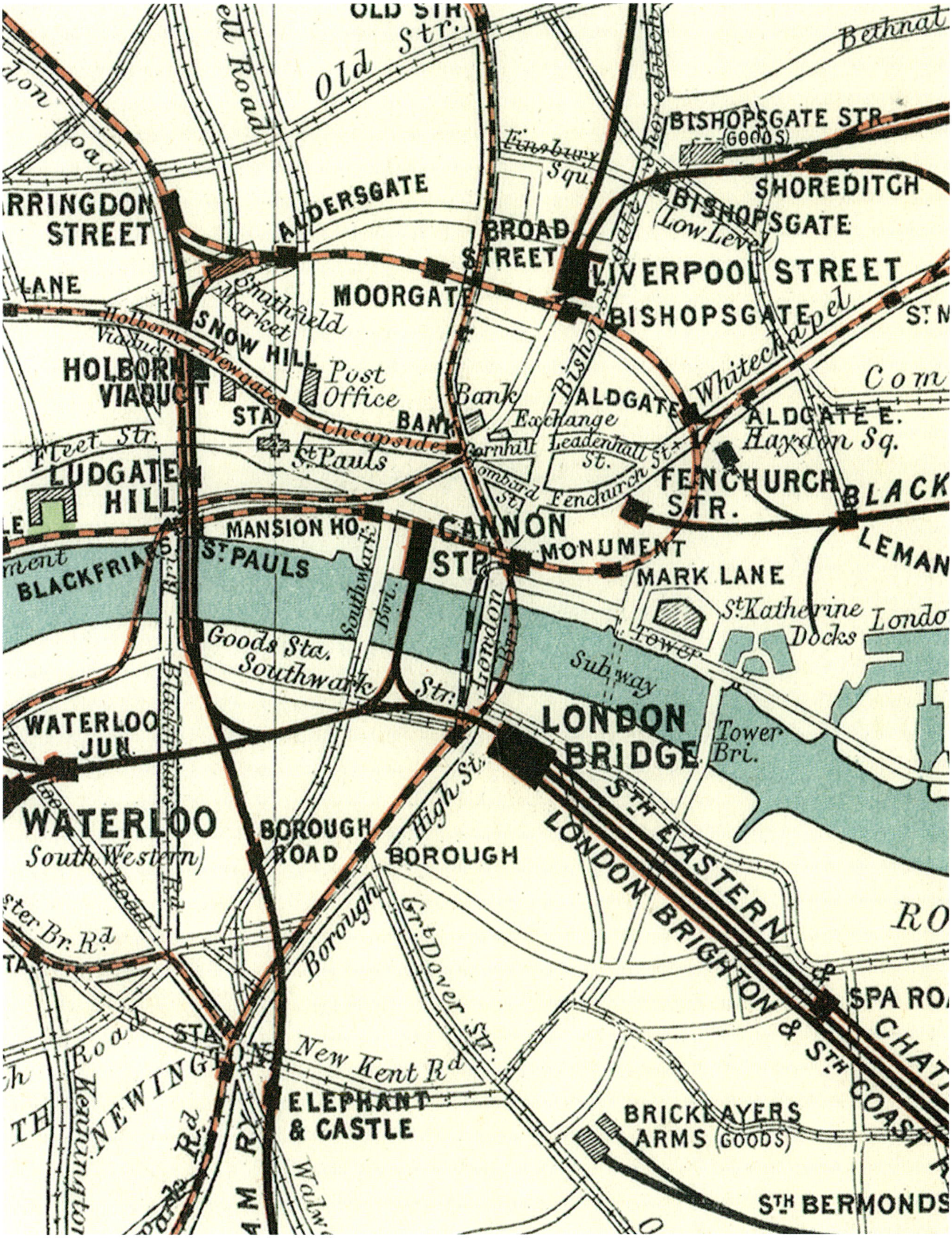
OLD STR
Old Str.
Bethnal
BISHOPSGATE STR
(GOODS)
Finsbury Squ.
ALDERSGATE
FARRINGDON STREET
SHOREDITCH
BISHOPSGATE (Low Level)
BROAD STREET
LIVERPOOL STREET
MOORGATE
BISHOPSGATE
Smithfield Market
SNOW HILL
HOLBORN VIADUCT
Post Office
Whitechapel
ALDGATE
ALDGATE E.
BANK
Bank
Exchange
Cheapside
Cornhill
Leadenhall St.
Haydon Sq.
Fleet Str.
St. Pauls
LUDGATE HILL
Lombard St.
Fenchurch St.
FENCHURCH STR.
MANSION HO.
CANNON STR.
MONUMENT
BLACKFRIARS
ST. PAULS
MARK LANE
LEMAN
St. Katherine Docks
Goods Sta. Southwark
Southwark Bri.
London Bri.
Subway
Tower
Southwark Str.
WATERLOO JUN.
LONDON BRIDGE
Tower Bri.
WATERLOO (South Western)
BOROUGH ROAD
BOROUGH
High St.
STH EASTERN
LONDON BRIGHTON & STH COAST
Borough
Gr. Dover Str.
SPA ROA
New Kent Rd.
NEWINGTON
ELEPHANT & CASTLE
BRICKLAYERS ARMS (GOODS)
STH BERMONDS

RAILWAY MAP OF CI

RAL LONDON - 1908

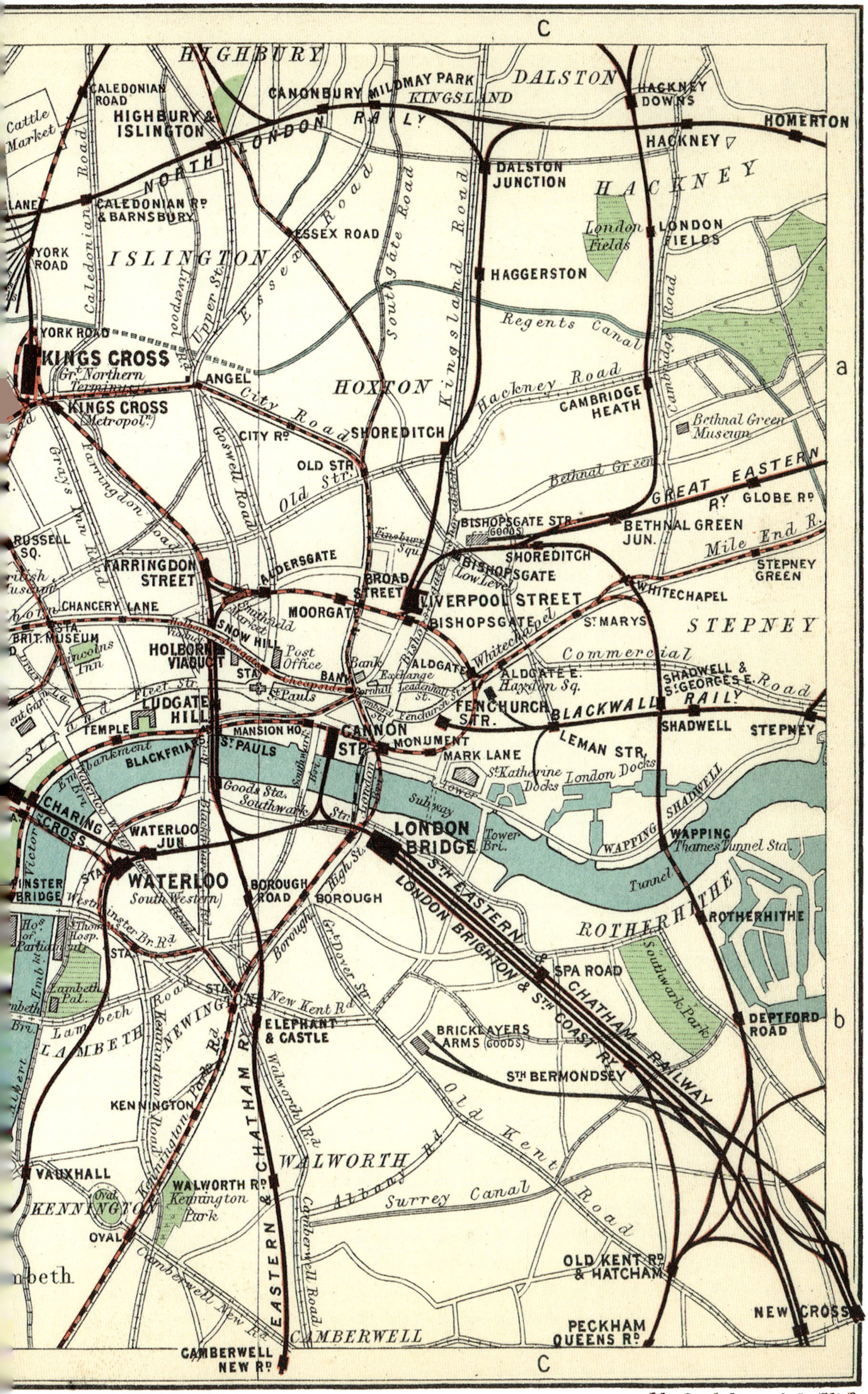

John Bartholomew & Co., Edin.

2 Miles

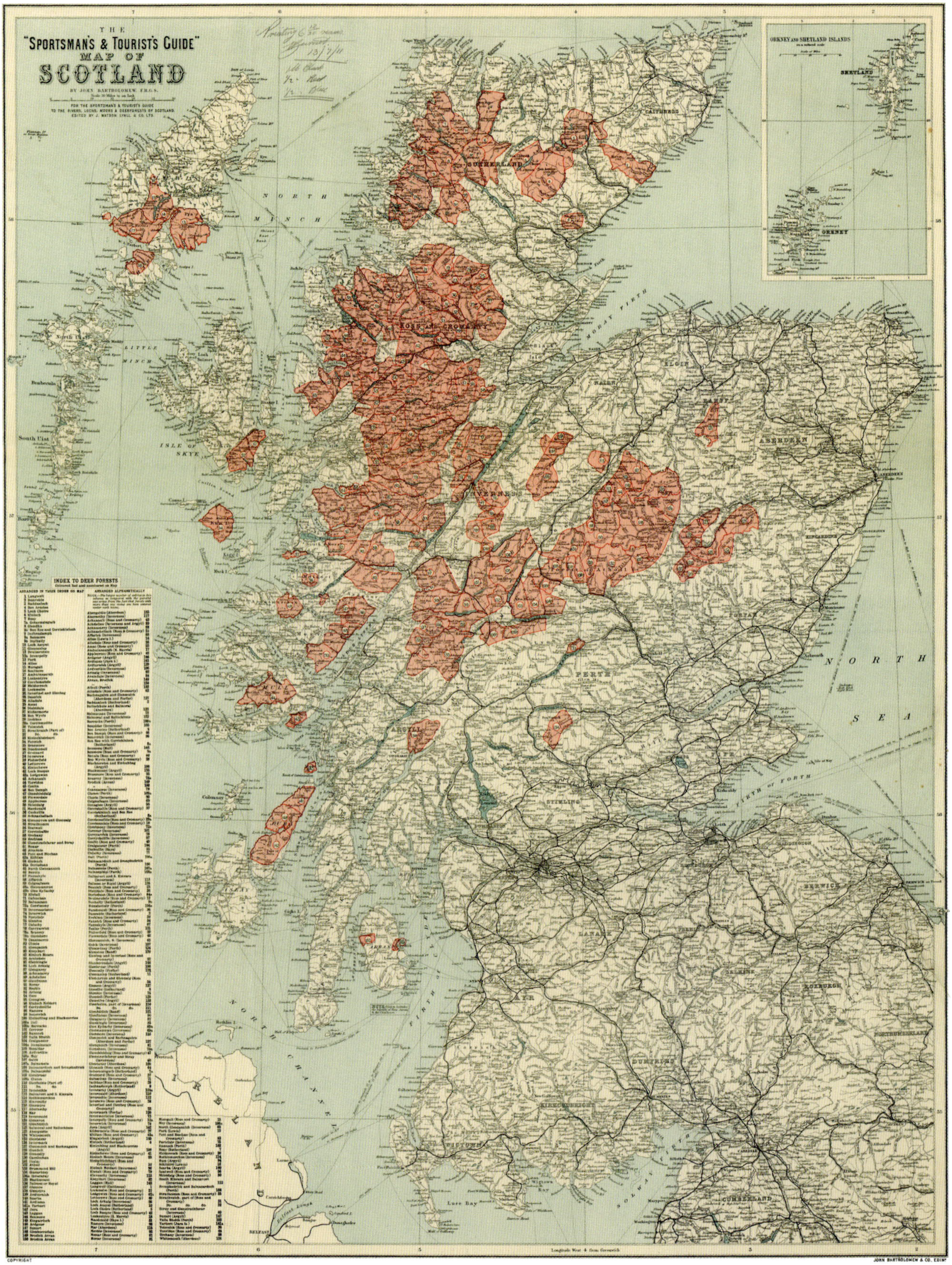
THE
"SPORTSMAN'S & TOURIST'S GUIDE"
MAP OF
SCOTLAND
BY JOHN BARTHOLOMEW, F.R.G.S.
ORKNEY AND SHETLAND ISLANDS
SHETLAND
ORKNEY
NORTH MINCH
LITTLE MINCH
ISLE OF SKYE
SUTHERLAND
CAITHNESS
ROSS AND CROMARTY
MORAY FIRTH
INVERNESS
ABERDEEN
PERTH
ARGYLL
STIRLING
FIRTH OF FORTH
NORTH SEA
NORTH CHANNEL
IRELAND
INDEX TO DEER FORESTS
COPYRIGHT
JOHN BARTHOLOMEW & CO. EDINR

Title	The "Sportsman's and Tourist's Guide" Map of Scotland
By	J Bartholomew, Edinburgh
Date	1911

The expansion of the rail network deep into the Highlands by the late 19th century opened up an enormous market for tourists and 'sportsmen' keen to indulge their passion for hunting, shooting and fishing in hitherto inaccessible country. In *The Highland Railway* (1963), H A Vallance reports a contemporary 1888 account of a train which had become the apotheosis of this traffic – the early morning service from Perth to Inverness over the Highland Main Line during the summer 'season':

> *In July and August this 7.50 train is the unique railway phenomenon. Passenger carriages, saloons, horseboxes and vans, concentrated at Perth from all parts of England, and intermixed to make an irregular caravan. Engines are attached fore and aft, and the procession toils pluckily over the Grampians.*

On 7th August, this train comprised rolling stock from ten companies – 37 carriages and two locomotives at the front, with a 'banker' loco attached to the rear at Blair Atholl for the stiff 17½ -mile climb at gradients of up to 1 in 70 to Britain's then highest railway summit at Druimuachdar, 1,484 ft above sea level.

As shown in the map extract below, with stations at Blair Atholl, Struan, Dalnaspidal, Dalwhinnie, Newtonmore, Kingussie, Kincraig and Aviemore, the Highland Main Line was well placed to serve this lucrative market. Together with the Kyle line, the Far North line to Caithness, the West Highland line from Glasgow to Fort William, its branch to Fort Augustus and the 'Mallaig Extension', the Highland and North British railway companies could legitimately claim to serve the overwhelming majority of the 150 'deer forests' identified on this 1911 map.

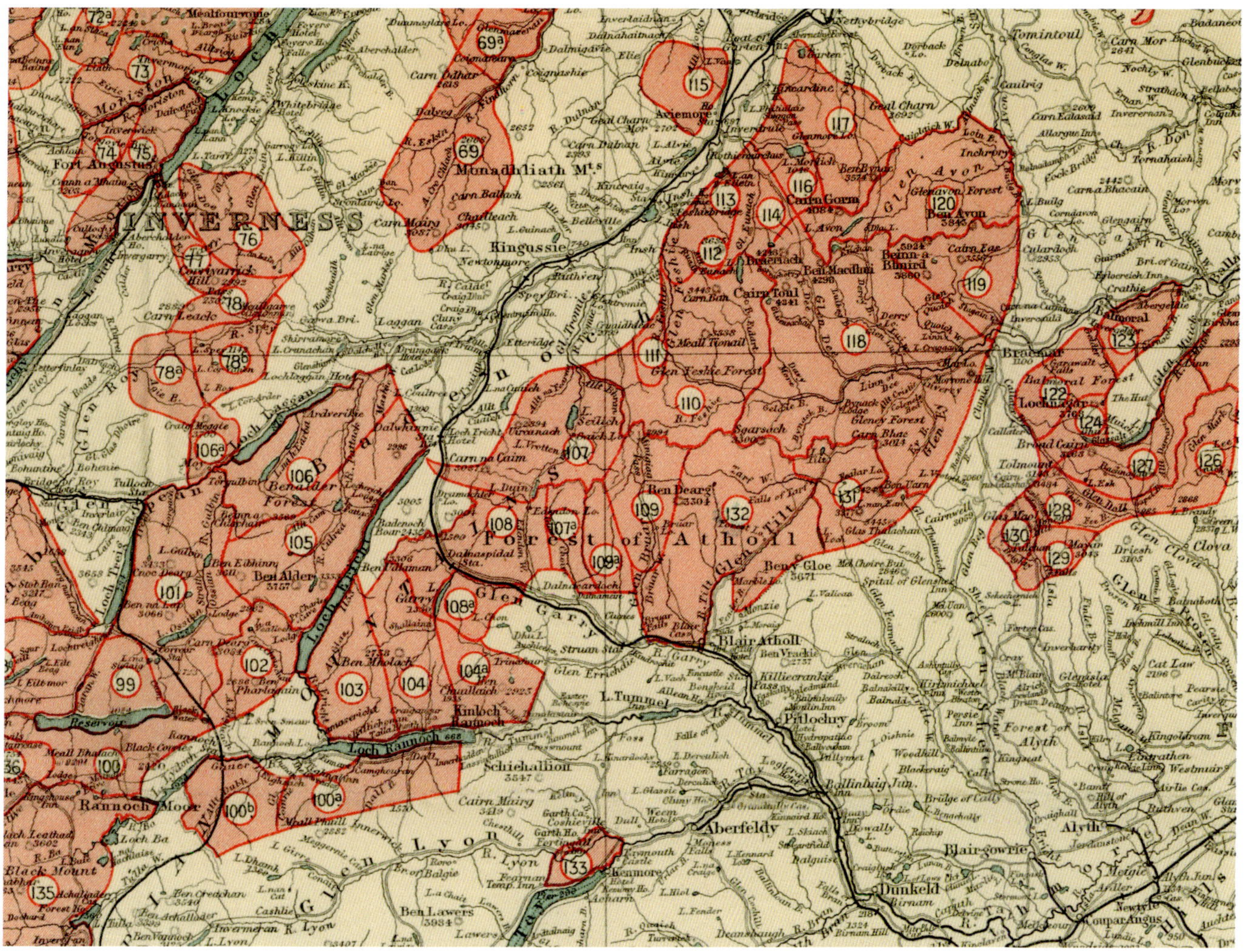

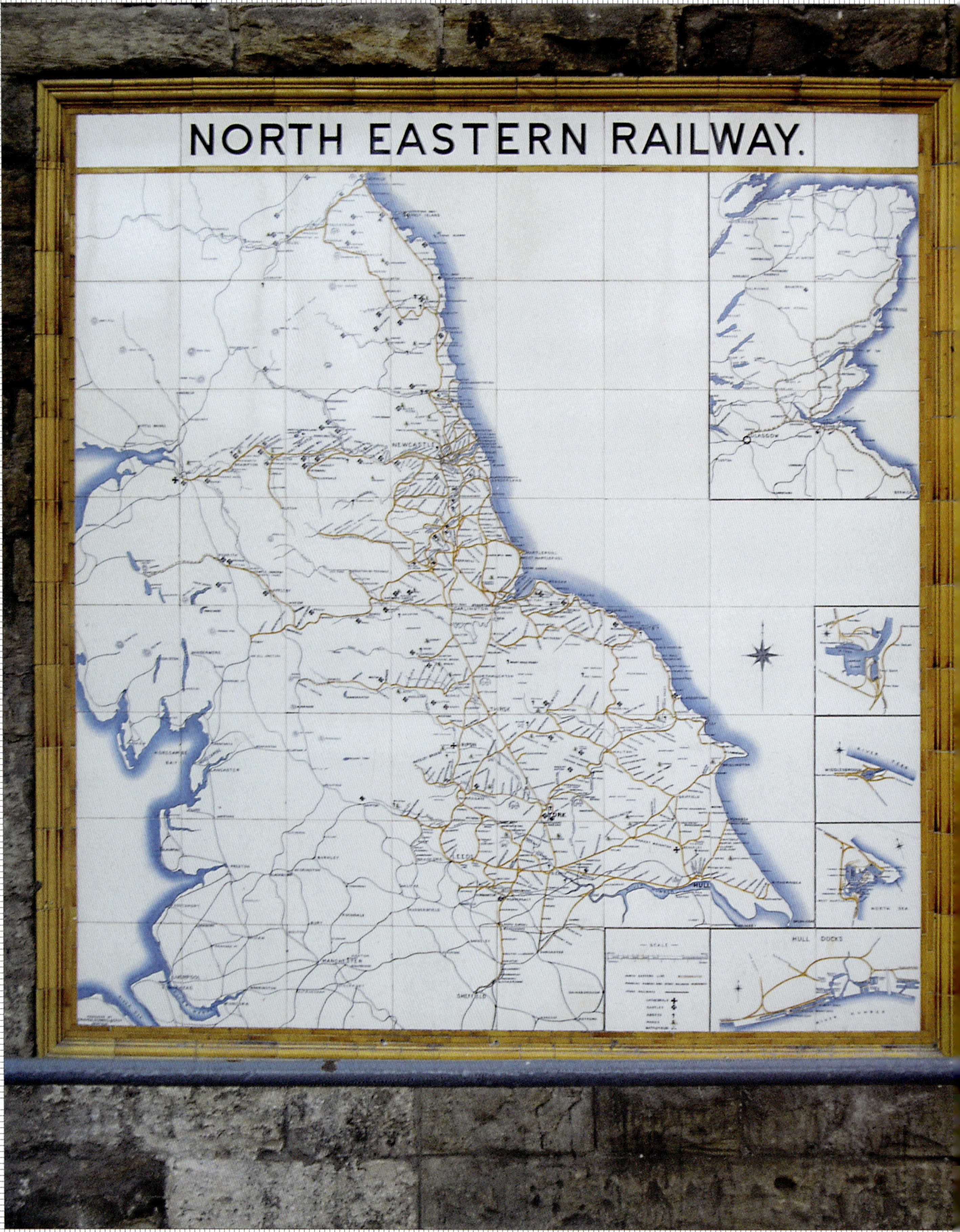
NORTH EASTERN RAILWAY.
NEWCASTLE
YORK
HULL
LEEDS
MANCHESTER
SHEFFIELD
GLASGOW
THIRSK
HULL DOCKS

Title	North Eastern Railway
By	Craven Dunnill & Co. Ltd, Jackfield (Shropshire)
Location	Morpeth railway station
Date	*c.*1900–10

In 1900 the North Eastern Railway (NER) began the commissioning of large glazed tile maps for display at a number of its stations, depicting the whole of the NER network, from Berwick and Carlisle in the north to southern outposts at Swinton, near Doncaster, and Withernsea on the North Sea coast.

It is thought that at least 25 maps – measuring 5 ft 8 in. by 5 ft 4 in. – were erected at stations, the last by 1910. A contemporary author, G W J Potter, wrote that the idea had 'attracted considerable attention, and its adoption has much to recommend it – being easily cleaned, very legible, practically everlasting'. The maps were indeed built to last, and 12 still exist, nine of them at their original stations – Beverley, Hartlepool, Middlesbrough, Morpeth, Saltburn, Scarborough, Tynemouth, York and Whitby. The map from King's Cross station – far from NER territory – is now preserved at the National Railway Museum in York. The example photographed here is from Morpeth station, beside the 'Up' (London-bound) platform.

After cheaper foreign tiles eliminated the domestic industry in Britain, the original factory in Shropshire which produced the NER tile maps was abandoned – but the building survived, and was eventually re-opened by the Ironbridge Gorge Museum Trust as a tile museum. By 1989 a demonstration tile works was established there, and in 2007 the wheel turned full circle when the North Eastern Tile Company commissioned Craven Dunnill to restart production of NER tiled maps, using original methods and in their original factory.

Intriguingly, as shown in the extract to the right, the tile map includes representation of a line that was never built – from Beverley to North Frodingham in East Yorkshire. This was a mistake which could not easily be corrected!

The tile map tucked away on Morpeth's 'Up' platform.

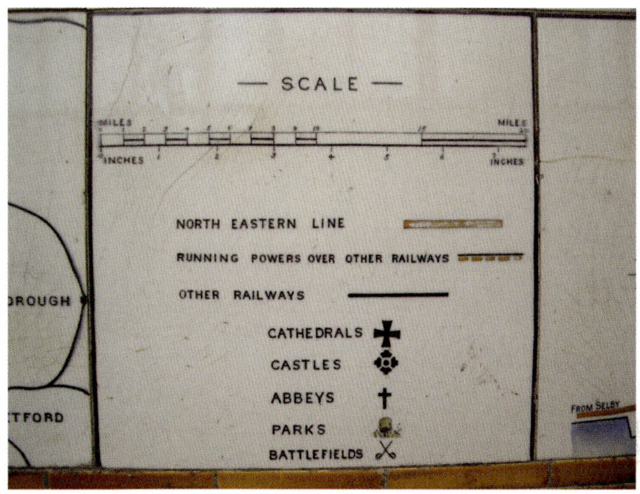

Buildings of historic interest are included in the map key.

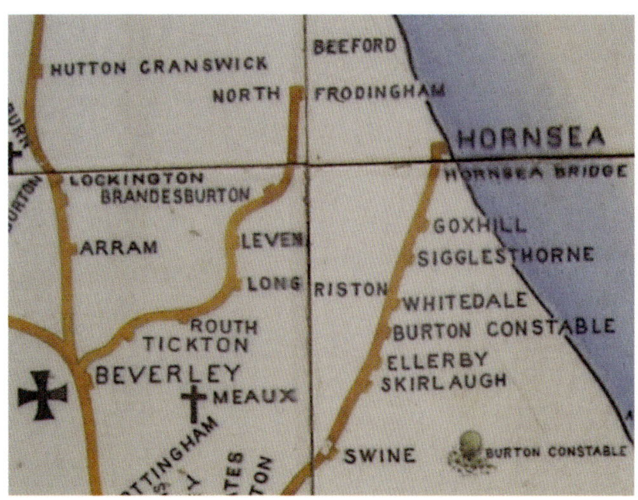

The line that never was – from Beverley to North Frodingham.

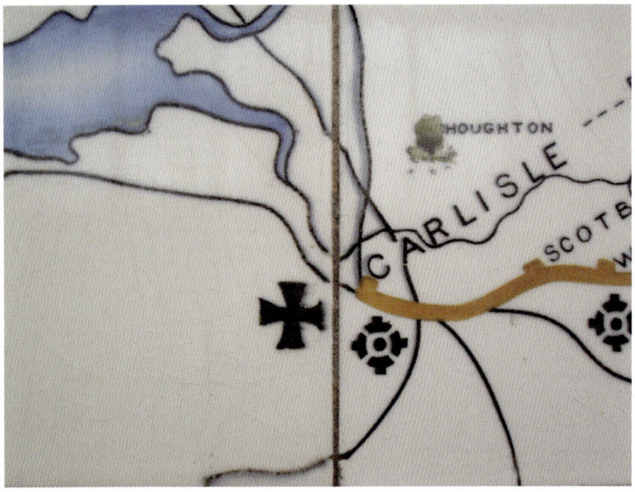

The NER's northwestern outpost at Carlisle.

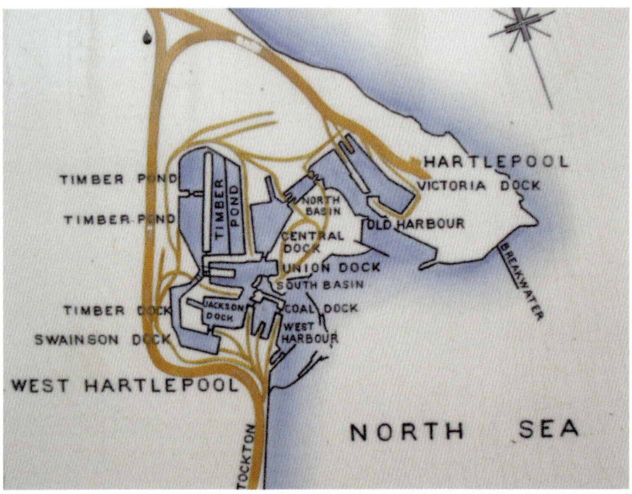

Goods traffic sidings at Hartlepool Docks.

The dense commuter network around Newcastle.

DURHAM
HARTLEPOOL
WEST HARTLEPOOL
FERRYHILL
BISHOP AUCKLAND
BARNARD CASTLE
DARLINGTON
STOCKTON
MIDDLESBROUGH
REDCAR
SALTBURN
WHITBY
ROBIN HOODS BAY
SCARBOROUGH
PICKERING
FILEY
RICHMOND
NORTHALLERTON
BEDALE
THIRSK
HELMSLEY
MALTON
BRIDLINGTON
FLAMBOROUGH
RIPON
PATELEY BRIDGE
HARROGATE
KNARESBOROUGH
DRIFFIELD
YORK
POCKLINGTON
MARKET WEIGHTON
BEVERLEY
HORNSEA
LEEDS
BRADFORD
SELBY
HULL
RIVER HUMBER
HALIFAX
HUDDERSFIELD
NORMANTON
PONTEFRACT
WAKEFIELD
GOOLE
BARNSLEY
DONCASTER
PENISTONE
ROTHERHAM
SHEFFIELD
GAINSBOROUGH
GRIMSBY
SCALE
NORTH EASTERN LINE
RUNNING POWERS OVER OTHER RAILWAYS
OTHER RAILWAYS
CATHEDRALS
CASTLES

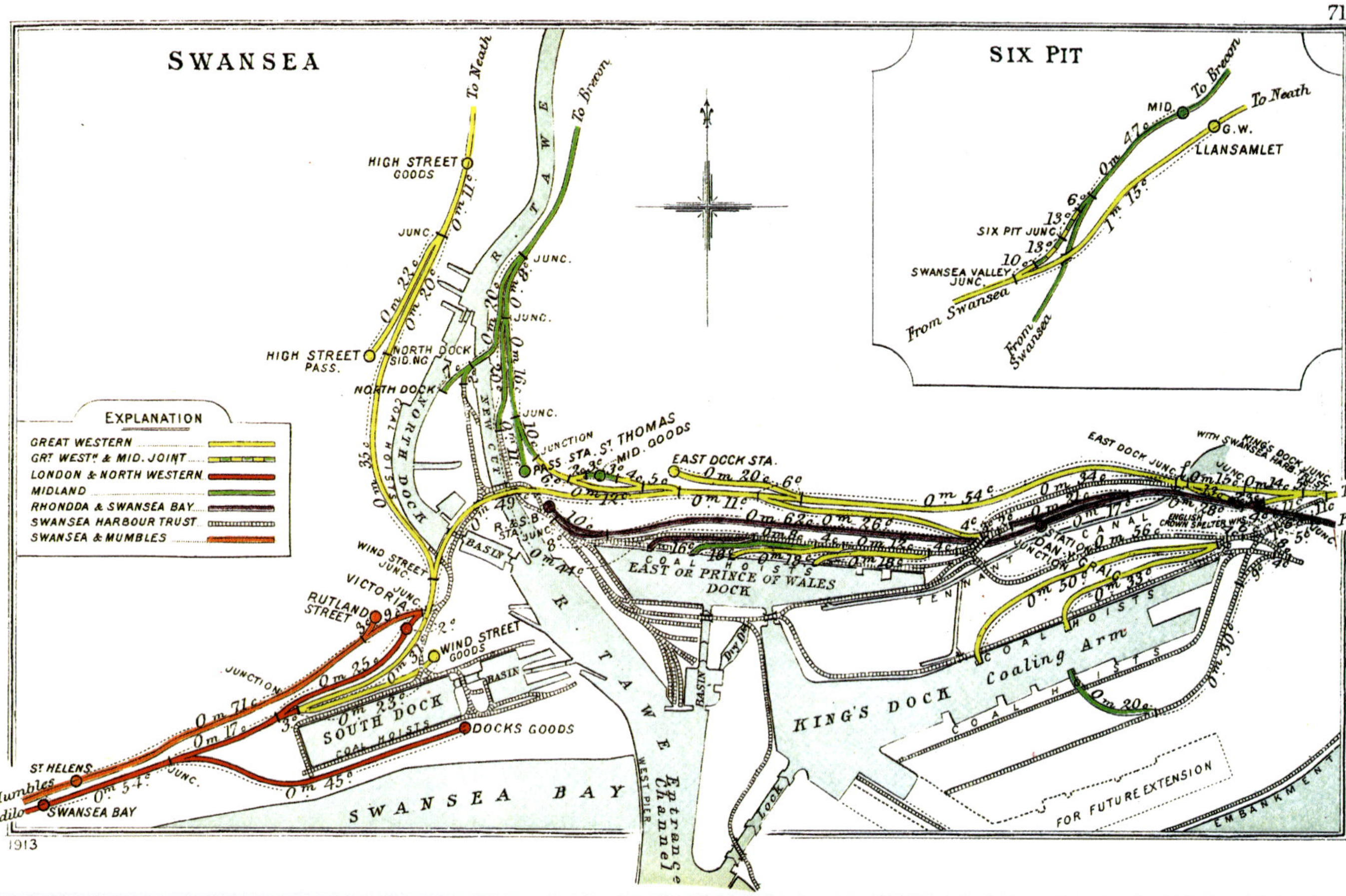

This 1913 Railway Clearing House map of Swansea shows seven different railway companies serving the town and docks.

Coal and steam power were the natural driving forces behind the Industrial Revolution in Britain. By the 18th century the increasing output of coal mines was already being transported on horse-drawn waggonways and tramways to rivers and harbours for onward distribution around Britain by boat – it was not by chance that the world's first railways were built to serve Britain's growing coal industry.

While coal deposits could be found in many parts of Britain, the main concentrations were in central Scotland, Teesside and Tyneside, South Yorkshire, Derbyshire, Nottinghamshire and South Wales and it was in these areas that railway development was the most rapid.

By far the largest coal-producing area in Britain was in South Wales. Its reserves are reckoned to cover around 1,000 square miles with much of it under water – consequently a large proportion has never been exploited. The region was one of the birthplaces of the Industrial Revolution in the late 18th century with a canal and tramway system feeding coal to nearly 100 blast furnaces.

At the height of coal production in the early 20th century nearly 37 million tons of coal were being exported around the world from ports along the coast of South Wales. At this time a total of 16 railway companies were involved in transporting the coal from the hundreds of mines in the Welsh Valleys down to the mechanical tips at Newport, Cardiff, Barry, Port Talbot, Swansea and Llanelli docks. There was so much coal being mined that the steep-sided valleys often had competing railway lines running along each side – morning, noon and night, they echoed to a continuous procession of loaded coal trains trundling down to the docks and rattling empty trains returning.

Such was the importance of the docks that two of the major railway companies, the Alexandra (Newport & South Wales) and the Barry were originally incorporated primarily as dock companies. Competition between the docks for the lucrative coal traffic was intense with some of them being served by no less than six railway companies – for example, Cardiff Docks was served by the Great Western Railway, London & North Western Railway, Taff Vale Railway, Rhymney Railway, Barry Railway and Cardiff Railway.

By far the most successful of these coal-carrying railway and docks companies was the Barry Docks & Railway Company which was incorporated in 1884 to build 68 miles of railway

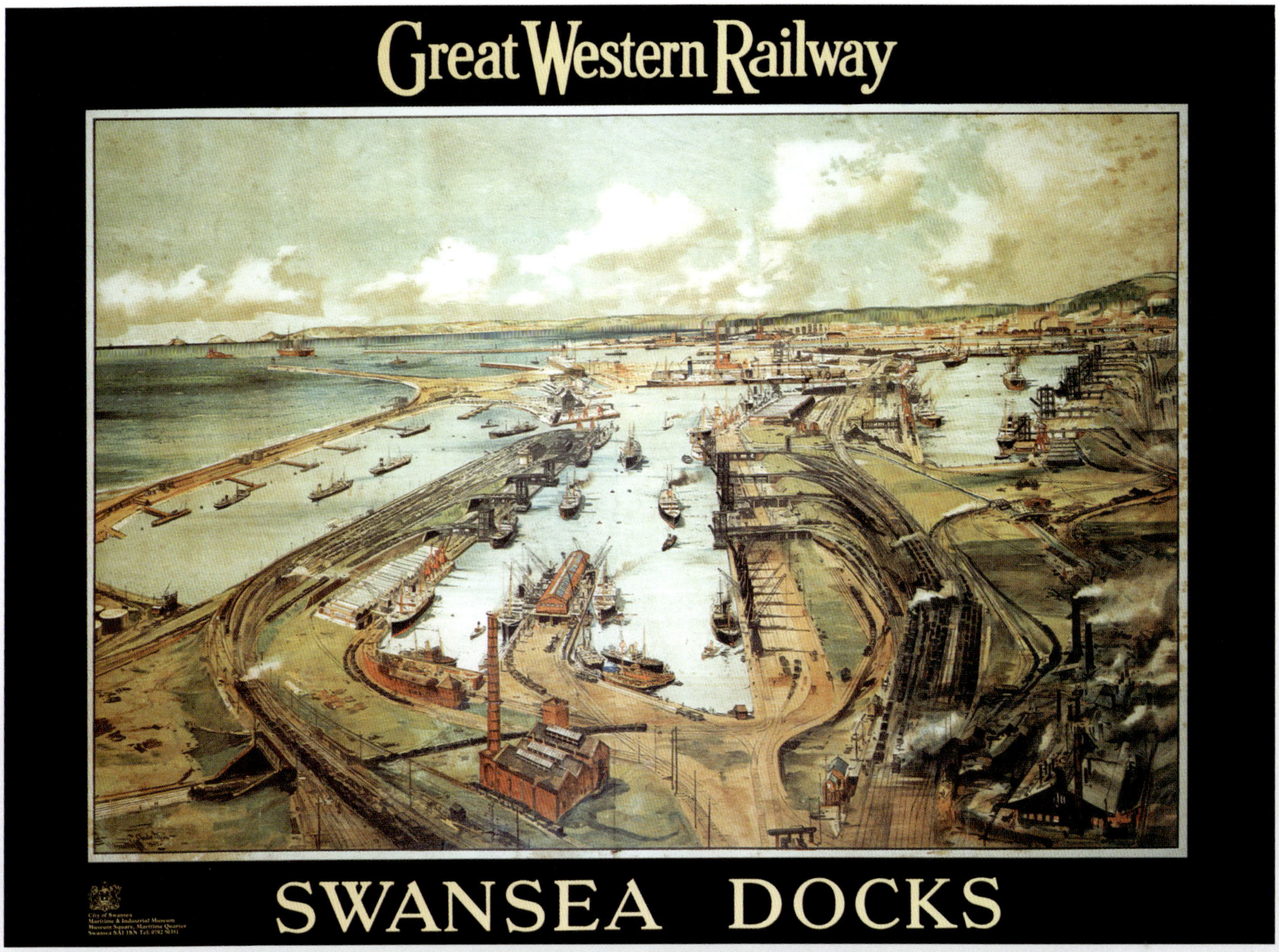

The Great Western Railway's docks at Swansea once exported vast amounts of Welsh coal to the far-flung corners of the British Empire.

between a new purpose-built 73-acre deep-water dock at Barry and the Rhondda and Aberdare Valleys. Engineered by Henry Marc Brunel (second son of Isambard Kingdom Brunel) and Sir John Wolfe Barry, the railway featured several long tunnels and three massive viaducts. With nearby Cardiff Docks already working to full capacity, the resourceful new railway was an instant success and on the day the line opened in 1889 several thousand invited guests were treated to a grandstand view of newly arrived coal trains discharging their loads into the first ship. With its conveniently located deep-water dock the railway delivered three million tons of coal to 1,700 ships in the first year of its existence and by 1913 was exporting 30 per cent of all of South Wales' coal.

One of the key innovations of the Beeching Report was the introduction of 'merry-go-round' coal trains – so named because the train both loads and unloads its cargo while moving at slow speed. Despite the massive changes in the coal industry since the 1960s, it is still closely linked to the railways, with trains of 1,000 tonnes and more providing a regular supply of coal from Britain's three remaining deep mines (in Nottinghamshire, the West Midlands and Yorkshire), from open cast sites in south Wales, the north of England and Scotland, and from major coal-importing ports such as Avonmouth, Hunterston and Immingham, to coal-fired power stations in all three countries. Coal remains by far the most important freight commodity on the railways.

One positive outcome of the Beeching Report was the introduction of 'merry-go-round' coal trains such as this one seen at Ratcliffe-on-Soar power station, Nottinghamshire, in 1993.

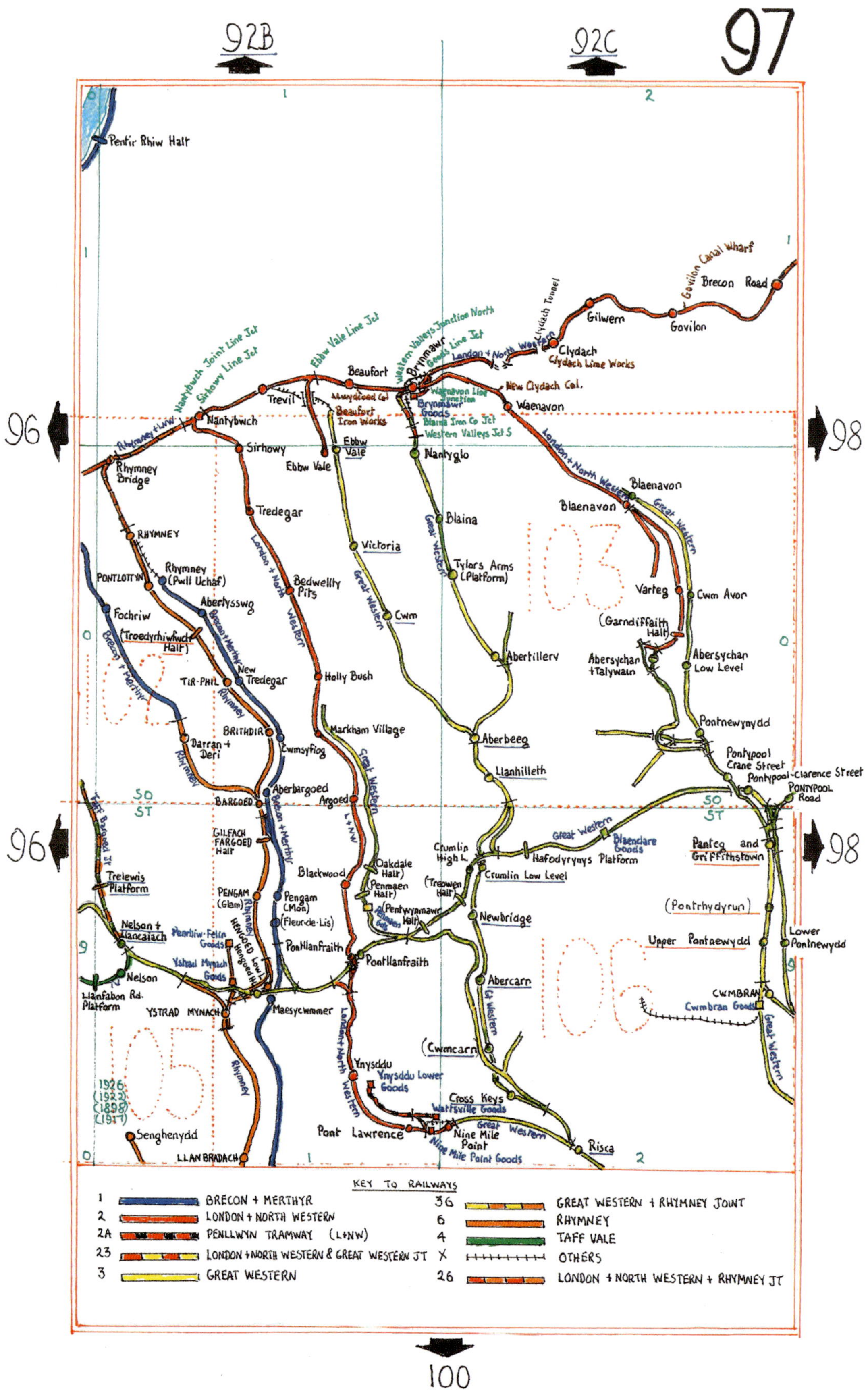
97
92B
92C
96
98
100
Pentir Rhiw Halt
Gilwern
Govilon
Brecon Road
Govilon Canal Wharf
Clydach
Clydach Lime Works
New Clydach Col.
Waenavon
Beaufort
Beaufort Iron Works
Brynmawr
Brynmawr Goods
Nantybwch
Trevil
Sirhowy
Ebbw Vale
Rhymney Bridge
Tredegar
RHYMNEY
Rhymney (Pwll Uchaf)
PONTLOTTYN
Fochriw
Abertysswg
(Troedyrhiwfwch Halt)
New Tredegar
TIR-PHIL
BRITHDIR
Darran + Deri
Cwmsyfiog
Aberbargoed
BARGOED
GILFACH FARGOED Halt
PENGAM (Glam)
Pengam (Mon)
(Fleur-de-Lis)
Bedwellty Pits
Holly Bush
Markham Village
Argoed
Blackwood
Oakdale Halt
(Penmaen Halt)
Pontllanfraith
Maesycwmmer
YSTRAD MYNACH
Nelson + Llancaiach
Nelson
Llanfabon Rd. Platform
Trelewis Platform
Ynysddu
Ynysddu Lower Goods
Pont Lawrence
Nine Mile Point
Nine Mile Point Goods
Cross Keys
Risca
Senghenydd
LLANBRADACH
Victoria
Cwm
Nantyglo
Blaina
Tylors Arms (Platform)
Abertillery
Aberbeeg
Llanhilleth
Crumlin High L.
Crumlin Low Level
Hafodyrynys Platform
Newbridge
Abercarn
(Cwmcarn)
Blaenavon
Varteg
Cwm Avon
(Garndiffaith Halt)
Abersychan + Talywain
Abersychan Low Level
Pontnewynydd
Pontypool Crane Street
Pontypool-Clarence Street
PONTYPOOL Road
Panteg and Griffithstown
(Pontrhydyrun)
Upper Pontnewydd
Lower Pontnewydd
CWMBRAN
Cwmbran Goods
London + North Western
Great Western
Brecon + Merthyr
Rhymney
1926
(1922)
(1898)
(1917)
103
102
105
106
KEY TO RAILWAYS
1 BRECON + MERTHYR
2 LONDON + NORTH WESTERN
2A PENLLWYN TRAMWAY (L+NW)
23 LONDON + NORTH WESTERN & GREAT WESTERN JT
3 GREAT WESTERN
36 GREAT WESTERN + RHYMNEY JOINT
6 RHYMNEY
4 TAFF VALE
X OTHERS
26 LONDON + NORTH WESTERN + RHYMNEY JT

Title	Jowett's Railway Atlas (page 97)
By	Alan Jowett
Publisher	Patrick Stephens Ltd, Cambridge
Date	1989

Jowett's Railway Atlas, published in 1989, is an historical atlas portraying Britain's rail network at or close to its zenith. Produced by the railway author Alan Jowett, the atlas is a meticulously hand drawn and written production. Like the map of the western valleys of South Wales on pages 170–173, Jowett's map of the upper eastern valleys shows the rail network at its peak – the culmination of heroic engineering and intensive competition for the big prize, coal. This sheet – at an original scale of 3 in. to 10 km – focuses on the northeast corner of the South Wales coalfield, where no less than 10 companies or joint lines vied for the potentially lucrative business of shifting the 'black diamonds' to markets near and far, not least the big export business through the docks at Barry, Cardiff and Newport.

The two major rail companies fighting for market share in this part of the valleys were the Great Western Railway (GWR), for which this was core territory, and the London & North Western Railway (LNWR) which had managed to penetrate far from its principal routes between London, Birmingham and northwest England. The GWR struck north from Cardiff, Newport and its east–west Pontypool–Hengoed–Aberdare route, while the LNWR pushed south from the east–west Abergavenny–Brynmawr–Merthyr Tydfil ('Heads of the Valleys') line. Such was the competition that on two routes, up the Afon Llwyd and Sirhowy valleys through Abersychan and Argoed respectively, both companies even managed to squeeze into the same narrow valley, as did the locally based Rhymney and Brecon & Merthyr railways through Bargoed.

There is stark contrast at the northern end of the map, where the physical barrier of the Brecon Beacons – and the absence of coal – meant that only one railway penetrated north into rural Mid Wales. This was the Brecon & Merthyr, part of whose route just meanders into the northwest corner of the map.

The decline of coal and the rise of road competition have denuded this area of most, but by no means all, of its rail network. Passenger services survive from Cardiff to Rhymney (along the old Rhymney Railway line); through Cwmbran and Pontypool on the inter-city route from Cardiff to Crewe; and on the Newport–Ebbw Vale Parkway line, whose passenger services were reinstated in 2008, following their earlier withdrawal in 1962. Coal traffic from northwest of Ystrad Mynach continues to move down the valley by train, while an isolated preserved line serves the 'Big Pit' colliery heritage site at Blaenavon.

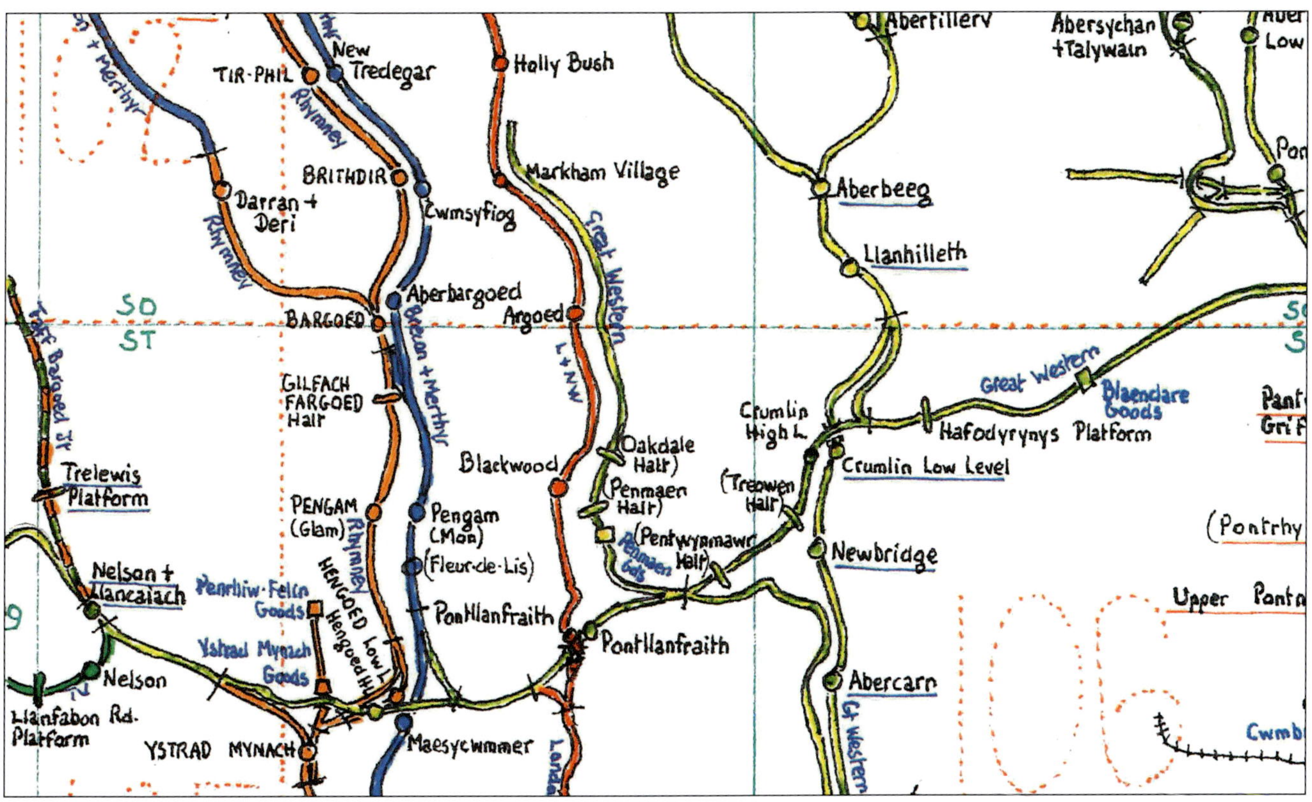

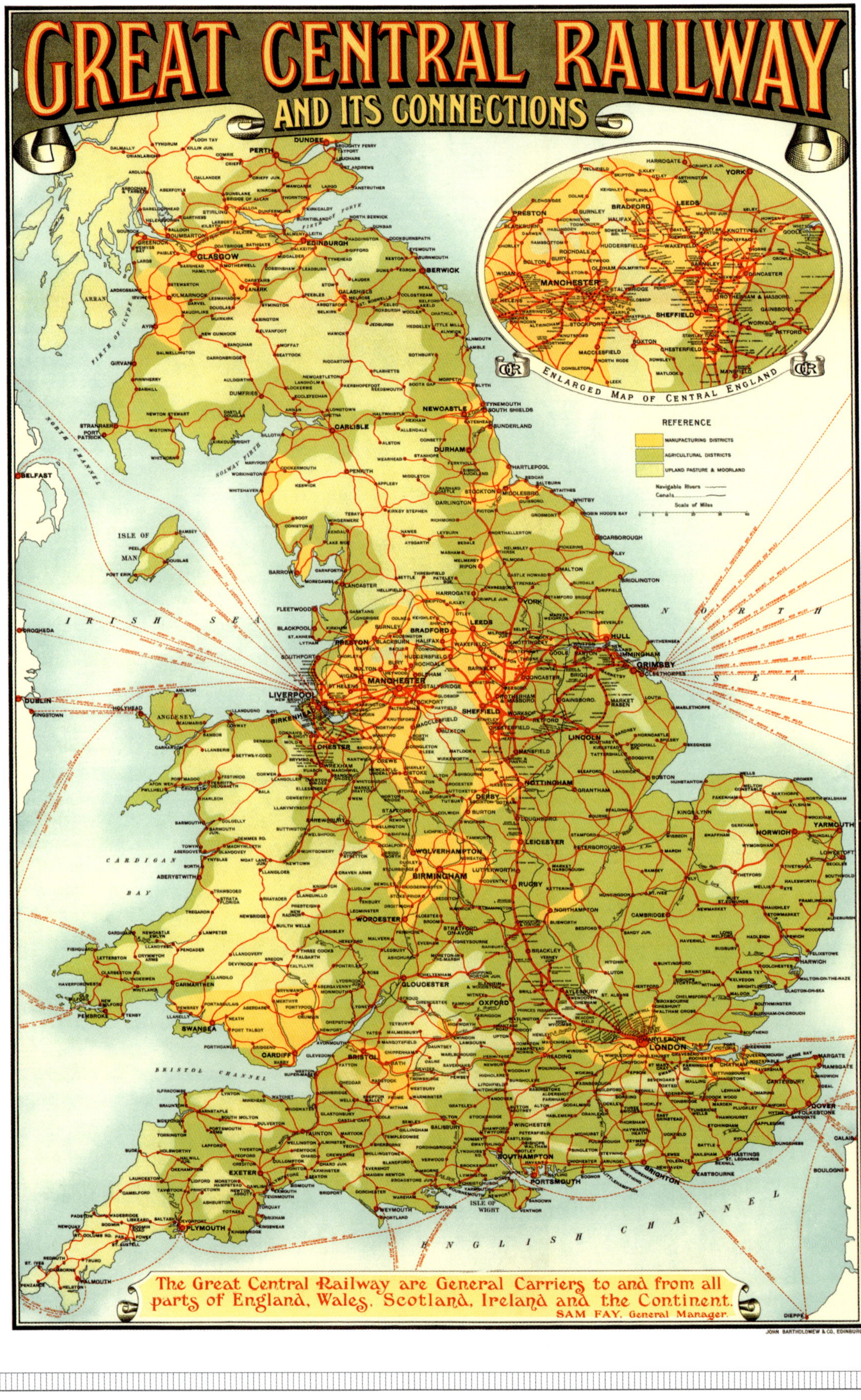
GREAT CENTRAL RAILWAY
AND ITS CONNECTIONS
ENLARGED MAP OF CENTRAL ENGLAND
REFERENCE
MANUFACTURING DISTRICTS
AGRICULTURAL DISTRICTS
UPLAND PASTURE & MOORLAND
IRISH SEA
NORTH SEA
CARDIGAN BAY
BRISTOL CHANNEL
ENGLISH CHANNEL
ISLE OF MAN
The Great Central Railway are General Carriers to and from all parts of England, Wales, Scotland, Ireland and the Continent.
SAM FAY, General Manager.
JOHN BARTHOLOMEW & CO. EDINBURGH

Title	Great Central Railway and its connections
Publisher	Great Central Railway
Mapmaker	J Bartholomew, Edinburgh
Date	1913

Map posters became a key tool of passenger information and railway marketing, with railway stations providing almost countless opportunities to proudly display each company's ability to transport the traveller the length and breadth of the country. A note handwritten in the margin of this poster for the Great Central Railway on 6 March 1913 records '26,550 copies'. The Bartholomew style of 'Great Central Railway and its connections' has close echoes with 'North British Railway and its communications' (see pages 120–123) and 'the L.&N.W.R. and its communications' (see pages 104–107) – all were produced by the Edinburgh mapmaker.

The Great Central, who built the last main line to reach London (in 1899), evidently felt sufficiently confident of its competitive strength not to downplay the extent of its rivals' routes on this map. The entire British rail network, as far as the map's northern limit beyond Central Scotland, is shown as the Great Central's 'connections' – and the Great Central's own lines are represented only marginally bolder than all the other railway companies'. Interestingly, navigable rivers and canals are also shown, while the map key distinguishes the underlying topography of the country as either manufacturing district, agricultural district or upland pasture and moorland.

The enlarged map of 'Central England', as shown in the extract overleaf, does not in fact show an area which would normally be regarded as the central part of England, but rather was central to the Great Central's core operations. Other than its long main line southwards to London, the railway's network was largely concentrated in the Lancashire/Yorkshire corridor from Manchester through Sheffield and Doncaster to Grimsby, with joint ownership and 'running powers' extending its influence beyond, for example west to Merseyside over the Cheshire Lines Committee tracks.

The critical importance of the Great Central's connectivity with the west and east coast shipping hubs at Liverpool and Grimsby respectively is graphically shown by the heavy concentration of shipping routes fanning out from these two ports.

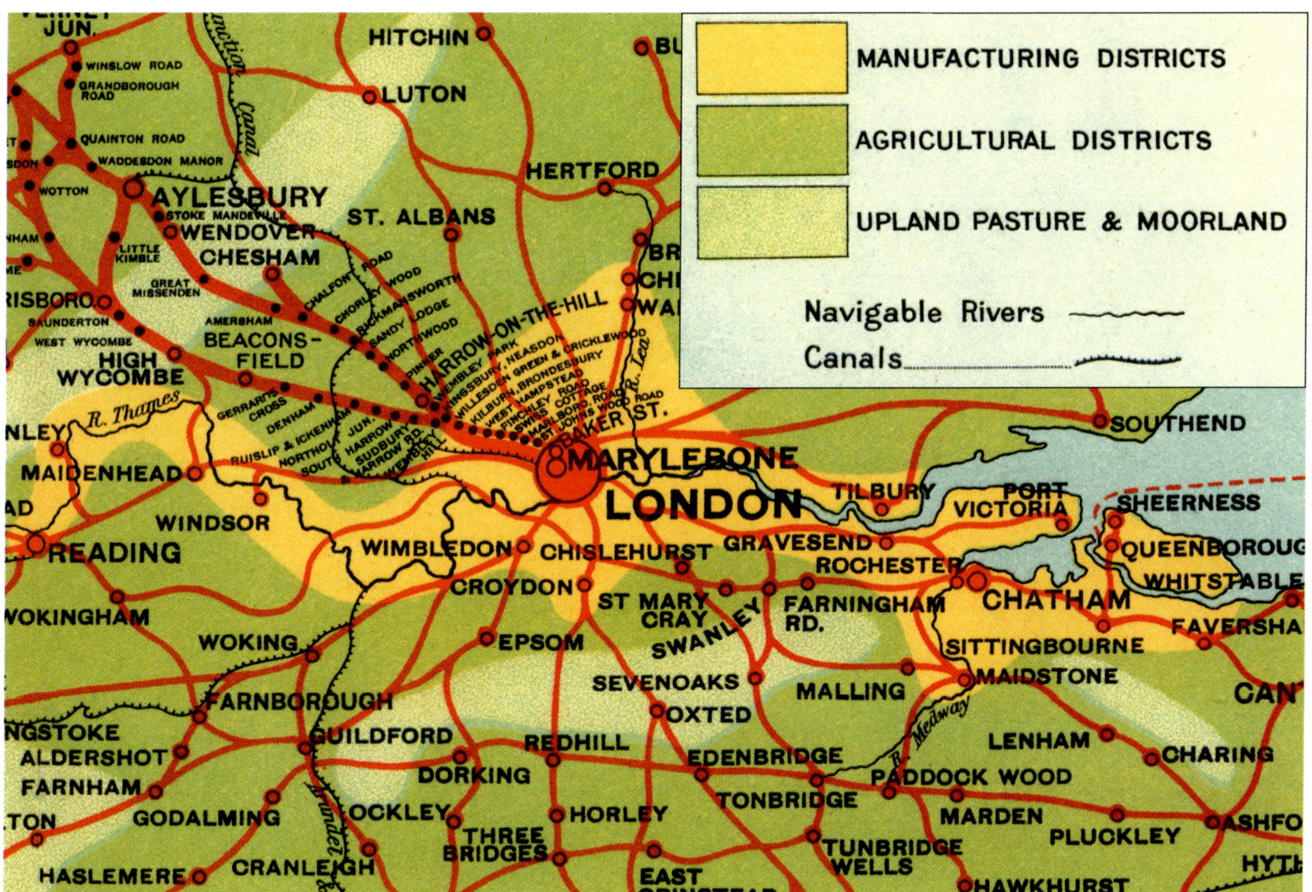

HELLIFIELD
SKIPTON
KEIGHLEY
COLNE
LONGRIDGE
PRESTON
BURNLEY
BLACKBURN
ACCRINGTON
HALIFAX
TODMORDEN
HASLINGDEN
DARWEN
BACUP
SOWERBY BRI.
GREETLAND
ELLAND
CHORLEY
RAMSBOTTOM
HUDDERSFIELD
ROCHDALE
BURY
BOLTON
HEYWOOD
OLDHAM
WIGAN
MIDDLETON
LOWER INCE
HINDLEY & PLATT BRIDGE
BICKERSHAW & ABRAM
WEST LEIGH & BEDFORD
LOWTON ST. MARY'S
GOLBORNE
HAYDOCK PARK
ASHTON-IN-MAKERFIELD
HAYDOCK
MANCHESTER
PARK BRIDGE
CENTRAL
LONDON RD.
ARDWICK
ASHBURY'S
GORTON
GUIDE BR.
ASHTON
STALYBRIDGE
CROWDEN
HADFIELD
DINTING
TRAFFORD PARK
URMSTON
FLIXTON
IRLAM
OLD TRAFFORD
BELLEVUE
HYDE RD.
FAIRFIELD
HYDE JUN.
NEWTON
HYDE
ST. HELENS
CULCHETH
GLAZEBROOK
STRETFORD
CHORLTON-CUM-HARDY
REDDISH
WOODLEY
MOTTRAM & BROADBOTTOM
GODLEY JUN.
PADGATE
CADISHEAD
PARTINGTON
SALE
BROOKLANDS
ALEXANDRA PARK
LEVENSHULME
FALLOWFIELD
BREDBURY
ROMILEY
MARPLE
FARNWORTH
SANKEY
HOUGH GREEN
HALEWOOD
WARRINGTON
TANHOUSE LANE
TIMPERLEY
W. TIMPERLEY
BASULEY
NORTHENDEN
CHEADLE
STRINES
BIRCH VALE
NEW MILLS
ROSE HILL
HUNTS CROSS
WIDNES
RUNCORN
ALTRINCHAM
HALE
STOCKPORT
HIGH LANE
MIDDLEWOOD
POYNTON
ASHLEY
MOBBERLEY
KNUTSFORD
LOSTOCK GRALAM
PLUMBLEY
NORTHWICH
BOLLINGTON
HARTFORD & GREENBANK
CUDDINGTON
MACCLESFIELD
NORTH RODE
CONGLETON
LEEK
ALNMOUTH
AMBLE
BLYTH
TYNEMOUTH
GCR
ENLARGED MAP

CRIMPLE JUN.
YORK
ARTHINGTON JUN.
LEEDS
HOLBECK
BEESTON
MILFORD JUN.
SELBY
HOWDEN
STADDLE-THORPE
ARDSLEY
CLECKHEATON
LIVERSEDGE
LOFTHOUSE
BATLEY
FERRY BR.
NORMANTON
KNOTTINGLEY
SALTMARSHE
WHITTON
GOOLE
WINTERINGHAM
WEST HALTON
WINTERTON & THEALBY
HECKMONDWIKE
DEWSBURY
PONTEFRACT
SANDAL
HARE PARK & CROFTON
MIRFIELD
WAKEFIELD
NORMANBY PARK
NOSTELL
RYHILL
HEMSWORTH
THORNE
MEDGE HALL
GUNNESS & BURRINGHAM
FRODINGHAM & SCUNTHORPE
APPLEBY
ALTHORPE
STOCKSMOOR
NOTTON & ROYSTON
SOUTH ELMSALL
HAMPOLE
STAINFORTH & HATFIELD
BRAMWITH
CROWLE
STAINCROSS
SHEPLEY & SHELLEY
BARNSLEY
CARCROFT & ADWICK-LE-STREET
BARNBY DUN
SUMMER LANE
STAIRFOOT
DODWORTH
SILKSTONE
WOMBWELL
DONCASTER
HAZLEHEAD BR.
DOVECLIFFE
MEXBORO.
SCAWBY & HIBALDSTOW
BIRDWELL & HOYLAND COMMON
WATH-ON-DEARNE
CONISBROUGH
SWINTON
PENISTONE
WORTLEY
WESTWOOD
KILNHURST
KIRTON LINDSEY
DEEPCAR
CHAPELTOWN & THORNCLIFFE
PARKGATE & ALDWARKE
TICKHILL & WADWORTH
NORTHORPE
ROTHERHAM ROAD
ECCLESFIELD
ROTHERHAM & MASBORO.
OUGHTIBRIDGE
GRANGE LANE
MALTBY
BLYTON
WADSLEY BRIDGE
MEADOW HALL & WINCOBANK
TINSLEY
DINNINGTON & LAUGHTON
NEEPSEND
BROUGHTON LANE
ATTERCLIFFE
GAINSBORO.
SHEFFIELD
WOODHOUSE
WALESWOOD
KIVETON PARK
DARNALL
STURTON
BEIGHTON
SHIREOAKS
WORKSOP
LEVERTON
KILLAMARSH
COTTAM
TORKSEY
ECKINGTON & RENISHAW
SPINK HILL
CHECKER HOUSE
RETFORD
STAVELEY TOWN
CLOWNE
WHITWELL
STAVELEY WORKS
SHEEPBRIDGE & BRIMINGTON
CRESWELL & WELBECK
ELMTON & CRESWELL
SCARCLIFFE
BOLSOVER
LANGWITH
CHESTERFIELD
LANGWITH JUN.
EDWINSTOWE
OLLERTON
BOUGHTON
TUXFORD
DUKERIES JUN.
ARKWRIGHT TOWN
WARSOP
SHIREBROOK
GRASSMOOR
HEATH
MANSFIELD WOODHOUSE
MANSFIELD
MATLOCK
PILSLEY
TIBSHELF TOWN
CENTRAL ENGLAND
GCR

At the start of the Second World War the GWR carried nearly 70,000 children from Paddington station to safe locations in the countryside. Here a large group of child evacuees has just arrived at Maidenhead station in 1940.

There is no doubt that Britain's railways played a vital role during the two World Wars. Even before the First World War, the London & South Western Railway had played a key part during the Second Boer War when it transported thousands of troops and their equipment from military camps in southern England to the new railway-owned docks at Southampton for embarkation on troopships to South Africa.

During the First World War, Britain's railways came under Government control and, with war raging in France and Belgium, they transported vast numbers of soldiers, their equipment and munitions to ports in southern England for embarkation across the English Channel. In the reverse direction came thousands of injured soldiers who were transported in ambulance trains to hospitals around the country. On the other side of the Channel the overloaded French railway system was soon in disarray so, in 1915, the Railway Operating Division (ROD) of the Royal Engineers was formed to operate railways in theatres of war. By 1916 the ROD had requisitioned around 600 freight locomotives and thousands of goods wagons from British railway companies for use not only in France but also in Italy, Palestine, the Balkans and Mesopotamia. With the war in northern France virtually at stalemate, the demand for heavy freight locomotives outstripped supply and so the ROD adopted the Great Central Railway's Class '8K' 2-8-0 as its standard freight locomotive. Over 500 of these powerful locomotives were built for service overseas, the majority returning home at the end of the war.

During the Second World War, the railways came under Government control for the second time in 20 years, but their task was made doubly difficult by the conditions under which they had to operate. Despite blackouts, major disruption caused by bombing, lack of maintenance and overworked staff and locomotives they kept on running. In 1939 the railways not only operated over 4,000 special trains carrying nearly 1.5 million passengers away from the capital during the first evacuation of London, but also carried almost 400,000 troops of the British Expeditionary Force (BEF) in over 1,000 trains to Channel ports for embarkation to France. The following May and June around 320,000 troops of the BEF were snatched from the Dunkirk beaches and carried to safety over the Channel in a flotilla of small boats – in a period of a fortnight the railways ran over 600 special trains to carry the traumatized troops away from the south coast ports.

The railways also suffered heavily during the German bombing of London, other cities and key installations – stations, junctions, engine sheds and railway works were all targeted but despite suffering considerable damage the railways kept running. Troop trains, workmen's trains and freight trains all had priority over passenger trains where delays and overcrowding were the norm. The war effort not only depended on the mining and railway distribution of coal, which by 1942 amounted to 160 million tons a year, but also the distribution of other raw materials, war equipment and munitions between rail-connected docks and war factories.

The lead up to the D-Day landings in 1944 saw US Army Transportation Corps locomotives shipped into Britain to assist with the largest mass movement of troops and military equipment in the history of warfare. By 3 June the railways had moved in great secrecy 500,000 British soldiers, 2,000,000 US soldiers and 1,500,000 tons of stores from their concentration areas to their ports of embarkation along the English south coast. The war left a terrible legacy for Britain's railways, their rundown state and lack of investment leading inevitably to nationalization in 1948.

During the Second World War, the railways were a major target for German bombers and were attacked over 10,000 times between 1939 and 1945. However, stations and permanent way damage was cleared up quickly and services resumed.

Title	Jowett's Railway Centres (pages 58, 62 and 63)
By	Alan Jowett
Publisher	Patrick Stephens Ltd, Cambridge
Date	1993

Jowett's Railway Centres, published in 1993, expands the coverage of *Jowett's Railway Atlas* by focusing on various snapshots in time, showing the development and decline of the railway network in key centres over 140 years and more. The same hand-drawn cartographic technique is used, graphically illustrating – with supporting commentary – the significant rail route changes seen over this period of time in a variety of British towns and cities.

The three selected snapshots of central Hull contrast the earliest railways in 1846, the zenith of the local network in 1915, and the truncated system of 1990. Jowett notes in his commentary that, 'Hull, in common with most railway centres, had a unique set of circumstances affecting its growth and

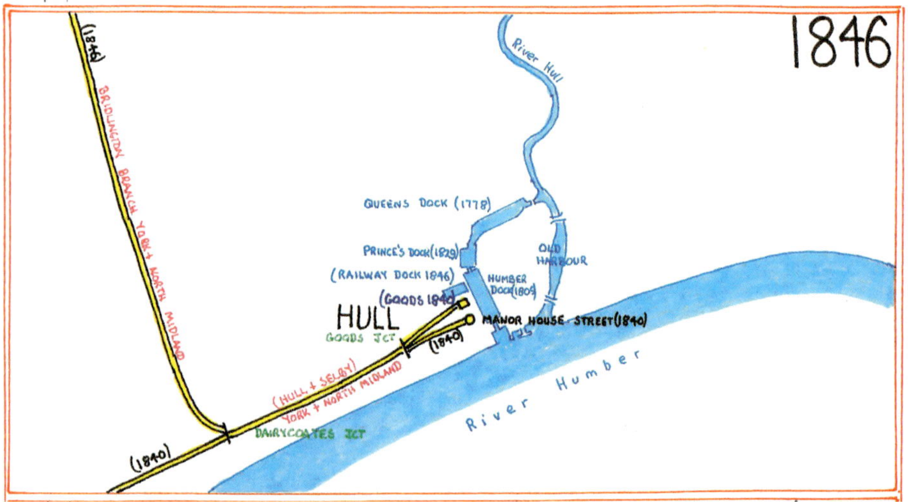

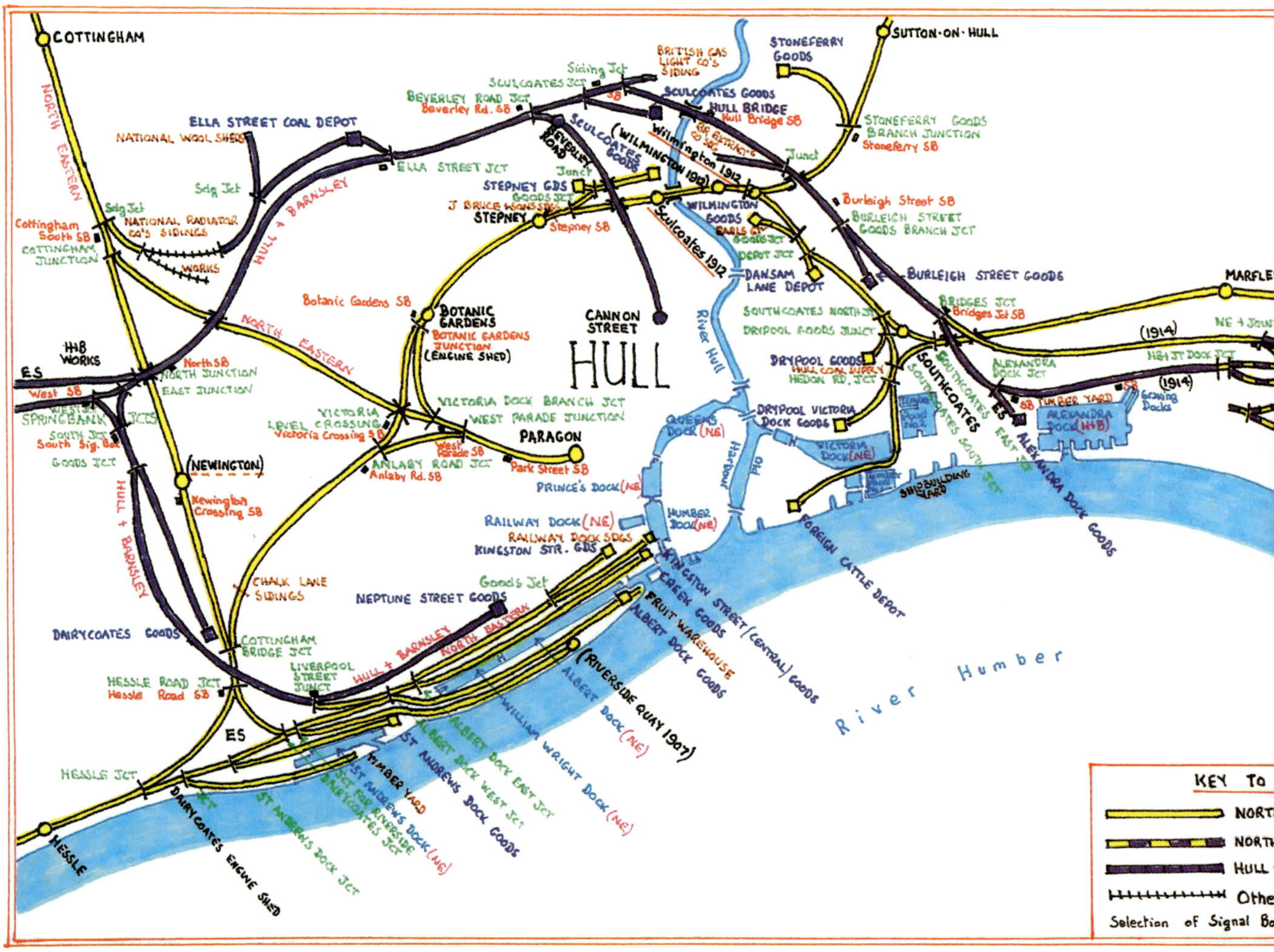

decline'. These were: its position on the Humber, serving important shipping trade routes; the barrier created by the Humber, which has tended to isolate Hull; its mainly agricultural hinterland, leaving the city 'at the end of the line'; and the relative absence of competition, as from the beginning Hull's rail network was effectively under the control of one company, the North Eastern Railway (NER). Jowett also compares and contrasts railway development with dock development:

> *With railways, usually the last railway to open is the first to close and quite often the present-day situation is very similar to that reached around 1850 and in very few places is this better illustrated than at Hull. . . . With dock development, however, exactly the opposite prevails, the first docks opened, being of shallow depth, are usually the first to close and the later deep-water docks that can accommodate the larger vessels at all states of the tide are the ones to remain.*

Following the 1923 Grouping only a small measure of rationalization was possible, closing the Cannon Street passenger terminus of the Hull & Barnsley Railway (HBR), which had been the one serious, but only partially successful, attempt to break the NER's monopoly in Hull.

Extensive rationalization took place following the Beeching Report, with all dock goods traffic concentrated on the former HBR route, sweeping round the north of the city centre on an embankment – and thereby eliminating road congestion in the city caused by the other lines' level crossings. Today the network is virtually identical to that of 1990, with Hull Paragon station handling local trains to York and Scarborough, trans-Pennine expresses to Leeds and beyond, and inter-city services to London. The single-track King George Dock branch line serves coal and steel terminals, and a chemicals plant at Saltend, now the very end of the line in east Yorkshire.

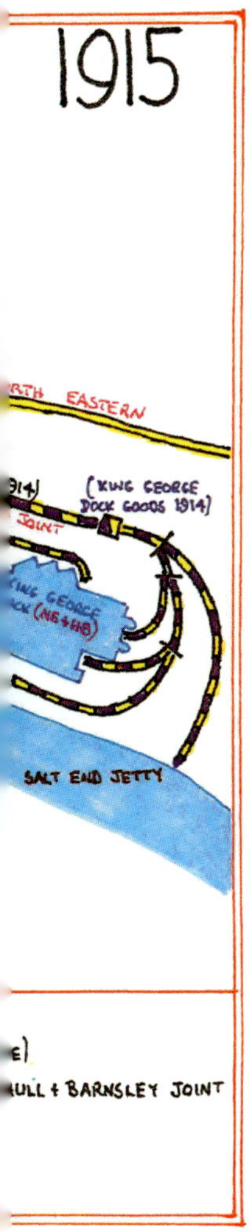

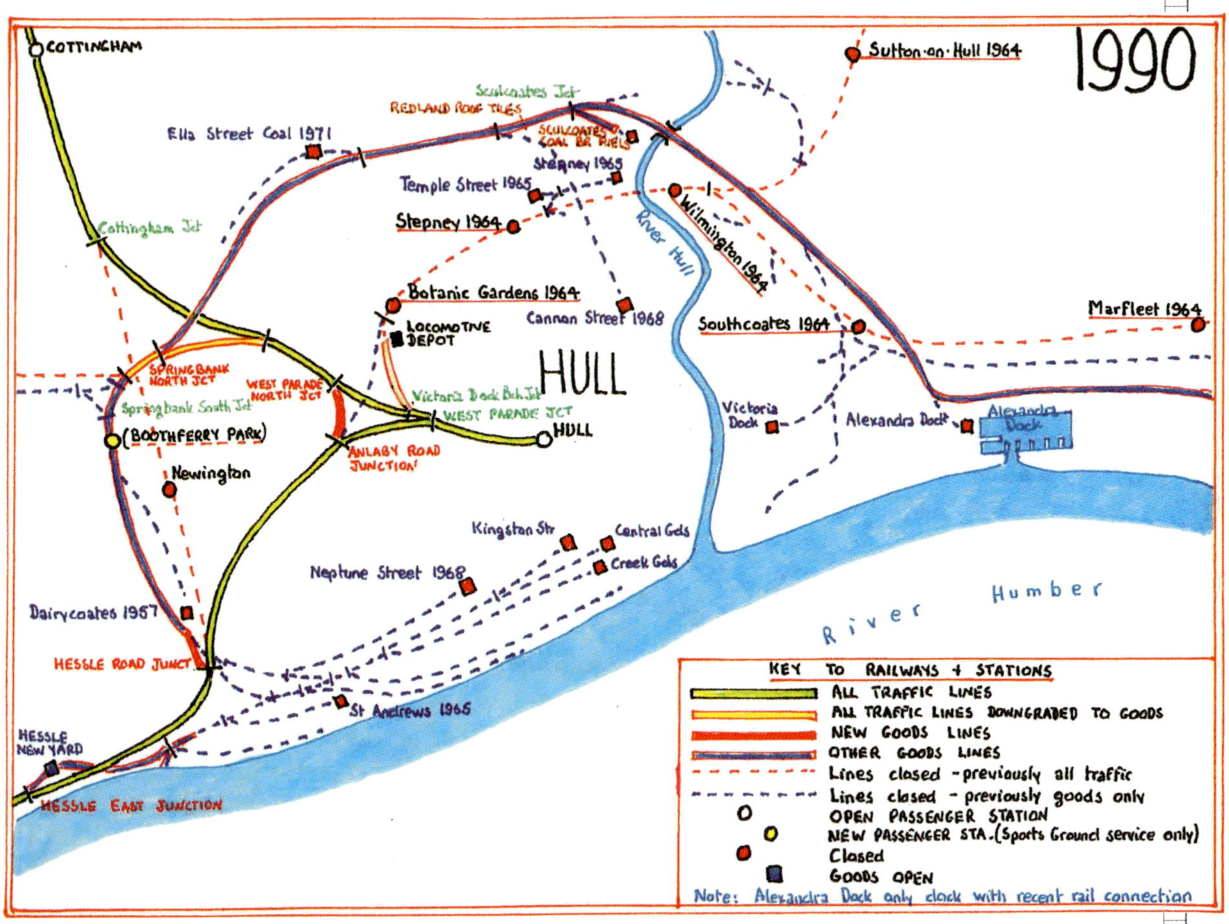

Prior to the Light Railways Act of 1896, all new railway lines in Britain required a specific Act of Parliament before they could be built. This was an expensive and time-consuming affair and the severe economic recession and associated high interest rates of the 1870s and 1880s had brought a halt to many new railway schemes in Britain. With the Light Railways Act in place railway companies could build new railways in sparsely populated regions of the country without requiring a specific Act of Parliament. While cutting out much of the red tape and expensive legislation, it did impose severe limitations including a weight limit of 12 tons per axle and a speed limit of 25 m.p.h. Cost-cutting measures included minimal earthworks, bridges and stations, lightly laid track spiked directly onto sleepers, the absence of level crossing gates, minimal signalling and the operation of 'One Engine in Steam' – i.e. only one locomotive was allowed to work on the line at any given time. Over 30 standard gauge and narrow gauge lines were built under the Act's provisions between 1896 and 1925.

Seeing a potential business opportunity, the arch proponent of light railways in England and Wales was the colourful Colonel H F Stephens who formed the Light Railway Syndicate a year before the Act was passed. By the time of his death in 1931 he had control of an eccentric railway empire of light railways across the land. Using second-hand equipment and with the minimum of maintenance these ranged from narrow gauge enterprises such as the Ffestiniog Railway, the Snailbeach District Railway and the Ashover Light Railway to meandering standard gauge rural lines such as the Kent & East Sussex Railway, the Shropshire & Montgomeryshire Railway and the East Kent Railway. That three of his ramshackle, standard gauge lines survived into British Railways ownership in 1948 can only be described as a miracle!

While the provisions of the Light Railway Act were perfect for the sparsely populated regions of Scotland, there were very few light railways actually built in the country. Of those that saw the light of day were the Dornoch Light Railway and the Leadhills & Wanlockhead Light Railway, both of which opened in 1902, and the Wick & Lybster Railway which opened in 1903. At the southern end of the Kintyre peninsula the narrow gauge Campbeltown & Macrihanish Railway (formerly a canal) was converted from a coal-carrying line to a light railway in 1906. North of Inverness construction of the Cromarty & Dingwall Railway was started but never completed.

By the end of the First World War the building of light railways in Scotland was back on the political agenda and the question of rural transport in Scotland was studied by a Parliamentary Committee which published its findings in

A train waits to depart from Lybster station on the short-lived Wick & Lybster Light Railway in 1928.

1919. In their report the Committee advocated the building of 382 miles of new railway, 85 miles of new road, road improvements and the introduction of new bus and steamer services. Schemes looked at were railways to serve the northwest fishing ports of Ullapool and Lochinver, narrow gauge lines on the islands of Lewis, Skye and Arran, an isolated railway from Dunoon to Strachur and a 28-mile line across the wilds of Galloway linking Dalmellington with the Portpatrick & Wigtownshire Joint Railway at Parton. As all of these proposed schemes would pass through sparsely populated and remote countryside the Committee pressed the point that Government funding would be necessary. Not surprisingly, during this period of austerity and increased competition from road transport none of these lines were built.

Light railway provisions enjoyed a new lease of life from the 1960s, enabling volunteer-run preserved railways to be operated without onerous legislative requirements. In the early years of the 21st century, an eagle-eyed civil servant in the then Scottish Executive spotted that the old legislation would allow a new two-mile branch line to be constructed to a open-cast coal site in East Ayrshire without an Act of Parliament – and so the Greenburn Light Railway (Scotland) Order 2003 came to pass, limiting coal trains to 20 m.p.h. and prohibiting the operation of passenger trains, albeit to an unlikely destination.

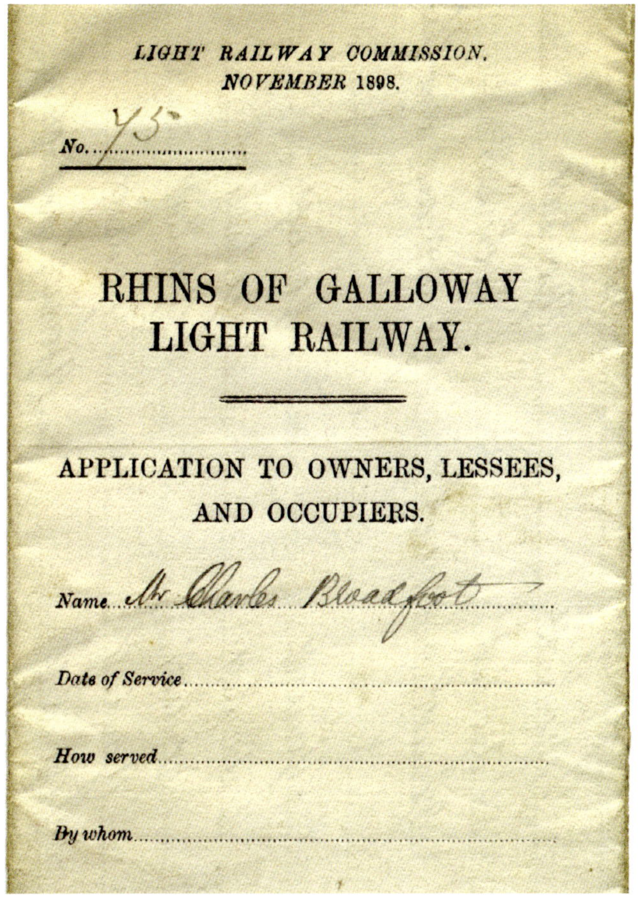

LIGHT RAILWAY COMMISSION.
NOVEMBER 1898.

No. 75

RHINS OF GALLOWAY
LIGHT RAILWAY.

APPLICATION TO OWNERS, LESSEES, AND OCCUPIERS.

Name Mr Charles Broadfoot

Date of Service

How served

By whom

A land purchase document for a light railway that was never built, the Rhins of Galloway Light Railway in southwest Scotland.

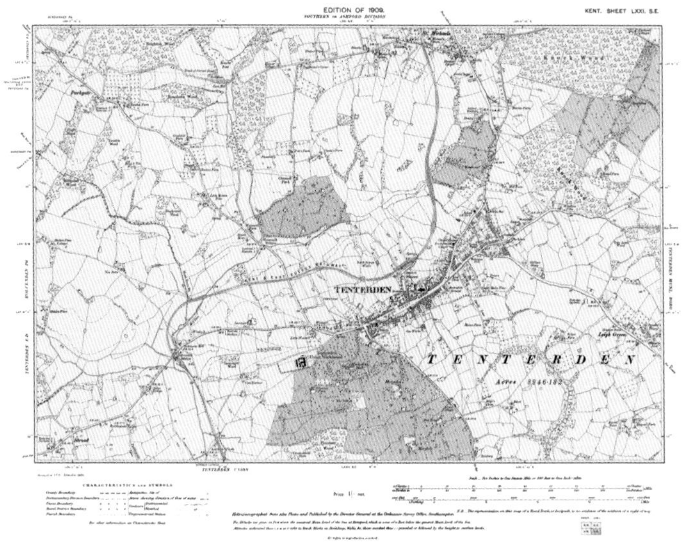

Series	Ordnance Survey Edition of 1909. 6 in. to 1 mile
Sheet	Kent LXXI, S.E.
Date	Surveyed 1870, revised 1906, printed 1909

The Rother Valley Railway, as it was initially known, was the first line to be built under the new Light Railway legislation of 1896. The first section, from the Hastings–London main line at Robertsbridge to Rolvenden, was opened in 1900, followed by a short extension to the present Tenterden Town station in 1903. In 1905 the final section to Headcorn on the Dover–Ashford–London main line was opened, providing a through cross-country route from East Sussex to Kent.

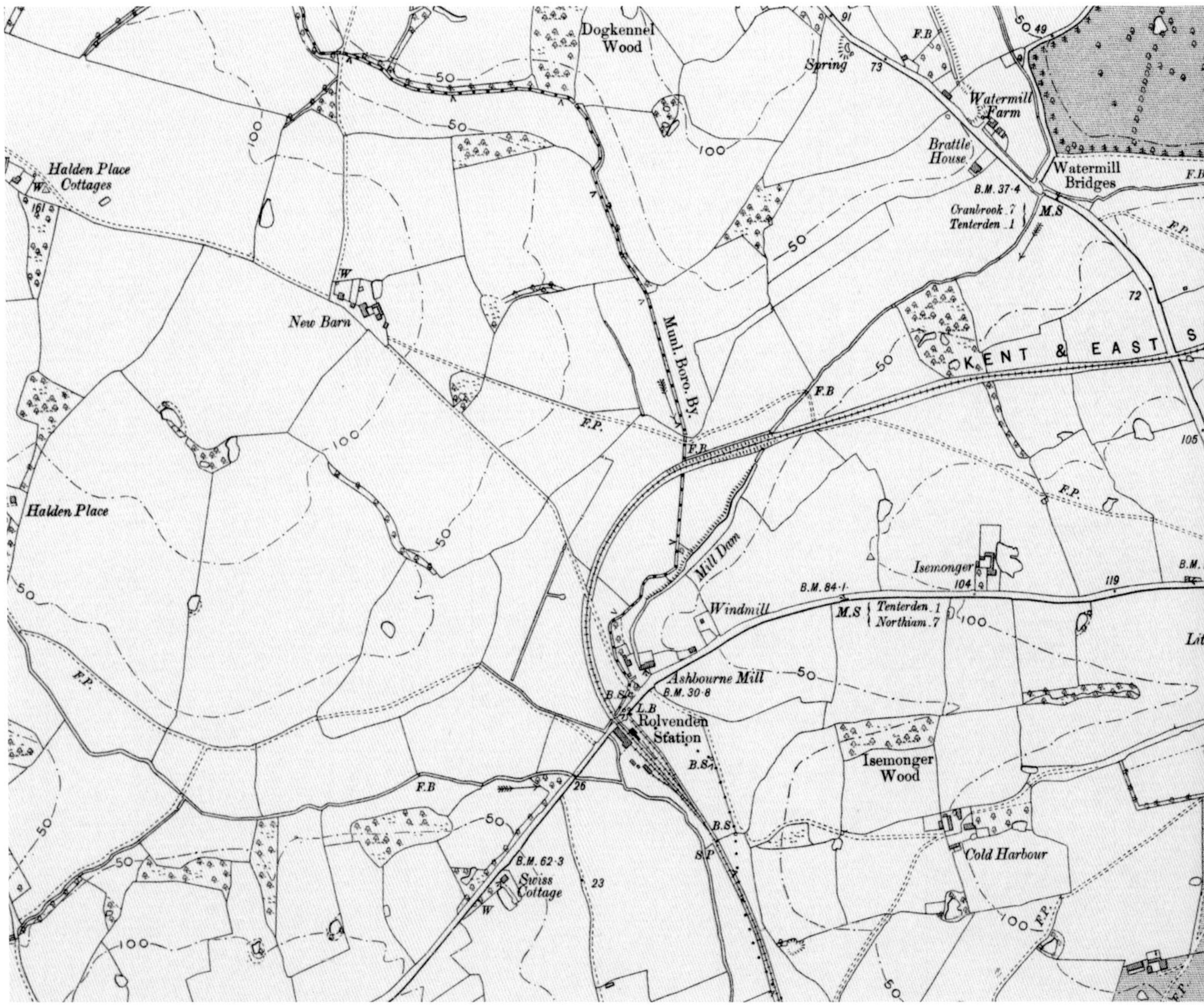

The Kent & East Sussex Railway (KESR), as it eventually became known, was built, owned and managed by Colonel H F Stephens, who went on to run a collection of light railways across England and Wales from his office in Tonbridge. He also managed the narrow gauge Ffestiniog Railway from 1923 until his death in 1931.

The map extract, from an Ordnance Survey 6 in. to 1 mile sheet of 1909, illustrates a number of aspects of the lower-cost engineering of light railways. From Tenterden Town station, despite having a destination to the south, the railway takes a wide sweep to the west, sticking to the contours and avoiding breaching the established fabric of the village. In a distance of just 1½ miles, the railways makes three level crossings with roads – whereas on lines built to more main-line standards, bridges might well have been constructed.

Unlike most railways, the KESR did not lose its independence at the 'Grouping' in 1923, and it was one of only three independent light railways which survived to be nationalized in 1948. However, growing road competition resulted in the closure of the passenger service in 1954; goods traffic lingered on south of Tenterden, and the occasional passenger train, particularly for hop-pickers and ramblers, operated in the summer; but the line closed completely in 1961. Soon after closure, a society was formed with the object of preserving the line as a volunteer-run steam railway, but it was not until 1974 that trains again ran over the first two miles south from Tenterden. In 2000, 100 years after it originally opened, the KESR finally reached its current terminus at Bodiam – and there are plans to once again make the main line connection at Robertsbridge.

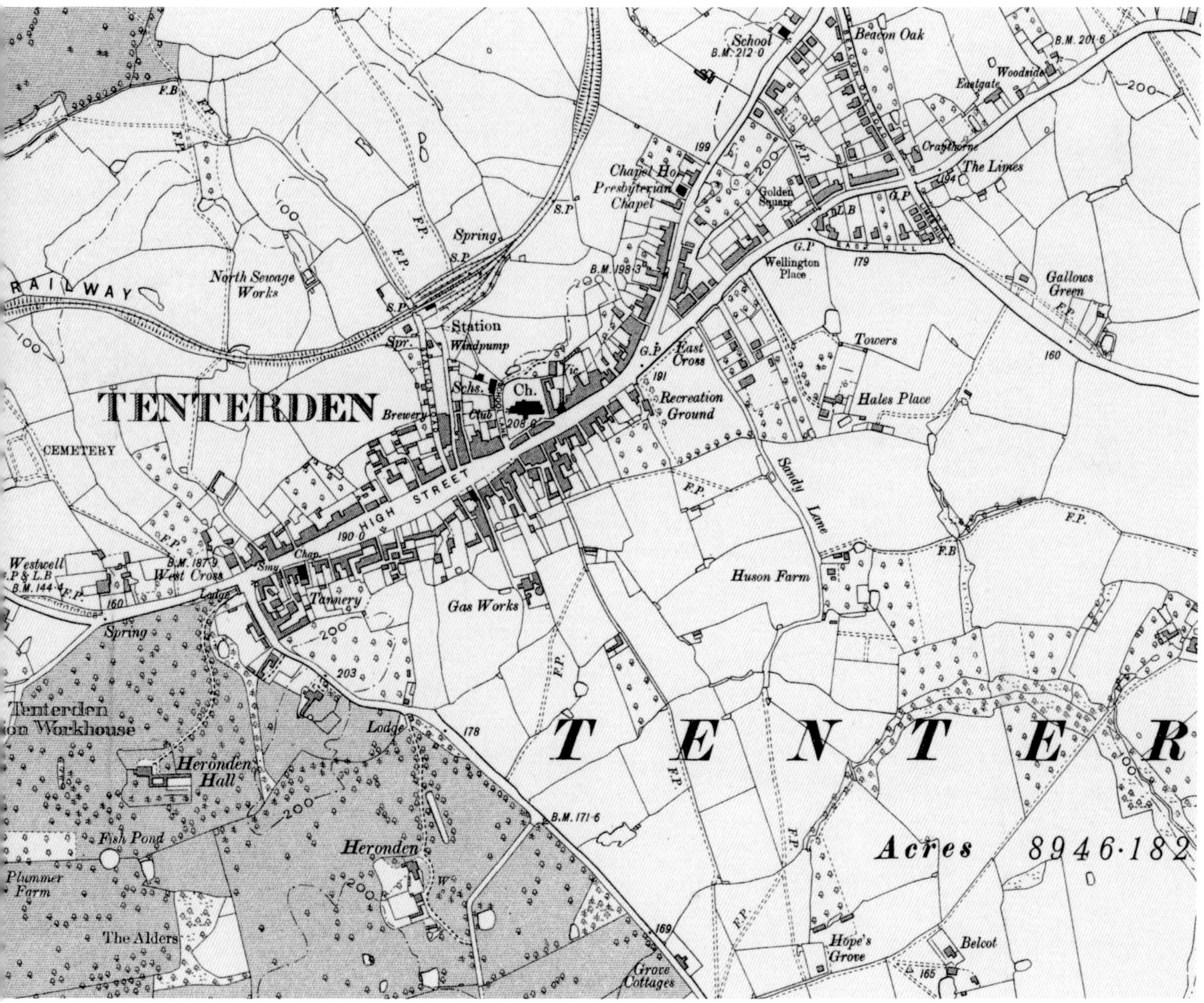

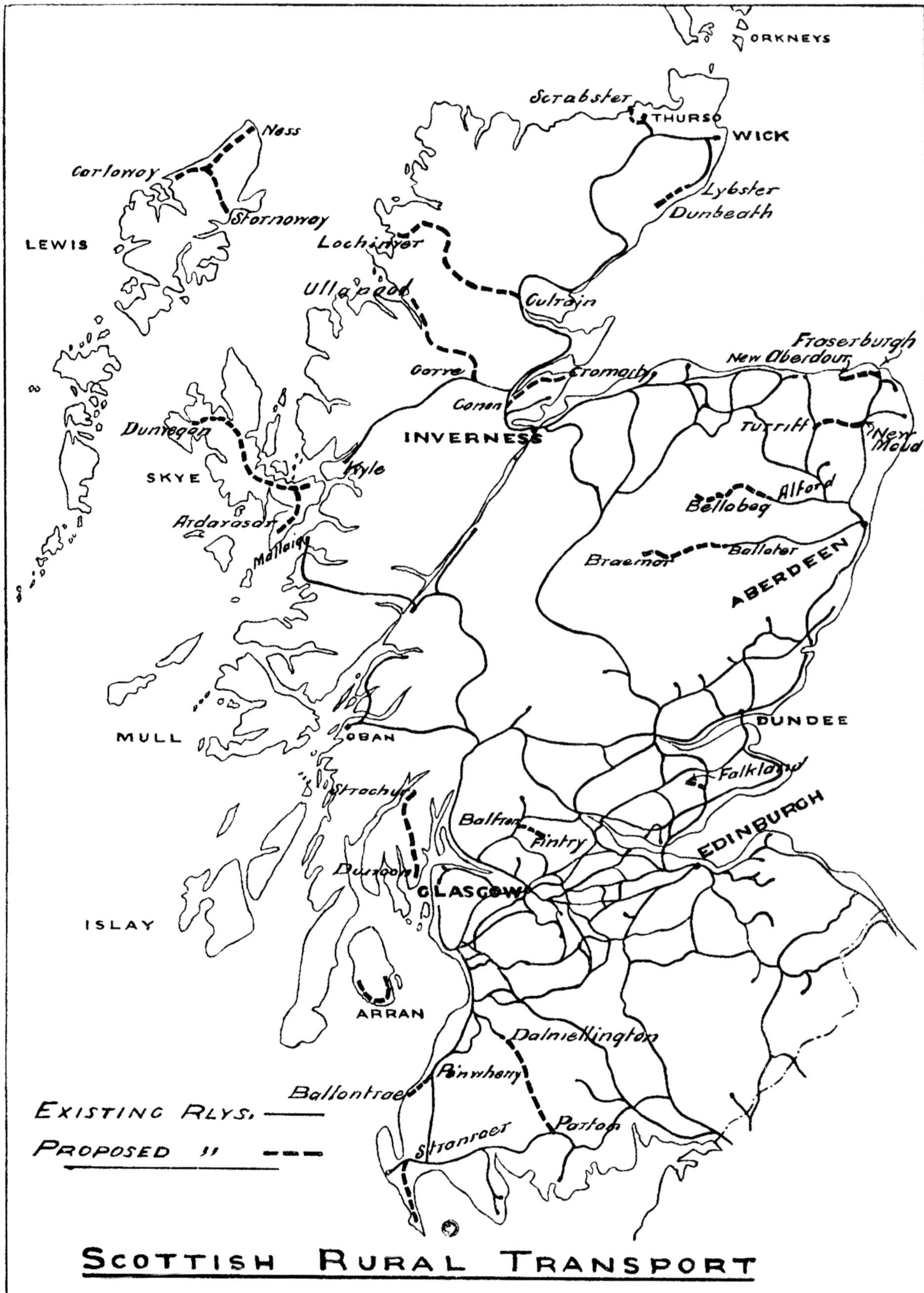

SCOTTISH RURAL TRANSPORT

Title	Scottish Rural Transport
Publisher	The Railway Gazette
Date	1919

Light railways were built to lower engineering standards than main lines – with more flexible operating arrangements for signalling, fencing, level crossings, etc., but lower speed limits – in order to open up rural areas at the lowest possible cost. There was great enthusiasm for this concept in the late Victorian and Edwardian era in Scotland, where a Parliamentary Committee in 1919 advocated extensive development of new lines from one end of the country to the other, as illustrated by this map from the *Railway Gazette*.

However, economic constraints – and the rise of the motor car and the lorry after the First World War – meant that not one of the proposed light railways shown here was ever opened. All were in areas of low population and little industrial activity, with particular concentrations in northeast Scotland and the Highlands and Islands. At the opposite end of the country, the suggested link from Stranraer to the isolated village of Drummore in the Rhins of Galloway would undoubtedly have struggled. Not much further north, the proposed route from the Ayr–Stranraer line to coastal Ballantrae would only have made strategic sense if pushed further south through the hills to Cairnryan (where all ferry services over the North Channel to Northern Ireland are now concentrated), but this would have required very heavy engineering works, well beyond the scope of a light railway. One port where engineering to light railway standards might have worked, and where direct rail–sea connection would have benefited both modes right through to modern times was Scrabster in Caithness, where the Committee proposed a short light-railway link from the Highland Railway terminus at Thurso, but it was not to be.

The *Oxford Companion to British Railway History* advances two reasons for the comparative failure of light railways in Britain:

> *Firstly, agricultural decline gathered pace in the 1880s, by which time much of the country was well served by a network of railways and branches built to main-line standards, so that the demand for better rural transport came from relatively few areas. Secondly, railway companies and the Board of Trade by then had entrenched ideas about how railways should be built and equipped. Reduced standards were considered to be retrograde. Had light railways developed earlier their history might have been very different.*

Opened throughout in 1902 to serve silver and lead mines, the Leadhills & Wanlockhead Railway in southwest Scotland was one of the first light railways built in Britain. A mixed train is seen here at Elvanfoot in 1931, only eight years before the line closed.

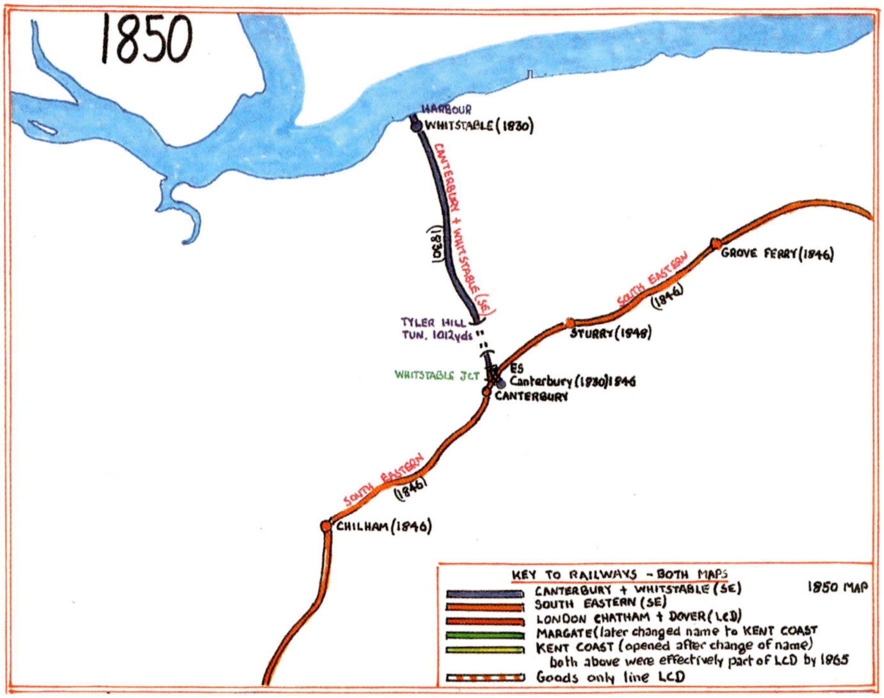
1850
HARBOUR
WHITSTABLE (1830)
CANTERBURY + WHITSTABLE (SE)
(1830)
TYLER HILL
TUN. 1012yds
WHITSTABLE JCT
ES
Canterbury (1830)1846
CANTERBURY
STURRY (1846)
SOUTH EASTERN
(1846)
GROVE FERRY (1846)
SOUTH EASTERN
(1846)
CHILHAM (1846)
KEY TO RAILWAYS - BOTH MAPS
CANTERBURY + WHITSTABLE (SE)
1850 MAP
SOUTH EASTERN (SE)
LONDON CHATHAM + DOVER (LCD)
MARGATE (later changed name to KENT COAST
KENT COAST (opened after change of name)
both above were effectively part of LCD by 1865
Goods only line LCD

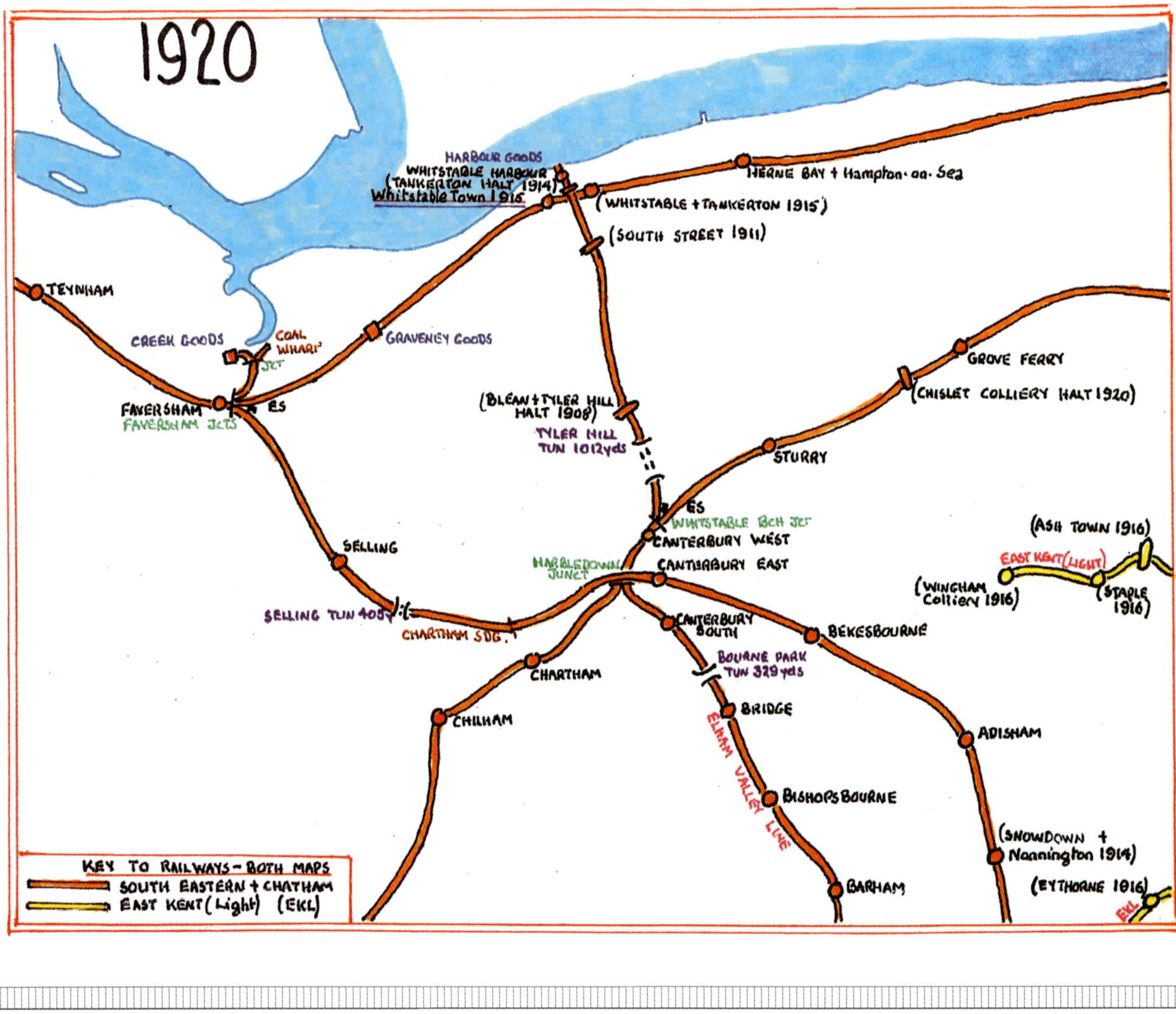
1920
HARBOUR GOODS
WHITSTABLE HARBOUR
(TANKERTON HALT 1914)
Whitstable Town 1915
HERNE BAY + Hampton-on-Sea
(WHITSTABLE + TANKERTON 1915)
(SOUTH STREET 1911)
TEYNHAM
CREEK GOODS
COAL WHARF
JCT
GRAVENEY GOODS
FAVERSHAM
FAVERSHAM JCTS
ES
(BLEAN + TYLER HILL HALT 1908)
TYLER HILL
TUN 1012yds
GROVE FERRY
(CHISLET COLLIERY HALT 1920)
STURRY
ES
WHITSTABLE BCH JCT
CANTERBURY WEST
SELLING
HARBLEDOWN JUNCT
CANTERBURY EAST
(ASH TOWN 1916)
EAST KENT (LIGHT)
(WINGHAM Colliery 1916)
(STAPLE 1916)
SELLING TUN 405y
CHARTHAM SDG.
CANTERBURY SOUTH
BEKESBOURNE
BOURNE PARK
TUN 329 yds
CHARTHAM
CHILHAM
BRIDGE
ELHAM VALLEY LINE
ADISHAM
BISHOPSBOURNE
(SNOWDOWN + Nonington 1914)
KEY TO RAILWAYS - BOTH MAPS
SOUTH EASTERN + CHATHAM
EAST KENT (Light) (EKL)
BARHAM
(EYTHORNE 1916)
EKL

Title	Jowett's Railway Centres (pages 234, 235 and 239)
By	Alan Jowett
Publisher	Patrick Stephens Ltd, Cambridge
Date	1850/1920/1991

These three selected snapshots in time from *Jowett's Railway Centres* – at an original scale of 2 miles to the inch – illustrate the growth and relative decline of the rail network around Canterbury in Kent between 1850, 1920 and 1991. In his commentary, Jowett notes that the rail network in the district has had a number of distinctive features. The first railway to be built in the district, the Canterbury & Whitstable (CWR), offered Britain's first fare-paying passenger service which was also steam-hauled (on 3 May 1830); it was also the first company to issue season tickets, in 1834; and it closed to passengers at an early date, in 1931.

The second route to open, the South Eastern Railway (SER), running southwest to northeast through what is now Canterbury West station, commenced operations in 1846, followed in 1860 by the London, Chatham & Dover Railway (LCDR), serving Canterbury East. There was no physical connection between the two lines, one crossing the other on a bridge, and Jowett notes that although both lines belonged to a single company by 1900, 'over 90 years later they remain unconnected'.

Railway development in the area was virtually complete by 1863, other than the Elham Valley Railway south to Folkestone via Canterbury South (1889) and, on the fringes of the area, the East Kent Light Railway, which, although largely completed by 1916, did not reach its ultimate terminus at Canterbury Road until 1925. This station had one of the shortest lives of any British railway station, succumbing to road competition just 23 years later. Unusually, all the passenger line closures in this area were implemented *before* the 1963 publication of the Beeching Report, the only casualties thereafter being two small intermediate stations on the former SER route northeast of Canterbury.

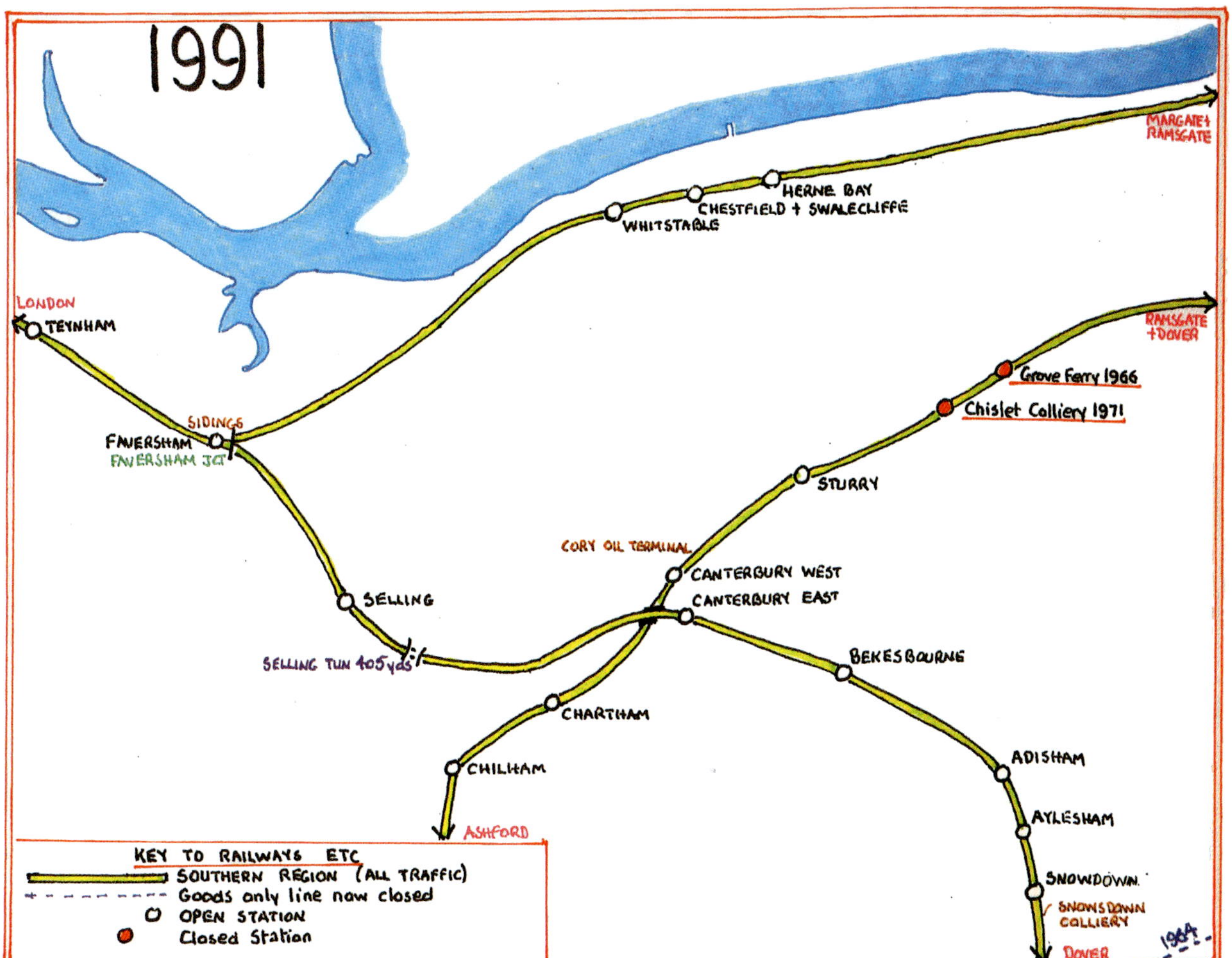

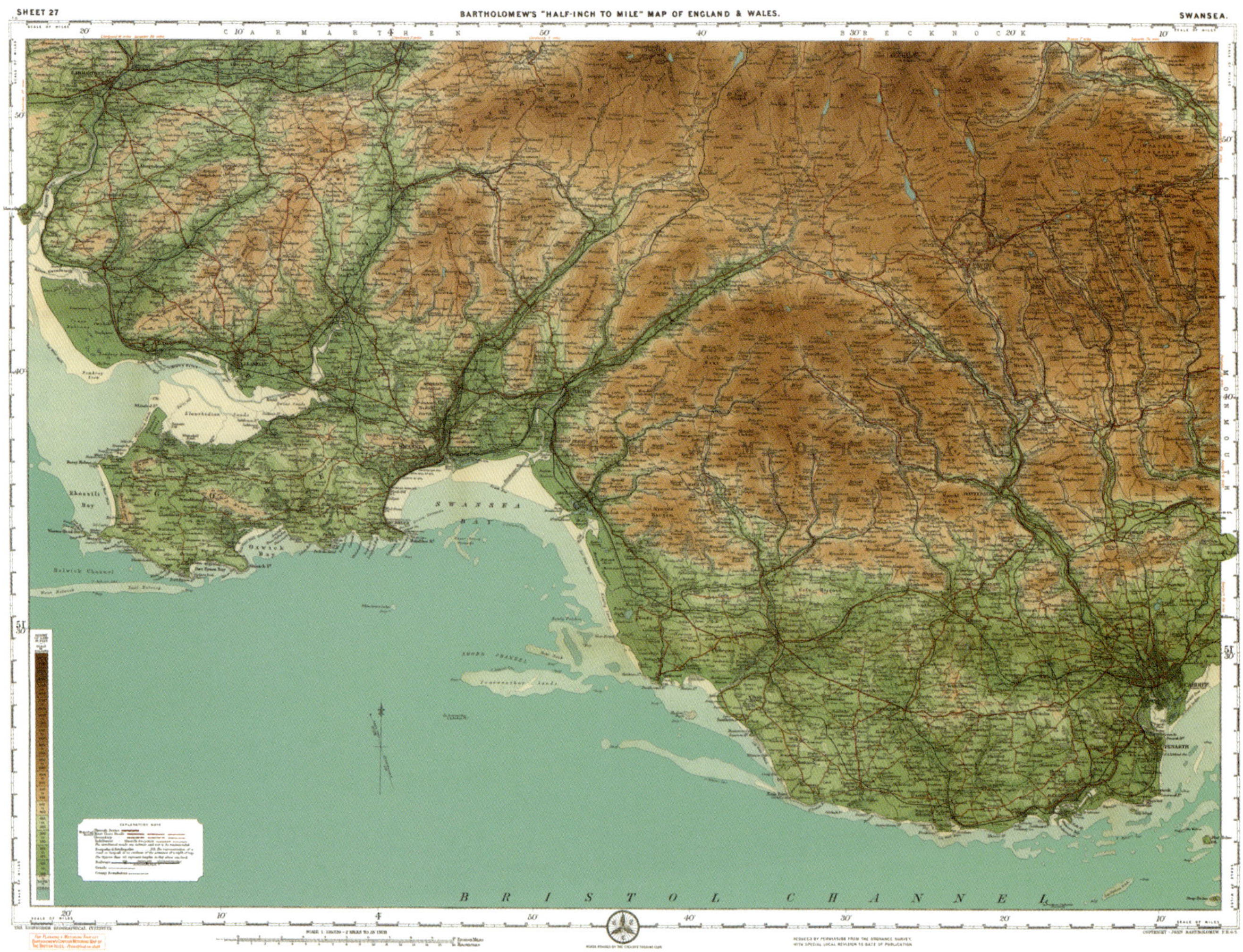

Title	Bartholomew's Half-Inch to Mile Map of England & Wales Sheet 27 Swansea
Mapmaker	J Bartholomew & Son Ltd, Edinburgh
Date	1921

The name of Bartholomew was synonymous with general mapping and with railway mapping from the 1830s to the 1980s. Their use of layer colouring on maps, in which low ground is shown in shades of green and higher ground in shades of brown is demonstrated to great advantage in this 1921 sheet covering Swansea, Cardiff and the Valleys. The widespread use of brown colouring underlines the difficult topography for railways in the north and east of the region – they were forced along narrow valleys in order to reach key sources of coal, and in some cases to penetrate into rural Wales around and beyond the Brecon Beacons.

While the half-inch scale meant that Bartholomew's could not replicate the level of detail of the classic one-inch maps produced by their Ordnance Survey competitors, they did manage to create some distinctive treatments of railways and roads. All railways are shown in black (like the Ordnance Survey representation of multiple-track railways at the time), but station names are shown in full where it is not obvious from the adjacent town or village. In an interesting twist – perhaps reflecting the intended appeal to tourists – there is a different symbol for 'stations with refreshment rooms', although this is so small as to be barely distinguishable.

While these maps were 'reduced [sic] by permission from the Ordnance Survey', the footnotes record that roads were revised by the Cyclists' Touring Club, which explains the quaint reference in the key to 'Indifferent Roads (Passable for Cyclists)'.

The extract overleaf of the area around Swansea – which had six passenger rail termini, more than any other British town outside London – shows the route of the Swansea & Mumbles Railway, originally a horse-drawn waggonway, which survived as an electrified tramway until 1960. Taking a wide sweep to the west and then north is the London & North Western Railway's Central Wales Line to Shrewsbury. The southern section of this route, from Swansea Victoria via Gowerton South to Pontardulais, was axed when the rest of the line was reprieved from the 'Beeching Axe' in 1964. Trains were diverted to run from the former Great Western Railway terminus at Swansea High Street to Llanelli, then reversing to reach the original routing at Pontardulais.

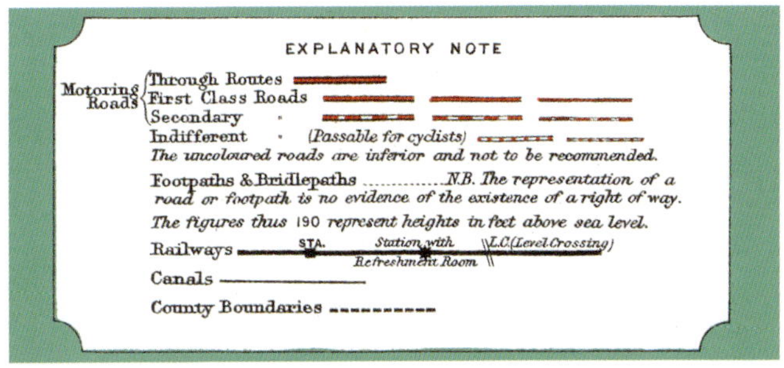

BARTHOLOMEW

The Edinburgh cartographic firm of Bartholomew was founded in 1826 by John Bartholomew (1805–1861). The second John Bartholomew (1831–1893) trained with his father and the eminent German geographer, August Petermann (1822–1878). Under John Junior, the business expanded from contract engraving into lithography and printing, with a particular emphasis on map production. John Junior and his son, John George (1860–1920), introduced the continental technique of layer colouring to Britain in *Baddeley's Guide to the Lake District*. The firm subsequently became synonymous with the renowned series of *Times* atlases.

Bartholomew remained a family business, passing from son to son, until 1980, when the business was sold to Reader's Digest Association, Inc., and then to News International, who later merged it with the Glasgow publisher Collins, as part of HarperCollins Publishers. The Bartholomew Archive – http://digital.nls.uk/bartholomew – is one of the most extensive cartographic archives available for research in a public institution. HarperCollins and John C Bartholomew and family donated the firm's business records, working maps and printing plates to the National Library of Scotland (NLS) between 1983 and 2008. Fittingly, the NLS Map Reading Room lies only a short walk from the site of Bartholomew's former Edinburgh base.

Llangain
River Towy
Ferryside
Broadlay
Llansaint
St. Ishmael
Kidwelly
Gwendraeth
Llandyfaelog
Llangyndeyrn
Llanddarog
Carmel
Llandybie
Mynydd Mawr
Ammanford
Cross Hands
Tumble
Pont Antwn
Pontyberem
Pont Henry
Pont Yates
Llannon
Llanedy
Hendy
Sylen
Five Roads
Pembrey
Burry Port
Pembrey Burrows
Towyn Burrows
Pembrey Scot
Burry River
Llanelly
Felin-foel
Dafen
Llangennech
Bynea
Loughor
River Loughor
Bacas Sands
Llanrhidian Sands
Whiteford Pt.
Whiteford Sands
Salthouse Pt.
Penclawdd
Crofty
Llanmorlais
Llanrhidian
Llanmadog
Cheriton
Landimore
Burry Holms
Broughton Bay
Llangennith
Rhossili Bay
Rhossili Downs
Rhossili
Worms Head
Helwick Channel
Middleton
Pitton
Reynoldston
Knelston
Llanddewi
Burry
Cefn Bryn
Oxwich
Oxwich Bay
Oxwich Pt.
Penrice
Nicholaston
Penmaen
Port Eynon
Port Eynon Bay
Port Eynon Pt.
Pennard
Parkmill
Ilston
Bishopston
Kittle
Fairwood Common
Clyne Common
Dunvant
Killay
Three Crosses
Gowerton
Pwll-du Head
Three Cliffs Bay
Langland Bay
Newton
East Helwick
5 Fathoms Line
G O W E R

HEIGHT OF LAND IN FEET
SCALE OF COLOURS
ABOVE 2750
2750 TO 2500
2500 TO 2250
2250 TO 2000
2000 TO 1750
1750 TO 1500
1500 TO 1250
1250 TO 1000
1000 TO 800
800 TO 600
600 TO 400
400 TO 300
300 TO 200
200 TO 100
100 TO SEA LEVEL
Low Tide TO 5 Fathoms
S W A N S E A
B A Y
Outer Green Grounds
Neath Bar
Breakwater
Baglan Burrows
Margam Moor
Margam Burrows
Low Water Mark
Old River
NEATH
MORRISTON
ABERAVON
BRITON FERRY
Port Talbot
Taibach
Margam
Mynydd Margam
Michaelston
Baglan
Skewen
Llansamlet
Clydach
Pontardawe
Ystalyfera
Ystradgynlais
Ynyswen
Glyn-neath
Resolven
Crynant
Seven Sisters
Onllwyn
Abercrave
Llangiwg
Cilybebyll
Mynydd March Hywel
Hir Pynydd
Penller fedwen
Mynydd y Drum
Cefn Morfydd
Cefn Mawr
Moel yr Hyrddod
Mynydd Caerau
Llangynwyd
Maes-teg
Cwm Afan
Cwm Amman
Brynamman
Garnant
Glanamman
Vale of Neath
Birchgrove
Crymlyn Bog
Crymlyn Burrows
Jersey Marine Sta.
Swansea Bay Sta.
G L A M

THE 'GROUPING' – AND A QUEST FOR SPEED

After wartime experience of the benefits of unified operation, the Government amalgamated 120 railway companies into just four large groups. The railway system as a whole stagnated, but on the principal main lines a short 'golden era' of ever faster and more luxurious services was launched.

MILEPOST 14	Railway Towns	176
1926	Oswestry Station and Railway Works	178
MILEPOST 15	The Dawn of High Speed Travel and the Age of the Streamliners	182
c.1923–8	Bartholomew's 'Post Grouping' British Isles	184
1923–47	Jowett's West Yorkshire	188
MILEPOST 16	Style Wars – The Classic Era of Railway Publicity	190
1929/35/38	George Dow's Diagrammatic Maps	194
1933	Harry Beck's London Underground	196
1850/1935/1990	Jowett's Yarmouth and Lowestoft	200

1923–1947

STIRLING
FIFE
Firth of Forth
North Berwick
Dunbar
EDINBURGH
Leith
Haddington
HADDINGTON
Lammermuir Hills
BERWICK
Berwick
Tweedmouth
Duns
Coldstream
Kelso
Greenlaw
GLASGOW
LINLITHGOW
Linlithgow
Bathgate
Airdrie
Penicuik
Peebles
PEEBLES
Selkirk
SELKIRK
Galashiels
Jedburgh
Hawick
ROXBURGH
Cheviot Hills
Lanark
LANARK
Biggar
Kilmarnock
Mauchline
Muirkirk
Douglas
Old Cumnock
AYR
Dalmellington
Sanquhar
Moffat
Thornhill
DUMFRIES
Dumfries
Lockerbie
Langholm
Annan
Riccarton Jun.
Alnwick
Belford
Wooler
Rothbury
Morpeth
NORTHUMBERLAND
Bellingham
Haltwhistle
Hexham
NEWCASTLE
Allendale
Alston
CARLISLE
Longtown
Brampton
Silloth
KIRKCUDBRIGHT
Kirkcudbright
Castle Douglas
Creetown
Solway Firth
Wigtown Bay
CUMBERLAND
Wigton
Aspatria
Maryport
Workington
Cockermouth
Keswick
Whitehaven
Penrith
Appleby
Shap
WESTMORLAND
Kirkby Stephen
Windermere
Coniston
Kendal
Sedbergh
Hawes
Richmond
Barnard Castle
Middleton
DURHAM
Stanhope
Crieff
Callander
Dunblane
Perth
St Andrews
Anstruther
Largo
Leslie
Kirkcaldy
Burntisland

An empty art room at the London & North Western Railway's Mechanics' Institute at Crewe in 1907.

The peace and quiet of England's green and pleasant land was rudely shattered by the coming of the railways in the mid-19th century – what were once just green pastures and sleepy villages were soon to become the throbbing workshops of the early railway companies. The railway town had arrived.

In the south of the country the village of Swindon was initially ignored by Brunel's broad-gauge Great Western Railway (GWR) when it partially opened in 1840 between Paddington and Bristol. This was soon to change when Swindon became the junction for the new railway to Gloucester and Cheltenham and by 1843 the GWR had established their locomotive repair depot on the land between the two railway lines. In the beginning only 200 men were employed here but by 1846 the works had started building new locomotives and within five years the workforce had soared to 2,000. Swindon expanded rapidly and by the early 20th century was home to over 13,000 railway workers plus their families and others employed in their support. The GWR took a paternalistic view of their employees and engendered a great deal of loyalty among them, providing housing, libraries, reading rooms, medical insurance, a hospital, entertainment venues, a sports and social club and annual outings. Swindon's output of steam locomotives was prolific and continued until 1960, when the last steam locomotive built for British Railways, 'Evening Star', emerged from the works. Although the building of diesels carried on until 1965, the works eventually closed in 1986. The buildings were converted into shops, a railway museum and the headquarters of English Heritage. Despite the loss of its railway works Swindon, with its excellent road and rail connections, has continued to be a boom town in the 21st century.

The opening of the Grand Junction Railway between Birmingham, Liverpool and Manchester in 1837 soon saw that company establishing their railway works on a greenfield site near the small village of Monks Coppenhall. Renamed Crewe after nearby Crewe Hall, the railway town, like Swindon, grew up around an important railway junction and in 1846 became the main locomotive works for the newly formed London & North Western Railway (LNWR). By the beginning of the 20th century the works and its steel-producing plant had a workforce of 8,000 and Crewe had become a railway town of 40,000 people. As with the GWR, the LNWR looked after its employees, providing housing, an educational institution, a hospital, church, school and recreational facilities. At its peak under the London, Midland & Scottish Railway in the 1930s, Crewe Works employed over 20,000 staff and the town's population had grown to around 60,000. It remained the most important railway works in Britain until 1990, by which time over 8,000 steam, diesel and electric locomotives had been built there. While Crewe still remains an important railway centre, with extensive freight and civil engineering

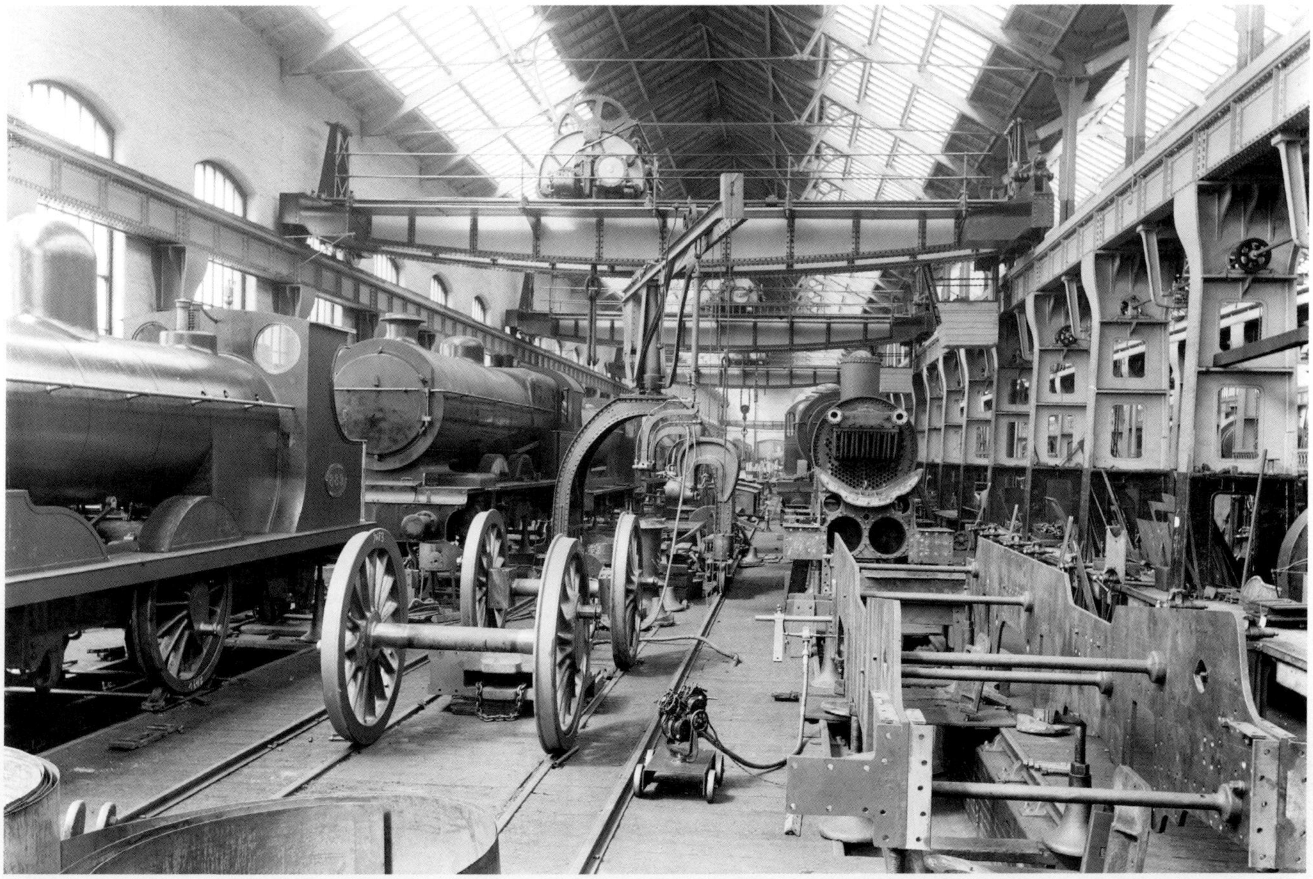

An early 20th-century scene inside the locomotive erecting shop at the Lancashire & Yorkshire Railway's works at Horwich, Lancashire.

yards to the south and west, what little remains of the once extensive railway works is now owned by Bombardier Transportation who carry out train maintenance and inspection.

Around Britain other railway towns also sprung up around locomotive works. Principal among these were the London & South Western Railway at Eastleigh, the South Eastern & Chatham Railway at Ashford, the Great Northern Railway at Doncaster, the North Eastern Railway at Darlington, the Lancashire & Yorkshire Railway at Horwich and the Glasgow & South Western Railway at Corkerhill. Today most of these railway works have long since disappeared with new locomotives now being imported from North America.

Men from the Horwich Railway Mechanics Institute play at the Lancashire & Yorkshire Railway's cricket ground, c.1892.

Series	Ordnance Survey 25 in. to 1 mile
Sheet	Shropshire Sheet XII.14 Oswestry
Date	Surveyed 1874, revised 1924, printed and published 1926

The first railway to reach Oswestry was a short branch from Gobowen, opened in January 1849 by the Shrewsbury & Chester Railway, which had bypassed the town to the north and east. Despite their Welsh-sounding names, both towns lie in England, a few miles from the border.

Independent cross-country lines were promoted southwest to Newtown and northeast to Whitchurch, opening in 1861 and 1864 respectively, but Oswestry was far from being a major railway junction. However, its status as a 'railway town' was assured in 1864 by the amalgamation of the Newtown and Whitchurch lines, together with the Newtown & Machynlleth and Llanidloes & Newtown Railways, to form the Cambrian Railway. The new concern decided to site its headquarters in the town, with facilities which included administrative offices in the substantial station building, a well-equipped locomotive, carriage and wagon works, and a locomotive running shed eventually expanded to six 'roads' (rail tracks).

In a small town whose population barely reached 10,000 even after the development stimulated by the railway, these topographical features of the railway were relatively dominant, as can be seen in the Ordnance Survey map on this page and overleaf, published four years after the Cambrian was absorbed into the Great Western Railway in 1922. The railway was kept on the edge of the town, where it had ample room for expansion as passenger and freight traffic and supporting activities grew. In addition to the works and locomotive ('engine') shed, and the passenger station, the map depicts goods sheds, goods stations, a coal depot, a timber yard and rail-connected cattle pens.

The railway at Oswestry saw little change until British Railways' 1955 Modernisation Plan and the 1963 Beeching Report resulted in not just a rapid decline, but the almost complete removal of the town from the railway map. After withdrawal of the remaining shuttle passenger services to Gobowen in 1966, all that survived were freight trains, latterly serving Blodwel quarry (five miles to the south), for which a single track through the former station sufficed.

This traffic lingered on until 1988, but more recent years have seen a mini railway revival in Oswestry with the establishment of the Cambrian Railways Society, which hopes to develop a heritage railway operation through the town from Gobowen to Blodwel.

Before – still multiple tracks to and through Oswestry just after the 1966 closure of the passenger station.

After – by 1978 only a single track was required through the station for the remaining freight branch line.

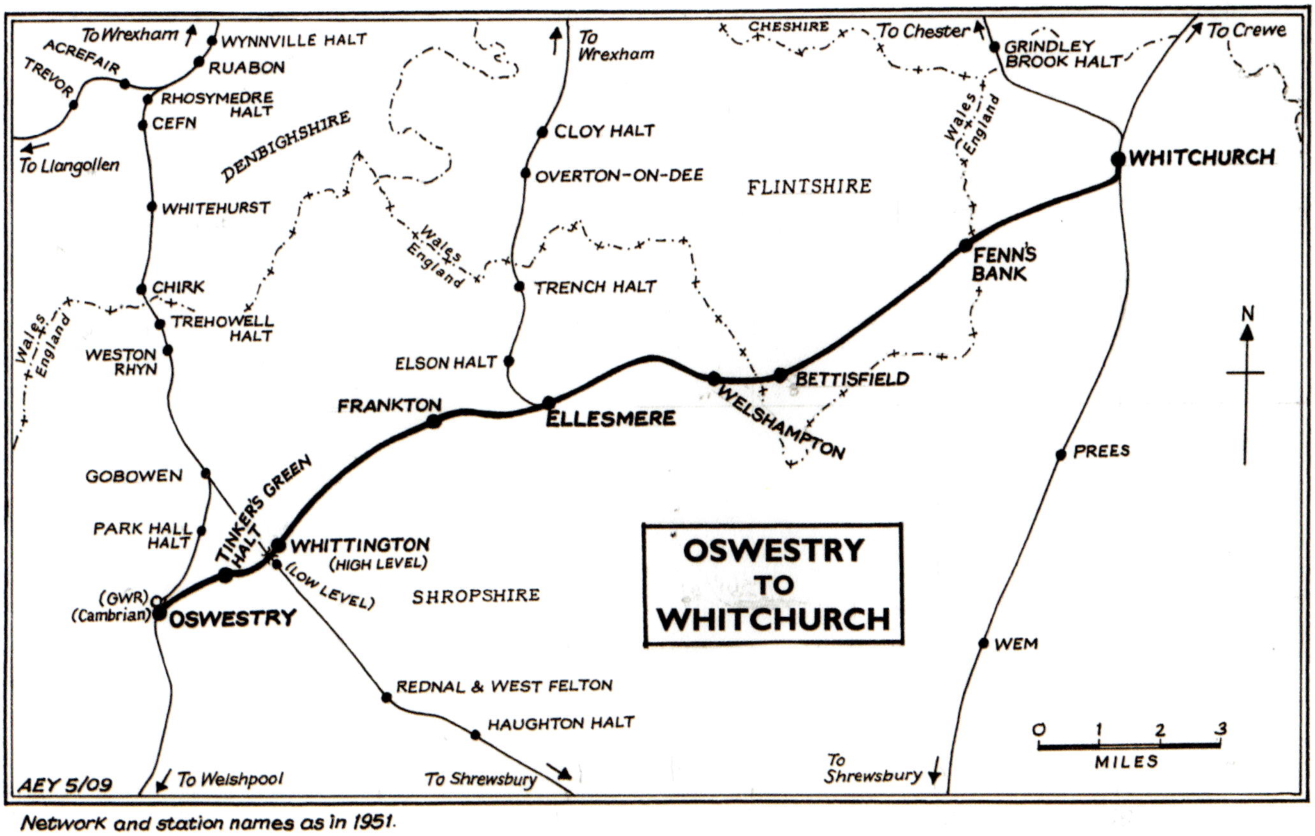

Alan Young's contemporary map shows Oswestry's position in the wider rail network in 1951.

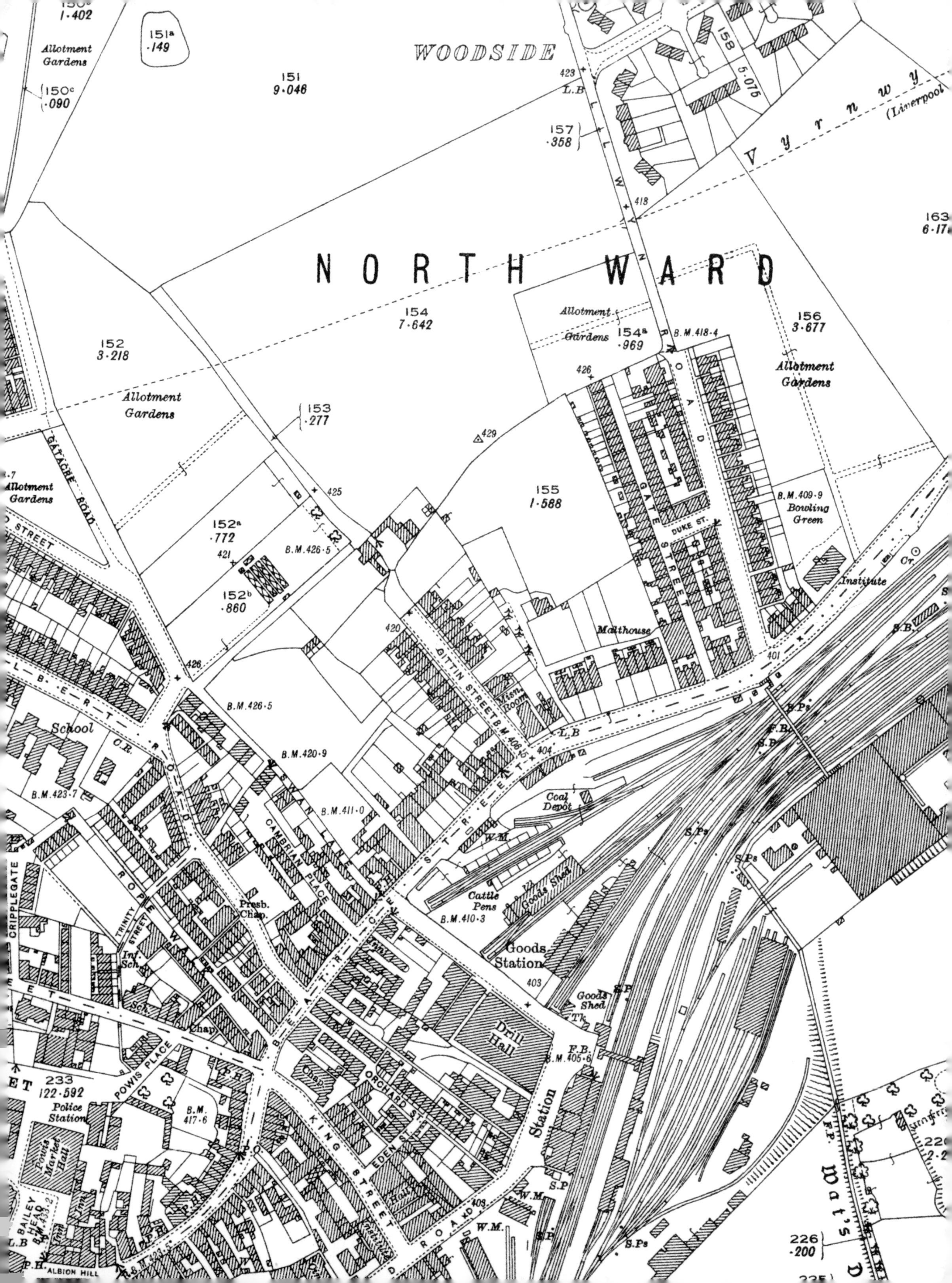
WOODSIDE
NORTH WARD
Allotment Gardens
Vyrnwy (Liverpool
School
Presb. Chap.
Police Station
Powis Market Hall
Drill Hall
Goods Station
Goods Shed
Cattle Pens
Coal Depot
Malthouse
Bowling Green
Institute
GATACRE ROAD
ROPE WALK
CAMBRIAN PLACE
SWAN LANE
KING STREET
ORCHARD ST.
EDEN ST.
BAILEY HEAD
ALBION HILL
POWIS PLACE
CRIPPLEGATE
TRINITY STREET
DUKE ST.
GATE STREET
STATION ROAD
Station
ROAD
CROSS STREET
GITTIN STREET
Misn. Room
Inf. Sch.
Chap.
Sch.
Hall

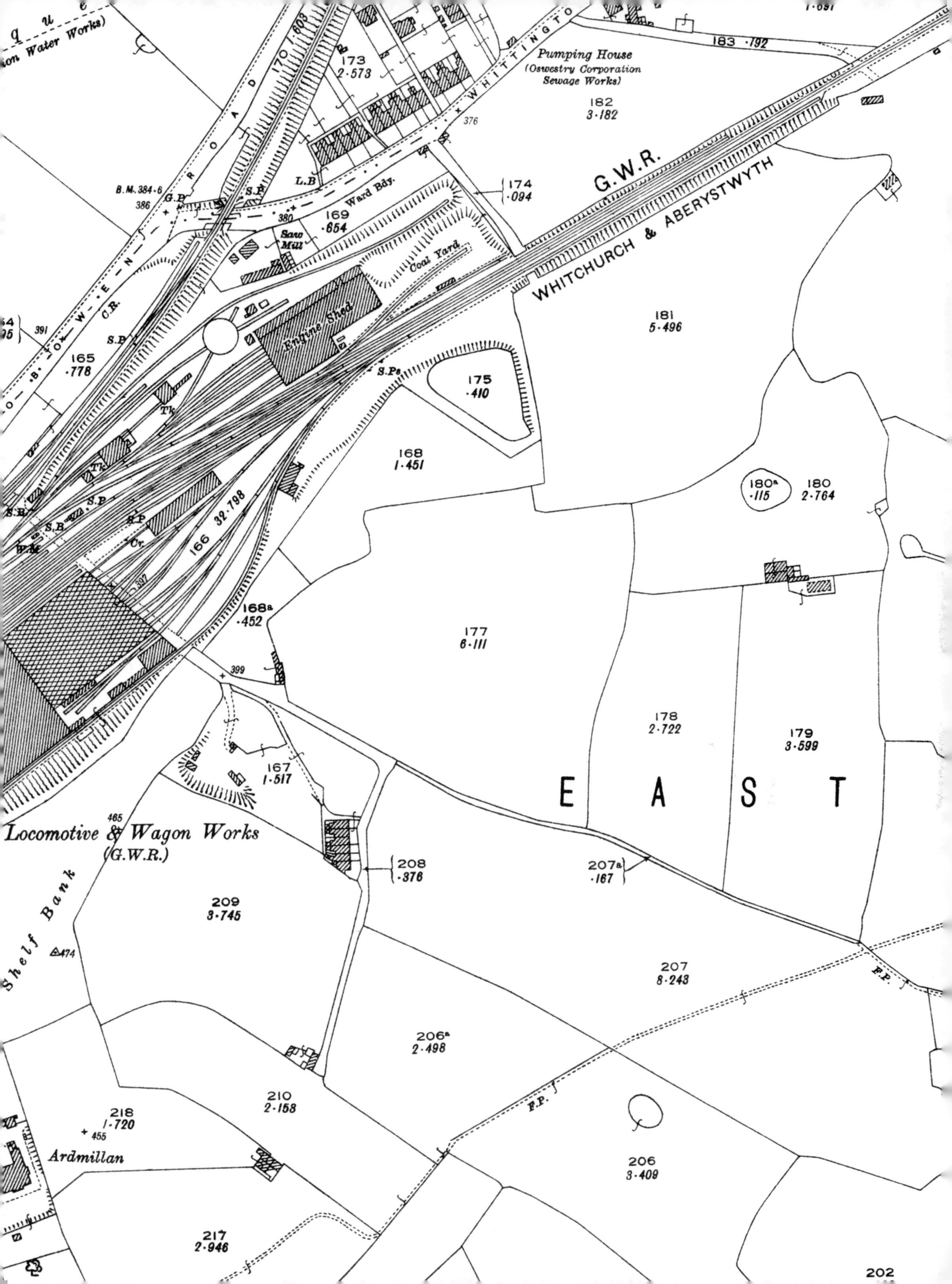

Pumping House
(Oswestry Corporation Sewage Works)
G.W.R.
WHITCHURCH & ABERYSTWYTH
Engine Shed
Coal Yard
Saw Mill
Ward Bdy.
Locomotive & Wagon Works
(G.W.R.)
Shelf Bank
Ardmillan
E A S T
F.P.
202

MILEPOST

15

By the end of the First World War, the 120 railway companies operating in Britain were facing a difficult future. Losses were mounting and in 1921 the Government passed the Railways Act which amalgamated these companies into four large regional companies. Commonly known as the Big Four Grouping, this took effect on 1 January 1923. Within a few years these four companies – Great Western Railway (GWR), London, Midland & Scottish Railway (LMS), London & North Eastern Railway (LNER) and Southern Railway (SR) – were getting their act together with the development of more powerful steam locomotives and modern coaching stock. Striving for faster and faster services, the rivalry between these publicity-seeking companies was intense.

Always at the forefront of railway technology, the GWR was first off the starting block with its world-beating 'Cheltenham Flyer'. Hauled by one of Collett's new 'Castle' Class 4-6-0 locomotives, this famous train gained the title 'The World's Fastest Train' in 1932 when 'Tregenna Castle' and its six coaches took just under 57 minutes to travel the

Introduced in 1935, the London & North Eastern Railway's 'Silver Jubilee' express between King's Cross and Newcastle was hauled by Nigel Gresley's streamlined 'A4' Pacific locomotives.

No. 5000 'Launceston Castle' speeds across the Thames at Maidenhead with the world's fastest scheduled train, the 'Cheltenham Flyer', in 1934.

77¼ miles between Swindon and Paddington at an average speed of 81.6 m.p.h. The train was soon rescheduled to take 65 minutes for the journey and became officially the fastest scheduled train service in the world.

Elsewhere in the world the German 'Flying Hamburger' and the American 'Burlington Zephyr' were setting new standards using streamlined high-speed diesels. After visiting Germany, Nigel Gresley, the Chief Mechanical Engineer of the LNER carried out trials with his 'A3' Pacific 'Papyrus', which set a speed record of 108 m.p.h., proving that steam was every bit as good as diesel. This success led to Gresley designing his world-famous 'A4' streamlined Pacifics, the first of which, 'Silver Link', reached a speed of 112.5 m.p.h. while hauling seven streamlined articulated coaches in September 1935. The new A4s were soon in regular service hauling the 'Silver Jubilee' express between King's Cross and Newcastle – the locomotives proved very reliable and were popular with their crews.

Not to be outdone by the LNER's success, the LMS introduced its 'Coronation Scot' streamlined train between Euston and Glasgow in 1937. Motive power was provided by William Stanier's new 'Coronation' Class Pacifics, the first ten members being fitted with streamlined casings, hauling matching coaches. During a test run in June 1937, No. 6220 'Coronation' achieved a speed of 114 m.p.h. just south of Crewe, narrowly avoiding a derailment as it approached the station at an excessive speed.

The competition was hotting up, and back on the LNER new streamlined expresses were soon introduced from King's Cross to Edinburgh and Leeds. However, the best was yet to come when, on 3 July 1938, Gresley's 'A4' Class 'Mallard' achieved a world speed record for steam traction of 126 m.p.h. while hauling a test train down Stoke Bank south of Grantham – this record still stands today.

Sadly, the onset of the Second World War in 1939 put a stop to these high jinks and the post-war austerity years never saw a return to this Golden Age. The LMS 'Coronations' lost their streamlined casings and the LNER 'A4's were shorn of their streamlined side valances, although the 'Elizabethan' non-stop express between King's Cross and Edinburgh provided a little glamour until it was withdrawn in 1961.

Fortunately six members of the 'A4' Class (including 'Mallard') have been preserved, including one each in museums in the USA and Canada. The preserved ex-LMS 'Coronation' Class Pacific 'Duchess of Hamilton' has recently had its streamlined casing reinstated, and the locomotive can be seen in all its glory at the National Railway Museum in York.

Three preserved 'A4' locomotives. In the foreground is 'Mallard' which attained 126 m.p.h. in 1938 – an unbeaten world record for steam locomotives.

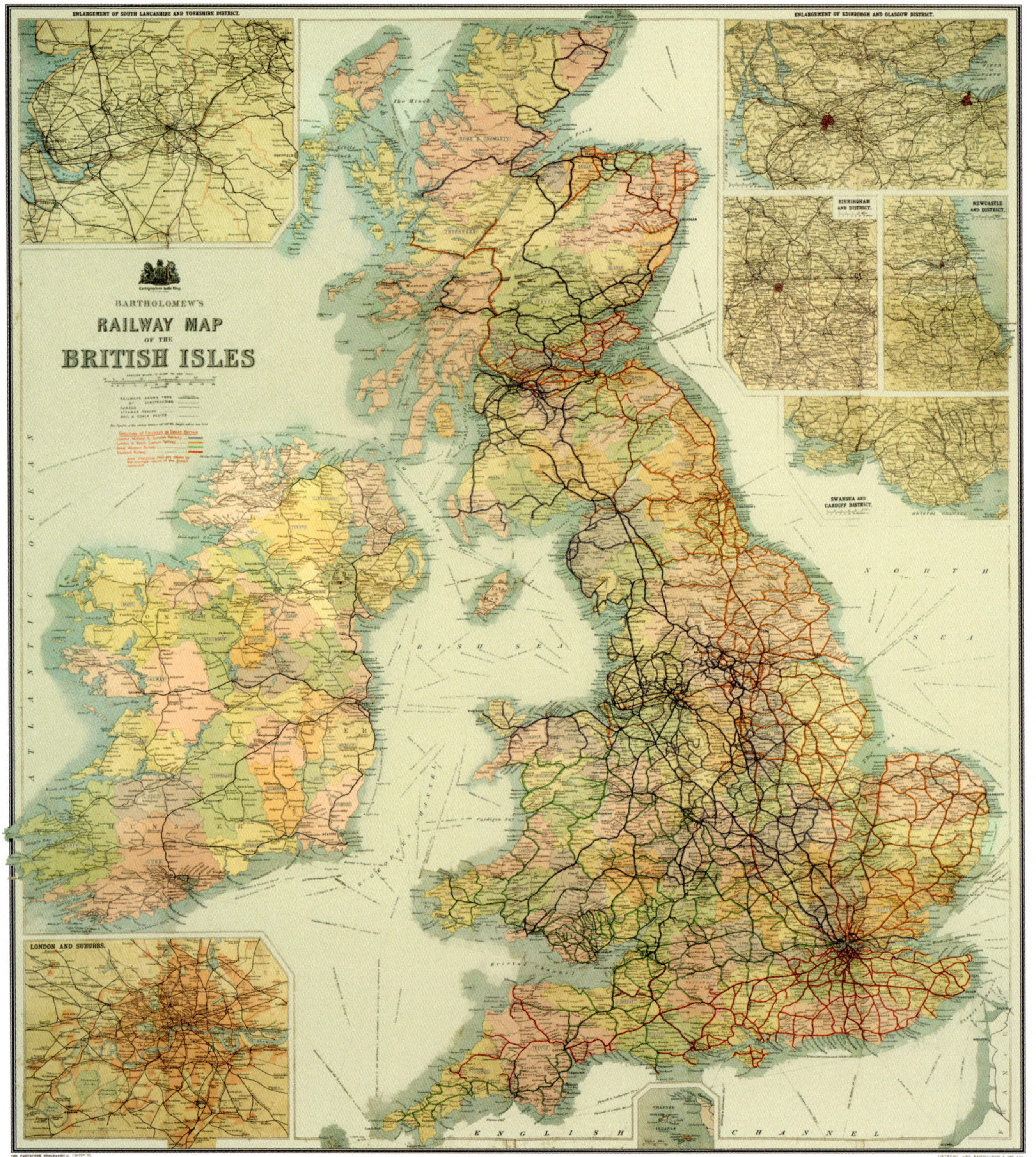

Title	Bartholomew's Railway Map of the British Isles
By	J Bartholomew, Edinburgh
Date	*c.*1923–8

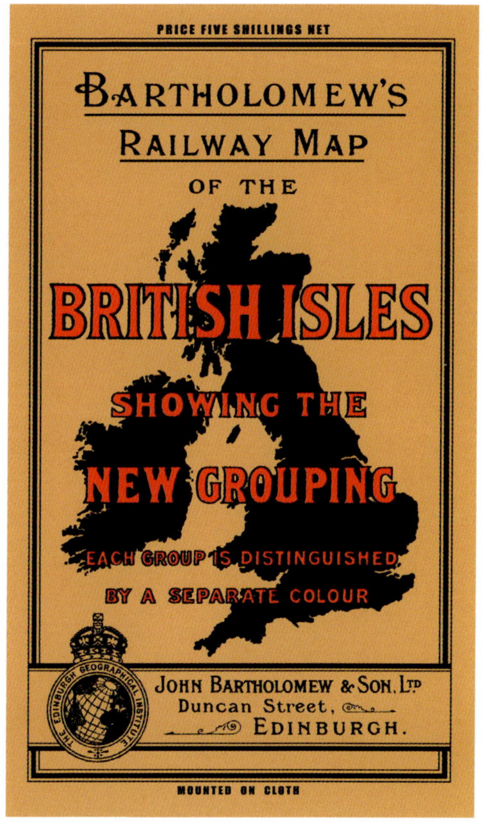

John Bartholomew & Son Ltd, the prestigious Edinburgh mapmaker, produced specialist railway maps not just on commission from the railway companies, but also as their own bespoke sheets showing the complete national network. In this particular case, an earlier map has been overprinted in colour to distinguish the routes of the 'Big Four' companies created by the Government's 1923 Grouping – the Great Western Railway (GWR), London, Midland & Scottish Railway (LMS), London & North Eastern Railway (LNER) and the Southern Railway (SR).

The base map was clearly produced some time before the Grouping, as it depicts a number of routes which had already closed by 1923, such as the Highland Railway's Keith to Buckie branch (closed in 1915), and the 'seventh Anglo-Scottish route' across the Solway Firth Viaduct, west of Port Carlisle (closed in 1922). The original key also has a symbol for 'railways constructing', but only a couple of very minor lines were built after the Grouping.

The overlaid key highlights a very British quirk which continued after the Grouping – 'joint railways', which were lines owned by more than one company. Designed to find a compromise between competing aims, in many ways they were a costly solution. Not only was there the cost of legal requirements to determine working arrangements, but in the case of the bigger joint railways there were separate headquarters and staff to be funded. A more sensible outcome would have been for a single company to take ownership and to charge the other company for train running rights, extending the principle of 'running powers' which was long established and accepted throughout the rail network.

The Grouping did reduce the number of joint lines from over 70 to just 16, perhaps the most famous of which were the Somerset & Dorset and the Midland & Great Northern.

The latter is shown in some detail in the extract on this page from another Bartholomew's map (of 1927), and was the longest of the joint railways, at 183 miles. It penetrated deep into Great Eastern Railway territory, designed to tap the Norfolk coast holiday market and fish traffic to the Midlands. As essentially a duplicate route for longer distance traffic and an early casualty of local road competition, it closed completely in 1959 – other than for a few freight stubs, and the Sheringham–Cromer link, which survives to this day.

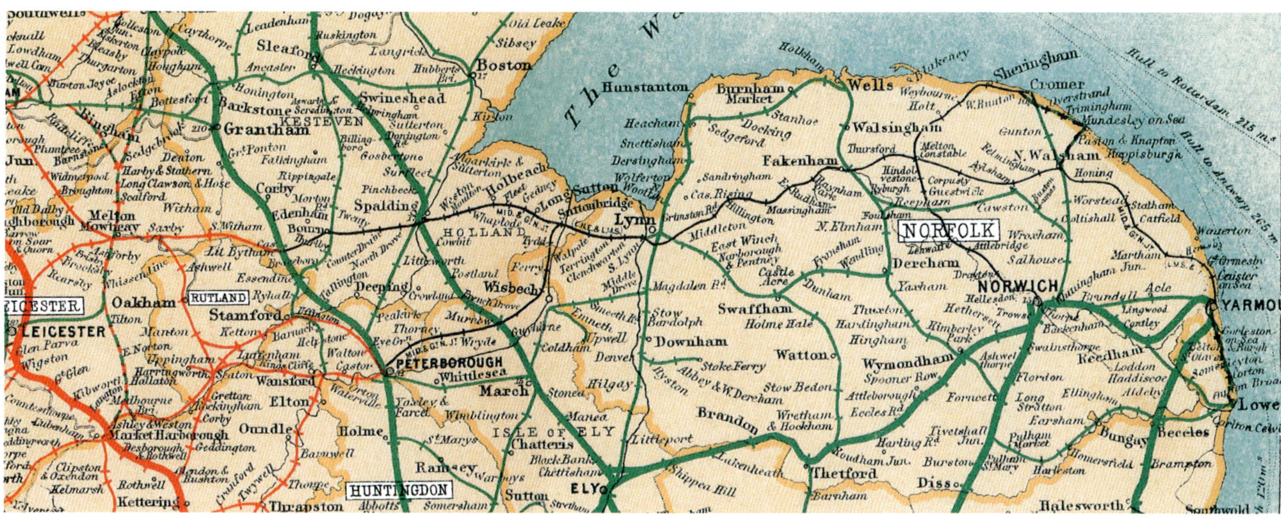

Cardigan Bay
ENGLISH MILES, 19 MILES TO ONE INCH
KILOMETRES
RAILWAYS SHOWN THUS
STATION
D° CONSTRUCTING
CANALS
STEAMER TRACKS
MAIL & COACH ROUTES
The figures at the railway stations indicate the height above sea level
GROUPING OF RAILWAYS IN GREAT BRITAIN
London, Midland & Scottish Railway
London & North Eastern Railway
Great Western Railway
Southern Railway
Joint intergroup lines are shown by the alternate colours of the groups concerned.

LINCOLN
NOTTINGHAM
LEICESTER
RUTLAND
HUNTINGDON
NORTHAMPTON
BEDFORD
CAMBRIDGE
NORFOLK
SUFFOLK
ESSEX
HERTFORD
OXFORD
BUCKINGHAM
BERKSHIRE
SURREY
KENT
SUSSEX
HAMPSHIRE
The Wash
Mouth of the River Thames
Strait of Dover
Isle of Wight
Hunstanton
Wells
Sheringham
Cromer
Walsingham
Fakenham
Lynn
Dereham
NORWICH
YARMOUTH
Lowestoft
Swaffham
Downham
Watton
Wymondham
Brandon
Thetford
Diss
Eye
Halesworth
Bungay
Beccles
Aldeburgh
Woodbridge
IPSWICH
Harwich
Hadleigh
Sudbury
Colchester
Walton on the Naze
Brightlingsea
Maldon
Chelmsford
Southminster
Southend
Shoeburyness
Sheerness
Whitstable
Herne Bay
Margate
Ramsgate
Sandwich
CANTERBURY
DEAL
DOVER
FOLKESTONE
Hythe
MAIDSTONE
Tonbridge
Seven Oaks
Tunbridge Wells
Ashford
Wye
Rye
Hastings
St. Leonards
Bexhill
Battle
Hailsham
Eastbourne
Lewes
Newhaven
Seaford
BRIGHTON
Shoreham by Sea
Worthing
Littlehampton
Chichester
Midhurst
Petworth
Pulborough
Horsham
Haslemere
Godalming
GUILDFORD
Dorking
Reigate
Farnham
Aldershot
Alton
Basingstoke
WINCHESTER
Eastleigh
Petersfield
PORTSMOUTH
Hayling I.
Selsey Bill
Ryde
Newport
Shanklin
Ventnor
READING
WINDSOR
Maidenhead
Marlow
High Wycombe
Henley
Wallingford
Abingdon
Bampton
OXFORD
Bicester
Banbury
Aylesbury
Tring
St. Albans
Hitchin
Baldock
Royston
Luton
Dunstable
Leighton
Bletchley
Buckingham
Towcester
Blisworth
Weedon
NORTHAMPTON
Wellingborough
Olney
BEDFORD
Ampthill
Biggleswade
St. Neots
HUNTINGDON
St. Ives
Thrapston
Oundle
Kettering
Market Harborough
RUGBY
COVENTRY
Hinckley
LEICESTER
Oakham
Stamford
Melton
Loughborough
NOTTINGHAM
Southwell
Newark
Grantham
Sleaford
Boston
Spalding
Holbeach
Sutton
Wisbech
PETERBOROUGH
March
Whittlesea
Ramsey
Chatteris
ELY
Littleport
Soham
Newmarket
CAMBRIDGE
Shelford
Saffron Walden
Audley
Bury St. Edmunds
Stowmarket
Needham
Framlingham
Halstead
Braintree
Witham
Dunmow
Bishop Stortford
Hertford
Ware
Epping
Ongar
Waltham
Hatfield
Uxbridge
Slough
LONDON
Greenwich
Gravesend
Chatham
Rochester
Strood
Faversham
Sittingbourne
Tuxford
Worksop
Retford
Gainsborough
Bawtry
LINCOLN
Market Rasen
Louth
Saltfleet
Mablethorpe
Sutton on Sea
Alford
Horncastle
Spilsby
Burgh
Wainfleet
Tattershall
Swineshead
Dungeness
New Romney
Winchelsea
Beachy Head
Walmer
The Downs
CALAIS
BOULOGNE
C. Grisnez
Hull to Rotterdam 216 m.
Grimsby to Rotterdam 180 m.
Hull to London 226 m.
London to Rotterdam 188 m.
Southampton to London 209. Portsmouth to London 193 m.
to Antwerp 191 m.
to Ostende 115 m.
to Flushing 120 m.
to Rotterdam 122 m.
to Antwerp 135 m.

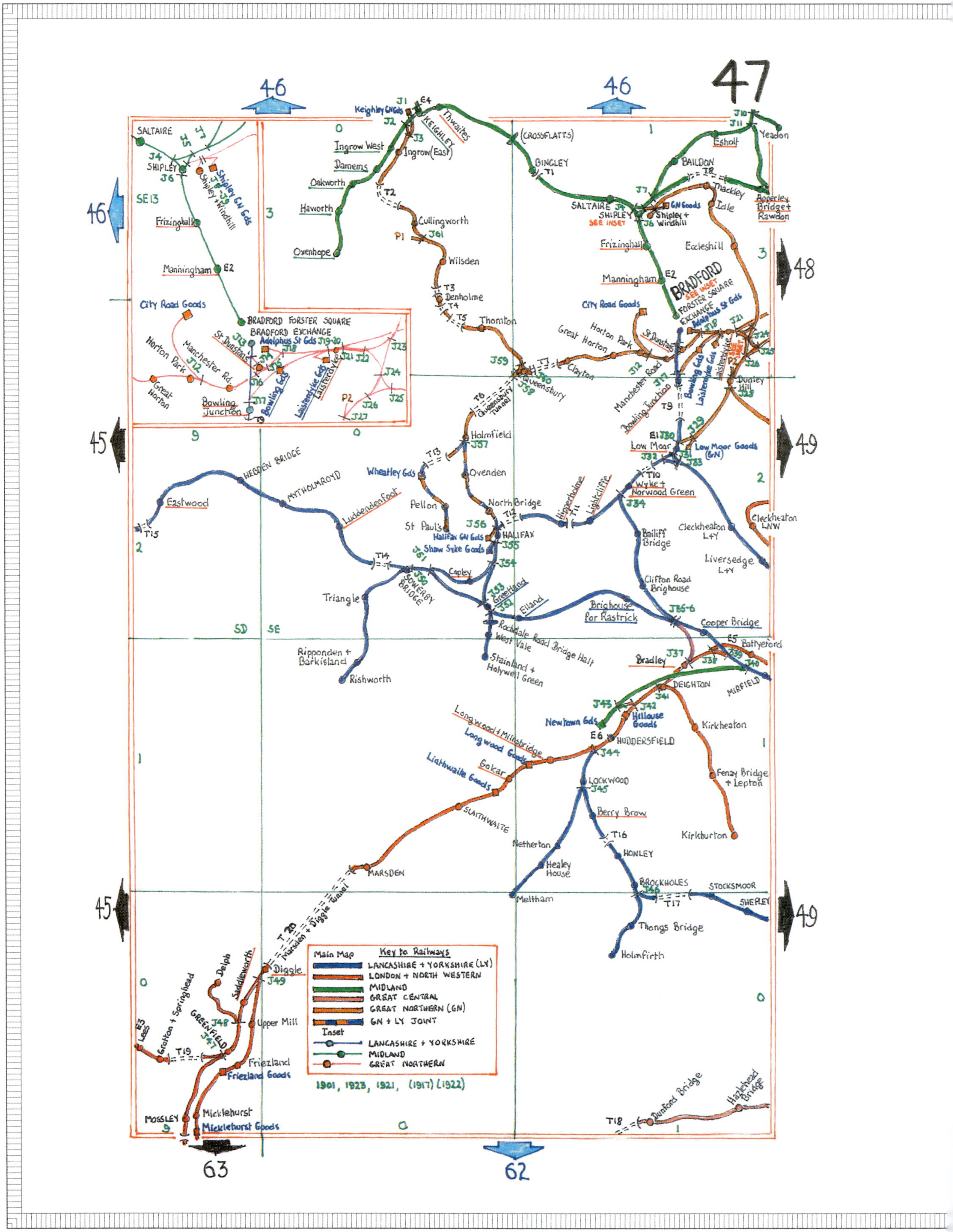
47
46
45
48
49
63
62
Key to Railways
Main Map
LANCASHIRE + YORKSHIRE (LY)
LONDON + NORTH WESTERN
MIDLAND
GREAT CENTRAL
GREAT NORTHERN (GN)
GN + LY JOINT
Inset
LANCASHIRE + YORKSHIRE
MIDLAND
GREAT NORTHERN
1901, 1923, 1921, (1917) (1922)
SALTAIRE
SHIPLEY
Frizinghall
Manningham
BRADFORD FORSTER SQUARE
BRADFORD EXCHANGE
City Road Goods
Adolphus St Gds
Horton Park
Great Horton
Manchester Rd.
St Dunstans
Bowling Junction
Bowling Gds
Laisterdyke Gds
Laisterdyke
Keighley GNGds
KEIGHLEY
Ingrow West
Ingrow (East)
Damems
Oakworth
Haworth
Oxenhope
Thwaites
(CROSSFLATTS)
BINGLEY
Cullingworth
Wilsden
Denholme
Thornton
Clayton
Queensbury
Queensbury Tunnel
Holmfield
Ovenden
North Bridge
HALIFAX
Halifax GN Gds
Shaw Syke Goods
Wheatley Gds
Pellon
St Paul's
Esholt
Yeadon
BAILDON
Thackley
Idle
Apperley Bridge + Rawdon
Shipley + Windhill
GN Goods
Eccleshill
BRADFORD
SEE INSET
Dudley Hill
Low Moor
Low Moor Goods (GN)
Wyke + Norwood Green
Lightcliffe
Hipperholme
Cleckheaton LNW
Cleckheaton L+Y
Liversedge L+Y
Bailiff Bridge
Clifton Road Brighouse
Brighouse for Rastrick
Cooper Bridge
Battyeford
MIRFIELD
Bradley
DEIGHTON
Kirkheaton
HUDDERSFIELD
Newtown Gds
Hillhouse Goods
Fenay Bridge + Lepton
Kirkburton
Longwood + Milnsbridge
Longwood Goods
Golcar
Linthwaite Goods
SLAITHWAITE
MARSDEN
Marsden + Diggle Tunnel
LOCKWOOD
Berry Brow
Netherton
Healey House
Meltham
HONLEY
BROCKHOLES
STOCKSMOOR
SHEPLEY
Thongs Bridge
Holmfirth
HEBDEN BRIDGE
MYTHOLMROYD
Eastwood
Luddendenfoot
Copley
SOWERBY BRIDGE
Triangle
Ripponden + Barkisland
Rishworth
Greetland
Elland
Rochdale Road Bridge Halt
West Vale
Stainland + Holywell Green
Diggle
Delph
Saddleworth
Upper Mill
GREENFIELD
Grotton + Springhead
Lees
Friezland
Friezland Goods
MOSSLEY
Micklehurst
Micklehurst Goods
Dunford Bridge
Hazlehead Bridge
SD
SE
SE13

Series	Jowett's Railway Atlas (page 47)
By	Alan Jowett
Publisher	Patrick Stephens Ltd, Cambridge
Date	1989

This map from *Jowett's Railway Atlas* portrays the zenith of the rail network on the West Yorkshire side of the Pennines to the southwest of Bradford. Originally mapped at a scale of 5 miles to the inch, it shows no less than six different railway companies competing for passenger and freight traffic with a dense network of lines in the industrialized and heavily populated valleys around Bradford, Halifax and Huddersfield. Further west, the topography of the Pennines and the relative absence of development thinned out the opportunities for the railways to just three main lines – the Midland Railway (MR) striking northwestwards from Leeds and Bradford through Keighley to Hellifield, then west to Lancaster and north to Carlisle via Settle; the Lancashire & Yorkshire Railway (LYR) through Halifax and Hebden Bridge to Manchester and Preston; and the London & North Western Railway through Huddersfield and the long Marsden and Diggle (Standedge) Tunnel to Manchester.

These three main lines are the principal survivors from that era, serving key inter-city, commuter and freight markets. The route southeastwards from Huddersfield via Stocksmoor to Penistone (and onwards to Barnsley and Sheffield) was reprieved from the 'Beeching Axe', while the previously freight-only lines from Huddersfield and Mirfield to Halifax and Sowerby Bridge via Brighouse re-opened to passenger traffic in 2000. The MR's scenic branch line from Keighley to Oxenhope closed in 1962, but in 1968 re-opened as the volunteer-run Keighley & Worth Valley Railway, famous for its scenes in the classic film *The Railway Children*.

The largest concentration of through routes that closed is the network of former Great Northern Railway lines connecting Bradford, Halifax and Keighley, centred on the triangular junction at Queensbury. This system was fully opened by 1884, both to serve local mill towns and villages and to compete with the MR and LYR. The lines were heavily engineered and steeply graded, with no less than 16 tunnels totalling over four and a half miles in length. Relatively slow through journeys, and intensifying bus competition after the First World War (and car competition after the Second World War), led to the early demise of passenger services in 1955. The last stretches, surviving for freight only, in Bradford (to City Road Goods) and in Halifax (to North Bridge) closed in 1972 and 1974 respectively.

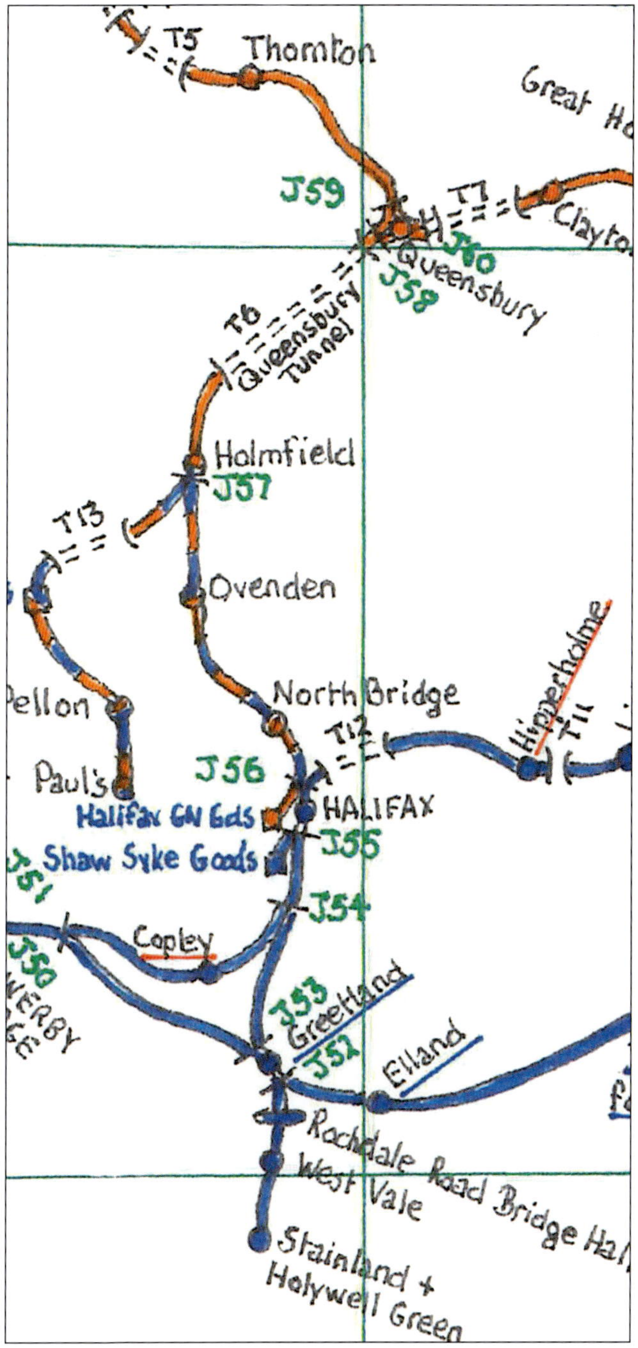

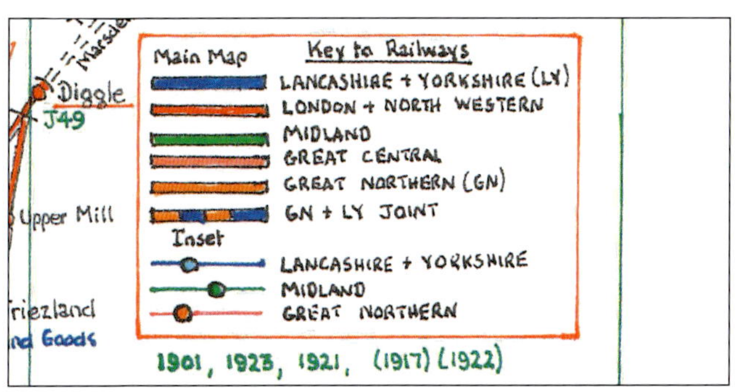

By the beginning of the 20th century Britain's railway companies had discovered the value of marketing their services, routes and destinations to the travelling public. Long before the First World War, railway publicity departments all over the land were extolling their companies' virtues with guidebooks, illustrated route maps, postcards and posters. Of note were the publicity departments of the North British Railway and the London & North Western Railway, but smaller companies such as the North Staffordshire Railway and the Furness Railway also led the way. In addition, the Great Northern Railway was in the forefront of publicity with its 'Skegness is So Bracing' poster – designed by the illustrator John Hassall in 1908, it is probably the most famous railway poster of all time.

However, by far the most prolific publicity department was that of the Great Western Railway (GWR) which, as early as 1903, was promoting its route to the southwest as 'The Holiday Line'. The GWR even commissioned its own film, 'The Story of the Holiday Line', which was first shown at the London Coliseum in 1914. By the 1920s the GWR was promoting the scenic delights of Devon and Cornwall as the 'Cornish Riviera', a marketing slogan which was used to good effect on posters, guide books and even on its premier express to Penzance. In the early 1930s the GWR's 'Cheltenham Flyer' – dubbed the 'World's Fastest Train' – was a tremendous coup for its publicity department, as was the centenary of the company in 1935. Posters by artists such as Murray Secretan, Leonard Richmond and Charles Mayo could be seen adorning stations the length and breadth of the GWR's system. Not content with its abundant output of posters and holiday guides, the GWR also produced over 40 jigsaw puzzles (manufactured by Chad Valley) and a series of books for railway enthusiasts.

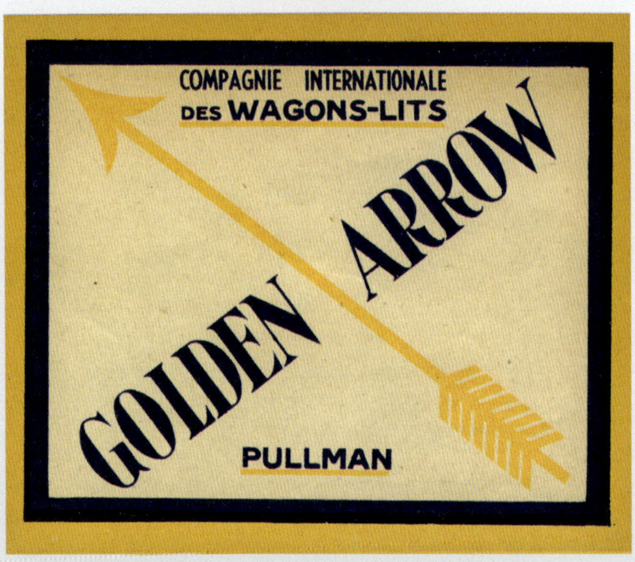

Even the luggage labels of the 1930s were stylishly designed.

Written by the prolific author, S P B Mais, the Southern Railway published a series of walking guide books in the 1930s.

The 1930s was the Golden Age of Railways in Britain and the publicity departments of the other three of the 'Big Four' railway companies – the London, Midland & Scottish (LMS), the London & North Eastern (LNER) and the Southern (SR) railways – were also hard at work glamourizing travel on their respective systems. The most prolific poster artist of this period was Norman Wilkinson, who produced a large number of classic large-format posters for both the LMS and the SR and was instrumental in introducing 17 Royal Academicians to work for the LMS publicity department. In addition to its classic holiday destination posters, the SR also produced a series of attractive railway walks books written by the author and broadcaster S P B (Stuart Petre Brodie) Mais.

Introduced in 1937, the London & North Eastern Railway's 'Coronation' express was the first streamlined train between London and Edinburgh.

Painted by Norman Wilkinson, this London Midland & Scottish Railway poster of 1937 depicts the streamlined 'Coronation Scot' express climbing up Shap in 1937.

CORPORATE IDENTITY

The post-war years brought nationalization of Britain's railways and a new period of corporate identity. A lion (nicknamed the 'Unicycling Lion') was chosen by British Railways (BR) to adorn the sides of locomotives while a 'sausage'-shaped totem was selected to appear on BR publicity material and for station nameboards. With some variations these remained as the BR corporate identity until the introduction of the famous double-arrow logo in 1965 – nicknamed the 'arrow of indecision' at the time, but subsequently proving to be one of the most enduring of logos, still an identifying sign for railway station locations today.

Introduced after nationalization in 1948, the 'Unicycling Lion' device was used to adorn the sides of British Railways' locomotives.

The 'sausage' logo was used by British Railways as their corporate identity on station signs and publicity material.

British Rail's famous double-arrow logo was introduced in 1965 and is still an identifying sign for railway station locations today.

Always ready to capitalize on a publicity coup, the LNER's department had a field day in 1935 when that company introduced Britain's first streamlined high-speed train, the 'Silver Jubilee', between King's Cross and Newcastle. Stylish locomotives and matching coaches ushered in a new age of modern travel, complemented by the company's stylish brochures and posters. By 1937 the LNER's new streamliner, 'Coronation', was competing with the LMS's streamlined 'Coronation Scot' for the important Anglo-Scottish traffic, with both companies vying with each other for the ultimate publicity coup – the LNER was the winner when one of their streamlined locomotives, 'Mallard', achieved a world record for steam traction in 1938. Just over a year later the Second World War broke out, and a new age of austerity descended on the railways.

By the early 1950s the newly formed British Railways (BR) publicity department had got back into full swing after the post-war austerity period and was producing posters, holiday guides, brochures, leaflets and regional timetables all carrying the corporate identity logo. The most famous

Illustrated by 'Michie', this stylish tariff for the 'Cornish Riviera Express' was produced by the Western Region of British Railways in the 1950s.

artist of the BR era was undoubtedly Terence Cuneo, who had started his career as a magazine and book illustrator before he was appointed an official war artist during the Second World War. From 1953, when he was also appointed official artist of Queen Elizabeth II's Coronation, Cuneo was commissioned by BR to produce a large number of oil paintings of the railways, their locomotives and infrastructure. Often putting his own personal safety at risk when sketching on location, Cuneo's masterpieces always included his personal trademark – a small mouse lurking in a corner of each painting. His output continued into the diesel and electric era and, today, his posters are prized and valuable possessions for collectors of railway ephemera.

Dating from the late 1950s, this elegant tariff for the Western Region's 'The Inter-City Express' contained a dinner menu and wine list.

Painted by Terence Cuneo, this BR poster celebrated the centenary of Brunel's Royal Albert Bridge at Saltash, Cornwall, in 1959. Inset: His trademark mouse can be seen in the bottom left of the poster.

As we have seen, simplified strip maps appeared at a relatively early stage of the development of Britain's rail network, but it was not until the early years of the 20th century that truly diagrammatic maps of railways were first produced. Around 1904 both the City & South London Railway and the Great Central Railway introduced maps with railways shown as straight lines with equal spacing between stations – designed to ease passenger understanding of how to use the dense and complex commuter networks which had developed, particularly in London. Over the next 25 years an increasing number of railways – notably the constituent routes of the London Underground – published maps with other essential diagrammatic characteristics such as coloured, vertical, horizontal and 45-degree lines.

However, it was not until George Dow produced his Great Eastern and Great Northern suburban lines maps for the London & North Eastern Railway (LNER) in 1929 that an entire *system* – as opposed to individual lines – was shown fully diagrammatically. Dow was also the first to introduce lines of common angle, and to show other companies' routes as pairs of thin lines – a double line known as 'casing'.

The end product – as seen in the Great Northern route diagram shown below – was two striking maps which in the words of the cartographer Doug Rose, 'straightened out the tangled geographical lines so that users could allow their eyes to flow effortlessly along each route on the map'. As described by George Dow's son, Andrew, in *Telling the Passenger Where to Get Off* (2005):

They were designed specifically to be placed in compartments of passenger stock, and they were therefore sized and proportioned to fit in the standard glazed frames above seatbacks, with the lettering of a size readable from any corner of the compartment.

Dow went on to produce a total of some 15 diagrammatic maps for the LNER, London Midland & Scottish Railway (LMS) and British Railways (BR), including the dramatically distinctive 1935 London Suburban map for the LMS in 1935, and the Tyneside Electric Lines Route Diagram created to coincide with the LNER's electrification of the Newcastle–South Shields line in 1938.

Unlike Harry Beck, the famed designer of the classic London Underground map (see overleaf), George Dow drew maps of a wide range of different rail networks, but as none of these would ever have the massive visitor base of the Underground – which ensured that Beck's design quickly became internationally recognizable – Dow's key role in the history of railway mapping was destined to be neglected for many years. Fortunately that is now being put right.

Title	Great Northern Suburban Lines Route Diagram
Mapmaker	George Dow
Publisher	London & North Eastern Railway
Date	1929

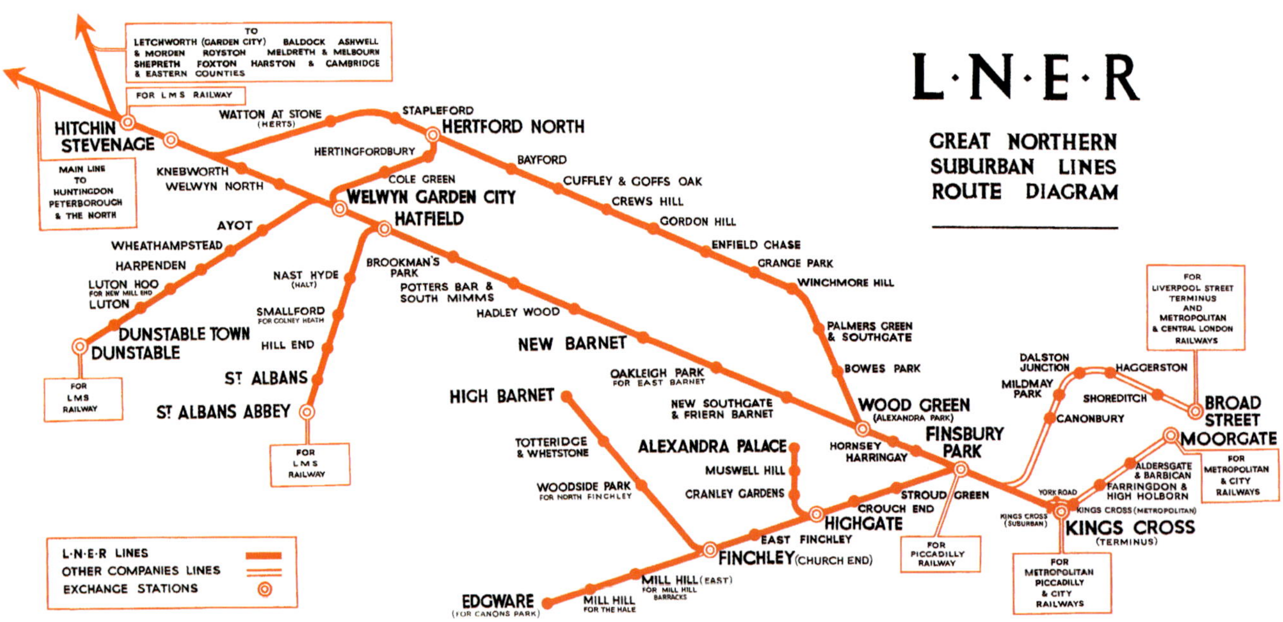

GEORGE DOW

George Dow (born 1907) was a grade five clerk who had worked for the LNER for two years when, in his own time and at his own initiative, he designed and drew the diagrammatic maps for the Great Eastern and Great Northern lines.

Having trained in engineering drawing, he became involved in preparing maps and diagrams for internal railway purposes – from which he extended the principles of simplified mapping to produce diagrams that provided passengers with easily identifiable information while they were *en route* on the train.

As Dow progressed upwards through railway management he inevitably had to withdraw from direct map making, but during his period as Public Relations & Publicity Officer for BR London Midland Region (1949–55) he commissioned award-winning posters, and under his guidance some 25 diagrammatic maps were produced by a member of his staff, Vic Welch.

Dow retired as BR Divisional Manager Stoke-on-Trent in 1968, but went on to have a very productive retirement – lecturing, railway modelling and writing many books on railway subjects, including the classic *Great Central* which includes 27 maps from his own pen. He died in 1987.

Title	LMS London Suburban Electric Lines Route Diagram
Mapmaker	George Dow
Publisher	London Midland & Scottish Railway
Date	1935

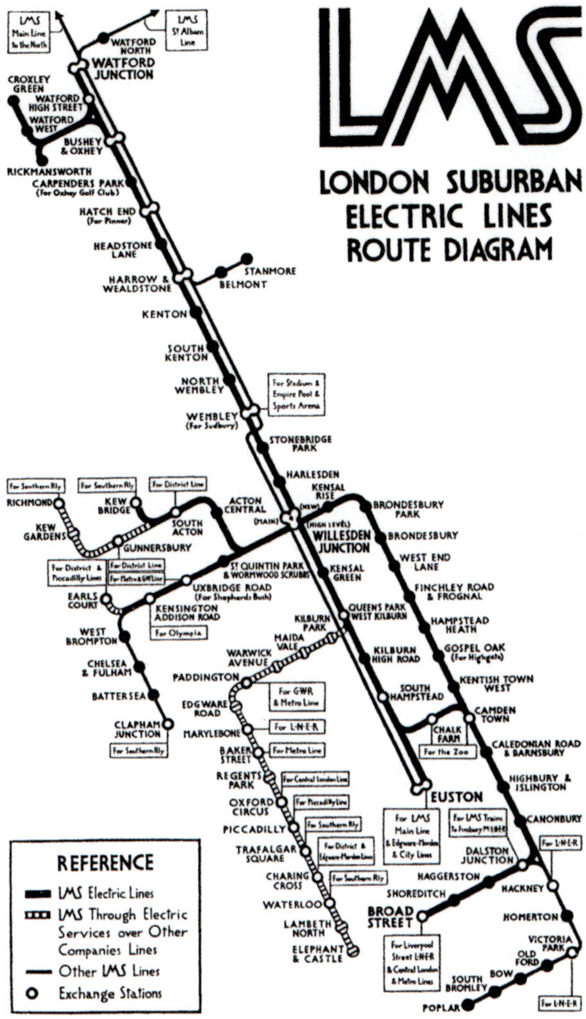

Title	Tyneside Electric Lines Route Diagram
Mapmaker	George Dow
Publisher	London & North Eastern Railway
Date	1938

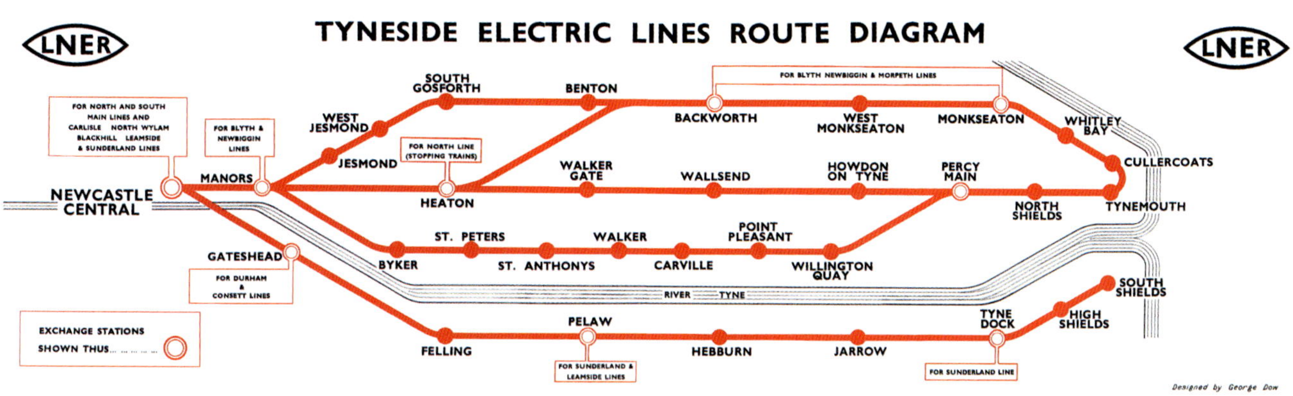

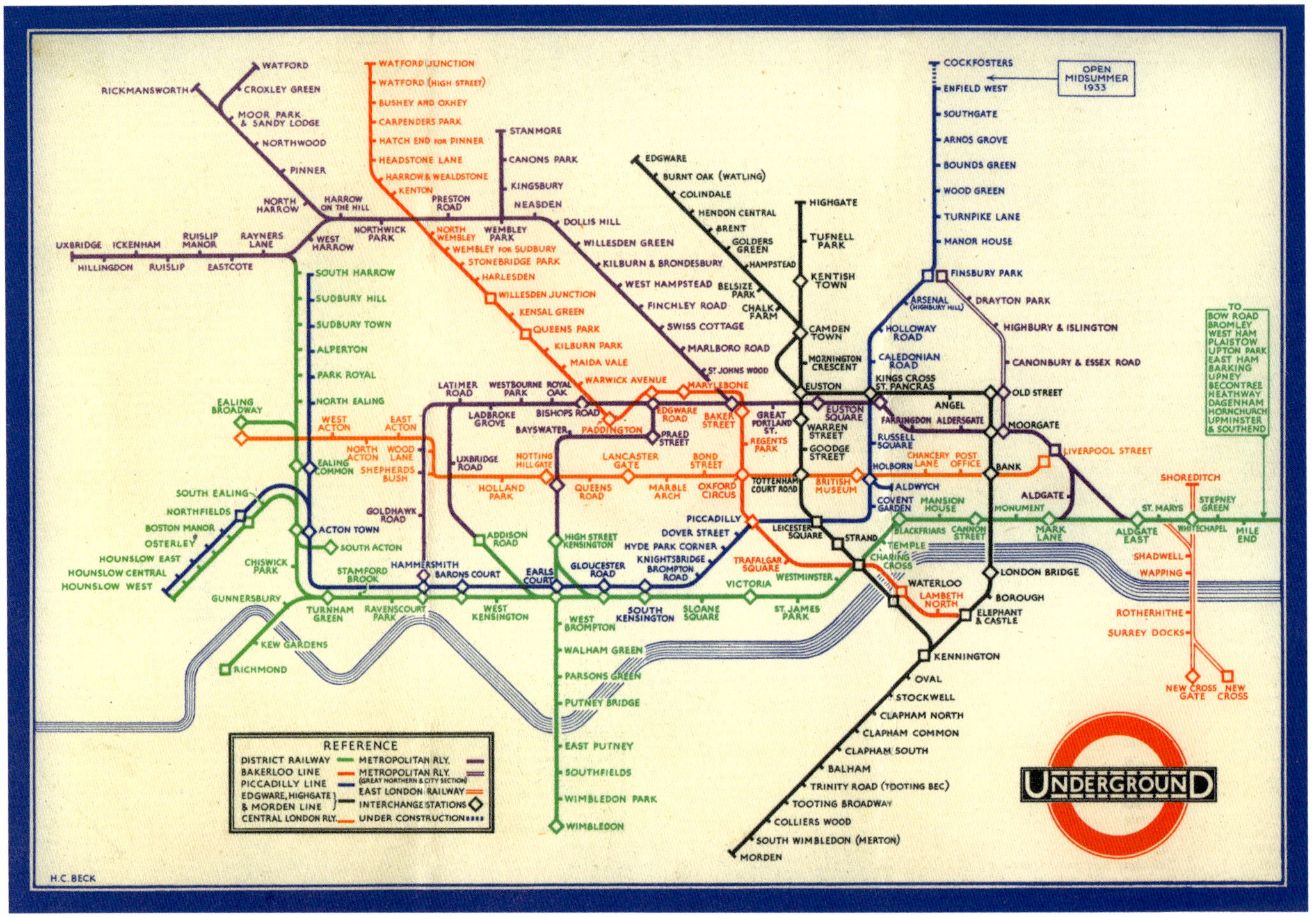

Series	The London Underground Diagram
Mapmaker	Henry C (Harry) Beck
Publisher	The Underground Group
Date	1933

The best known railway map in Britain – and indeed the world – is based on Harry Beck's classic diagrammatic map of the London Underground, first published in 1933, and which quickly became an iconic representation of Britain's capital city. Today's London Underground map is instantly recognizable as a direct descendant of Beck's abstract design of 80 years ago, a map concept which has largely stood the test of time by regular adaptation to changing circumstances. While the current map, as shown overleaf, includes far more information than Beck had to incorporate – inevitably affecting the ease of understanding – the essence of his simplified design, which made it the bench mark for underground system maps worldwide, is still evident.

As we have seen, Beck was far from being the first designer to deploy diagrammatic representation of railways, and as Andrew Dow (George Dow's son) notes in *Telling the Passenger Where to Get Off* (2005), '...it is clear that most or all of the eight lines that made up the Underground system [by 1933] were already represented in individual diagrammatic route maps'. Beck must have been familiar with at least some of them, and interestingly, as a resident of Highgate on the London & North Eastern Railway's Great Northern suburban system, it seems highly likely that he would have seen George Dow's diagrammatic map (see page 194) on the trains in which he travelled, and would have been influenced by them.

However, Beck's good fortune – although in light of his later dealings with London Transport management, he might not have seen it this way – was to pull together coherently a variety of elements of diagrammatic mapping to create a unique map design for one of the busiest rail networks in the world, increasingly heavily used by visitors, and thereby destined to achieve international recognition. Beck's map can be seen as part of a wider drive to apply consistent design standards across the growing London transport system, much of it led by Frank Pick – as Managing Director of the Underground Group from 1928 and from 1933 of the new London Passenger Transport Board, which managed underground railway, bus and tram transport across London after the system was brought into public ownership.

Beck sketched out his first design concept in 1931, initially rejected by the Underground Group, before they recognized that his schematic representation of reality would transform understanding of the system's links and connections – both for Londoners and its many visitors. The key features of the 1933 map and its modern successors are:

- simplification of the route lines to verticals, horizontals or diagonals;
- enlargement of the central area relative to the suburbs;
- elimination of all surface detail except the River Thames.

Beck had to accommodate many enforced changes in the design – some of them ill-advised management interference – as well as making his own improvements. A regular challenge was the extension and augmentation of existing 'tube' lines from 1933 through to 1948, with just one closure in his time – the withdrawal of the shuttle service from Acton Town to South Acton in 1959.

Early design changes, in 1934, involved a change of colouring for the Central Line – always intended as a strong horizontal base for the map – from orange to red, and for the Bakerloo Line from red to brown. The colour representation of the routes open in 1934 has remained largely unchanged ever since, but there have been a number of important design changes since Beck's first vision – such as the use of rings instead of diamonds to denote interchange stations, and the identification of station names in lower case, with an initial capital, rather than upper case.

Since Beck's last map, many Underground network changes have had to be incorporated into 'the diagram', notably the opening of the Victoria and Jubilee lines, and the closure of the Aldwych

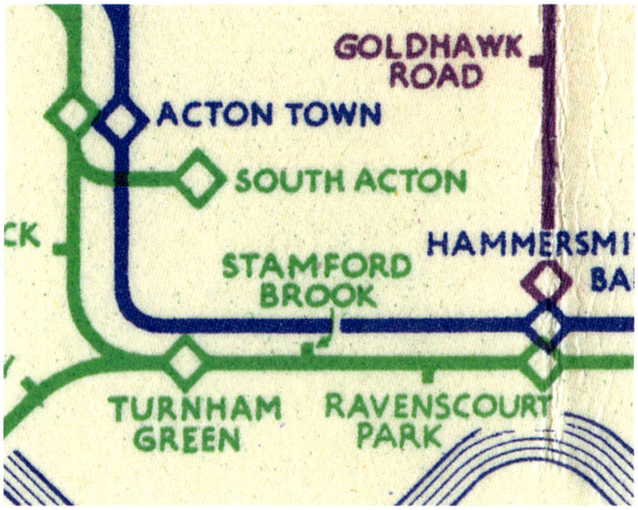

and Epping to Ongar branches. The current version of the diagram shown overleaf has had to accommodate too much additional detail to maintain the essential simplicity of the original concept, notably through the incorporation of the various Docklands Light Railway and 'London Overground' routes. Some modern critics have even suggested that too strict an adherence to Beck's 'rules' – horizontal lines, vertical lines, and diagonal lines at 45 degrees – has inhibited the clarity and utility of the design. However, it seems more likely that these rules have in fact ensured that the essential coherence of the map has been protected from the visual clutter which a large variety of angles would have introduced.

While Beck may have been less of an innovator than typically credited in the last two decades since his 'rediscovery', no doubt the majority of the countless millions of people who have used the Underground map would concur with the sentiments of Ken Garland, in *Mr Beck's Underground Map* (1994), where he concludes that this seminal and enduring map 'achieved both visual distinction and proven usefulness in equal measure'.

HARRY BECK

Henry (Harry) Beck (born 1902) was an out-of-work engineering draughtsman – formerly an employee of the Underground Group – when he first sketched out his design concept in 1931. He returned to the employ of the Group on the eve of the almost immediate success of his first published design in 1933, but amazingly remained a 'temporary' employee until 1937. Being both an employee and a contracted freelance created a status problem for Beck as a designer, and he had an uneasy relationship with management for most of his career. On a number of occasions changes were made to the map design without Beck even being consulted, and 'the diagram' became something of an obsession in his life. As Ken Garland notes, 'he was determined to guard and nurture his creation against all opposition.'

After his 1959 version of the map, other designers were brought in, and increasingly angry exchanges with what was by then London Transport did nothing to improve the relationship. All of Beck's subsequent ideas – including a simpler way of representing the new Victoria Line than that which was adopted – were rejected. He died in 1974, aged 72, a somewhat embittered man, still working on the map design – and it would not be until the 1990s that London Transport fully acknowledged his central role in creating the world's best-known railway map.

Chesham
Chalfont & Latimer
Amersham
Chorleywood
Watford
Croxley
Rickmansworth
Moor Park
Northwood
Northwood Hills
Pinner
North Harrow
Harrow-on-the-Hill
West Harrow
Northwick Park
West Ruislip
Hillingdon
Ruislip
Ruislip Manor
Eastcote
Uxbridge
Ickenham
Rayners Lane
Ruislip Gardens
South Ruislip
Northolt
Greenford
Perivale
Hanger Lane
South Harrow
Sudbury Hill
Sudbury Town
Alperton
Park Royal
North Ealing
Ealing Broadway
Ealing Common
South Ealing
Northfields
Boston Manor
Osterley
Hounslow East
Hounslow Central
Hounslow West
Hatton Cross
Heathrow Terminals 1, 2, 3
Heathrow Terminal 4
Heathrow Terminal 5
Watford Junction
Watford High Street
Bushey
Carpenders Park
Hatch End
Headstone Lane
Harrow & Wealdstone
Kenton
Preston Road
South Kenton
North Wembley
Wembley Central
Stonebridge Park
Harlesden
Willesden Junction
Wembley Park
Kensal Rise
Brondesbury
Kensal Green
Queen's Park
Kilburn Park
Maida Vale
Warwick Avenue
Royal Oak
Westbourne Park
Ladbroke Grove
Latimer Road
Stanmore
Canons Park
Queensbury
Kingsbury
Neasden
Dollis Hill
Willesden Green
Kilburn
Brondesbury Park
West Hampstead
Finchley Road
Swiss Cottage
St. John's Wood
Kilburn High Road
South Hampstead
Edgware
Burnt Oak
Colindale
Hendon Central
Brent Cross
Golders Green
Hampstead
Finchley Road & Frognal
Belsize Park
Chalk Fa
Camden
Mornington Cresc
Hampst Heat
Paddington
Edgware Road
Marylebone
Baker Street
Great Portland Street
Euston
Warren Street
Regent's Park
Goodge Street
Edgware Road
Bayswater
Notting Hill Gate
Lancaster Gate
Bond Street
Oxford Circus
Tottenham Court Road
Queensway
Marble Arch
Holland Park
East Acton
White City
Shepherd's Bush
West Acton
North Acton
Wood Lane
Acton Central
Shepherd's Bush Market
South Acton
Goldhawk Road
Kensington (Olympia)
Acton Town
Hammersmith
Barons Court
High Street Kensington
Hyde Park Corner
Knightsbridge
Green Park
Piccadilly Circus
Gloucester Road
Sloane Square
St. James's Park
Westminster
Chiswick Park
Turnham Green
Stamford Brook
Ravenscourt Park
West Kensington
Earl's Court
South Kensington
Victoria
West Brompton
Gunnersbury
Kew Gardens
Richmond
Fulham Broadway
Parsons Green
Putney Bridge
Imperial Wharf
East Putney
Southfields
Wimbledon Park
Wimbledon
Pimlico
Waterloo
River Thames
Vauxhall
Clapham Junction
Kennington
Clapham North
Stockw
Clapham Common
Clapham South
Balham
Tooting Bec
Tooting Broadway
Colliers Wood
South Wimbledon
Morden
Bakerloo
Central
Circle
District
District open weekends, public holidays and some Olympia events
Hammersmith & City
Jubilee
Metropolitan
Northern
Piccadilly
Victoria
Waterloo & City
DLR
London Overground
Emirates Air Line
Interchange stations
Step-free access from street to train
Step-free access from street to platform
National Rail
Riverboat services
Tramlink
Airport
Emirates Air Line (opening summer 2012)
MAYOR OF LONDON
Website
tfl.gov.uk
24 hour travel information
0843 222 123
© Transport for London
Reg. user No. 12/2182/P
Improvement works may affe

u pay no more than 5p per minute if calling
n a BT landline. There may be a connection charge.
rges from mobiles or other landline providers may vary.

Transport for London

UNDERGROUND

ourney, please check before you travel

Correct at time of going to print

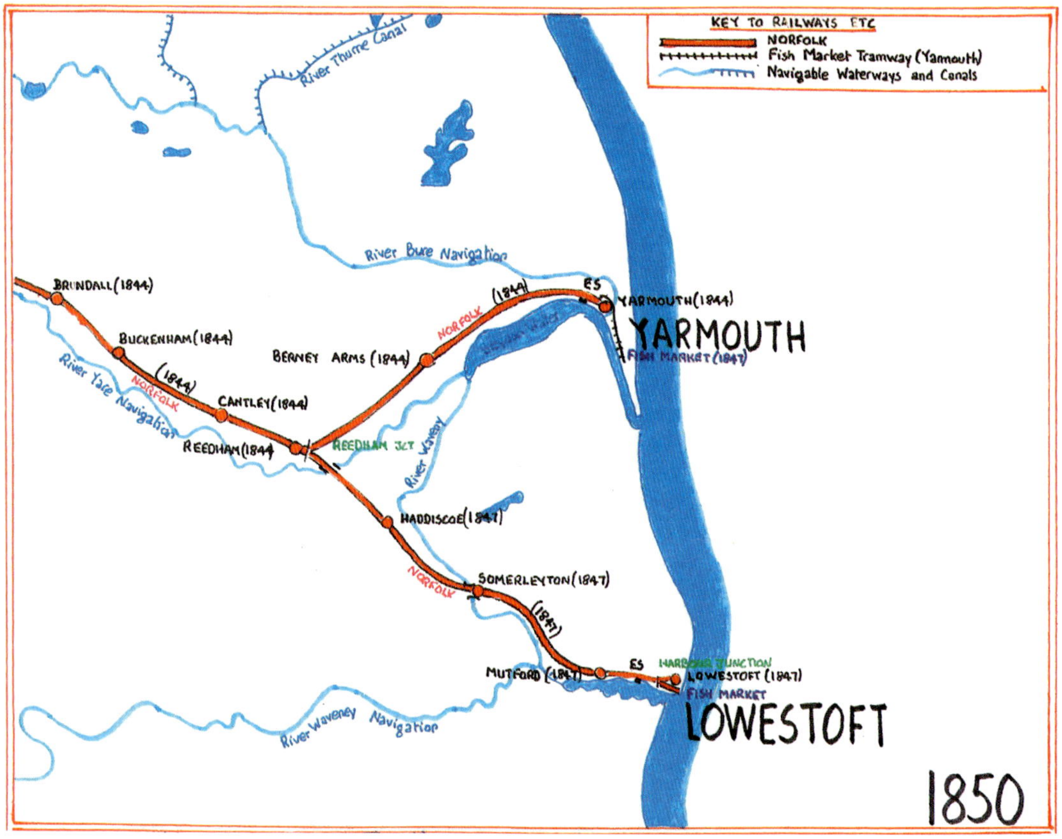
KEY TO RAILWAYS ETC
NORFOLK
Fish Market Tramway (Yarmouth)
Navigable Waterways and Canals
River Thurne Canal
River Bure Navigation
BRUNDALL (1844)
BUCKENHAM (1844)
(1844)
NORFOLK
CANTLEY (1844)
River Yare Navigation
REEDHAM (1844)
REEDHAM JCT
BERNEY ARMS (1844)
NORFOLK
(1844)
ES
YARMOUTH (1844)
YARMOUTH
FISH MARKET (1847)
River Waveney
HADDISCOE (1847)
NORFOLK
SOMERLEYTON (1847)
(1847)
MUTFORD (1847)
ES
HARBOUR JUNCTION
LOWESTOFT (1847)
FISH MARKET
LOWESTOFT
River Waveney Navigation
1850

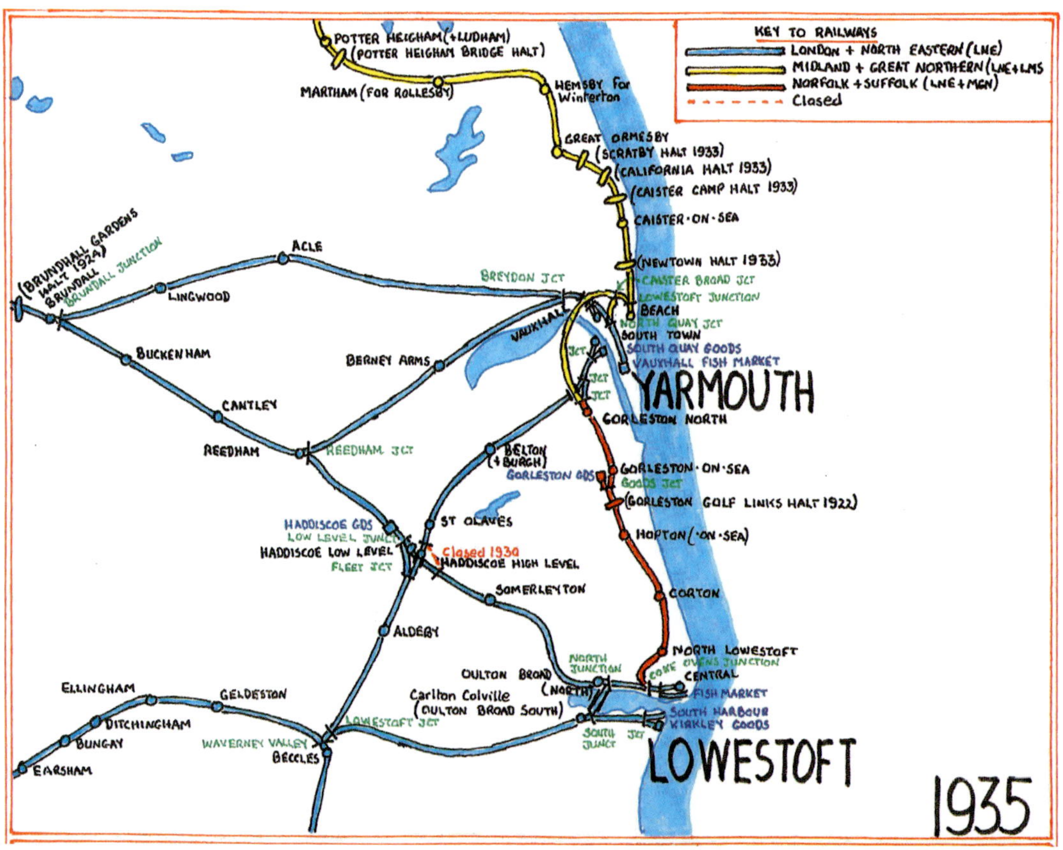
KEY TO RAILWAYS
LONDON + NORTH EASTERN (LNE)
MIDLAND + GREAT NORTHERN (LNE+LMS)
NORFOLK + SUFFOLK (LNE+MGN)
Closed
POTTER HEIGHAM (+LUDHAM)
(POTTER HEIGHAM BRIDGE HALT)
MARTHAM (FOR ROLLESBY)
HEMSBY for Winterton
GREAT ORMESBY
(SCRATBY HALT 1933)
(CALIFORNIA HALT 1933)
(CAISTER CAMP HALT 1933)
CAISTER-ON-SEA
(NEWTOWN HALT 1933)
BREYDON JCT
CAISTER BROAD JCT
LOWESTOFT JUNCTION
BEACH
NORTH QUAY JCT
SOUTH TOWN
SOUTH QUAY GOODS
VAUXHALL FISH MARKET
VAUXHALL
JCT
JCT
JCT
YARMOUTH
GORLESTON NORTH
(BRUNDALL GARDENS HALT 1924)
BRUNDALL
BRUNDALL JUNCTION
LINGWOOD
ACLE
BUCKENHAM
BERNEY ARMS
CANTLEY
REEDHAM
REEDHAM JCT
BELTON (+BURGH)
GORLESTON GDS
GORLESTON-ON-SEA
GOODS JCT
(GORLESTON GOLF LINKS HALT 1922)
ST OLAVES
HOPTON (-ON-SEA)
HADDISCOE GDS
LOW LEVEL JUNCT
HADDISCOE LOW LEVEL
FLEET JCT
Closed 1930
HADDISCOE HIGH LEVEL
SOMERLEYTON
CORTON
ALDEBY
NORTH LOWESTOFT
NORTH JUNCTION
COKE OVENS JUNCTION
OULTON BROAD (NORTH)
CENTRAL
FISH MARKET
ELLINGHAM
GELDESTON
Carlton Colville (OULTON BROAD SOUTH)
SOUTH HARBOUR
KIRKLEY GOODS
DITCHINGHAM
BUNGAY
LOWESTOFT JCT
SOUTH JUNCT
JCT
WAVERNEY VALLEY
BECCLES
EARSHAM
LOWESTOFT
1935

Title	Jowett's Railway Centres (pages 150, 154 and 155)
By	Alan Jowett
Publisher	Patrick Stephens Ltd, Cambridge
Date	1993

These 1850, 1935 and 1990 snapshots of Yarmouth and Lowestoft – at an original scale of 2 miles to the inch – focus on a very short stretch of Norfolk and Suffolk coast which was the object of intensive competition between three railway operators for more than 50 years. The railways opened up this area of East Anglia for two key traffics – holidaymakers and fish – and while the Great Eastern Railway (GER) was pre-eminent for flows from and to London through its core territory to the south, the spoils were shared on the corridor west to the major Midlands markets.

The two competitors for the longer distance traffic were the GER, with routes from Yarmouth and Lowestoft to Norwich then onwards via Thetford and March to Peterborough and the Midlands, and the Midland & Great Northern (joint) Railway (M&GNR) which meandered from Yarmouth through north Norfolk to Wisbech, Spalding and Peterborough. The M&GNR's wish to get to Lowestoft, and competition for tourist traffic at small coastal resorts, led to the construction of yet another joint line, the Norfolk & Suffolk Joint Railway (NSJR), which was the creation of one joint railway (the M&GNR) and the GER.

The 1935 map shows that both joint railways survived the Grouping of 1923, despite being deep in London & North Eastern Railway (LNER) territory. No less than seven new halts – most of them for coastal holiday traffic – were opened between 1922 and 1933. The M&GNR and the NSJR came entirely under the control of the LNER in 1936, but other than the 1959 loss of the M&GNR north of Yarmouth and the short link south from Yarmouth Beach station over Breydon Water to the Yarmouth–Lowestoft line, everything in the map area survived until the Beeching Report in 1963. By 1977, however, the local rail network had shrunk back to its minimum extent in modern times – with only the East Suffolk line from Lowestoft south to Ipswich surviving the threat of closure.

Berney Arms station, on the more minor of the two routes between Norwich and Yarmouth, has the unusual distinction of being one of just three stations on the national rail network to which there is no road access, the others being Corrour on the West Highland Line, and Dovey Junction on the Cambrian Coast Line.

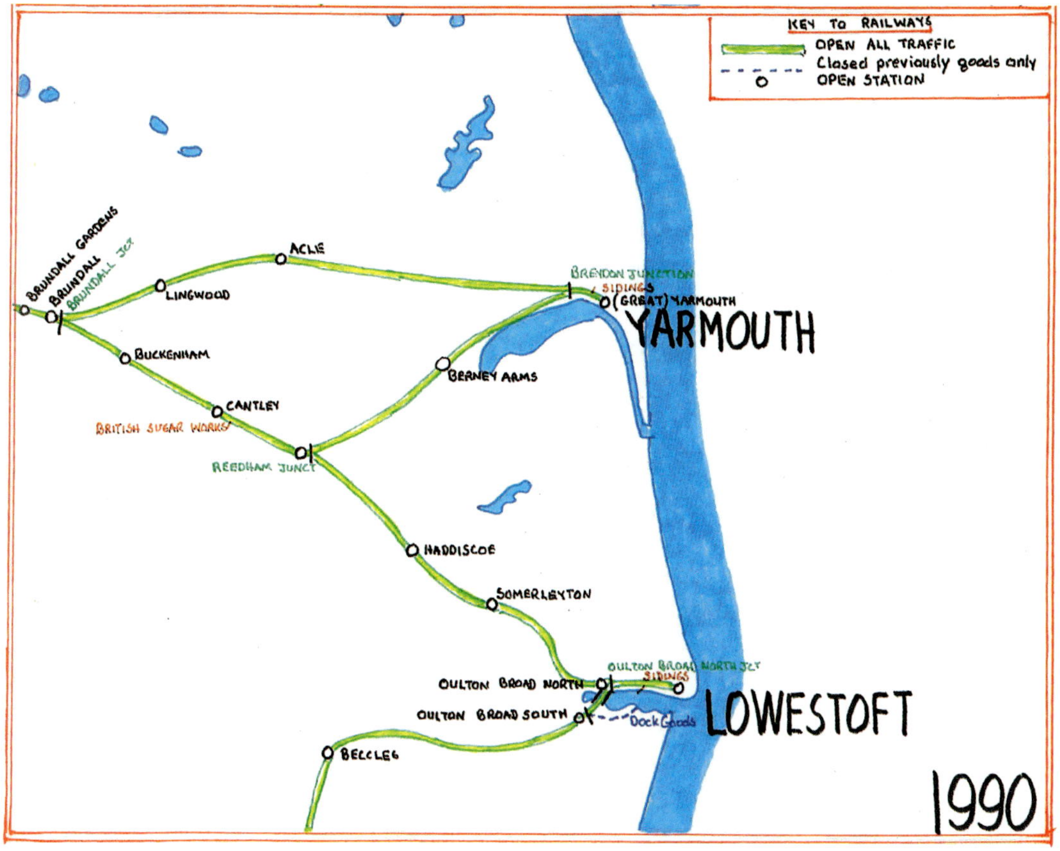

NATIONALIZATION AND MODERNIZATION

Another exhausting war effort was followed by nationalization of the railways under British Railways. The process of upgrading the largely worn-out system culminated in the 1955 Modernisation Plan – which failed to stem the steady tide of traffic losses to fast-growing road competition.

MILEPOST 17	The Decline and Fall of Steam	204
1948	Crewe Junction	206
1949	Pictorial Map Poster of Yorkshire	210
MILEPOST 18	Railway Accidents and Safety	212
1951	British Railways 'Travel By Train' Map	214
1909/53	Ffestiniog Railway	218
MILEPOST 19	Diesel and Electric Dawn	222
1961	Strip Map of the East Coast Main Line	224
1958/66	Millerhill (Edinburgh) Marshalling Yard	226
1963	Eastern England Pictorial Map Brochure	228

1948–1963

Darlington
Richmond
Eaglescliffe
Picton
Battersby
Grosmont
Robin Hood's Bay
Ravenscar
Cloughton
SCARBOROUGH
Catterick Bridge
Bedale
Northallerton
Helmsley
Aysgarth
Leyburn
Thirsk
Pickering
Filey
Speeton
Pilmoor
Gilling
Flamborough Head
Ripon
Alne
Malton
Bridlington
Pateley Bridge
Knaresborough
Driffield
Skipton
Harrogate
YORK
Wetherby
Hornsea
Ilkley
Keighley
Arthington
Church Fenton
Market Weighton
Beverley
Shipley
LEEDS
Selby
South Howden
HULL
BRADFORD
HALIFAX
Dewsbury
Castleford
Withernsea
Goole
New Holland
Knottingley
Pontefract
Spurn Head
Wakefield
Normanton
GRIMSBY
Hemsworth
Rochdale
Barnsley
Scunthorpe
Barnetby
MANCHESTER
Doncaster
Cleethorpes
Penistone
Stalybridge
Swinton
Market Rasen
Rotherham
Louth
SHEFFIELD
Gainsborough
Mablethorpe
Stockport
Sutton-on-
Buxton
Miller's Dale
Worksop
Retford
LINCOLN
Horncastle
Willoughby
Bakewell
Chesterfield
Macclesfield
Tuxford
Bardney
Firsby
Skegness
Congleton
Matlock
Clay Cross
Mansfield
Woodhall Junc.
Matlock Bath
Kirkby-in-Ashfield
Southwell
Newark
Leek
Alfreton
Sleaford
The Wash
Ambergate
Hunstanton
STOKE
Ilkeston
NOTTINGHAM
Ashbourne
Boston
Cheadle
Grantham
DERBY
Wolferton
Uttoxeter
Trent
Harby and Stathern
Repton
Kegworth
Spalding
Burton
Castle Bytham
Ashby-de-la-Zouch
Loughborough
Rugeley
Bourne
King's Lynn
Melton Mowbray
Oakham
Wisbech
LEICESTER
Syston
Stamford
Lichfield
Tamworth
Manton
Walsall
Wansford
March
Sutton Coldfield
Wigston
Uppingham
Peterborough
Seaton
Oundle
Nuneaton
Hinckley
Market Harborough
Holme

On 1 January 1948, the newly nationalized railways of Britain inherited a vast array of steam engine types, some dating to the 1870s. Despite the introduction of standardized parts such as boilers and driving wheels by G. J. Churchward and Charles Collett of the Great Western Railway (GWR) and their protégé, William Stanier of the London, Midland & Scottish Railway (LMS), many of Britain's steam locomotives belonged to a bygone age.

Diesels were still in their early stages of development and it would be some years before suitable designs could be mass-produced in sufficient numbers to replace the tried and tested steam locomotive. The progress of steam locomotive development by the Big Four companies in the 1930s had been stopped dead in its tracks by the onset of the Second World War, and by 1948 steam motive power and the railway system as a whole were in a very rundown state. Despite this, some proven and some groundbreaking designs from the Big Four continued to be built into the British Railways' era – notable among the former were Collett's 'Castle' Class 4-6-0s, Stanier's 'go-anywhere' Class '5MT' (Black Five) 4-6-0s and Peppercorn's Class 'A1' 4-6-2s. Of the latter, Oliver Bulleid's highly innovative 'Merchant Navy' and the lighter 'West Country' and 'Battle of Britain' 4-6-2s, and Hawksworth's 'County' Class 4-6-0s, are worthy of mention.

Meanwhile, in 1947, with nationalization looming, the newly formed Railway Executive appointed Robert Riddles as the first Chief Mechanical & Electrical Engineer of British Railways (BR). His job was to introduce a series of new and efficient steam locomotives using interchangeable standardized parts, thus allowing for the wholesale withdrawal of out-of-date steam types and the consequential savings for the British taxpayer. Drawing on his considerable

British Railways Standard 'Britannia' Class 4-6-2 heads a parcels train up Shap on the West Coast Main Line in June 1966.

experience while working under the LMS's William Stanier and for the Ministry of Supply during the war, Riddles went on to design 12 different classes of locomotives ranging from the humble Class '2MT' 2-6-2Ts right through to the powerful '9F' 2-10-0s.

By 1968 thousands of withdrawn steam locomotives had ended their lives awaiting the cutter's torch at scrapyards around the country.

By 1951 Riddles' new standard class locomotives were starting to emerge from the main locomotive works around the country – notable among these were the 'Britannia' Class '7P6F' 4-6-2s which were soon putting in superlative performances on the Great Eastern main line out of Liverpool Street; the Class '4' and Class '5' 4-6-0s; and the Class '4' 2-6-4Ts. By 1954 all of the 12 new classes had been introduced, culminating in the heavy freight Class '9F' 2-10-0s which mark the high-point of British steam locomotive design. However, a year later the writing was on the wall for steam traction with the publication of the British Transport Commission's Modernisation Plan (see Milepost 18) – this ill-conceived and ill-timed document announced the death sentence for steam and the rapid introduction of, at that time, many types of untried main line diesel.

Construction of new steam locomotives continued until 18 March 1960 when the last example (and the 999th Standard locomotive) to be built by BR, Class '9F' 2-10-0 No. 92220 'Evening Star', emerged from Swindon Works. Just over eight years later the last standard gauge steam locomotives were withdrawn from service on BR – many of Riddles' fine locomotives had had a working life of less than ten years.

The last steam locomotive to be built for British Railways, Standard Class '9F' 2-10-0 'Evening Star', was named at Swindon in 1960.

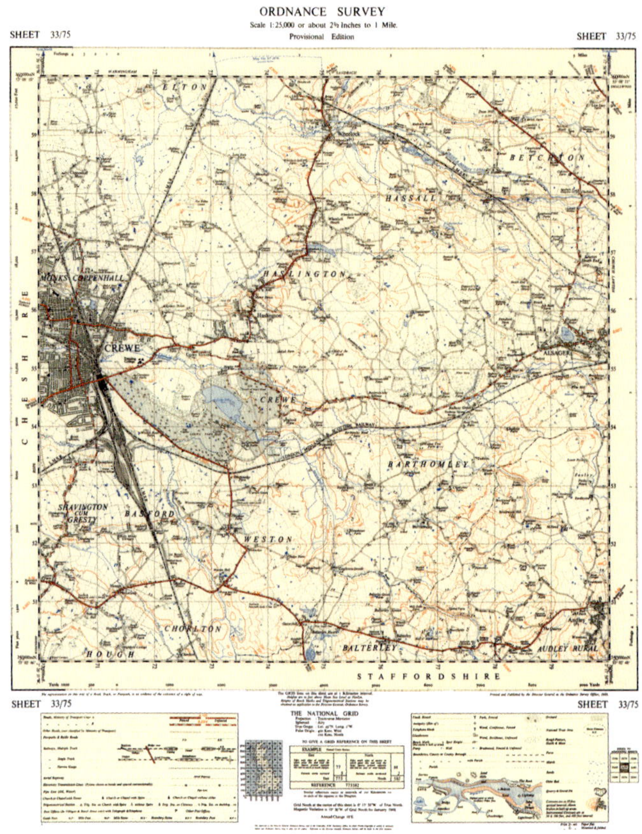

Series	Ordnance Survey 1:25,000
Sheets	33/65 and 33/75
Date	1948

Crewe was a classic railway town, shown in the composite Ordnance Survey map overleaf from the year of nationalization, not long after its heyday. In 1948 all the key components of the railway presence at Crewe were still very much in evidence – a junction of six important lines (as it is to this day), a large works (which, as we have seen in Milepost 9, employed over 20,000 staff during its peak of activity in the 1930s) and extensive freight marshalling yards marked as 'Sorting Sidings'. North and west of the passenger station were the (un-named) steam locomotive running sheds, two belonging to the London, Midland & Scottish Railway (LMS) and a third for the exclusive use of the Great Western Railway, which reached Crewe using running powers over the LMS from nearby Nantwich, and extended onwards to Manchester.

The map demonstrates that despite being a railway town, Crewe's substantial station was by 1948 some distance away from the geographical centre of the town, with two of the rail routes effectively acting as barriers to development to the east. Virtually all the built-up area remained west of the line running northeast to Manchester and north of the line running southwest to Shrewsbury.

Crewe would later benefit directly from the 1955 British Railways Modernisation Plan, with the first stages of overhead electrification of the West Coast Main Line (WCML) concentrated on Crewe. The initial phase to Manchester was electrified in 1960, to Liverpool in 1962, to Euston in 1966 and finally to Birmingham in 1967. Although the completion of the WCML electrification in 1974 – from Weaver Junction, 17 miles to the north, to Glasgow – did not impact on the railway infrastructure at Crewe, there was some consequent downgrading of its passenger interchange function, with fewer Glasgow–London trains calling at Crewe and the principal cross-country 'hub' role shifting to Birmingham's grimly modernized New Street station.

Crewe remains an important railway centre today, with diesel and electric locomotive depots active beside the station; a continuing works presence under the ownership of Bombardier Transportation; and Freightliner container trains re-marshalled day and night at Basford Hall sidings to the south – many of these trains still using the freight-only lines which tunnel under Crewe station.

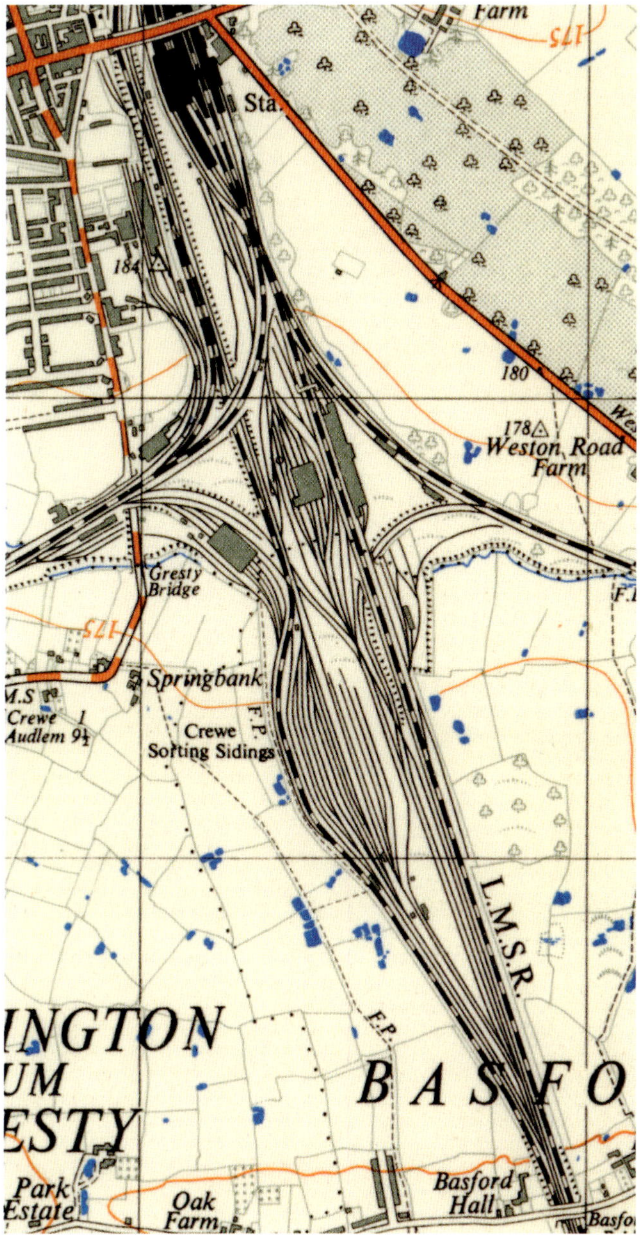

CREW JUNCTION

Until 1965, there was another Crewe which served as a British railway junction. The Caledonian Railway's branches from Granton and Leith joined at Crewe Toll in northwest Edinburgh, and to avoid confusion with its much bigger namesake, the location was described as 'Crew Junction', as shown in this photo taken a few months after closure.

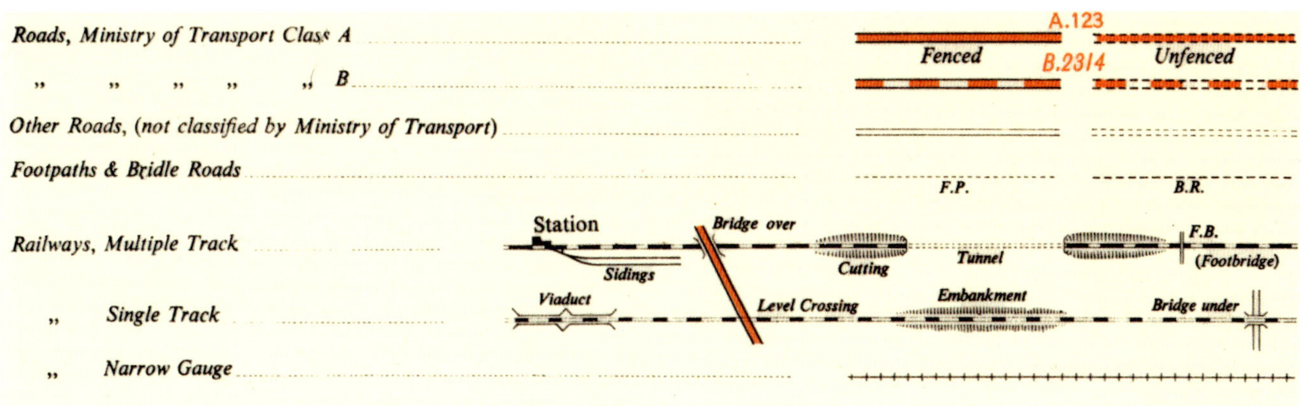

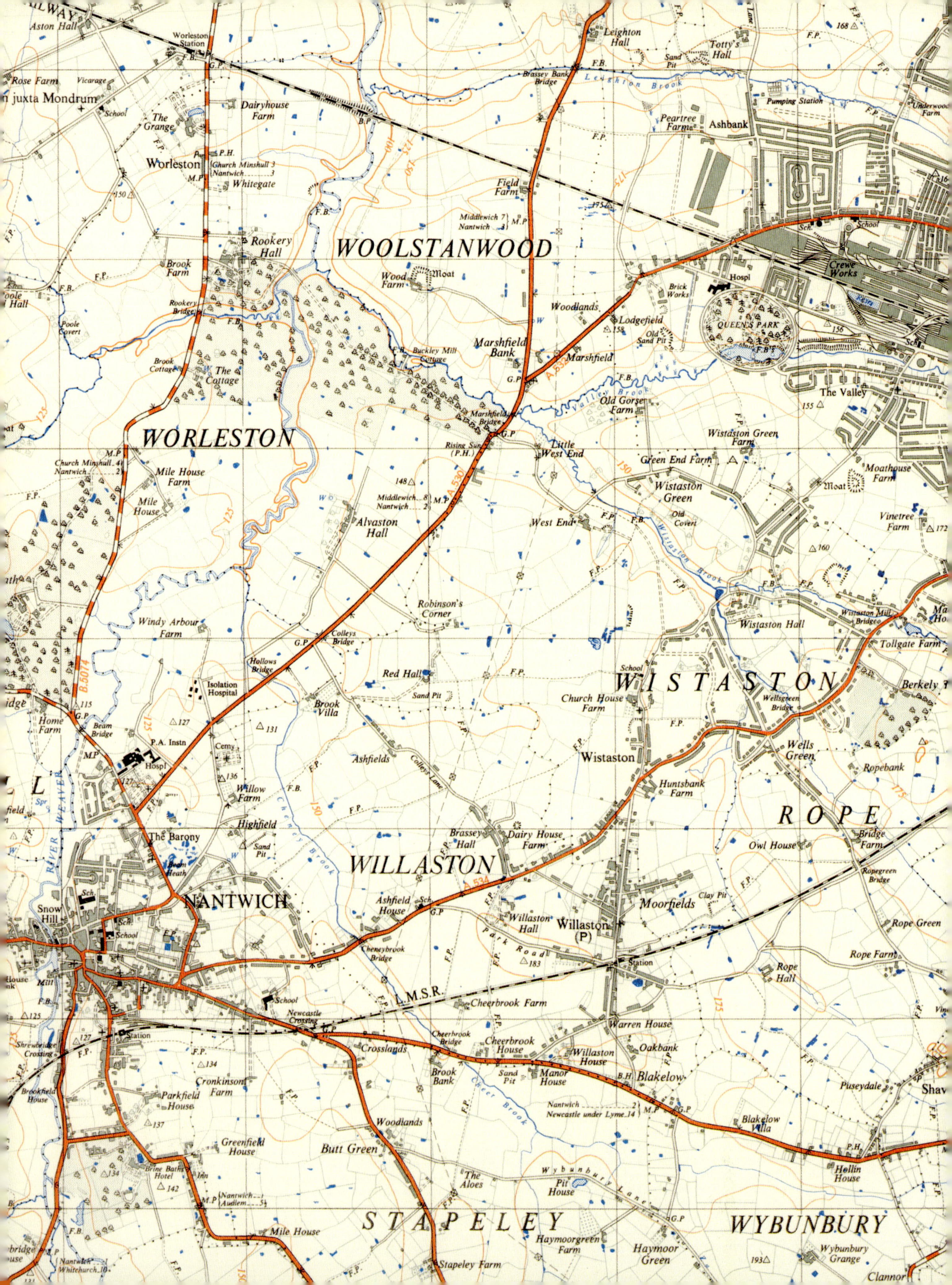

WOOLSTANWOOD
WORLESTON
WISTASTON
WILLASTON
NANTWICH
ROPE
STAPELEY
WYBUNBURY
Aston Hall
Worleston Station
Rose Farm
Vicarage
Dairyhouse Farm
The Grange
Worleston
Whitegate
Leighton Hall
Totty's Hall
Leighton Brook
Brassey Bank Bridge
Pumping Station
Peartree Farm
Ashbank
Field Farm
Rookery Hall
Brook Farm
Wood Farm
Moat
Crewe Works
Hospl
Brick Works
Queen's Park
Woodlands
Lodgefield
Marshfield Bank
Marshfield
Buckley Mill Cottage
Rookery Bridge
Poole Covert
Brook Cottage
The Cottage
Valley Brook
Old Gorse Farm
The Valley
Marshfield Bridge
Rising Sun (P.H.)
Little West End
Wistaston Green Farm
Green End Farm
Mile House Farm
Mile House
Alvaston Hall
West End
Wistaston Green
Old Covert
Moathouse Farm
Vinetree Farm
Wistaston Brook
Robinson's Corner
Windy Arbour Farm
Colleys Bridge
Hollows Bridge
Wistaston Hall
Wistaston Mill Bridge
Tollgate Farm
Red Hall
Sand Pit
Isolation Hospital
Brook Villa
Church House Farm
Wellsgreen Bridge
Berkely
Home Farm
Beam Bridge
P.A. Instn
Cemy
Ashfields
Colleys Lane
Wistaston
Wells Green
Ropebank
Willow Farm
Highfield
Huntsbank Farm
River Weaver
The Barony
Cheney Brook
Brassey Hall
Dairy House Farm
Owl House
Bridge Farm
Ropegreen Bridge
Snow Hill
Ashfield House
Willaston Hall
Willaston
Moorfields
Clay Pit
Rope Green
Park Road
Cheneybrook Bridge
Station
Rope Farm
Rope Hall
L.M.S.R.
Cheerbrook Farm
Newcastle Crossing
School
Warren House
Shrewbridge Crossing
Cheerbrook Bridge
Cheerbrook House
Crosslands
Willaston House
Oakbank
Brookfield House
Parkfield House
Cronkinson Farm
Brook Bank
Manor House
Blakelow
Puseydale
Cheer Brook
Woodlands
Blakelow Villa
Greenfield House
Butt Green
Hollin House
Brine Baths Hotel
The Aloes
Wybunbury Lane
Pit House
Mile House
Haymoorgreen Farm
Haymoor Green
Wybunbury Grange
Stapeley Farm
Clannor

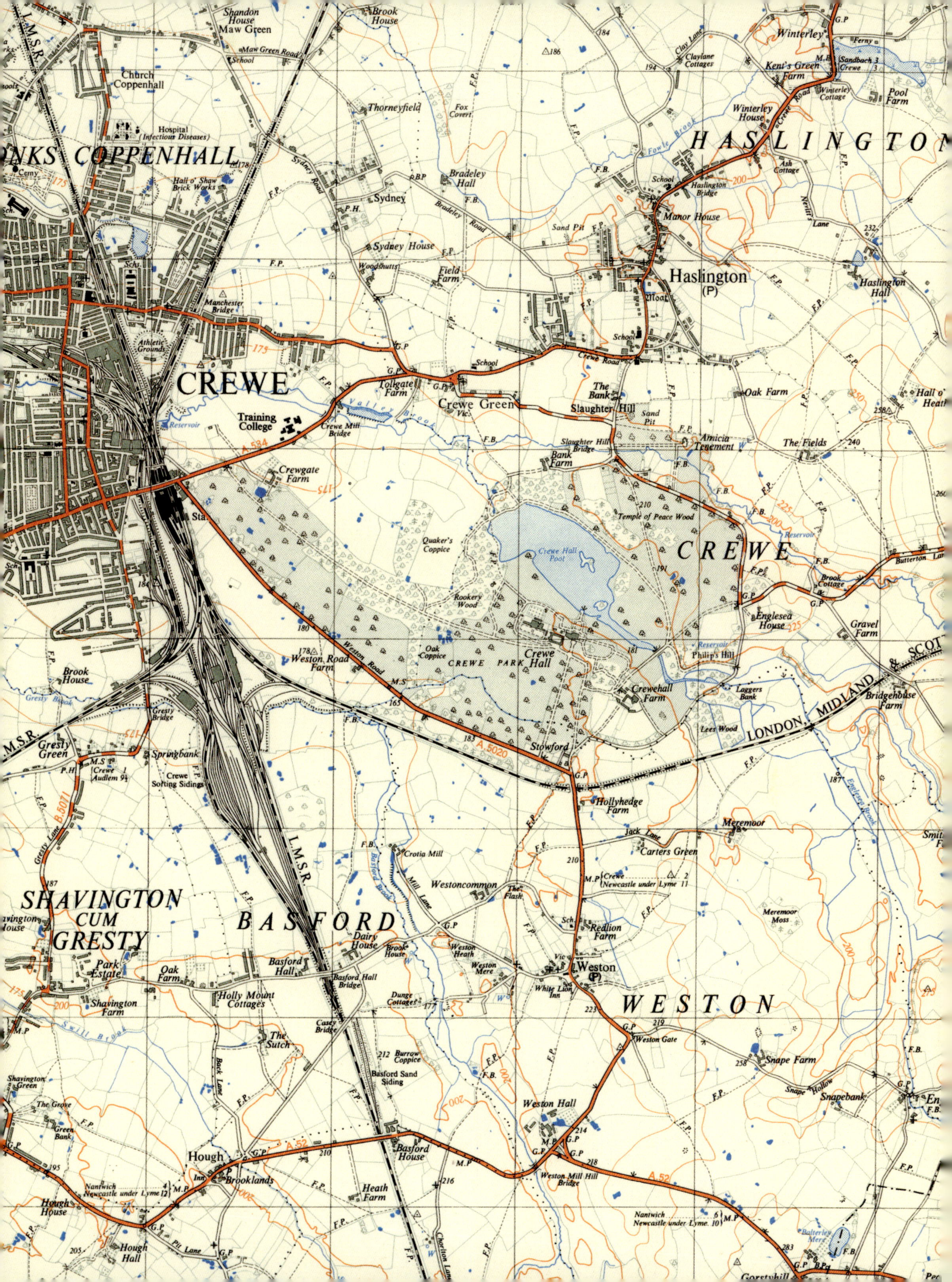

CREWE
HASLINGTON
Haslington (P)
CREWE
CREWE PARK Hall
Crewe Hall Pool
Temple of Peace Wood
Quaker's Coppice
Rookery Wood
Oak Coppice
Crewe Green
Slaughter Hill
Tollgate Farm
Crewegate Farm
Training College
Crewe Mill Bridge
Valley Brook
Weston Road Farm
Weston Road
Stowford
Crewehall Farm
Philips Hill
Englesea House
Gravel Farm
Oak Farm
The Fields
Amicia Tenement
Bank Farm
The Bank
Sydney
Sydney House
Bradeley Hall
Thorneyfield
Fox Covert
Field Farm
Manor House
Haslington Hall
Winterley
Kent's Green Farm
Winterley House
Winterley Cottage
Pool Farm
Claylane Cottages
Church Coppenhall
Shandon House
Maw Green
Brook House
COPPENHALL
Hospital (Infectious Diseases)
Hall o' Shaw Brick Works
Manchester Bridge
Athletic Grounds
L.M.S.R.
LONDON MIDLAND & SCOT
Bridgehouse Farm
Laggers Bank
Lees Wood
Hollyhedge Farm
Meremoor
Carters Green
Meremoor Moss
Crotia Mill
Westoncommon
The Flash
Redlion Farm
Weston (P)
WESTON
White Lion Inn
Weston Gate
Snape Farm
Snapebank
Weston Hall
Weston Mill Hill Bridge
BASFORD
Basford Hall
Basford Hall Bridge
Dairy House
Brook House
Weston Heath
Weston Mere
Dunge Cottages
Casey Bridge
Burrow Coppice
Basford Sand Siding
Basford House
Heath Farm
SHAVINGTON CUM GRESTY
Gresty Green
Gresty Bridge
Springbank
Crewe Sorting Sidings
Brook House
Park Estate
Oak Farm
Holly Mount Cottages
Shavington Farm
The Sutch
Swill Brook
Back Lane
Shavington Green
The Grove
Green Bank
Hough
Brooklands
Hough House
Hough Hall
Pit Lane
Gorstyhill
Balterley Mere
A.534
A.5020
A.52
B.5071

Title	A Map of Yorkshire
Mapmaker	Estra Clark
Publisher	British Railways North Eastern Region
Date	1949

Pictorial maps – where drawings or paintings are used to illustrate locations within the map – represent a distinctive sub-genre of railway mapping, often associated with tourist or leisure markets, and in the more modern era typically designed in a more stylized way than conventional topographical or rail network mapping.

This map – the original of which measures some 50 in. by 40 in. – was created by Estra Clark for British Railways in the early period of nationalization for display on station poster boards. It very effectively combines a slightly simplified geographic representation of all the principal rail routes, at a relatively small scale, with charming painted illustrations of more than a hundred locations throughout Yorkshire – on a backcloth of layered colouring to depict lowland and upland, with important land and sea features also marked and named.

This is not a stereotypical tourist map, in that major industries and services are proudly illustrated beside the towns with which they were associated. The map extract opposite features town sketches extolling Bradford wool, Halifax carpet making and the 'fine worsted fabrics, chemicals & dyestuffs' of Huddersfield – whose past sporting glories are evoked by the illustration of a footballer in the distinctive blue and white stripes of Huddersfield Town AFC. However, it was the Yorkshire Dales, the North Sea coast, and towns like Skipton, Scarborough and Whitby which continued to draw tourists by train in Yorkshire, as they had done since Victorian times.

At its full size, this map poster is a feast of revelations about the settlements, landscape features, history and folk tales that make up Yorkshire – as well as providing an attractive picture of just how much of the county was accessible by rail in 1949.

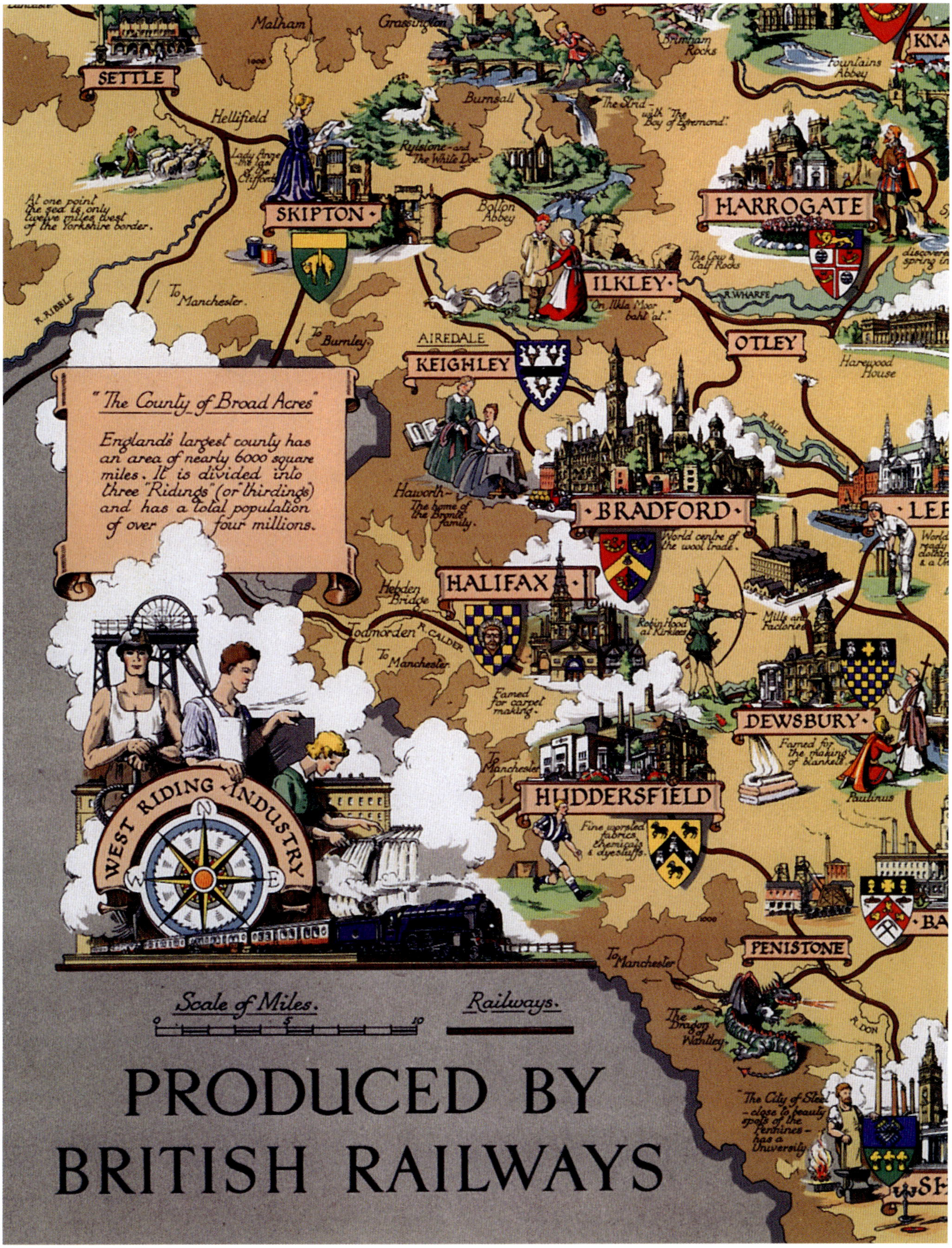
SETTLE
Malham
Hellifield
Burnsall
Rylstone – and The White Doe.
SKIPTON
At one point the sea is only twelve miles west of the Yorkshire border.
To Manchester.
R. RIBBLE
To Burnley.
Bolton Abbey
ILKLEY
On Ilkla Moor baht 'at.
HARROGATE
Fountains Abbey
The Cow & Calf Rocks
R. WHARFE
OTLEY
Harewood House
AIREDALE
KEIGHLEY
"The County of Broad Acres"
England's largest county has an area of nearly 6000 square miles. It is divided into three "Ridings" (or thirdings) and has a total population of over four millions.
Haworth – The home of the Brontë family.
BRADFORD
World centre of the wool trade.
HALIFAX
Hebden Bridge
Todmorden
R. CALDER
To Manchester
Famed for carpet making.
Robin Hood at Kirklees
Mills and Factories
DEWSBURY
Famed for the making of blankets.
Paulinus
WEST RIDING INDUSTRY
HUDDERSFIELD
Fine worsted fabrics, chemicals & dyestuffs.
To Manchester
PENISTONE
The Dragon of Wantley.
R. DON
Scale of Miles.
Railways.
PRODUCED BY BRITISH RAILWAYS

As events in the last two decades at Ladbroke Grove, Hatfield and Potters Bar have shown, railway safety continues to be a major issue in the modern era of the high-speed train. However, compared to other forms of land transport the railways are by far the safest way to travel. Since the early days of the railways many lessons have been learnt and put into practice, with new safety technology continually evolving.

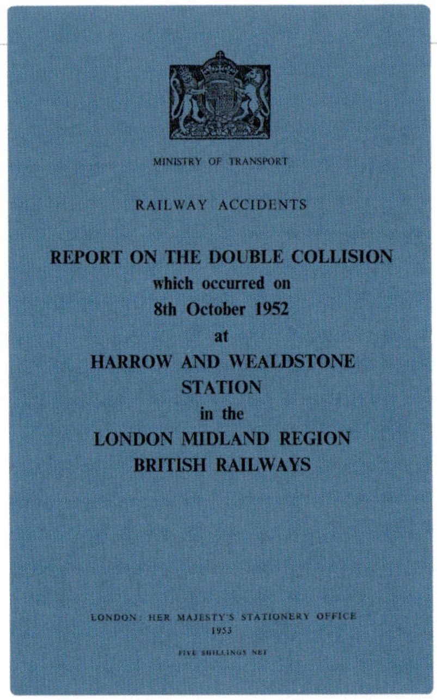

MINISTRY OF TRANSPORT

RAILWAY ACCIDENTS

REPORT ON THE DOUBLE COLLISION
which occurred on
8th October 1952
at
HARROW AND WEALDSTONE STATION
in the
LONDON MIDLAND REGION
BRITISH RAILWAYS

LONDON: HER MAJESTY'S STATIONERY OFFICE
1953
FIVE SHILLINGS NET

The first known railway accident to a passenger occurred on 15 September 1830 when the unfortunate politician, William Huskisson, was run over by George Stephenson's 'Rocket' at Parkside station during the opening ceremony of the Liverpool & Manchester Railway – he later died in hospital from his injuries. Safety on these early railways was either non-existent or at best extremely basic with collisions, runaway trains, mechanical failure, excessive speed, derailments and human error all common events. The first attempts at early signalling – a vital ingredient in railway safety – were rudimentary to say the least, with signalmen holding flags or lamps positioned along the line. The introduction of the electric telegraph in 1848, the token system for single line railways – to avoid head-on collisions only one driver was allowed to be in possession of the metal token that was issued for each stretch of track between passing stations – and John Saxby's invention of interlocking signals and points in 1856 were all key aspects in railway safety. The gradual introduction in the 1850s and 1860s of block signalling – this ensured that trains travelling in the same direction on double-tracked railway lines were kept far enough apart to avoid collisions – was a major step forward, and became mandatory in 1889 following the Armagh disaster in Ireland when a runaway train killed 78 passengers.

Britain's worst railway disaster occurred during the First World War at Quintinshill, ten miles north of Carlisle, when correct signalling practice was not followed. On 22 May

The triple train crash at Harrow & Wealdstone station in 1952, when 112 people died, was Britain's worst peacetime rail disaster.

1915, a speeding southbound troop train collided head-on with a stationary local train. The wooden carriages of the troop train, lit by acetylene gas, completely disintegrated but worse was to follow when a northbound express not only ploughed into the wreckage but also hit a stationary goods train in the adjoining loop. Hot cinders from the derailed locomotives ignited the acetylene gas and an inferno ensued – 227 people were killed including 215 soldiers from the Royal Scots Regiment. Because the accident happened during the war it was not publicized, but the two signalmen on duty were found guilty of gross neglect and culpable homicide.

Britain's worst peacetime railway disaster occurred at Harrow & Wealdstone station on 8 October 1952, when a stationary local train was hit in the rear by a southbound Perth to Euston sleeping car train. Seconds after the crash a northbound express hit the wreckage – 112 people were killed and 340 injured.

The second worst peacetime railway disaster happened in thick fog at St John's, Lewisham, on 4 December 1957, when a suburban electric train was hit in the rear by a steam-hauled Cannon Street to Ramsgate express – 90 people were killed and 173 injured. Driver error (the steam locomotive driver had passed a red signal) and the lack of an Automatic Warning System in the steam locomotive were to blame.

In more recent times the head-on crash between a High Speed Train (InterCity 125) express and a diesel multiple unit at Ladbroke Grove outside Paddington on 5 October 1999 killed 31 people and injured 523. The accident occurred after the driver of the diesel multiple unit had passed a red signal – poorly positioned signals, poor driver training procedures and driver error were all to blame.

Since Ladbroke Grove, Hatfield and Potters Bar, safety on Britain's railways has seen a marked improvement. In the most recent statistics (2009/10) there were no fatalities for either passengers or workforce in train accidents while only five passengers died in accidents at stations and there were only three workforce fatalities. However, by far the most dangerous place on the railways is at level crossings where the number of deaths to pedestrians and motorists has been rising steadily since 2003 – in 2009/10 there were 12 public fatalities while in the first four months of 2011 alone a similar number had already been killed in this way. Of course, the vast majority of level crossing accidents are caused by road users disobeying the rules. Sadly the highest mortality rate on the railways are suicides or suspected suicides with a staggering 236 being recorded in 2009/10.

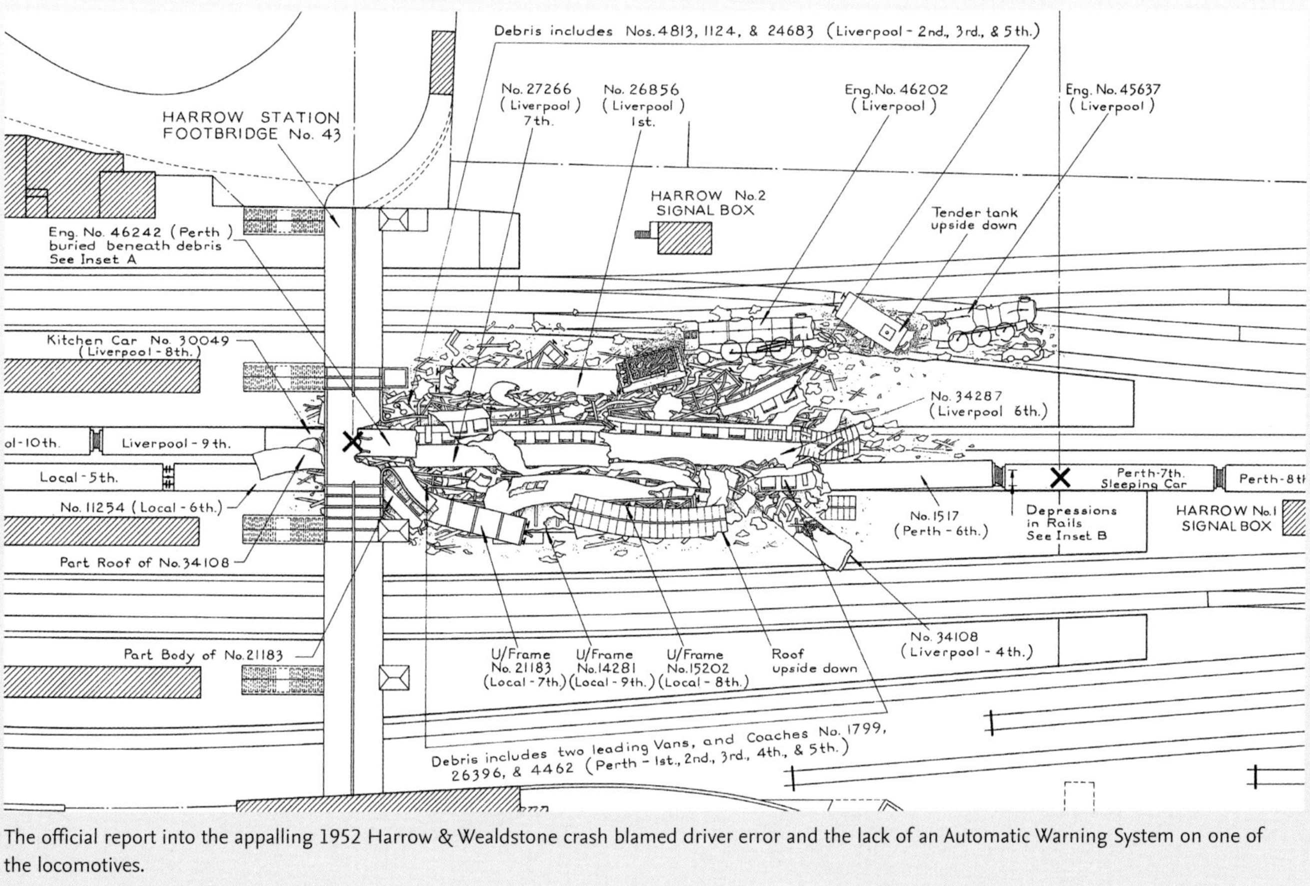

The official report into the appalling 1952 Harrow & Wealdstone crash blamed driver error and the lack of an Automatic Warning System on one of the locomotives.

Title	British Railways – travel by train: Passenger facilities and Map
Publisher	The Railway Executive
Date	1951

While rail closures – due to both road competition and industrial change – had become a growing phenomenon from the 1920s onwards, a very substantial rail network remained in 1951 throughout rural and urban Britain. But the newly nationalized railway was generally in poor shape after a massive war effort, and the big investment of the 1955 Modernisation Plan had yet to come. Many lines were in their last years – unbeknown to staff or passengers.

Every part of Britain suffered from the closure of rail routes after 1951, and the pain was not spread evenly. A striking feature of this map and its extract, is the extent to which the Home Counties north of London lost so many cross-country routes which could have been playing a crucial role in today's transport system. These generally more affluent areas were among the first to benefit from the 'never had it so good' era of growing household incomes – and rising car ownership – which sealed the fate of so many railway routes. The irony is that 30 years later – when the routes of most closed lines had been breached by roads, industry and housing development – road congestion was becoming a big problem which cross-country railways could have alleviated, but it was generally too late to retrieve the situation.

As comparison with page 240 demonstrates, many routes in the 'commuter belt' north of London had succumbed *before* the Beeching Report. Very few were terminating branch lines, and most provided lateral connections between the many main routes radiating out of London to the rest of the country. A notable pre-Beeching loss was the Northampton–Bedford line which could have formed part of an important east–west cross-country route, but the key cut came with a route which even the Beeching Report had not proposed for closure – the Oxford–Cambridge line, built by the London & North Western Railway. During the Second World War, the Government chose Bletchley Park as the location for its crucial base for monitoring German military signals, in part because Bletchley station was virtually equidistant between the two university centres of intellectual excellence.

Other than the reprieved Bletchley–Bedford section, all passenger services between Oxford and Cambridge were withdrawn in 1968 by a Labour Government which had opposed most of Beeching's cuts while in opposition. Freight services survived on the western section only, and passenger trains were reinstated by British Rail between Oxford and Bicester in 1987. Local authorities and development agencies have been campaigning for complete reinstatement of the 'East West Rail' route, initially between Bletchley and Bicester, so that at long last travellers who wish to make direct journeys across England north of London and south of Birmingham can once again do so by rail. To the delight of campaigners, in late 2011 the Chancellor of the Exchequer announced that the western section of the projected route would now be part of the National Infrastructure Plan – re-establishing a strategic railway linking Oxford, Reading and Aylesbury with Milton Keynes and Bedford.

The map was to remain as the standard design of British Railways' network map until 1967. The version shown overleaf has been enhanced to show the boundaries of the new British Railways' regions.

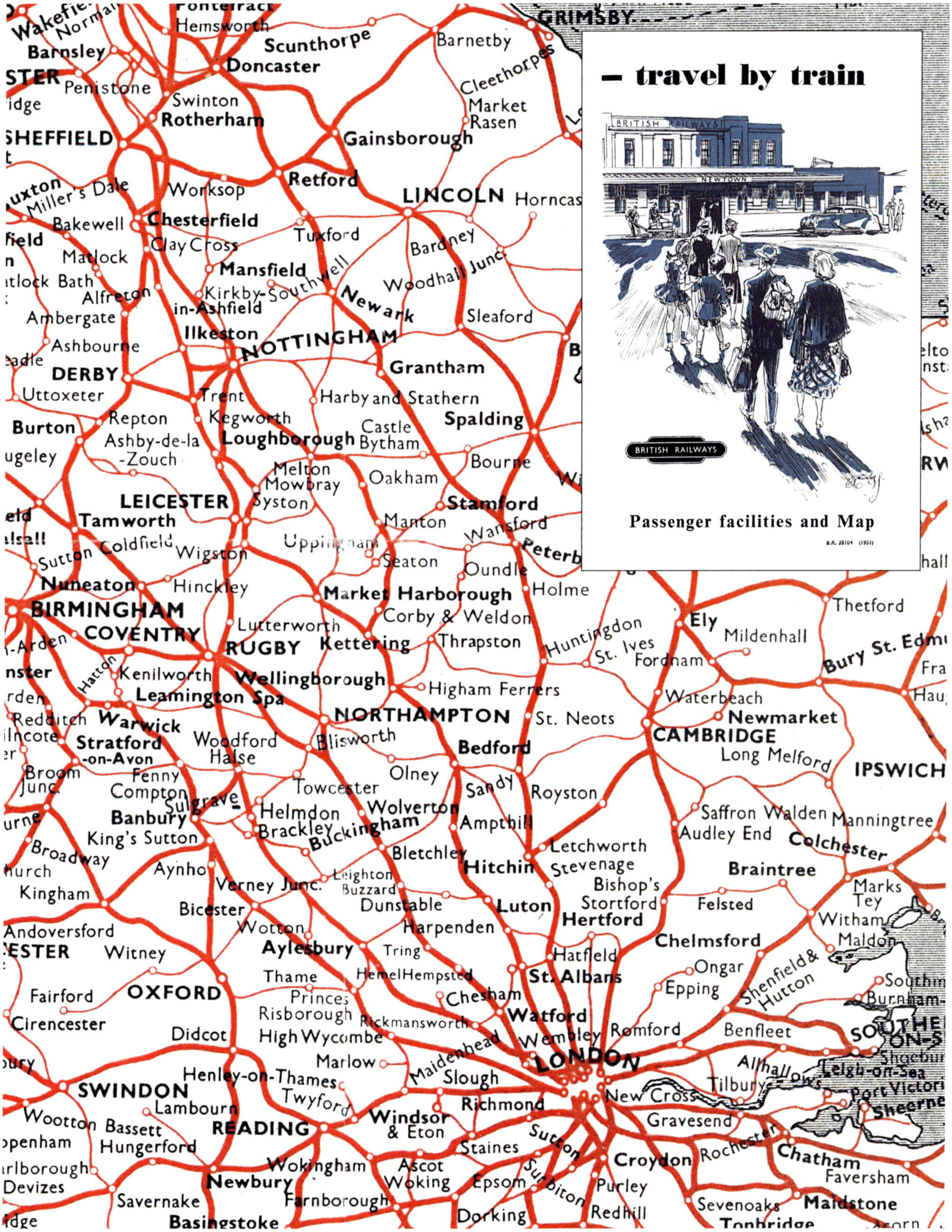
GRIMSBY
Hemsworth
Barnsley
Scunthorpe
Barnetby
Doncaster
Penistone
Cleethorpes
Swinton
Market Rasen
Rotherham
SHEFFIELD
Gainsborough
Miller's Dale
Worksop
Retford
LINCOLN
Bakewell
Chesterfield
Tuxford
Clay Cross
Bardney
Matlock
Mansfield
Woodhall Junc.
Matlock Bath
Kirkby-in-Ashfield
Southwell
Newark
Alfreton
Ambergate
Sleaford
Ilkeston
Ashbourne
NOTTINGHAM
DERBY
Grantham
Uttoxeter
Trent
Harby and Stathern
Burton
Repton
Kegworth
Castle Bytham
Spalding
Ashby-de-la-Zouch
Loughborough
Melton Mowbray
Bourne
Oakham
LEICESTER
Syston
Stamford
Tamworth
Manton
Wansford
Sutton Coldfield
Wigston
Seaton
Oundle
Nuneaton
Hinckley
Holme
Market Harborough
BIRMINGHAM
Lutterworth
Corby & Weldon
Thetford
COVENTRY
Ely
RUGBY
Kettering
Thrapston
Huntingdon
Mildenhall
St. Ives
Fordham
Bury St. Edm
Hatton
Kenilworth
Wellingborough
Higham Ferrers
Leamington Spa
Waterbeach
Warwick
NORTHAMPTON
St. Neots
Newmarket
Redditch
CAMBRIDGE
Stratford-on-Avon
Woodford Halse
Blisworth
Bedford
Long Melford
IPSWICH
Broom Junc.
Fenny Compton
Olney
Towcester
Sandy
Royston
Sulgrave
Wolverton
Banbury
Helmdon
Saffron Walden
Manningtree
Brackley
Buckingham
Ampthill
Audley End
Colchester
King's Sutton
Letchworth
Broadway
Bletchley
Hitchin
Stevenage
Braintree
Aynho
Leighton Buzzard
Bishop's Stortford
Marks Tey
Kingham
Verney Junc.
Bicester
Dunstable
Luton
Felsted
Witham
Hertford
Maldon
Wotton
Harpenden
Andoversford
Chelmsford
Witney
Aylesbury
Tring
Hatfield
Ongar
Thame
Hemel Hempsted
St. Albans
Shenfield & Hutton
Fairford
OXFORD
Princes Risborough
Chesham
Epping
Cirencester
Rickmansworth
Watford
Didcot
High Wycombe
Wembley
Romford
Benfleet
Marlow
Maidenhead
LONDON
Allhallows
Henley-on-Thames
Slough
Leigh-on-Sea
SWINDON
Tilbury
Port Victoria
Lambourn
Twyford
Richmond
New Cross
Sheerness
Wootton Bassett
Windsor & Eton
Gravesend
Hungerford
READING
Sutton
Rochester
Chatham
Staines
Croydon
Wokingham
Ascot
Surbiton
Faversham
Devizes
Newbury
Woking
Epsom
Purley
Savernake
Farnborough
Sevenoaks
Maidstone
Dorking
Redhill
Basingstoke
Tonbridge
– travel by train
BRITISH RAILWAYS
NEWTOWN
BRITISH RAILWAYS
Passenger facilities and Map

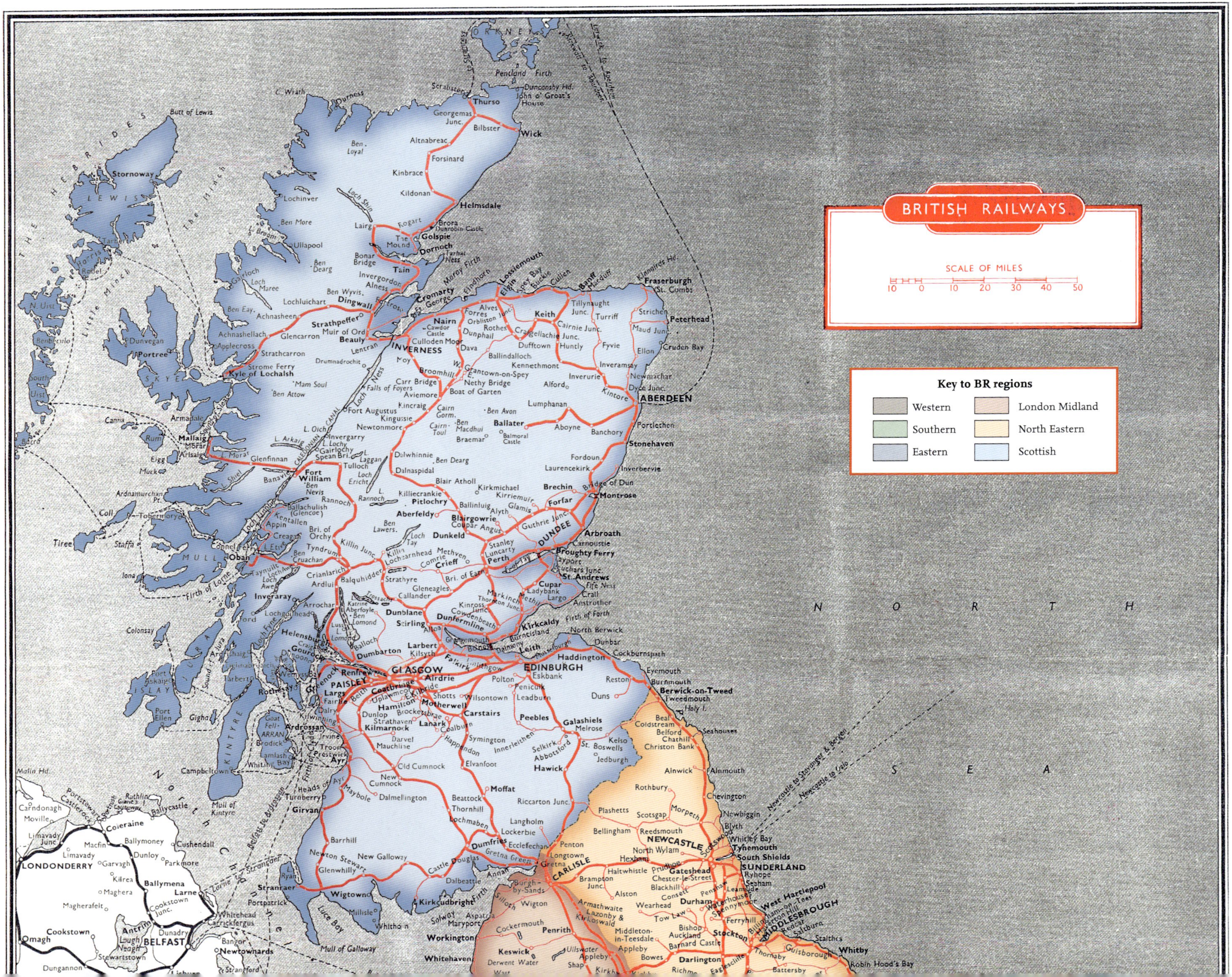

BRITISH RAILWAYS
SCALE OF MILES
10 0 10 20 30 40 50
Key to BR regions
Western
Southern
Eastern
London Midland
North Eastern
Scottish

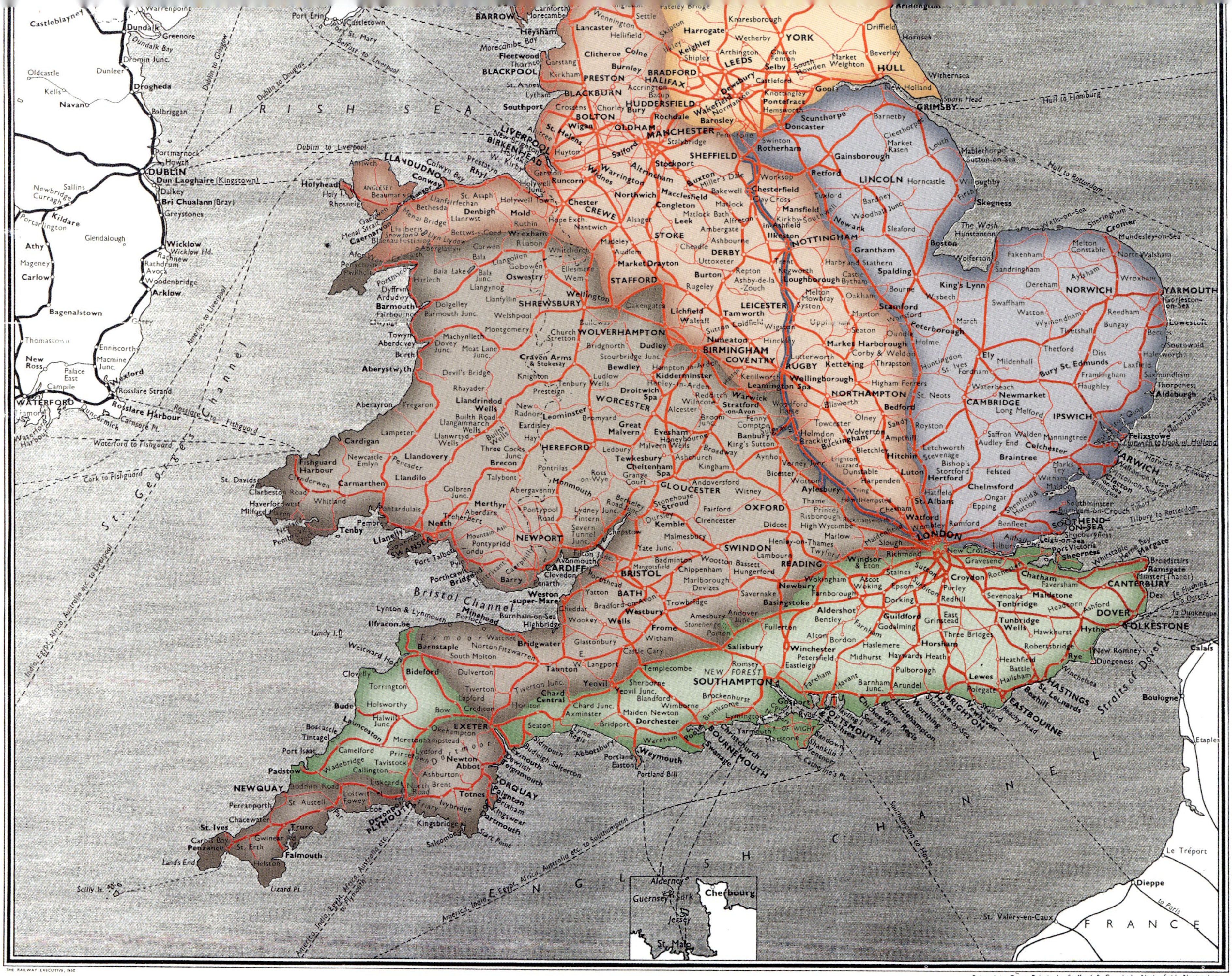

THE RAILWAY EXECUTIVE, 1950

Printed in Great Britain by Stafford & Co., Ltd., Netherfield, Nottingham.

The Ffestiniog (FR) was a railway of firsts – the first British narrow gauge line (using horse power for its first 27 years), the first successful application of the double Fairlie-type articulated steam locomotives (1870), and the first British railway to use bogie carriages (1873). Narrow gauge – in this case, just 1 ft 11½ in. between the rails – was much better suited to the difficult mountain terrain, using much sharper curves than was possible with standard (4 ft 8½ in.) gauge, and allowing construction at much lower capital cost.

Although built for slate traffic, the FR ran passenger trains from 1864 until 1939, the outbreak of war in that year curtailing the tourist traffic which had by then become its mainstay. By the time the Ordnance Survey (OS) produced the 1953 1 in. map, from which an extract is reproduced overleaf, the FR had lain dormant for seven years. The OS used a barely discernible 'ladder' representation for narrow gauge railways from the 1930s onwards – and this hardly does justice to the prodigious efforts made by the railway, over the 110 years from 1836, in moving slate from the Blaenau quarries down to the harbour at Porthmadog for export (133,000 tons moved by rail in the peak year of 1894). One needs to go to more detailed maps to appreciate the scale of the undertaking involved.

At the 6 in. scale shown overleaf, the above-ground impact of the slate industry around Blaenau Ffestiniog can be clearly seen, not least in the extensive spoil heaps which (still) so dominate the landscape there. Also evident are the standard gauge lines which reached the town from Llandudno Junction (to the north) and Bala Junction (to the south), in 1879 and 1883 respectively – thereby ending the FR's monopoly of the slate traffic. Down at Porthmadog (then anglicized as Portmadoc), the extract opposite of the 1909 25 in. OS map shows in great detail the rail-served wharves on each bank of the river, at a time when slate production was still near its peak volume, albeit shortly to go into long-term decline.

Following complete closure in 1946, the line was saved for preservation and a modest tourist service started across the Cob causeway between Porthmadog and Boston Lodge in 1955, later extended to allow direct connections from British Railways' Cambrian Coast Line trains at Minffordd. Since that time the railway has gone from strength to strength, with re-opening through to Blaenau Ffestiniog – and direct connection with BR services north to Llandudno Junction – taking place in 1982.

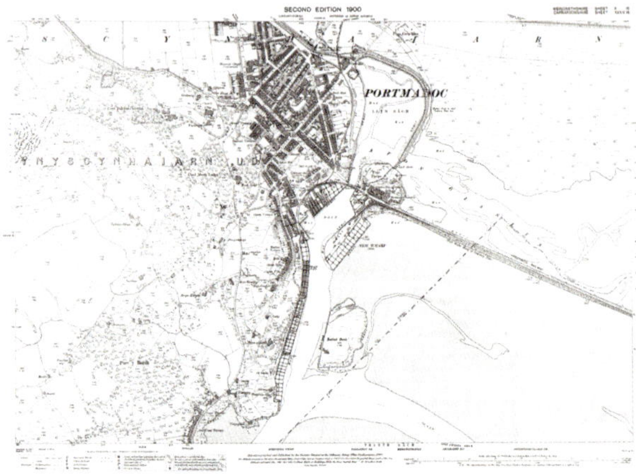

Ordnance Survey 1909 (Carnarvonshire) – see extract opposite

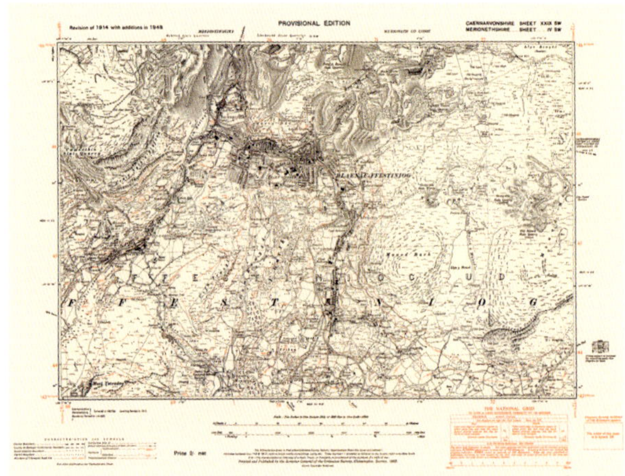

Ordnance Survey 1953 (Caernarvonshire) – see extract overleaf

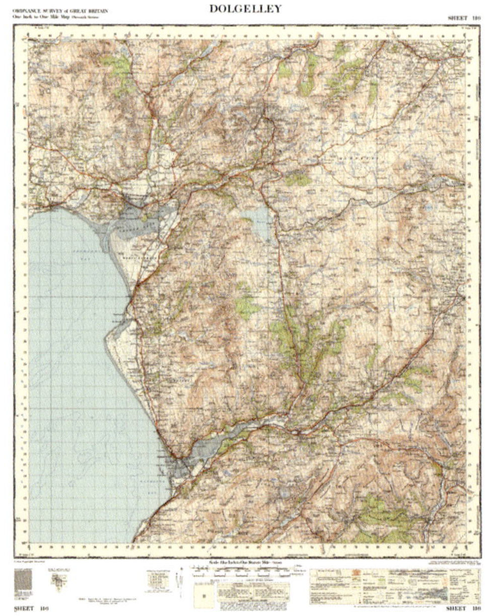

Ordnance Survey 1953 (Dolgelley) – see extract overleaf

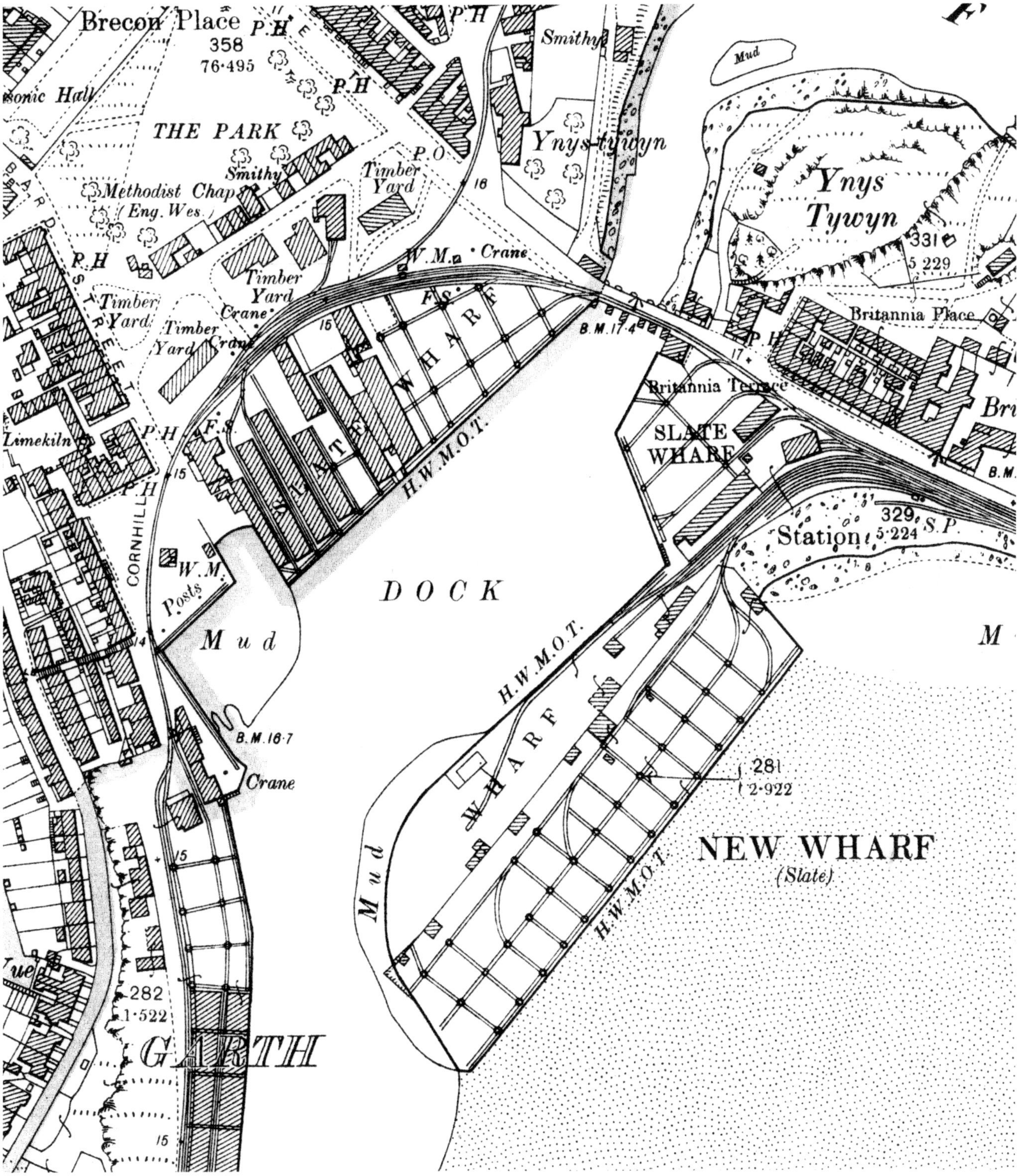

Series	Ordnance Survey Second Edition 1900. 25 in. to 1 mile
Sheet	Carnarvonshire [sic] XXXIV.16
Date	Surveyed 1877, revised 1899, reprinted 1909

Series	Ordnance Survey revision of 1914 with additions in 1949. 6 in. to 1 mile
Sheet	Caernarvonshire XXIX SW
Date	Surveyed 1887–8, levelling revised 1913, printed and published 1953

Series	Ordnance Survey Seventh Series. 1 in. to 1 mile
Sheet	116 Dolgelley
Date	1953

Apart from the London suburban and Brighton line electrification on the Southern Railway (SR), a few diesel shunters, some Great Western Railway railcars and two main line diesel prototypes being developed by the London, Midlands & Scottish Railway (LMS) and the SR, the railway scene in Britain on 1 January 1948 was 100 per cent steam powered. Lagging behind the USA by at least a decade, the advent of diesel haulage on British Railways was slow in coming and, when it did, was ill-conceived and poorly implemented.

The war and its aftermath had killed off any hopes of main-line electrification and, apart from the ex-London & North Eastern Railway's Woodhead route across the Pennines which was 'switched on' in 1954, steam power continued to rule the rails. This was soon to change.

Published on 24 January 1955, the British Transport Commission's *Modernisation and Re-Equipment of British Railways*, a 35-page 'blueprint' document, recommended a £1.24 billion (£24.8 billion at today's prices) investment over a period of 15 years to modernize and re-equip British Railways. The document outlined five key areas for modernization, including £345 million for replacement of steam locomotives by diesel and electric traction.

Of these two types of traction, diesel won the day, as the cost of implementing widespread electrification was enormous. While diesels were more expensive to build than steam locomotives, their benefits compared to steam were obvious – they were cleaner, their running costs were lower and their range was much greater.

This was all well intentioned – albeit involving vast amounts of taxpayers' money – but the authors of the Modernisation Plan had failed not only to take into account the almost complete lack of expertise in building diesel locomotives in Britain, but also the selfish and autonomous aspirations of some regions of British Railways – in fact they were still being run more-or-less on the same lines as the old Big Four prior to nationalization.

While Ivatt's two ex-LMS prototype mainline diesel-electrics, Nos. 10000 and 10001, and Bulleid's three examples, Nos 10201–10203, belched their fumes up and down the West Coast Main Line (WCML), the drawing offices in Britain's railway industry were working overtime to design some of the biggest calamities ever to run on Britain's railways. Despite the lessons to be learnt from steam locomotive design, standardization was never part of the master plan and within a few years perfectly good and, in many cases, nearly new steam locomotives were being sent to the scrapyard to make way for a motley crew of first generation diesel-hydraulic and diesel-electric locomotives. These locomotives

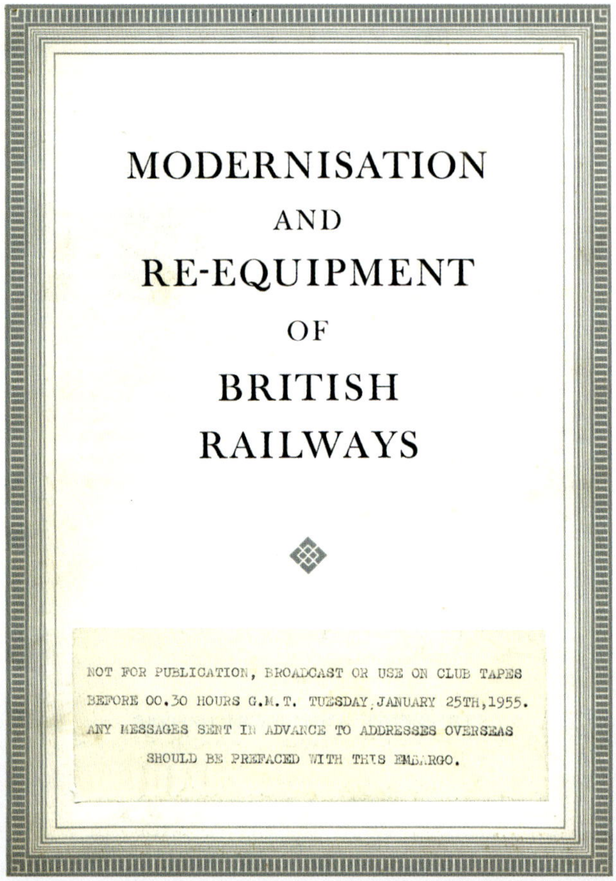
MODERNISATION
AND
RE-EQUIPMENT
OF
BRITISH
RAILWAYS

NOT FOR PUBLICATION, BROADCAST OR USE ON CLUB TAPES
BEFORE 00.30 HOURS G.M.T. TUESDAY, JANUARY 25TH, 1955.
ANY MESSAGES SENT IN ADVANCE TO ADDRESSES OVERSEAS
SHOULD BE PREFACED WITH THIS EMBARGO.

The 1955 Modernisation Plan for British Railways recommended the replacement of steam locomotives by diesel and electric traction.

were categorized according to power output, with Type 1 the least powerful and Type 5 the most powerful.

By far the biggest culprit was the Western Region which decided to opt for lightweight diesel-hydraulic transmission based on a successful German design. But there the similarity ended, as the early British examples (Type 4 'Warships' and Type 2 'Baby Warships'), built by the North British Locomotive Company in Glasgow and by BR at Swindon, were unreliable, expensive to maintain and of a non-standard design. Introduced in 1961, the Beyer-Peacock Type 3 'Hymeks' were a vast improvement but they and the more powerful 'Westerns' had a short working life, with the last being withdrawn in 1977.

Opting for heavier diesel-electric transmission, other regions had more success than Swindon with the London Midland Region's (LMR) Sulzer-engined Type 4 'Peaks' and the Eastern Region's (ER) English Electric Types 1 and 4 and Brush Type 2s being reasonably successful (the latter when re-engined), and some of which still survive today. Introduced in 1961, the Type 5 'Deltics' were impressive machines which transformed East Coast Main Line services, but even they only had a working life of 20 years. Probably the most successful design was the humble diesel shunter (now classified 08) which was introduced in 1953 and is still a familiar sight the length and breadth of the network.

To be fair to Swindon, other regions also had their spectacular failures, notable among them being the Scottish Region's Type 2s whose chronic unreliability brought about the collapse of their builders, the world-renowned North British Locomotive Company; the same region's Clayton Type 1s; the LMR's Metrovick Type 2s; and the ER's English Electric Type 2 'Baby Deltics'. All in all, this was a sad story.

Fortunately BR eventually got it right with many of their second-generation diesels, some of which, such as the English-Electric Type 3s (or Class 37s in current parlance) and Brush Type 4s (Class 47s) can still be seen in service today, but by the late 1980s the writing was on the wall for British diesel locomotive manufacturers. Now Britain buys its locomotives from the North American companies of General Electric and General Motors.

The introduction of modern electric services was much slower despite the obvious benefits such as energy efficiency, absence of pollution, unlimited range and rapid acceleration. Unfortunately the Government of the day was not prepared to invest in the initial high cost of widespread electrification of BR and because of this only a limited, but successful, programme was embarked upon, with new 25 kV schemes for the WCML and suburban lines out of Liverpool Street being some of the earliest. Starting in 1959, electrification of the WCML was not completed through from Euston to Glasgow until 1974, while full electrification of the former Great Eastern main line from Liverpool Street to Norwich was completed in 1987. Elsewhere in the country the Glasgow electric suburban 'Blue Trains' were 'switched on' between 1960 and 1967 and the Southern Region third-rail system was extended to Bournemouth in 1967. Electrification of the East Coast Main Line was completed in 1991, but only a few short 'infill' schemes have been undertaken since then – other than in Scotland where re-opened lines to Larkhall (2005) and from Airdrie to Bathgate (2010) have been electric since services resumed.

Built by the North British Locomotive Company, the short-lived 'Warship' diesel-hydraulic locomotives were introduced on the Western Region in 1958.

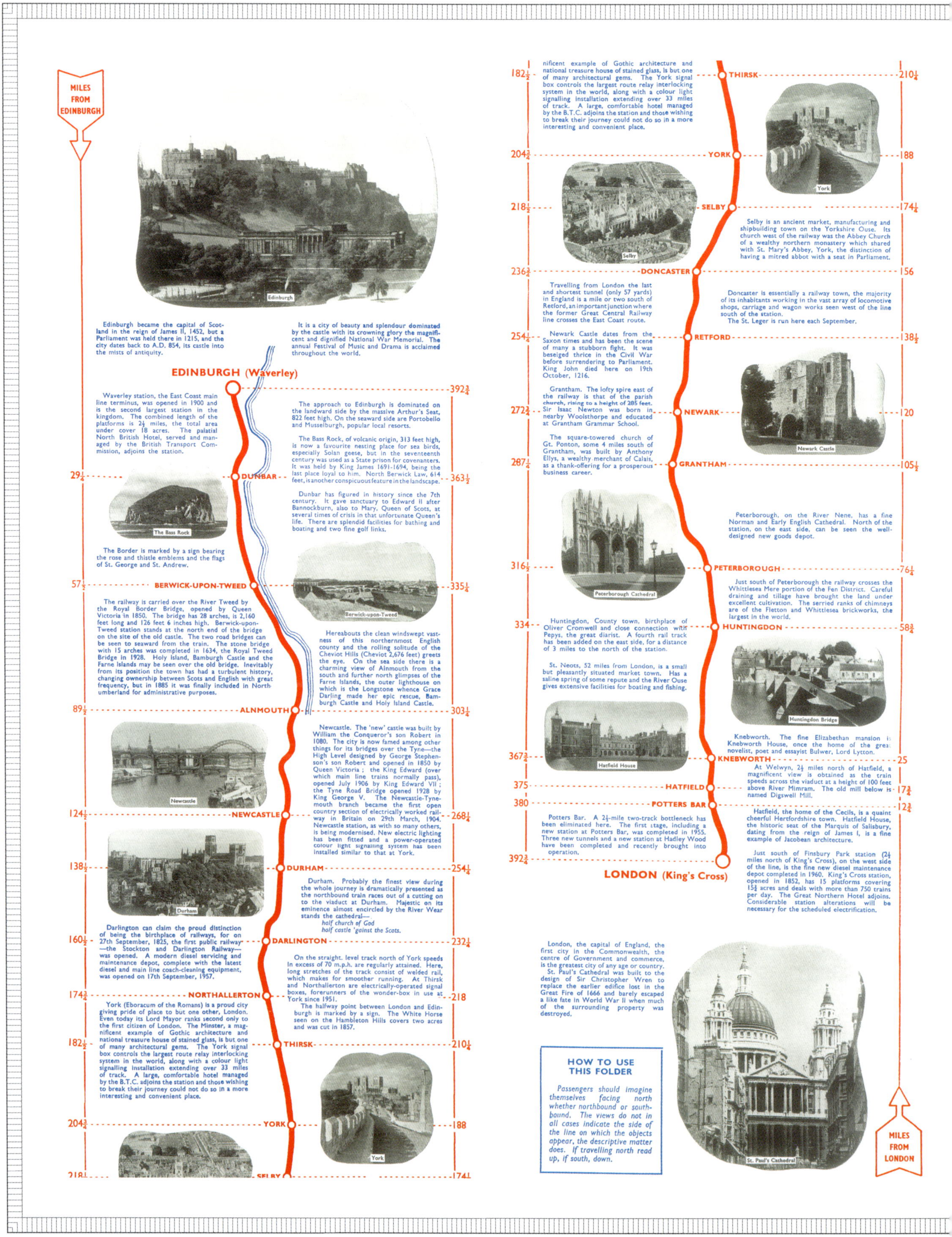

MILES FROM EDINBURGH

Edinburgh

Edinburgh became the capital of Scotland in the reign of James II, 1452, but a Parliament was held there in 1215, and the city dates back to A.D. 854, its castle into the mists of antiquity.

It is a city of beauty and splendour dominated by the castle with its crowning glory the magnificent and dignified National War Memorial. The annual Festival of Music and Drama is acclaimed throughout the world.

EDINBURGH (Waverley) 392¾

Waverley station, the East Coast main line terminus, was opened in 1900 and is the second largest station in the kingdom. The combined length of the platforms is 2¼ miles, the total area under cover 18 acres. The palatial North British Hotel, served and managed by the British Transport Commission, adjoins the station.

The approach to Edinburgh is dominated on the landward side by the massive Arthur's Seat, 822 feet high. On the seaward side are Portobello and Musselburgh, popular local resorts.

The Bass Rock, of volcanic origin, 313 feet high, is now a favourite nesting place for sea birds, especially Solan geese, but in the seventeenth century was used as a State prison for covenanters. It was held by King James 1691-1694, being the last place loyal to him. North Berwick Law, 614 feet, is another conspicuous feature in the landscape.

29¼ DUNBAR 363½

The Bass Rock

Dunbar has figured in history since the 7th century. It gave sanctuary to Edward II after Bannockburn, also to Mary, Queen of Scots, at several times of crisis in that unfortunate Queen's life. There are splendid facilities for bathing and boating and two fine golf links.

The Border is marked by a sign bearing the rose and thistle emblems and the flags of St. George and St. Andrew.

57½ BERWICK-UPON-TWEED 335¼

Berwick-upon-Tweed

The railway is carried over the River Tweed by the Royal Border Bridge, opened by Queen Victoria in 1850. The bridge has 28 arches, is 2,160 feet long and 126 feet 6 inches high. Berwick-upon-Tweed station stands at the north end of the bridge on the site of the old castle. The two road bridges can be seen to seaward from the train. The stone bridge with 15 arches was completed in 1634, the Royal Tweed Bridge in 1928. Holy Island, Bamburgh Castle and the Farne Islands may be seen over the old bridge. Inevitably from its position the town has had a turbulent history, changing ownership between Scots and English with great frequency, but in 1885 it was finally included in Northumberland for administrative purposes.

Hereabouts the clean windswept vastness of this northernmost English county and the rolling solitude of the Cheviot Hills (Cheviot 2,676 feet) greets the eye. On the sea side there is a charming view of Alnmouth from the south and further north glimpses of the Farne Islands, the outer lighthouse on which is the Longstone whence Grace Darling made her epic rescue, Bamburgh Castle and Holy Island Castle.

89½ ALNMOUTH 303¼

Newcastle

Newcastle. The 'new' castle was built by William the Conqueror's son Robert in 1080. The city is now famed among other things for its bridges over the Tyne—the High Level designed by George Stephenson's son Robert and opened in 1850 by Queen Victoria; the King Edward (over which main line trains normally pass), opened July 1906 by King Edward VII; the Tyne Road Bridge opened 1928 by King George V. The Newcastle-Tynemouth branch became the first open country section of electrically worked railway in Britain on 29th March, 1904. Newcastle station, as with so many others, is being modernised. New electric lighting has been fitted and a power-operated colour light signalling system has been installed similar to that at York.

124½ NEWCASTLE 268¼

Durham

138½ DURHAM 254¼

Durham. Probably the finest view during the whole journey is dramatically presented as the northbound train races out of a cutting on to the viaduct at Durham. Majestic on its eminence almost encircled by the River Wear stands the cathedral—

half church of God
half castle 'gainst the Scots.

Darlington can claim the proud distinction of being the birthplace of railways, for on 27th September, 1825, the first public railway—the Stockton and Darlington Railway—was opened. A modern diesel servicing and maintenance depot, complete with the latest diesel and main line coach-cleaning equipment, was opened on 17th September, 1957.

160½ DARLINGTON 232¼

On the straight, level track north of York speeds in excess of 70 m.p.h. are regularly attained. Here, long stretches of the track consist of welded rail, which makes for smoother running. At Thirsk and Northallerton are electrically-operated signal boxes, forerunners of the wonder-box in use at York since 1951.

174¾ NORTHALLERTON 218

The halfway point between London and Edinburgh is marked by a sign. The White Horse seen on the Hambleton Hills covers two acres and was cut in 1857.

York (Eboracum of the Romans) is a proud city giving pride of place to but one other, London. Even today its Lord Mayor ranks second only to the first citizen of London. The Minster, a magnificent example of Gothic architecture and national treasure house of stained glass, is but one of many architectural gems. The York signal box controls the largest route relay interlocking system in the world, along with a colour light signalling installation extending over 33 miles of track. A large, comfortable hotel managed by the B.T.C. adjoins the station and those wishing to break their journey could not do so in a more interesting and convenient place.

182½ THIRSK 210¼

York

204¾ YORK 188

218½ SELBY 174¼

nificent example of Gothic architecture and national treasure house of stained glass, is but one of many architectural gems. The York signal box controls the largest route relay interlocking system in the world, along with a colour light signalling installation extending over 33 miles of track. A large, comfortable hotel managed by the B.T.C. adjoins the station and those wishing to break their journey could not do so in a more interesting and convenient place.

182½ THIRSK 210¼

York

204¾ YORK 188

Selby

218½ SELBY 174¼

Selby is an ancient market, manufacturing and shipbuilding town on the Yorkshire Ouse. Its church west of the railway was the Abbey Church of a wealthy northern monastery which shared with St. Mary's Abbey, York, the distinction of having a mitred abbot with a seat in Parliament.

236¾ DONCASTER 156

Travelling from London the last and shortest tunnel (only 57 yards) in England is a mile or two south of Retford, an important junction where the former Great Central Railway line crosses the East Coast route.

Doncaster is essentially a railway town, the majority of its inhabitants working in the vast array of locomotive shops, carriage and wagon works seen west of the line south of the station.

The St. Leger is run here each September.

254¼ RETFORD 138½

Newark Castle dates from the Saxon times and has been the scene of many a stubborn fight. It was beseiged thrice in the Civil War before surrendering to Parliament. King John died here on 19th October, 1216.

Grantham. The lofty spire east of the railway is that of the parish church, rising to a height of 285 feet. Sir Isaac Newton was born in nearby Woolsthorpe and educated at Grantham Grammar School.

272¾ NEWARK 120

Newark Castle

The square-towered church of Gt. Ponton, some 4 miles south of Grantham, was built by Anthony Ellys, a wealthy merchant of Calais, as a thank-offering for a prosperous business career.

287¼ GRANTHAM 105½

Peterborough Cathedral

Peterborough, on the River Nene, has a fine Norman and Early English Cathedral. North of the station, on the east side, can be seen the well-designed new goods depot.

316½ PETERBOROUGH 76¼

Just south of Peterborough the railway crosses the Whittlesea Mere portion of the Fen District. Careful draining and tillage have brought the land under excellent cultivation. The serried ranks of chimneys are of the Fletton and Whittlesea brickworks, the largest in the world.

334 HUNTINGDON 58¾

Huntingdon, County town, birthplace of Oliver Cromwell and close connection with Pepys, the great diarist. A fourth rail track has been added on the east side, for a distance of 3 miles to the north of the station.

Huntingdon Bridge

St. Neots, 52 miles from London, is a small but pleasantly situated market town. Has a saline spring of some repute and the River Ouse gives extensive facilities for boating and fishing.

Hatfield House

Knebworth. The fine Elizabethan mansion is Knebworth House, once the home of the great novelist, poet and essayist Bulwer, Lord Lytton.

367¾ KNEBWORTH 25

At Welwyn, 2¼ miles north of Hatfield, a magnificent view is obtained as the train speeds across the viaduct at a height of 100 feet above River Mimram. The old mill below is named Digswell Mill.

375 HATFIELD 17¾

380 POTTERS BAR 12¾

Potters Bar. A 2¼-mile two-track bottleneck has been eliminated here. The first stage, including a new station at Potters Bar, was completed in 1955. Three new tunnels and a new station at Hadley Wood have been completed and recently brought into operation.

Hatfield, the home of the Cecils, is a quaint cheerful Hertfordshire town. Hatfield House, the historic seat of the Marquis of Salisbury, dating from the reign of James I, is a fine example of Jacobean architecture.

392¾

LONDON (King's Cross)

Just south of Finsbury Park station (2¼ miles north of King's Cross), on the west side of the line, is the fine new diesel maintenance depot completed in 1960. King's Cross station, opened in 1852, has 15 platforms covering 15¾ acres and deals with more than 750 trains per day. The Great Northern Hotel adjoins. Considerable station alterations will be necessary for the scheduled electrification.

London, the capital of England, the first city in the Commonwealth, the centre of Government and commerce, is the greatest city of any age or country.

St. Paul's Cathedral was built to the design of Sir Christopher Wren to replace the earlier edifice lost in the Great Fire of 1666 and barely escaped a like fate in World War II when much of the surrounding property was destroyed.

St. Paul's Cathedral

HOW TO USE THIS FOLDER

Passengers should imagine themselves facing north whether northbound or southbound. The views do not in all cases indicate the side of the line on which the objects appear, the descriptive matter does. If travelling north read up, if south, down.

MILES FROM LONDON

Title	Named Trains on the East Coast Main Line
Publisher	British Railways (North Eastern Region)
Date	1961

As we have seen on pages 48–51, the concept of strip maps – diagrammatically transforming the representation of rail routes into roughly straight lines, and thereby facilitating reproduction in booklet or leaflet form – has a long tradition on Britain's railways. So, 112 years after the *Reid's Railway Ride* map illustrated the route for travellers at a time when the East Coast Main Line (ECML) had still not been completed south of York, British Railways (BR) published this 1961 map with commentaries on the towns and landscape en route, accompanied by black and white photographs.

On the reverse of the map is a timetable for all 11 'named' trains operating between Edinburgh (Waverley) and London (King's Cross) on weekdays during the period 12 June to 9 September 1961. This was an epochal period for the ECML, with the steady replacement of the former LNER A1, A3 and streamlined A4 Pacific steam locomotives by diesels – in particular BR's iconic Type 5 'Deltics', which cut typical end-to-end journey times from seven to six hours.

The named trains conjure up nostalgic memories of a golden era of rail travel – of particular note are 'The Elizabethan', the non-stop Edinburgh–London train (which did not out-last steam haulage, as the ability to change drivers en route without stopping depended on the through corridor connections incorporated in the A4 locomotive design), and 'The Queen of Scots', an all-'Pullman' luxury train, which left the ECML at Northallerton to travel over the now closed line through Ripon to Harrogate, then onwards via Leeds to rejoin the traditional route to London at Doncaster.

The commentary notes that, 'On the straight track north of York speeds in excess of 70 m.p.h. are regularly attained' – a reference to the disappearing steam era, as the new Deltic diesels would regularly attain their maximum 100 m.p.h. speed on this section and on many other stretches of the ECML. An interesting feature of the timetable is the reference, *'The figures shown in italics, e.g. 1 34½, indicate passing times and that the train does not call'* – allowing attentive passengers to assess whether or not their train was on time.

At the time of writing, only one of these named trains is still so named – the world-renowned 'The Flying Scotsman', which, in its modern incarnation, departs from Edinburgh at 05.40 and arrives at King's Cross at 09.40, a far cry from the 7 hours 5 minutes of summer 1961.

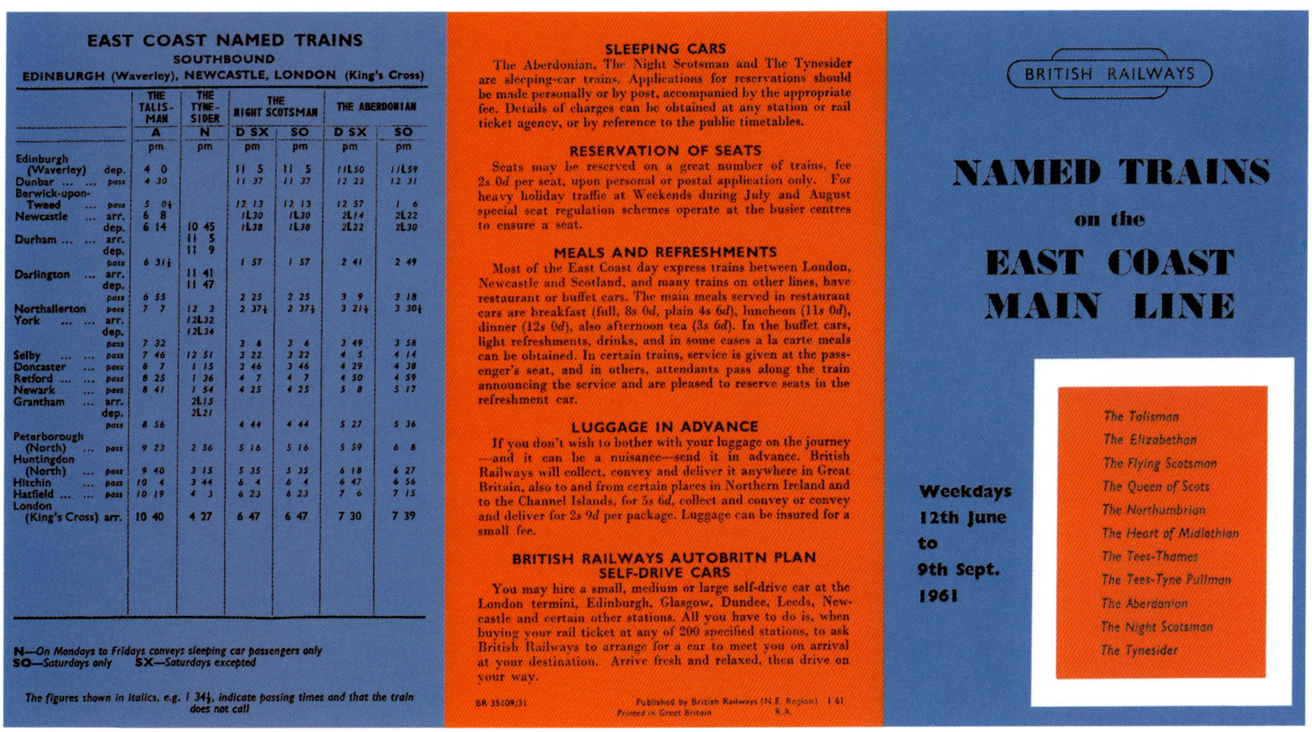

EAST COAST NAMED TRAINS
SOUTHBOUND
EDINBURGH (Waverley), NEWCASTLE, LONDON (King's Cross)

		THE TALISMAN	THE TYNESIDER	THE NIGHT SCOTSMAN		THE ABERDONIAN	
		A	N	D SX	SO	D SX	SO
		pm	pm	pm	pm	pm	pm
Edinburgh (Waverley)	dep.	4 0		11 5	11 5	11L50	11L59
Dunbar	pass	4 30		11 37	11 37	12 22	12 31
Berwick-upon-Tweed ...	pass	5 0½		12 13	12 13	12 57	1 6
Newcastle ...	arr.	6 8		1L30	1L30	2L14	2L22
	dep.	6 14	10 45	1L38	1L38	2L22	2L30
Durham	arr.		11 5				
	dep.		11 9				
	pass	6 31½		1 57	1 57	2 41	2 49
Darlington ...	arr.		11 41				
	dep.		11 47				
	pass	6 55		2 25	2 25	3 9	3 18
Northallerton	pass	7 7	12 3	2 37½	2 37½	3 21½	3 30½
York	arr.		12L32				
	dep.		12L34				
	pass	7 32		3 6	3 6	3 49	3 58
Selby	pass	7 46	12 51	3 22	3 22	4 5	4 14
Doncaster ...	pass	8 7	1 15	3 46	3 46	4 29	4 38
Retford	pass	8 25	1 36	4 7	4 7	4 50	4 59
Newark ...	pass	8 41	1 54	4 25	4 25	5 8	5 17
Grantham ...	arr.		2L15				
	dep.		2L21				
	pass	8 56		4 44	4 44	5 27	5 36
Peterborough (North) ...	pass	9 23	2 56	5 16	5 16	5 59	6 8
Huntingdon (North) ...	pass	9 40	3 15	5 35	5 35	6 18	6 27
Hitchin ...	pass	10 4	3 44	6 4	6 4	6 47	6 56
Hatfield	pass	10 19	4 3	6 23	6 23	7 6	7 15
London (King's Cross)	arr.	10 40	4 27	6 47	6 47	7 30	7 39

N—*On Mondays to Fridays conveys sleeping car passengers only*
SO—*Saturdays only* **SX**—*Saturdays excepted*

The figures shown in italics, e.g. 1 34½, indicate passing times and that the train does not call

SLEEPING CARS

The Aberdonian, The Night Scotsman and The Tynesider are sleeping-car trains. Applications for reservations should be made personally or by post, accompanied by the appropriate fee. Details of charges can be obtained at any station or rail ticket agency, or by reference to the public timetables.

RESERVATION OF SEATS

Seats may be reserved on a great number of trains, fee 2*s* 0*d* per seat, upon personal or postal application only. For heavy holiday traffic at Weekends during July and August special seat regulation schemes operate at the busier centres to ensure a seat.

MEALS AND REFRESHMENTS

Most of the East Coast day express trains between London, Newcastle and Scotland, and many trains on other lines, have restaurant or buffet cars. The main meals served in restaurant cars are breakfast (full, 8*s* 0*d*, plain 4*s* 6*d*), luncheon (11*s* 0*d*), dinner (12*s* 0*d*), also afternoon tea (3*s* 6*d*). In the buffet cars, light refreshments, drinks, and in some cases a la carte meals can be obtained. In certain trains, service is given at the passenger's seat, and in others, attendants pass along the train announcing the service and are pleased to reserve seats in the refreshment car.

LUGGAGE IN ADVANCE

If you don't wish to bother with your luggage on the journey—and it can be a nuisance—send it in advance. British Railways will collect, convey and deliver it anywhere in Great Britain, also to and from certain places in Northern Ireland and to the Channel Islands, for 5*s* 6*d*, collect and convey or convey and deliver for 2*s* 9*d* per package. Luggage can be insured for a small fee.

BRITISH RAILWAYS AUTOBRITN PLAN SELF-DRIVE CARS

You may hire a small, medium or large self-drive car at the London termini, Edinburgh, Glasgow, Dundee, Leeds, Newcastle and certain other stations. All you have to do is, when buying your rail ticket at any of 200 specified stations, to ask British Railways to arrange for a car to meet you on arrival at your destination. Arrive fresh and relaxed, then drive on your way.

BR 35109/31 Published by British Railways (N.E. Region) 1 61
Printed in Great Britain R.A.

BRITISH RAILWAYS

NAMED TRAINS on the EAST COAST MAIN LINE

Weekdays 12th June to 9th Sept. 1961

The Talisman
The Elizabethan
The Flying Scotsman
The Queen of Scots
The Northumbrian
The Heart of Midlothian
The Tees-Thames
The Tees-Tyne Pullman
The Aberdonian
The Night Scotsman
The Tynesider

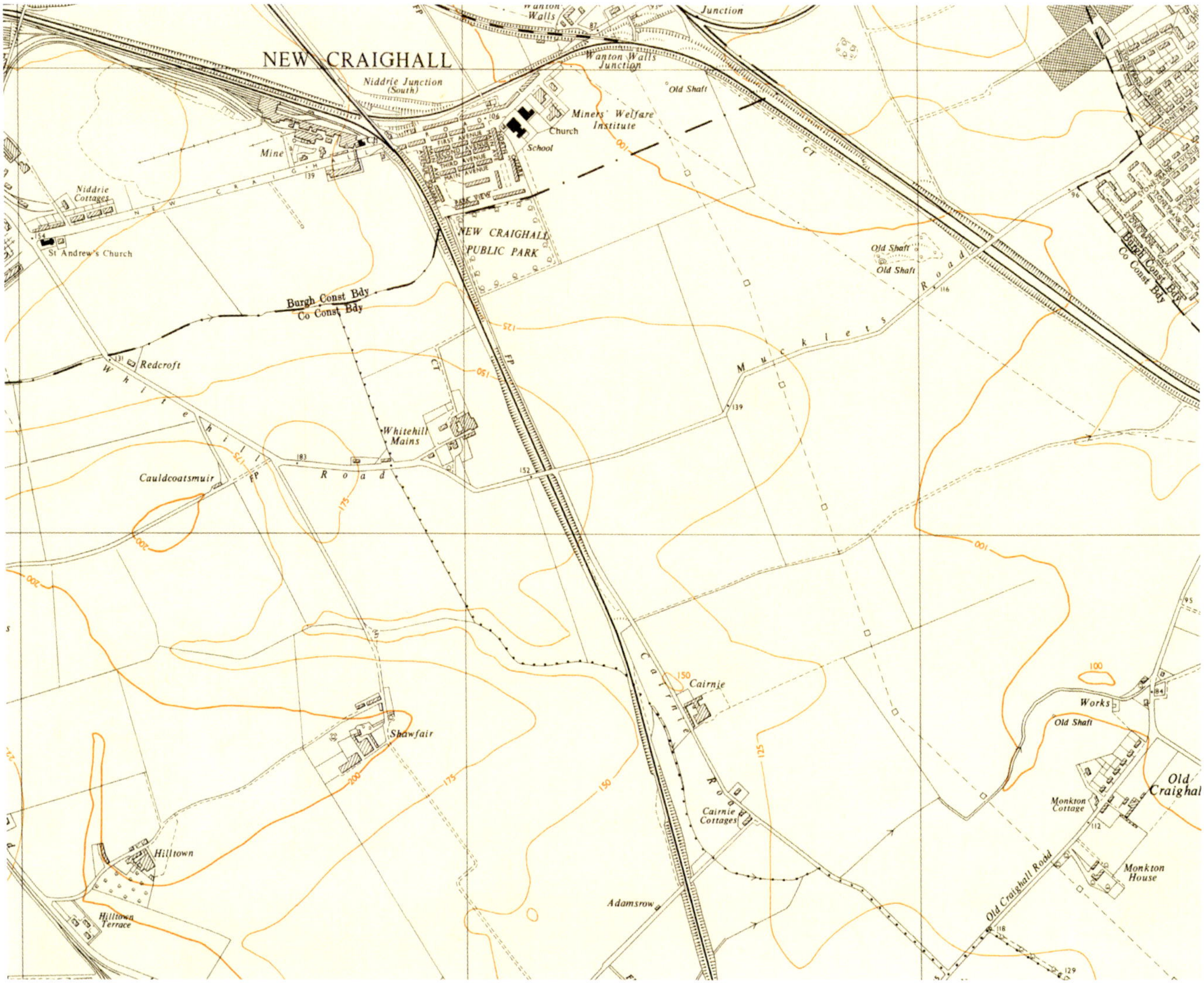

Series	Ordnance Survey 6 in. to 1 mile
Sheets	NT 37 SW (extracts, reduced)
Dates	1958 and 1966

Marshalling yards were developed at an early stage of the growth of the rail network, allowing individual freight wagons from a local catchment area to be sorted and re-marshalled into longer distance trains destined for yards closer to the wagons' ultimate destination.

In response to the operational inefficiencies of much freight traffic being handled at a multiplicity of smaller yards of pre-Grouping origin, the 1955 British Railways Modernisation Plan had allocated £85 million for the construction of between 20 and 30 large mechanized marshalling yards. The yards which were built incorporated: shunting over a hump from which wagons gravitated into the appropriate siding to form a train; remote-controlled points; and power-operated retarders decelerating wagons descending from the hump.

While Edinburgh in the 1950s was a relatively modest industrial centre, it did have extensive local goods traffic from breweries, collieries, and the docks in Granton and Leith, as well as long distance traffic in transit between the east of Scotland and eastern England. This traffic was handled at six unmechanized yards scattered throughout the eastern outskirts of the capital – an inefficient operation, which the new yard on a green field site at Millerhill, astride the Waverley Route to Carlisle via Hawick, was designed to largely replace.

However, by 1960 some senior railway managers were beginning to doubt the rationale for many of these expensive new yards – arguing that the future largely lay in cost-effective full trainloads running direct from A to B, and in

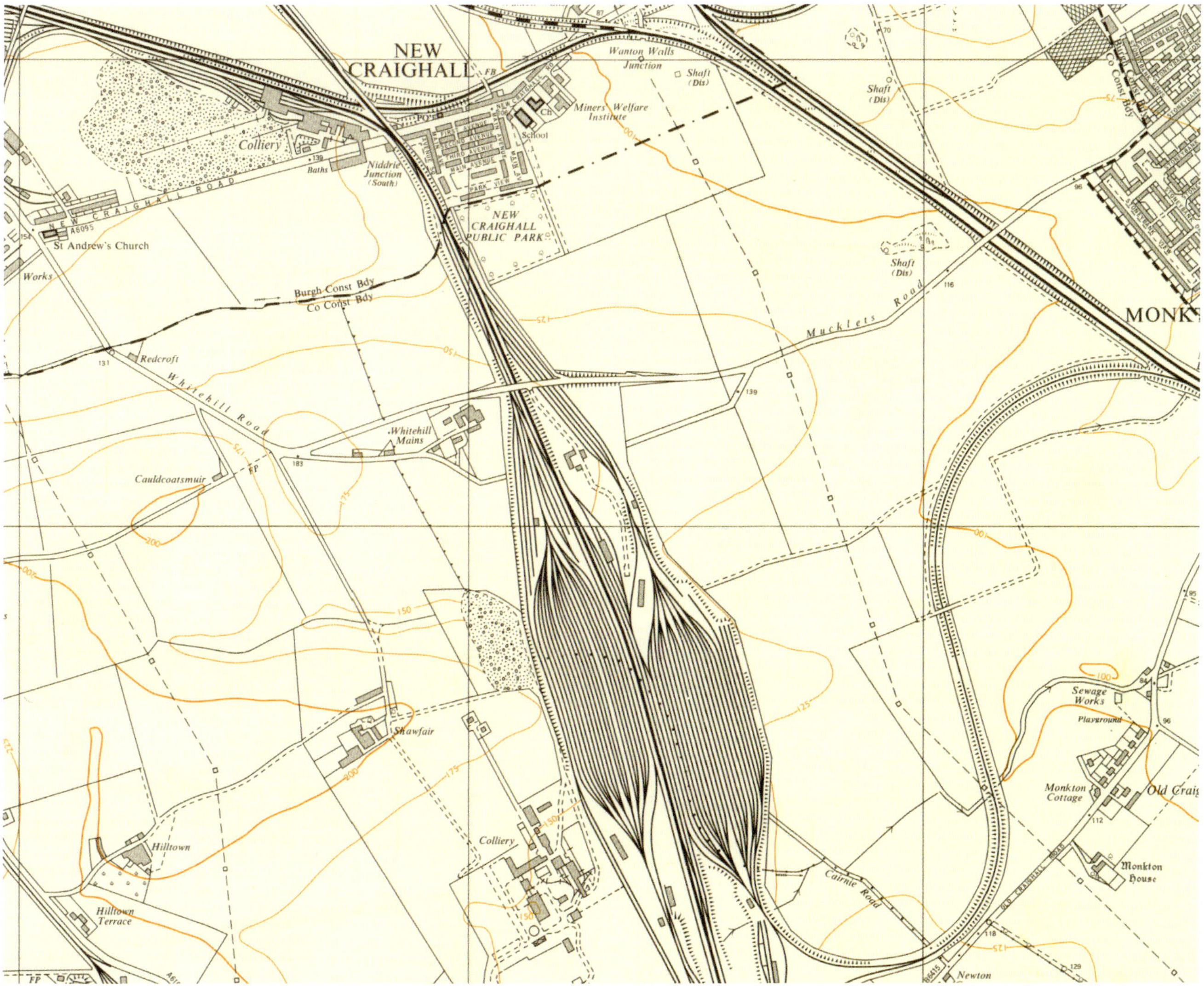

containerization. But it was too late to stop the Edinburgh scheme, which was already under construction, transforming the landscape of the Millerhill area, as shown in the contrast between these 1958 and 1966 Ordnance Survey map extracts. When fully completed in 1963, Millerhill's 100 sorting sidings had the capacity to handle up to 5,000 wagons daily, and to optimize its 'hub' role a new connecting line was built on a sweeping reverse curve to the East Coast Main Line. A further change in the landscape was the construction of the Monktonhall deep coal mine immediately west of the yard, with its own rail sidings. This colliery would generate substantial freight movements for the railway over the next three decades, but ironically most of it would be in full trainloads which had no need for Millerhill's wagonload sorting facilities.

By the end of the 1960s the marshalling yard had suffered from the continuing nationwide decline in wagonload rail freight traffic – through increased road competition, industrial change, and BR's ongoing switch of remaining traffic to trainload operation. The complete closure of the Waverley Route in 1969 led to significant rationalization at Millerhill, and with further traffic decline the entire 'Down Yard' – the fan of sidings between the old Waverley Route and Monktonhall colliery – was closed completely in 1983. The loop line running through the yard from the East Coast Main Line was, however, electrified in 1991; a new passenger station opened at Newcraighall in 2002; and in recent years aggregates and open cast coal have been loaded to rail at sidings within the remaining shrunken Millerhill complex. Advance works for the planned re-opening of the Borders Railway to Galashiels and Tweedbank in 2014 have been undertaken at the north end of the old Down Yard, now a forest of birch trees, but which has been proposed as the site for a new rail-connected waste transfer station.

EASTERN ENGLAND
NORTH SEA
GRIMSBY
CLEETHORPES
GAINSBOROUGH
The Old Hall Gainsborough
Louth Ch.
Market Rasen
LOUTH
Tennyson's Birthplace, Somersby
LINCOLN
MABLETHORPE
SUTTON-ON-SEA
Bardney
NEWARK
Somersby
WOODHALL JUNCTION
WILLOUGHBY
Tattershall
Tattershall Castle
Clock Tower
SKEGNESS
The Pottergate, Lincoln
GRANTHAM
SLEAFORD
BOSTON
The Angel and Royal Grantham
BOURNE
HUNSTANTON
WELLS-NEXT-THE-SEA
Holbeach
Sandringham Church
SPALDING
Sandringham
SHERINGHAM
STAMFORD
The Three-way Bridge, Crowland
FAKENHAM
CROMER
Crowland
KING'S LYNN
MELTON CONSTABLE
Aylsham
MUNDESLEY-ON-SEA
Swaffham Village
Wisbech
PETERBOROUGH
NORTH WALSHAM
SWAFFHAM
MARCH
DEREHAM
NORWICH
Watton
Chatteris
Cromwell's Residence, Ely
Market Cross
WYMONDHAM
Gt. YARMOUTH
Peterborough Cathedral
ELY
Thetford
GORLESTON
The Guildhall, Norwich
HUNTINGDON
St. IVES
BECCLES
Bury St. Edmunds
MILDENHALL
The Angel,
Diss
The Bridge St. Ives
LOWESTOFT
CAMBRIDGE
NEWMARKET
Eye
SOUTHWOLD
BURY St. EDMUNDS
The Swan Lavenham
HAUGHLEY
Stowmarket
Framlingham
Framlingham Castle
Cambridge, The Bridge of Sighs
Royston
SAXMUNDHAM
HAVERHILL
Lavenham
The Guildhall, Thaxted
Wickham Market
Thorpeness
Long Melford
HITCHIN
Saffron Walden
Kersey
ALDEBURGH
Sudbury
Finchingfield
IPSWICH
WOODBRIDGE
Nayland
Thaxted
Dedham
Bishop's Stortford
Halstead
Chappel & Wakes Colne
Manningtree
FELIXSTOWE
HERTFORD
BRAINTREE
Harlow Town
Gt. Dunmow
MARKS TEY
Parkeston Quay
Dovercourt Bay
HARWICH
COLCHESTER
Broxbourne
WITHAM
BRIGHTLINGSEA
CHELMSFORD
WALTON-ON-THE-NAZE
Frinton-on-Sea
CLACTON-ON-SEA
MALDON
Shenfield & Hutton
SOUTHMINSTER
KING'S CROSS
LIVERPOOL STREET
FENCHURCH STREET
ROMFORD
Burnham-on-Crouch
LONDON
Barking
UPMINSTER
Great Northern Lines
Great Eastern Lines
SOUTHEND-ON-SEA
TILBURY
Grays
SHOEBURYNESS
BLAKE
LINCOLN
NORTHAMPTON
HUNTINGDON
CAMBRIDGE
NORFOLK
SUFFOLK
ESSEX

Title	Eastern England
Publisher	British Railways (Eastern Region)
Date	1963

This map brochure captures the last year in which travellers could use the rail network to reach almost every significant holiday destination in a wide swathe of eastern England from Lincolnshire through Norfolk and Suffolk to the borders of Essex and London. The text notes that, 'fast and comfortable trains serve most of the places mentioned in this folder', and – in an evocation of a service concept soon to disappear from the railways – reminds readers that:

> *Luggage can be sent in advance – it can be collected from your home and delivered to the holiday address for 5/6 or, for 2/9, it can be collected from your home and delivered to your holiday station; alternatively you can hand it in at your local station and have it delivered to your holiday address for 2/9.*

Although this is a pictorial map, it does not have the accomplished art work and attention to detail of the Yorkshire map on pages 210–11, published just 14 years previously by British Railways. The small paintings of buildings and monuments of note are rather two-dimensional, and collectively over-dominate the map – however, as an illustration for a brochure, rather than a large poster for display, it serves its purpose. The charming postcard-style snapshot photos certainly underline the sheer range of different types of towns and villages which the holidaymaker could still visit by rail in 1963.

The brochure – timed to promote the 1963 holiday season – must have been published within months, if not weeks, of the 27 March publication of the Beeching Report which was to seal the fate of so many of the routes on the map. By the end of that year, the branch line from Fakenham to the charmingly named Wells-next-the-Sea had gone, to be followed in 1964 by Fakenham's own connection to the rest of the network at Dereham; branch lines to Maldon, Melton Constable, and Mundesley-on-Sea; and minor through routes from Swaffham to Thetford and through Saffron Walden.

Other communities put up stiffer opposition to the closure programme, and it was not until 1967 that the Sudbury–Haverhill–Cambridge cross-country line succumbed – deemed to be just too uneconomic to be salvageable by the innovative 'Basic Railway' and 'Paytrain' concepts, which as described in Milepost 20 saved the nearby Ipswich–Lowestoft 'East Suffolk Line'. Much of the rail network in east and south Lincolnshire finally succumbed under Dr Beeching's prescription in 1970, but perhaps the most draconian cuts came in north Cambridgeshire and northwest Norfolk between the late 1960s and early 1980s – as we shall see on pages 248–9.

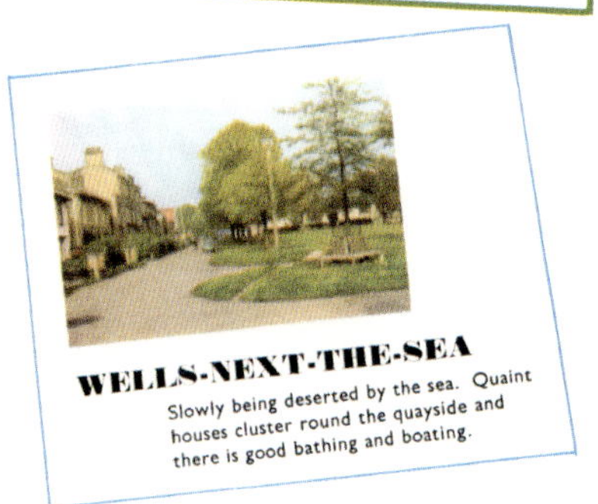

DR BEECHING'S PRESCRIPTION

Rising railway deficits led to a drastic re-appraisal of the network, with extensive closure of branch lines, cross-country routes and even some main lines. While these cuts made relatively little impression on overall railway finances, some of Beeching's innovations on the freight railway had a lasting and positive impact on the rail system.

MILEPOST 20	The Beeching Report	232
1963	Beeching Report: Passenger Station Receipts	236
1963	Beeching Report: Passenger Service Withdrawal Proposals	240
1963	Beeching Report: Liner Train Network Proposals	244
1965	Beeching's Trunk Routes Report	246
1964–82	Beeching Reprieves and Post-Beeching Closures	248
1966	East and West Coast Main Lines: Gradient Profiles	250
1968	British Rail Passenger Network Map	252

1963–1970

NEWCASTLE
MIDDLESBROUGH
SCARBOROUGH
YORK
LEEDS
HULL
GRIMSBY
MANCHESTER
SHEFFIELD
LINCOLN
DERBY
STAFFORD

MILEPOST

20

Dr Richard Beeching holds a copy of his infamous report, *The Reshaping of British Railways*, at a press conference on 27 March 1963.

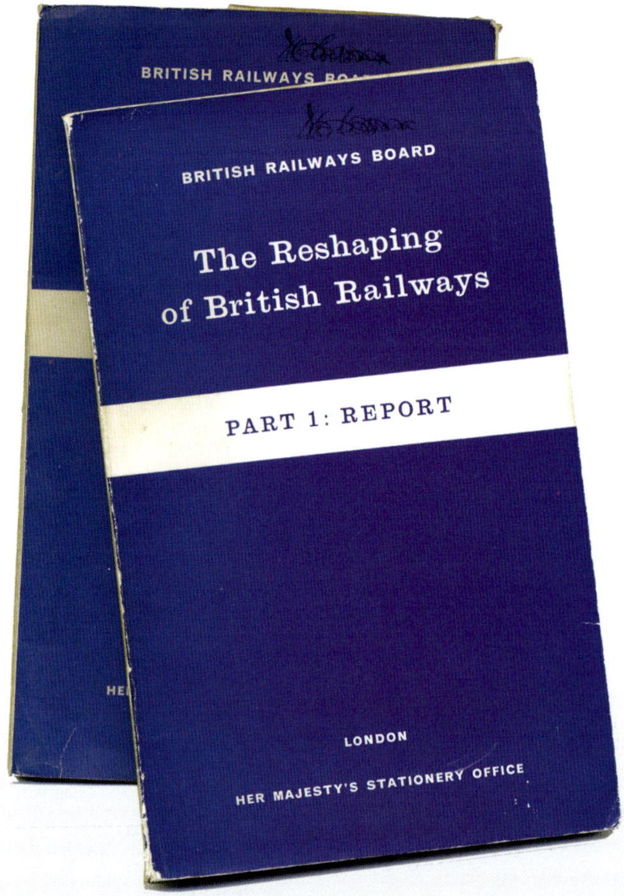

Price one shilling the Beeching Report was published in two parts – the second part consisting of 13 maps showing traffic flows, proposed closures and new liner-train routes.

While closure of unprofitable railway lines had been a small part of the railway scene until the Second World War, the austerity years that followed were a prelude to the biggest shake-up of Britain's railways ever witnessed. In the immediate years after the war the railways were in a run-down, lamentable state, and nationalization seemed the only answer. So, in 1948, the Labour Government did nationalize the railways at the same time as the canals, road haulage, bus services and airlines. The newly formed British Railways (BR) was faced with the enormous task of modernizing its vast network while turning it into a profitable concern. Modernization plans were slow to evolve (see Milepost 19), but cost-cutting, in the shape of line closures, was fairly rapid, and by the beginning of the 1960s about 3,300 miles of uneconomic lines had already disappeared. Probably the most famous of these was the Midland & Great Northern Joint Railway's network of lines in East Anglia which closed in 1959 – a leftover from the Victorian age, its duplication of other lines in this rural area being its final undoing.

Despite these closures and the implementation of the 1955 £1.24 billion Modernisation Plan, BR was continuing to lose taxpayers' money hand over fist – drastic action needed to be taken to counter ever-increasing competition from road transport, out-of-date working practices and mainly 19th-century technology.

Harold Macmillan's Conservative Government soon turned its attention to the near-bankrupt railways which by 1960 were losing £67.7 million a year (£1.15 billion at today's prices). In March of that year Macmillan had given this message in the House of Commons:

> *First the industry must be of a size and pattern suited to modern conditions and prospects. In particular, the railway system must be remodelled to meet current needs, and the Modernisation Plan must be adapted to this new shape.*

With this ringing in his ears and rail losses escalating, the Transport Minister, Ernest Marples, set up in 1960 an advisory group, known as the Stedeford Committee, to report on the state of the railways, and to come up with some solutions. One of the committee's members was Dr Richard Beeching who was then the Technical Director

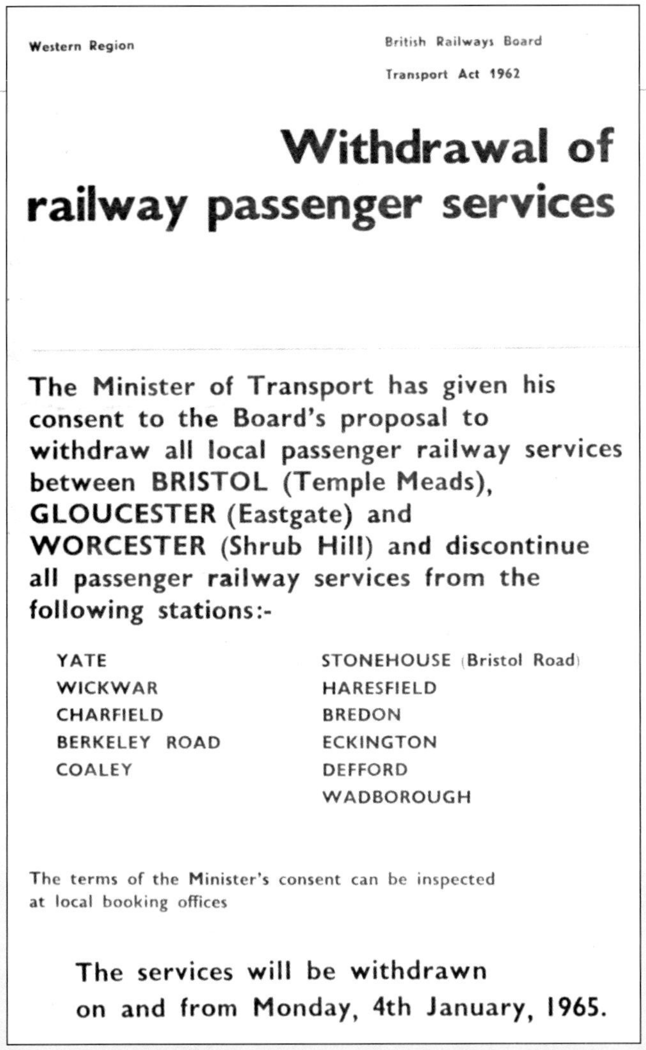

Western Region

British Railways Board

Transport Act 1962

Withdrawal of railway passenger services

The Minister of Transport has given his consent to the Board's proposal to withdraw all local passenger railway services between BRISTOL (Temple Meads), GLOUCESTER (Eastgate) and WORCESTER (Shrub Hill) and discontinue all passenger railway services from the following stations:-

YATE	STONEHOUSE (Bristol Road)
WICKWAR	HARESFIELD
CHARFIELD	BREDON
BERKELEY ROAD	ECKINGTON
COALEY	DEFFORD
	WADBOROUGH

The terms of the Minister's consent can be inspected at local booking offices

The services will be withdrawn on and from Monday, 4th January, 1965.

Notices such as this one announcing withdrawal of stopping trains between Bristol, Gloucester and Worcester were soon appearing at railway stations all over Britain.

of ICI, renowned for his analytical mind and undoubted skill in solving business problems. Beeching suggested that the railways should not provide a public service at any cost, and should instead be run as a profitable concern. While Beeching's ideas were not initially implemented, they certainly were noted by Marples, who appointed him as Chairman of the British Transport Commission (BTC) in June 1961, and then as Chairman of the British Railways Board which replaced the BTC in late 1962.

With Beeching now in the driving seat, the outlook for Britain's rural railways was grim. Within days of taking over, Beeching commissioned a detailed survey of both freight and passenger traffic on British Railways – often misquoted, the date of these surveys actually took place during the week of 17–23 April 1961 inclusive. In his eventual report, *The Reshaping of British Railways*, Beeching and his team acknowledged that 'it was impossible to continue the massive recording effort involved for a longer period.' Even the most loyal of his lieutenants admitted to doubts about the way they went about their analysis and decision-making on lines to be put up for closure. In his 1989 biography, *Beeching: Champion of the Railway?*, former senior BR manager R H N Hardy conceded that:

> *. . . there were innumerable critics who said, not without some justification, that our figures were suspect, that we based everything on one single week's traffic – which was true – that we closed on crude statistics – they had a point – and that both our staff and the public were bulldozed into submission.*

Nearly two years then elapsed before the publication of his 'notorious' report on 27 March 1963. Priced one shilling and available from Her Majesty's Stationery Office, *The Reshaping of British Railways* was a two-volume collection of graphs, statistics, analysis and maps that became commonly known as the 'Beeching Report'. Part 1 consisted of 148 pages of statistics and their analysis, the majority of which were effectively a death sentence for around 5,000 route miles of railway and the closure of over a third of the country's stations. With the 41-page Appendix 2 to the Report listing lines and stations throughout England, Scotland and Wales proposed for closure, there was little room for optimism.

According to the report one third of the route mileage carried only 1 per cent of the total passenger miles and 1 per cent of freight ton miles on BR – but of course, a similar analysis of Britain's road network would have produced much the same picture. Beeching was in fact commissioned by the 1964 Labour Government to produce such a study, examining the co-ordination of passenger and freight transport by road and rail – but the Prime Minister, Harold Wilson, under pressure from road haulage interests, soon concluded that Beeching was too pro-rail and too anti-road! He decided that Beeching must have 'assessors' imposed on him, but Beeching was unwilling to be shackled, declined the consultancy, and left BR to return to ICI earlier than expected, in May 1965.

An important flaw in Beeching's approach was identified by the charismatic and controversial General Manager of BR Eastern Region, Gerry Fiennes, who demonstrated one of the key omissions of the Report by showing that 'Paytrains' and 'Basic Railways' could cut costs dramatically on loss-making lines, while still retaining and even increasing patronage. Paytrains operated like buses where passengers paid their fare to a conductor on board. Basic railways had unstaffed stations so doing away with booking offices, etc. In this way the whole line could be run with a minimum of staff and facilities, thus reducing costs. In his 1967 book, *I Tried to Run a Railway* he commented that:

> *[the Marples/Beeching axis] took no account of the new techniques; either coming into operation like diesel traction;*

or just round the corner like automatic level crossings, mechanized track maintenance, tokenless block signalling, and 'bus stop' operation which can cut the cost of rural railways by more than a half. . . . They laid it down in general that rural railways did not pay, which was true; and could never pay, which was false.

Part 2 of the Beeching Report consisted of 13 large fold-out maps that had been produced to illustrate graphically the statistical information gathered for Part 1. Ranging from 'Density of passenger traffic' and 'Density of freight traffic' to 'Proposed withdrawal of passenger services' and 'Proposed modification of passenger train services', they nearly all gave bad news for the railway industry. However, the map of 'Liner train routes and terminals under consideration' was an important pointer to one of the positive outcomes of the report – and the Freightliner train service (using rail for the trunk haul and road for local collection and delivery) is today the market leader for movement of export and import containers to and from England's big five Deep Sea ports. Another positive legacy of Beeching was the widespread introduction of merry-go-round trains – so called because they load and unload coal while still moving – from collieries to power stations, a practice which continues to this day.

Protesters against the proposed Beeching cuts parade through the streets of Manchester in 1963.

As a fig leaf for the proposed passenger service closures the final route map in Part 2 showed 'Bus Services in Britain'.

Dubbed the 'Beeching Axe' by the press, the report's predominantly negative conclusions sparked a nationwide outcry, especially from the rural communities who were most at risk. Despite this, Beeching's report claimed to have all the answers – 'most areas of the country are already served by a network of buses' and that 'it appears that hardship will arise on only a very limited scale.' The consideration of hardship was the special responsibility of Transport Users' Consultative Committees where objections to closures could be lodged. If no bus service already existed then, in some cases, replacement bus services had to be laid on. At the end of the day it was for the Government – having also consulted its regional economic planning councils – to decide whether a line closed or not.

Beeching did not altogether get his own way, although over the next ten years 4,065 route miles succumbed to closure. While strong public and political pressure saved the Scottish lines from Inverness to Wick and Thurso and to Kyle of Lochalsh and in the southwest from Ayr to Stranraer, the closure of the Waverley Route between Edinburgh and Carlisle in January 1969 left the Borders region as by far the largest population grouping in Britain with no accessible railway services (although it is now planned to reopen the northern third of the line, to Edinburgh, in 2014). Conversely, its English equivalent, the Settle to Carlisle line, was reprieved as late as 1989 and is now an important freight route and alternative to the West Coast Main Line. Poor roads and the lack of bus services (but above all the fact that it passed through a number of marginal parliamentary constituencies) saved the Central Wales line, along with several Cornish branch lines, while in East Anglia the East Suffolk line from Ipswich to Lowestoft benefited from clever cost cutting, inspired by Gerry Fiennes. Despite the latter success story, other lines in East Anglia were not so lucky, including several that had not originally been listed for closure by Beeching, around King's Lynn and Peterborough. Ironically it was the 1964–70 Labour Government which wielded the axe more swingeingly than any other administration, despite their criticism of Beeching while in opposition.

In retrospect, many of the Beeching closures were very short-sighted – closing a third of the network brought savings of only £30 million, and the railways continued to run at an enormous loss. Cutting off the branches only starved the main lines of feeder traffic, causing an even greater downward spiral for rail finances. Fortunately a second Beeching report published in 1965, *The Development of the Major Railway Trunk Routes*, was not implemented – otherwise Britain's rail network would have shrunk even further.

Protesters halted the passage of the last train on the Waverley Route at Newcastleton on the night of 5–6 January 1969.

Title	Map No.3 – British Railways: distribution of passenger traffic station receipts
From	*The Reshaping of British Railways* (Part 2: Maps)
Publisher	British Railways Board, London
Date	1963

With its breathtaking plan for swingeing line closures the length and breadth of Britain, the Beeching Report did not hide its light under a bushel. Some critics felt that Dr Beeching was naïve in not foreseeing the political storm that such a transparent announcement of drastic surgery would cause. Part of that transparency came in Part 2 of the report – in a fascinating portfolio of 13 detailed maps showing line traffic densities, station receipts, services proposed for closure, and other aspects of the programme for reshaping the rail system. Beeching's Map 3 of passenger traffic station receipts in theory shows every passenger station still operational at the time of the report's research, other than those in the congested conurbations, but in practice some routes – such as March–Spalding–Lincoln – were completely omitted, demonstrating that mapmakers also make mistakes.

Stations were allocated to three categories – those earning up to £5,000 per annum; those between £5,000 and £25,000, and those of £25,000 and over – marked by coloured 'blobs' (solid circles). By far the most striking feature of the map (shown in detail overleaf) is its demonstration that the overwhelming majority of Britain's stations fell into the lowest revenue category, thereby visually underpinning the finding of Part 1 of the report that one third of the railway route mileage carried just 1 per cent of the total passenger miles.

A comparison of the station receipts map with Beeching's Map 9 of lines proposed for passenger service withdrawal (see pages 240–43) shows that very few stations in the highest revenue category were threatened with closure – but there were notable exceptions in East Anglia and Scotland. North of the Border, Stranraer and Thurso – at the southwestern and northern extremities respectively of the Scottish network – faced closure, as did Galashiels and Hawick on the Waverley Route from Edinburgh to Carlisle through the Scottish Borders, perhaps the most controversial of all Beeching's line closure proposals. In East Anglia, the planned closure of the East Suffolk line from Ipswich to Lowestoft would remove rail services from two stations which fell into the highest revenue category – Beccles and Saxmundham.

Following strong local protests, the East Suffolk line was saved, while in Scotland the Far North Line to Thurso (and the Kyle of Lochalsh line) benefited from the vociferous 'MacPuff' campaign which traded on political sensitivities about Highland depopulation. The line to Stranraer was reprieved in order to maintain the rail–ferry connection from Scotland to Northern Ireland, but the entire Waverley Route was closed by a Labour Government in 1969.

The contrast with rural mid-Wales was stark, as illustrated in the map extract. The Central Wales line service from

Shrewsbury to Swansea served more than 30 intermediate stations, none of which fell into Beeching's highest revenue category, and just three of which (at the southern end) earned between £5,000 and £25,000 annually. While the provision of replacement bus services was undoubtedly problematic, the line served a small declining population – but crucially it passed through three marginal constituencies, and as a result was reprieved twice, in 1964 and again in 1969. By contrast, the Waverley Route corridor had no key marginals, and the Borders lacked the political resonance of mid-Wales or the Highlands; its 98-mile railway closed on 6 January 1969, leaving the region it served as by far the largest population grouping in Britain with no accessible rail services. Such was the rough justice of British railway closures in the 1960s.

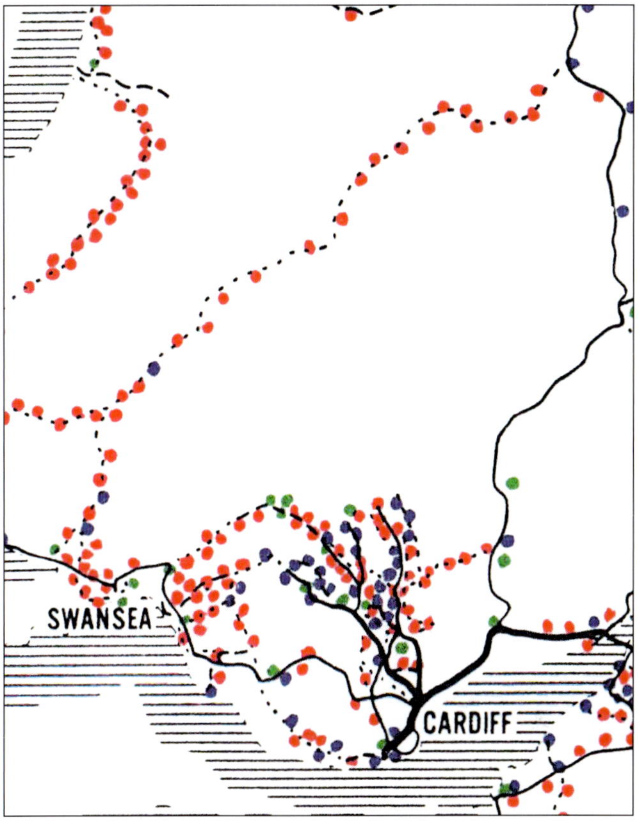

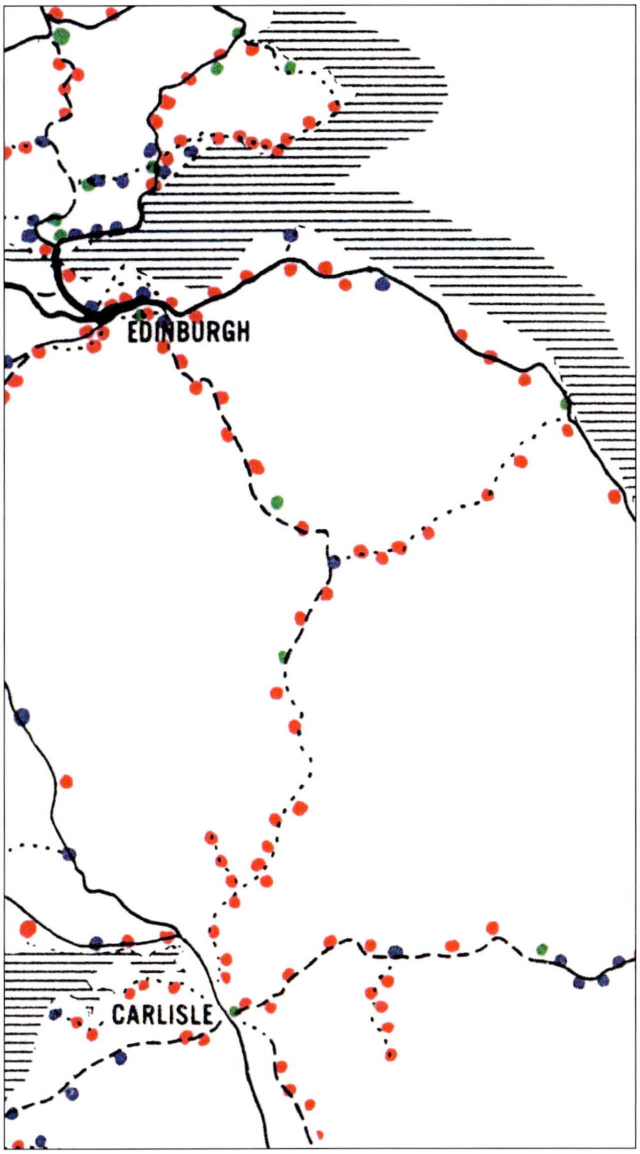

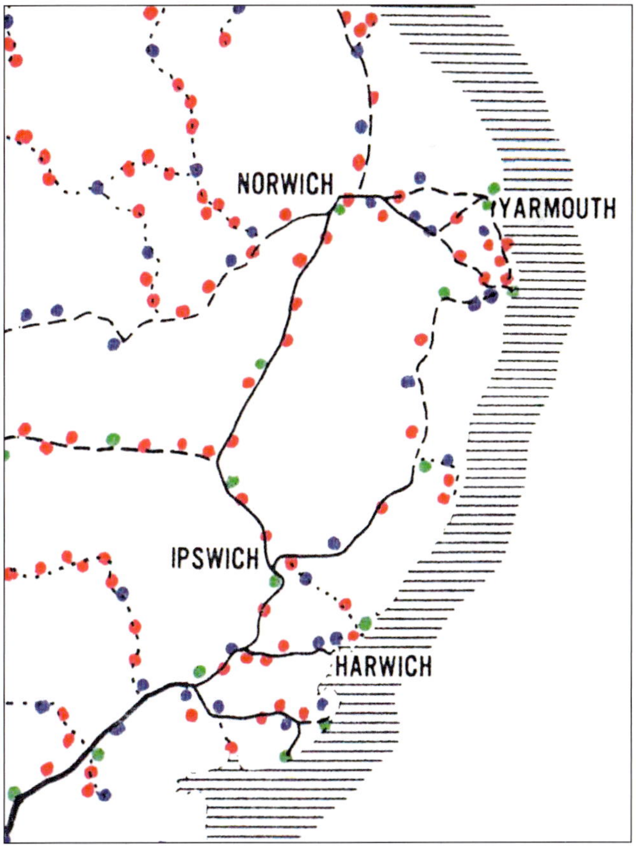

Map No.3

BRITISH RAILWAYS DISTRIBUTION OF PASSENGER TRAFFIC STATION RECEIPTS

Red • £0 to 5,000 per annum
Blue • £5,000 to 25,000 per annum
Green • £25,000 and over per annum
Stations in certain congested areas are omitted

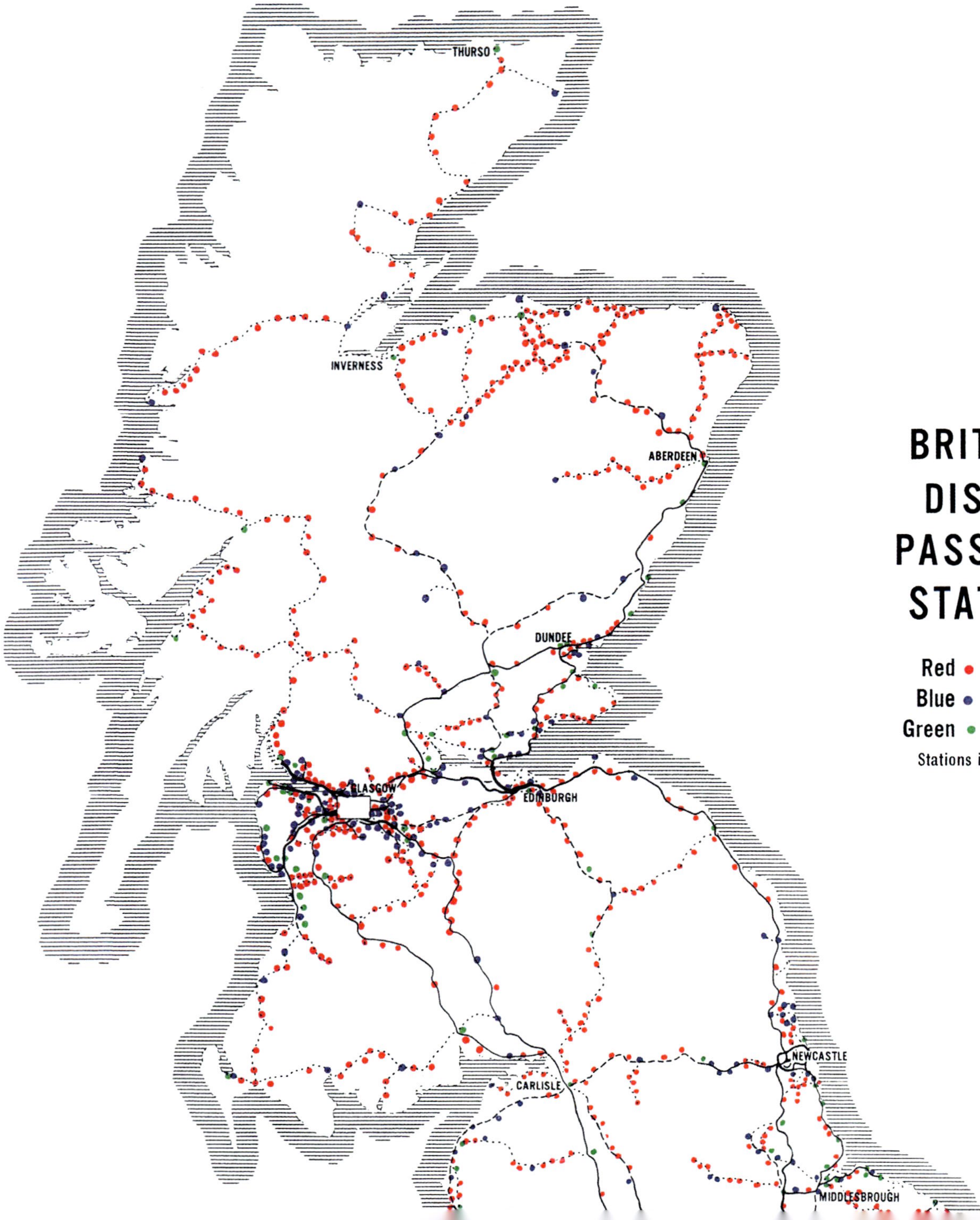

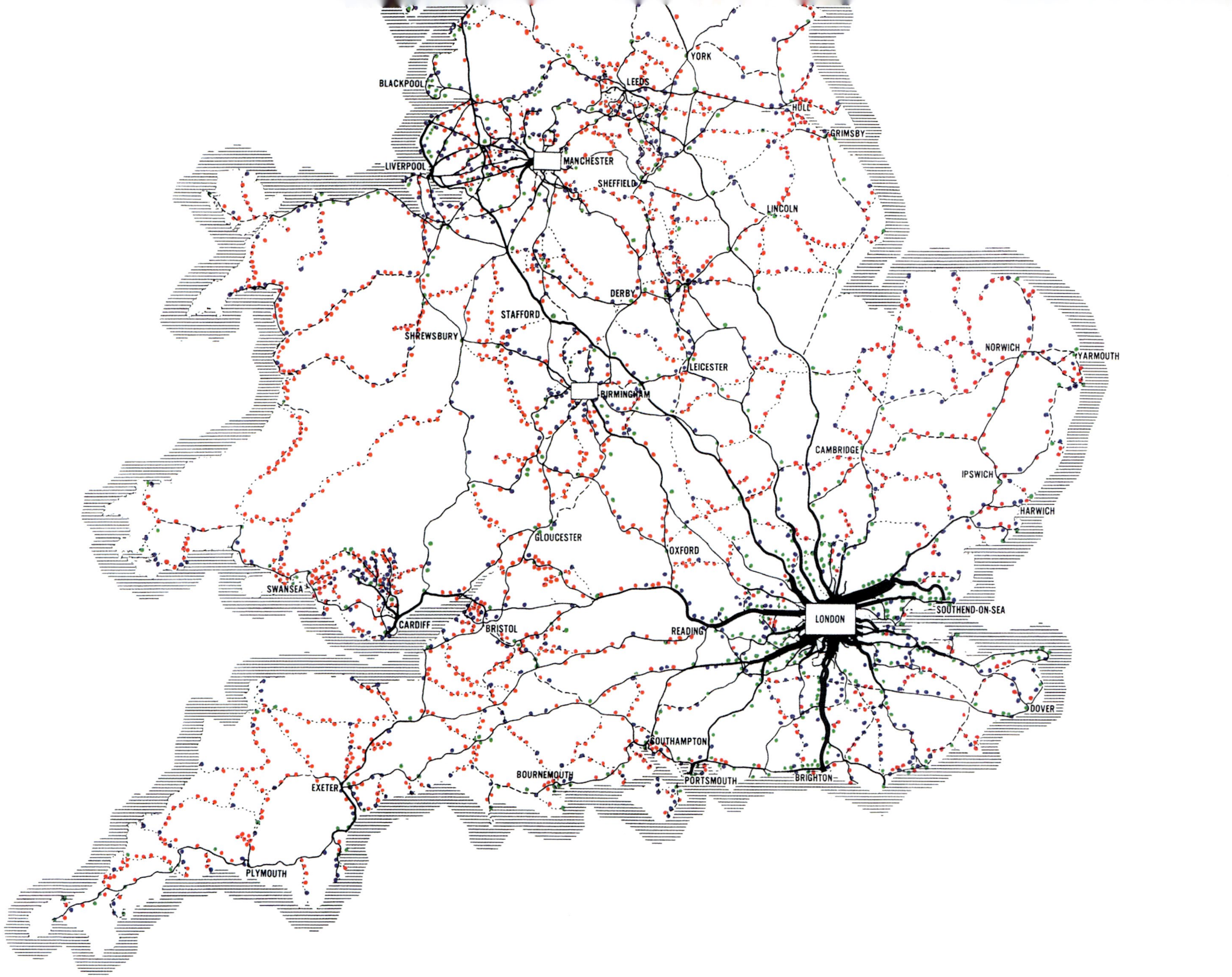
YORK
BLACKPOOL
LEEDS
HULL
GRIMSBY
LIVERPOOL
MANCHESTER
SHEFFIELD
LINCOLN
DERBY
STAFFORD
SHREWSBURY
NORWICH
YARMOUTH
LEICESTER
BIRMINGHAM
CAMBRIDGE
IPSWICH
HARWICH
GLOUCESTER
OXFORD
SWANSEA
SOUTHEND-ON-SEA
LONDON
CARDIFF
BRISTOL
READING
DOVER
SOUTHAMPTON
BOURNEMOUTH
PORTSMOUTH
BRIGHTON
EXETER
PLYMOUTH

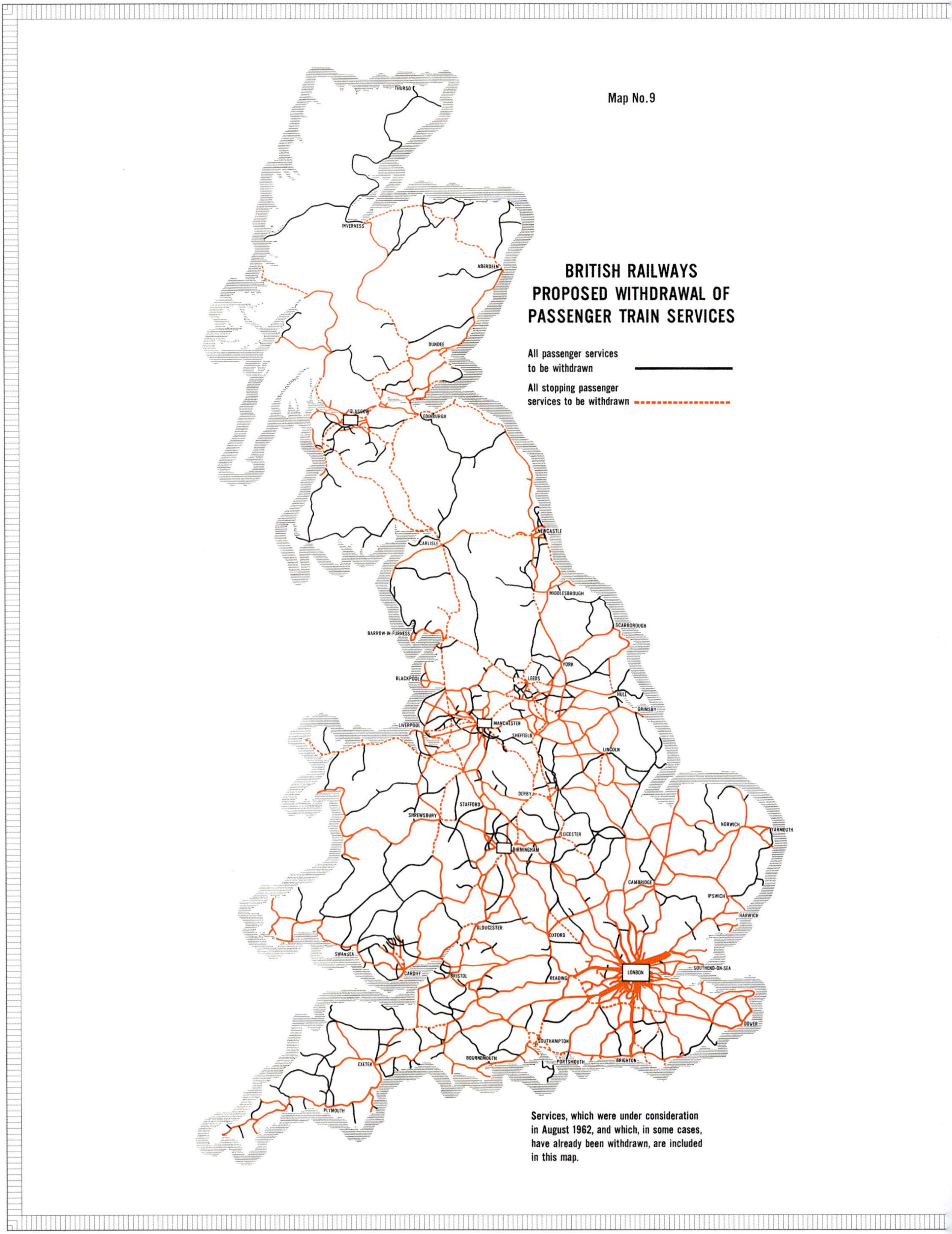
Map No.9
BRITISH RAILWAYS
PROPOSED WITHDRAWAL OF
PASSENGER TRAIN SERVICES
All passenger services
to be withdrawn
All stopping passenger
services to be withdrawn
THURSO
INVERNESS
ABERDEEN
DUNDEE
GLASGOW
EDINBURGH
NEWCASTLE
CARLISLE
MIDDLESBROUGH
SCARBOROUGH
BARROW-IN-FURNESS
YORK
BLACKPOOL
LEEDS
HULL
GRIMSBY
LIVERPOOL
MANCHESTER
SHEFFIELD
LINCOLN
DERBY
STAFFORD
SHREWSBURY
LEICESTER
BIRMINGHAM
NORWICH
YARMOUTH
CAMBRIDGE
IPSWICH
HARWICH
GLOUCESTER
OXFORD
SWANSEA
CARDIFF
BRISTOL
LONDON
SOUTHEND-ON-SEA
READING
DOVER
SOUTHAMPTON
PORTSMOUTH
BRIGHTON
BOURNEMOUTH
EXETER
PLYMOUTH
Services, which were under consideration
in August 1962, and which, in some cases,
have already been withdrawn, are included
in this map.

Title	Map No.9 – British Railways: proposed withdrawal of passenger train services. Map No.9a – British Railways: proposed withdrawal of passenger train services London area.
From	*The Reshaping of British Railways* (Part 2: Maps)
Publisher	British Railways Board, London
Date	1963

For most readers of the Beeching Report, the nub of its message lay in Maps 9 and 9a which set out the routes over which passenger services were proposed for withdrawal, and the surviving routes over which all local stopping passenger services were to suffer a similar fate. Every region of the country was faced with the consequences of closing 5,000 route miles of railway nationwide, the only exception being the majority of Kent – admittedly part of the London-focused commuter network, but very much unscathed compared to other parts of the Home Counties rail system.

In practice, due to the vagaries of politics, only a relatively few areas experienced the full implementation of Beeching's line closure proposals – northeast Scotland, Fife, the Scottish Borders, the West Midlands, north Cornwall, north Devon, Somerset, Dorset and the threatened branches of the 'outer suburban' commuter network west and southwest of London.

As noted earlier, London's passenger rail network survived Beeching almost untouched. The only inner London line to succumb was the Palace Gates branch, while in the suburbs just the short Belmont branch from Harrow & Wealdstone, and the Uxbridge and Staines branches from West Drayton, were closed. The threatened branches from Watford Junction to Croxley Green and St Albans City, and from Woodside to Sanderstead in south London, were all reprieved.

We explore the lines which were reprieved nationwide on pages 248–249 – but which were the worst Beeching passenger closures? The longest route to be closed was the Great Central Railway (GCR) from Aylesbury, north of London, through the East Midlands to Sheffield, a total route length of 126 miles. The GCR was unquestionably a main line of strategic value, not least through its continental 'loading gauge' for freight traffic, which nowadays could have been playing a key role in intermodal transport of the modern generation of tall containers. The Great Central was also a major loss in terms of cross-country connectivity, but the small towns which lost their train service were virtually all left within 10 miles of the nearest surviving railheads. The closure of the 86 miles of the 'Somerset & Dorset' network removed train services from a number of significant towns – including Glastonbury – and although all were within 20 miles of the nearest stations on the continuing 'main line' network, it left an area of fast growing population completely dependent on road transport.

Of all the areas that lost their rail services, in terms of population or distance from the surviving rail network the most drastic cuts came in Scotland. The northeast fishing towns of Fraserburgh (population 11,000) and Peterhead (17,000) were left just over 40 miles and 30 miles respectively from the rail network. The severity of loss was even greater in the Scottish Borders served by the Waverley Route, with Galashiels (population 13,000) left 33 miles from the nearest railhead and Hawick (16,000) 45 miles. The closure of the East Lincolnshire Line from Spalding to Grimsby left one town of comparable size to Galashiels or Hawick – Louth – cut off from the rail network, but by less than 20 miles.

It is clear that while a substantial proportion of Beeching's closure proposals may have made sense in an era of growing car ownership, a more strategic vision of Britain's transport future would have spared some very significant routes from the fate of crude closure.

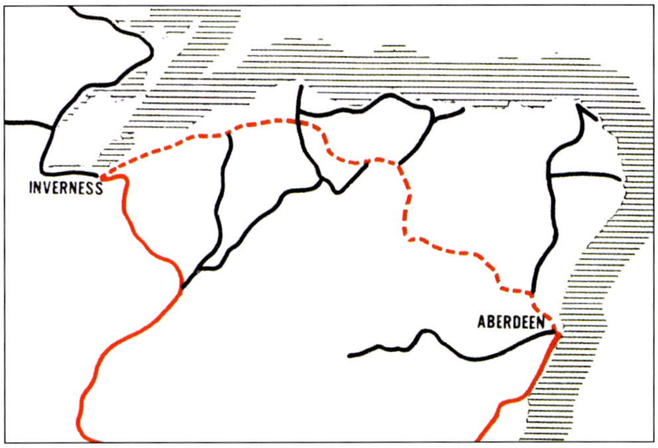

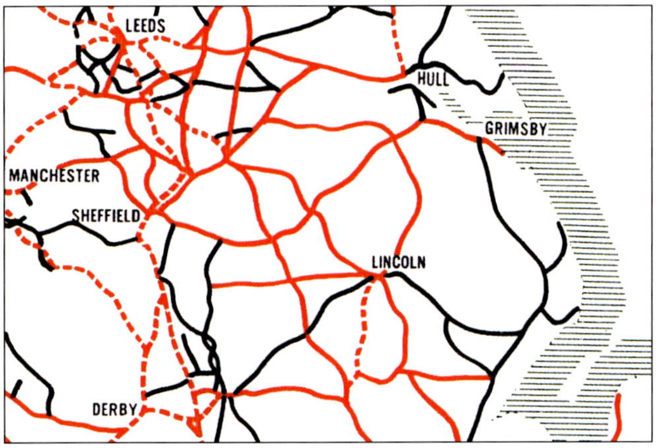

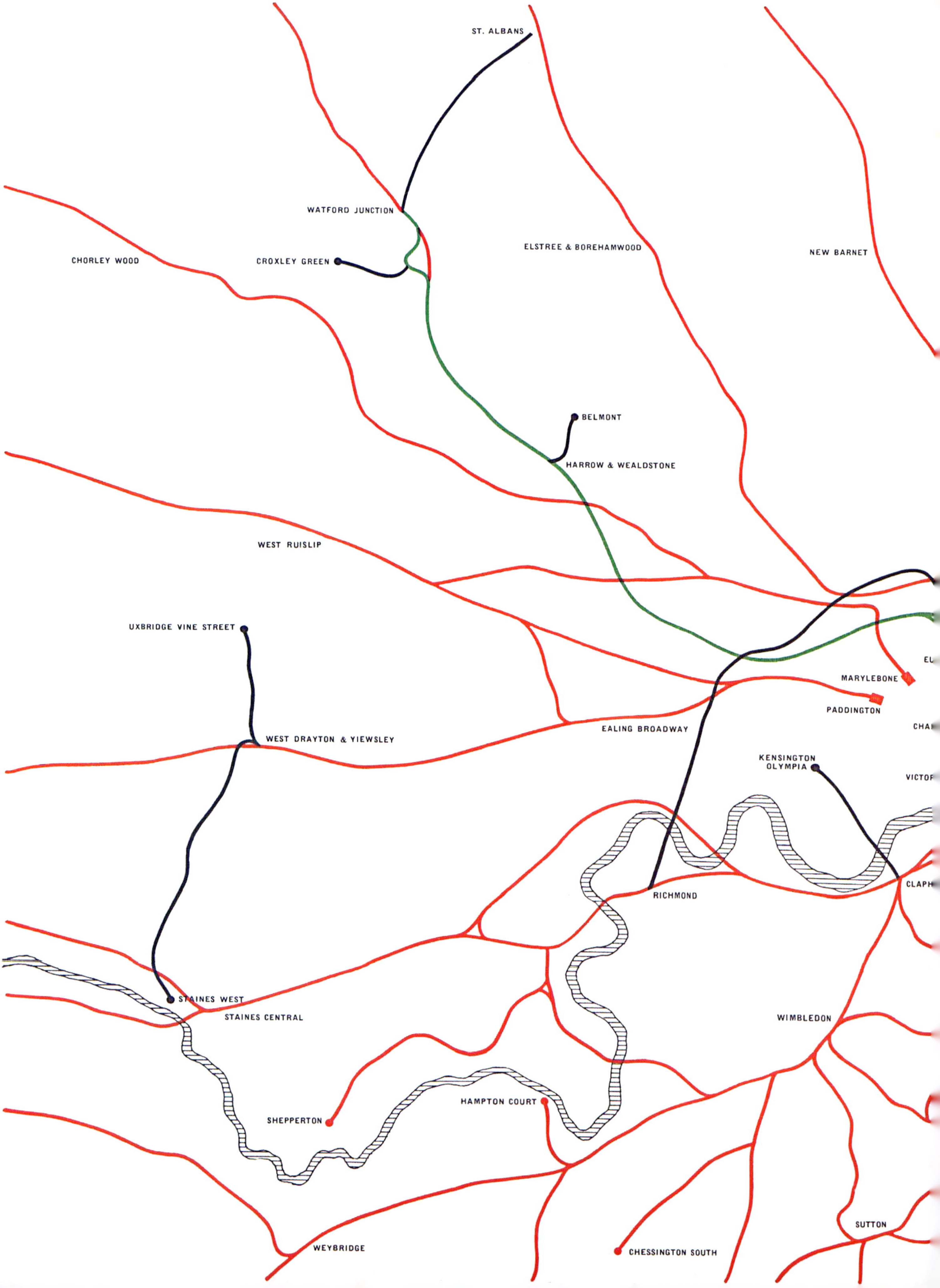
ST. ALBANS
WATFORD JUNCTION
CHORLEY WOOD
CROXLEY GREEN
ELSTREE & BOREHAMWOOD
NEW BARNET
BELMONT
HARROW & WEALDSTONE
WEST RUISLIP
UXBRIDGE VINE STREET
MARYLEBONE
PADDINGTON
WEST DRAYTON & YIEWSLEY
EALING BROADWAY
KENSINGTON OLYMPIA
RICHMOND
STAINES WEST
STAINES CENTRAL
WIMBLEDON
HAMPTON COURT
SHEPPERTON
SUTTON
WEYBRIDGE
CHESSINGTON SOUTH

Map No. 9A

BRITISH RAILWAYS
PROPOSED WITHDRAWAL OF PASSENGER TRAIN SERVICES
LONDON AREA

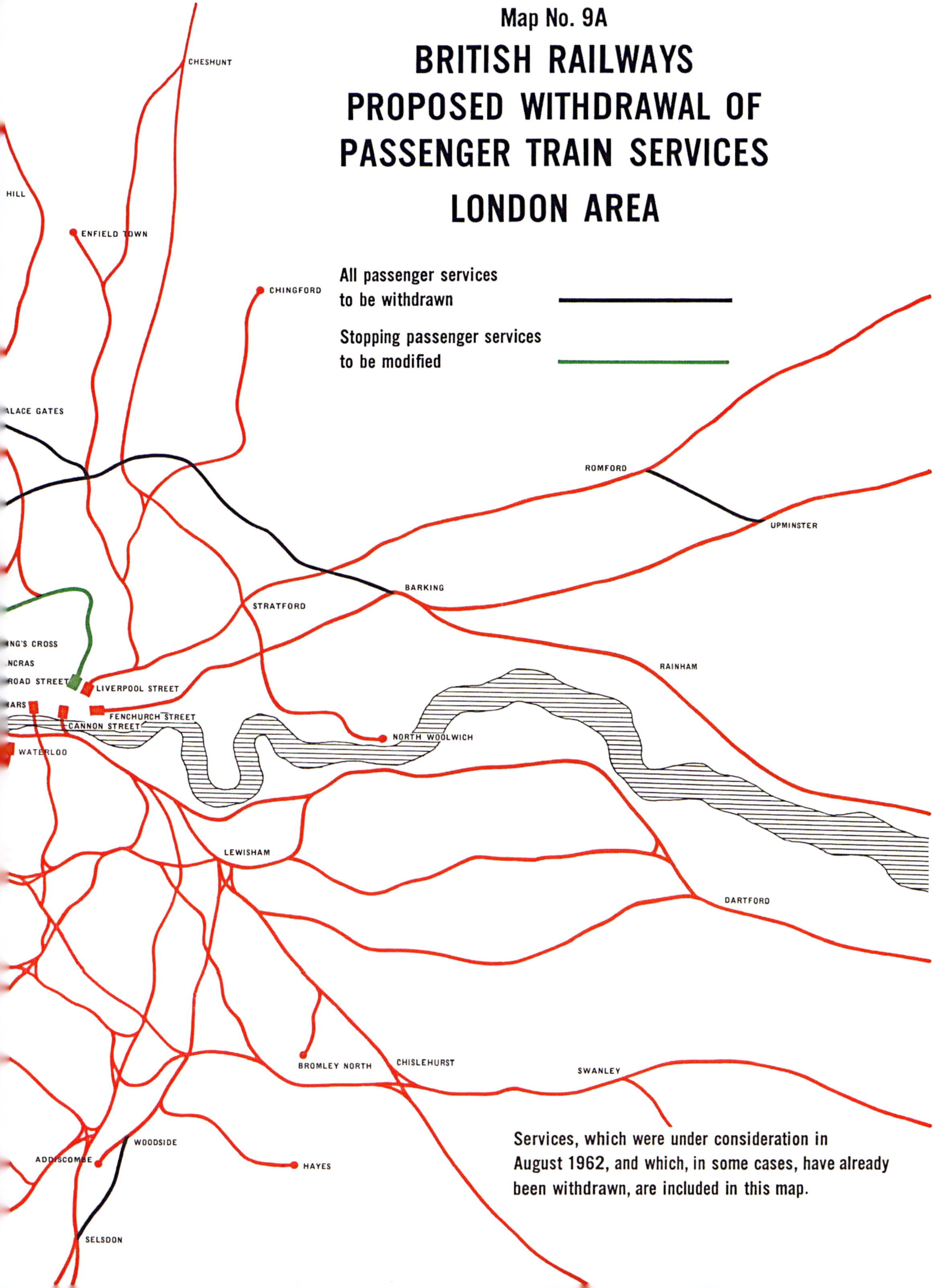

Services, which were under consideration in August 1962, and which, in some cases, have already been withdrawn, are included in this map.

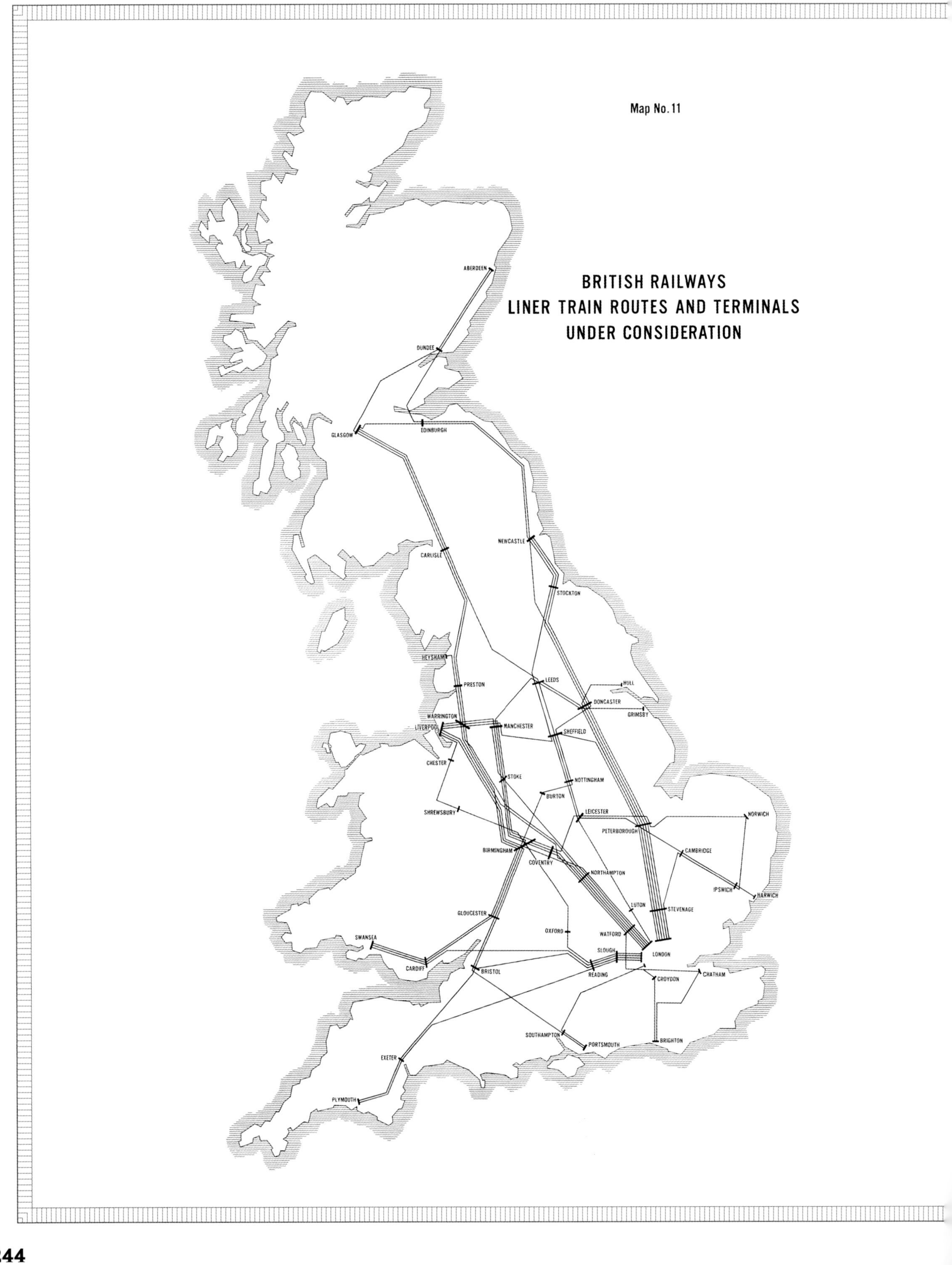
Map No. 11
BRITISH RAILWAYS
LINER TRAIN ROUTES AND TERMINALS
UNDER CONSIDERATION
ABERDEEN
DUNDEE
GLASGOW
EDINBURGH
NEWCASTLE
CARLISLE
STOCKTON
HEYSHAM
PRESTON
LEEDS
HULL
DONCASTER
GRIMSBY
WARRINGTON
LIVERPOOL
MANCHESTER
SHEFFIELD
CHESTER
STOKE
NOTTINGHAM
BURTON
SHREWSBURY
LEICESTER
NORWICH
PETERBOROUGH
BIRMINGHAM
COVENTRY
CAMBRIDGE
NORTHAMPTON
IPSWICH
HARWICH
LUTON
STEVENAGE
GLOUCESTER
OXFORD
WATFORD
SWANSEA
SLOUGH
LONDON
CARDIFF
BRISTOL
READING
CROYDON
CHATHAM
SOUTHAMPTON
PORTSMOUTH
BRIGHTON
EXETER
PLYMOUTH

Title	Map No.11 – British Railways: liner train routes and terminals under consideration
From	*The Reshaping of British Railways* (Part 2: Maps)
Publisher	British Railways Board, London
Date	1963

The wholesale closure programme which was the centrepiece of *The Reshaping of British Railways* – and the controversy which it provoked – have ensured that the name 'Beeching' will always have negative connotations. As Christian Wolmar concludes in *Fire & Steam* (2007), it was not all Beeching's fault, however:

> *His terms of reference had been too narrow: he was only supposed to find a way of returning the industry to profitability as soon as possible rather than consider the social and economic value of the railway or examine the economics of other modes of transport.*

The widespread condemnation of Beeching has tended to obscure the positive aspects of the report, where his fiercely intellectual analysis of problems yielded solutions from which the rail system has benefited right through to the present day. This was particularly the case in terms of the freight market – which inevitably was of less interest to the media, politicians or the public. Here Beeching came up with innovative ideas to equip the railway to do what it was inherently best at, in competition with a road haulage industry which was making major inroads into rail traffic, on the back of the expanding motorway network and bigger and faster lorries.

The biggest challenge was in the non-bulk market where the flexibility of the lorry – and its door-to-door capability – was causing a haemorrhaging of wagonload freight from rail to road. Part 1 of the report set out the thinking behind 'liner' trains as the key future method of moving 'general merchandise' traffic by rail:

> *The basic [intermodal] idea is to combine road and rail movement in such a way as to take advantage of the low cost of fast through-train movement as the means of providing trunk haulage over medium to long distances, for flows of traffic which, though dense are composed of consignments too small in themselves to justify through-train operation, and to do so without the disadvantages of the costly assembly of trains by wagon-load movements on rail, or costly transfer of merchandise between road and rail vehicles.*

Based on the intermodal concept – using high-capacity containers – being cheaper door-to-door than road haulage for distances of 100 miles or more, Beeching envisaged about 55 terminals being required. In practice, the relative economics of road haulage improved dramatically, pushing the 'break-even' point between road and rail to ever higher distances, which effectively shrunk the profitable market available to rail. More than half of the terminals 'under consideration' in this map were never built, but the new Freightliner division of British Rail (BR) did play a pioneering global role in containerization, in parallel with the American shipping company Sealand. It was the presence of a BR official at a 1966 conference in Moscow which ensured that the future international ('ISO') standard for containerization would be based on an 8 ft cross-section – if the agreed height had been any greater the restricted 'loading gauge' on most of Britain's rail routes would have severely diminished Freightliner's future ability to serve the maritime market at major ports.

Freightliner was eventually privatized with the rest of BR's freight businesses, and today continues to be the market leader for overland movement of containers to Britain's big Deep Sea ports at Felixstowe, Liverpool, Southampton, Thamesport and Tilbury. Lord Beeching (who died in 1985, aged 71) would have been proud of that legacy.

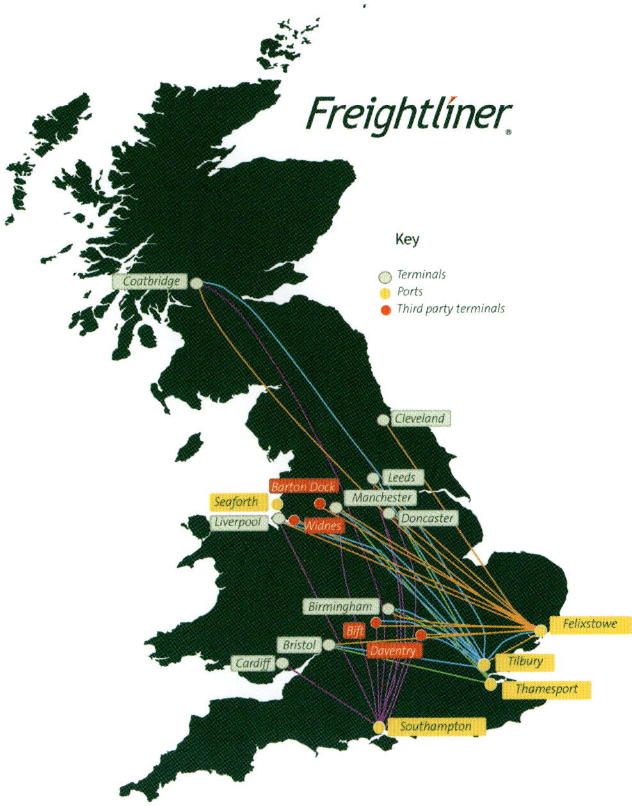

The Freightliner network in 2011, focused on Britain's big Deep Sea ports.

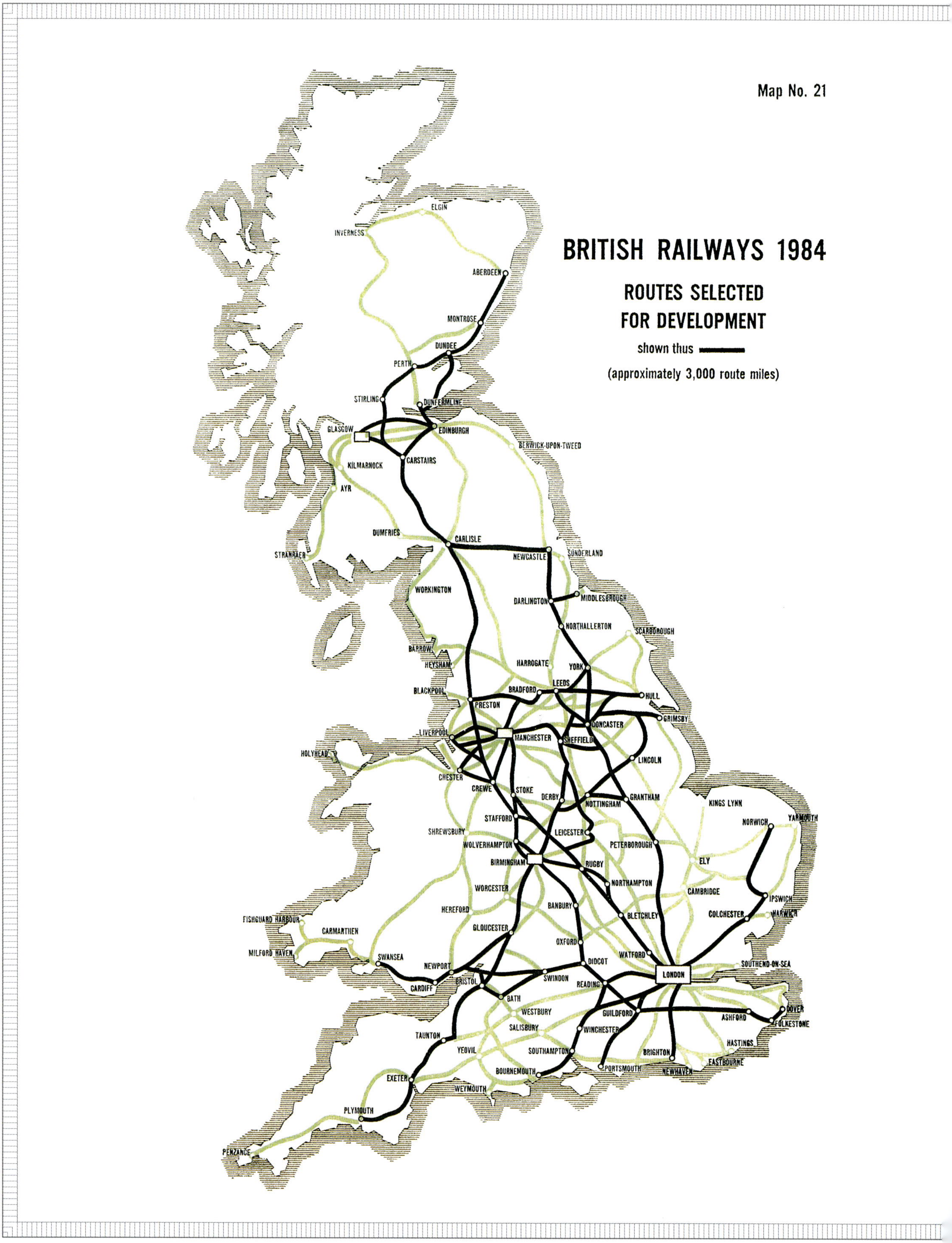
Map No. 21
BRITISH RAILWAYS 1984
ROUTES SELECTED FOR DEVELOPMENT
shown thus
(approximately 3,000 route miles)
ELGIN
INVERNESS
ABERDEEN
MONTROSE
DUNDEE
PERTH
STIRLING
DUNFERMLINE
GLASGOW
EDINBURGH
BERWICK-UPON-TWEED
CARSTAIRS
KILMARNOCK
AYR
DUMFRIES
STRANRAER
CARLISLE
NEWCASTLE
SUNDERLAND
WORKINGTON
DARLINGTON
MIDDLESBROUGH
NORTHALLERTON
SCARBOROUGH
BARROW
HEYSHAM
HARROGATE
YORK
BLACKPOOL
BRADFORD
LEEDS
HULL
PRESTON
GRIMSBY
DONCASTER
LIVERPOOL
MANCHESTER
SHEFFIELD
HOLYHEAD
LINCOLN
CHESTER
CREWE
STOKE
DERBY
GRANTHAM
NOTTINGHAM
KINGS LYNN
NORWICH
YARMOUTH
STAFFORD
SHREWSBURY
LEICESTER
WOLVERHAMPTON
PETERBOROUGH
ELY
BIRMINGHAM
RUGBY
NORTHAMPTON
WORCESTER
CAMBRIDGE
IPSWICH
HEREFORD
BANBURY
HARWICH
COLCHESTER
BLETCHLEY
FISHGUARD HARBOUR
GLOUCESTER
CARMARTHEN
OXFORD
MILFORD HAVEN
WATFORD
SWANSEA
DIDCOT
NEWPORT
LONDON
SOUTHEND-ON-SEA
SWINDON
BRISTOL
CARDIFF
READING
BATH
WESTBURY
GUILDFORD
DOVER
ASHFORD
FOLKESTONE
SALISBURY
WINCHESTER
TAUNTON
YEOVIL
SOUTHAMPTON
BRIGHTON
HASTINGS
PORTSMOUTH
EASTBOURNE
NEWHAVEN
BOURNEMOUTH
EXETER
WEYMOUTH
PLYMOUTH
PENZANCE

Title	Map No. 21 – British Railways 1984: routes selected for development
From	*The development of the major railway trunk routes*
Publisher	British Railways Board, London
Date	1965

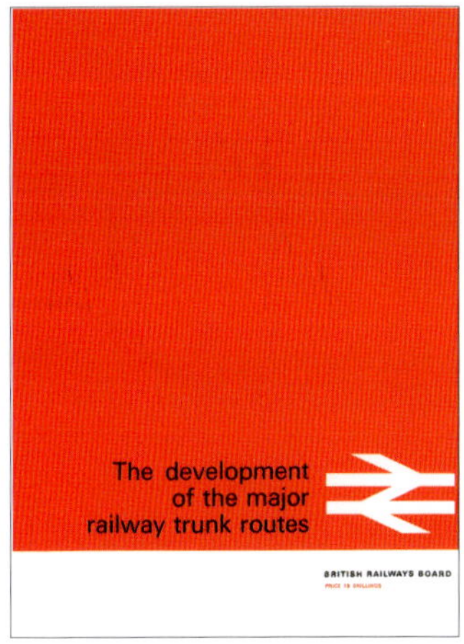

Dr Beeching's final major report for British Railways (BR) before he returned to ICI was *The development of the major railway trunk routes*, published in February 1965. This report focused on trunk routes as the key element of that part of the rail system, 'which is clearly sound now, and potentially sound under any foreseeable future conditions', with the crucial caveat that, 'Not all existing trunk routes can be included, however, because of the duplication which has resulted from the competitive building of the past.'

The report examined the pattern of traffic between main centres at that time, likely changes in the economy over the next 20 years, and forecasts of the pattern of transport demand between main centres in 1984, illustrated by no less than 27 maps. It concluded that:

> *The railways must reshape and redeploy their assets and service, so as to concentrate on cheap bulk movement, if the decline [in railway traffic] is to be arrested and reversed.*

The network of through routes was underutilized, resulting in higher than necessary costs of track and signalling per unit of traffic, and to get these costs down to about half their then level it was:

> *demonstrably practical to channel the through movement over a reduced network of lines amounting to some 3,000 route miles rather than the 7,500 miles provided today.*

It was argued that once the lines most likely to form the trunk rail network during the next 20 years had been determined, it would be possible to concentrate investment on the selected routes in the confidence that the essential dense flows of traffic would be maintained. As one would expect from Beeching, this was in many ways a logical argument – playing to the railways' strengths rather than weaknesses – but to the average passenger Map 21 implied, almost incredibly, a much greater threat to the network than even the 1963 report. To name but a few of the shocks, the 'routes selected for development' penetrated no further than Swansea in Wales, only the London–Norwich route would be developed in East Anglia, and Scotland faced the prospect of just one Anglo-Scottish line securing significant investment – echoing the 1839 Royal Commission, which believed that the traffic could support only one route, and Gladstone's 1889 comment to the House of Commons that 'as late as 1842, it was firmly believed to be absolutely impossible that there should ever be more than one railway to Scotland.'

The report's clarification that:

> *non-selection of lines for intensive development does not necessarily mean that they will be abandoned in the foreseeable future, nor even that money will not be spent upon some of them to improve their suitability for their continuing purpose*

would provide little comfort for most readers. But Beeching left BR just three months later; in three years' time a radical Transport Act introduced specific line-by-line grants in belated recognition of the railways' wider social value; and BR steadily developed their world-leading InterCity brand across a far wider range of routes than the shrunken network for development envisaged by Beeching.

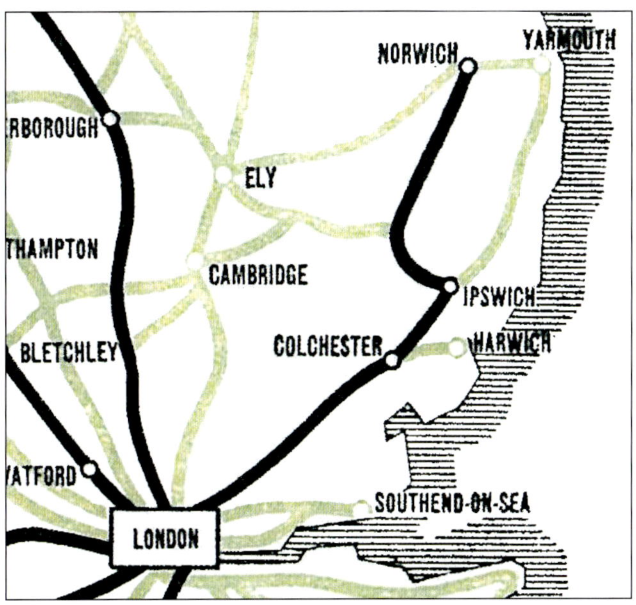

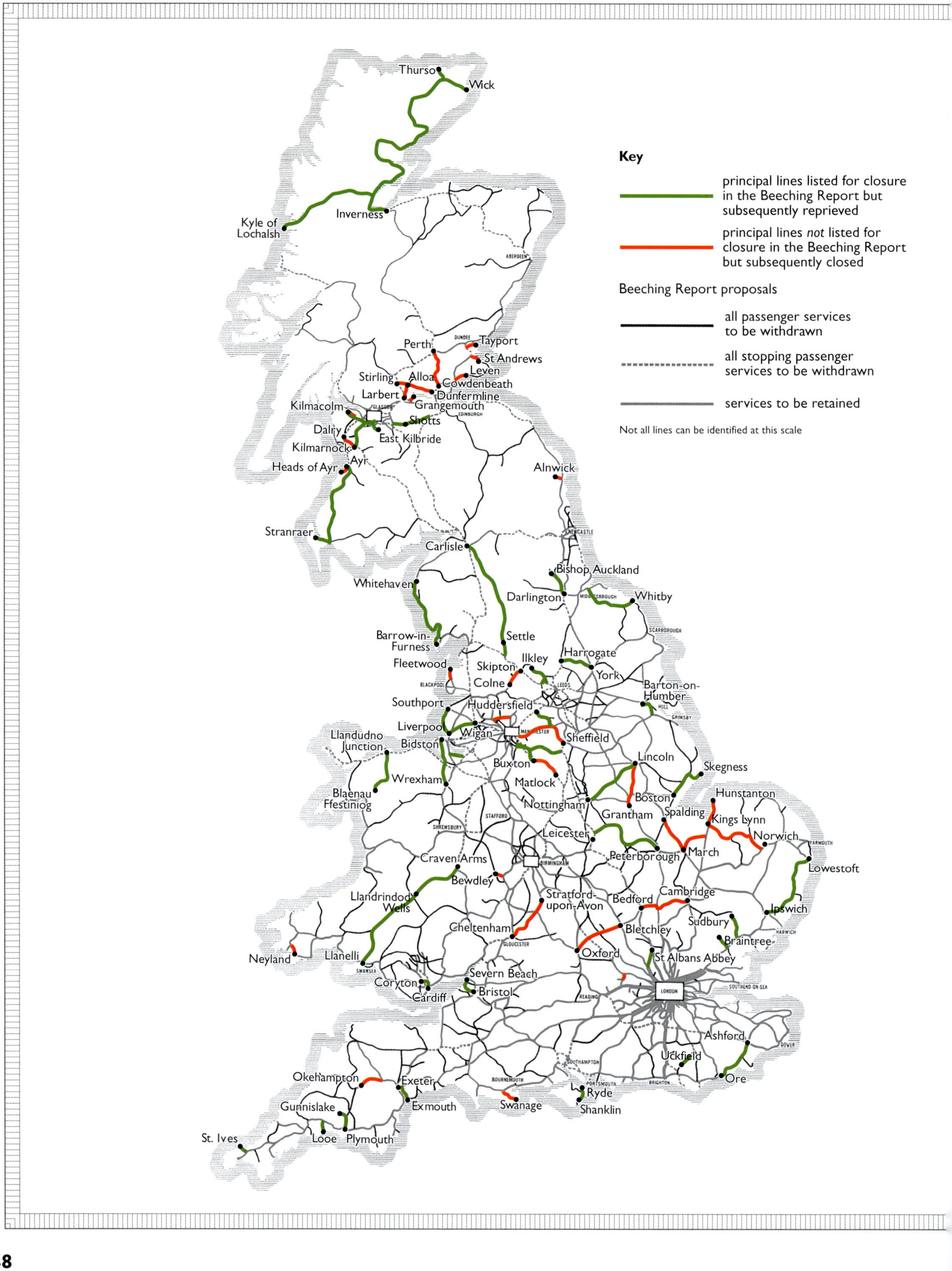
Key
principal lines listed for closure in the Beeching Report but subsequently reprieved
principal lines *not* listed for closure in the Beeching Report but subsequently closed
Beeching Report proposals
all passenger services to be withdrawn
all stopping passenger services to be withdrawn
services to be retained
Not all lines can be identified at this scale
Thurso
Wick
Inverness
Kyle of Lochalsh
ABERDEEN
Perth
DUNDEE
Tayport
St Andrews
Leven
Stirling
Alloa
Cowdenbeath
Larbert
Dunfermline
Grangemouth
Kilmacolm
GLASGOW
EDINBURGH
Shotts
Dalry
East Kilbride
Kilmarnock
Ayr
Heads of Ayr
Alnwick
Stranraer
NEWCASTLE
Carlisle
Bishop Auckland
Whitehaven
Darlington
MIDDLESBROUGH
Whitby
SCARBOROUGH
Barrow-in-Furness
Settle
Fleetwood
Skipton
Ilkley
Harrogate
York
BLACKPOOL
Colne
LEEDS
Barton-on-Humber
HULL
GRIMSBY
Southport
Huddersfield
Liverpool
Wigan
MANCHESTER
Sheffield
Llandudno Junction
Bidston
Lincoln
Skegness
Buxton
Wrexham
Matlock
Blaenau Ffestiniog
Boston
Nottingham
Hunstanton
STAFFORD
Grantham
Spalding
Kings Lynn
SHREWSBURY
Leicester
Norwich
YARMOUTH
March
Peterborough
Craven Arms
BIRMINGHAM
Lowestoft
Bewdley
Stratford-upon-Avon
Bedford
Cambridge
Llandrindod Wells
Ipswich
Sudbury
Cheltenham
Bletchley
HARWICH
Braintree
GLOUCESTER
Neyland
Llanelli
Oxford
St Albans Abbey
SWANSEA
Severn Beach
Coryton
Bristol
LONDON
SOUTHEND-ON-SEA
Cardiff
READING
Ashford
DOVER
Uckfield
SOUTHAMPTON
Okehampton
Exeter
BOURNEMOUTH
PORTSMOUTH
BRIGHTON
Ore
Ryde
Exmouth
Swanage
Shanklin
Gunnislake
St. Ives
Looe
Plymouth

Title	Map of (a) principal lines listed for closure in the Beeching Report but subsequently reprieved, and (b) principal lines *not* listed for closure in the Beeching Report but subsequently closed.
By	Collins Bartholomew (using data compiled by David Spaven)
Date	2011

Strong local campaigns and party political factors ensured that a significant number of routes threatened by the 'Beeching Axe' won reprieves from Ministers of Transport, the ultimate arbiters of what did and did not close. Because local issues were paramount, it is difficult to identify common underlying factors across the wide variety of services saved – from the 4-mile St Ives branch in Cornwall to the 168 miles of the Far North line from Inverness to Wick and Thurso. The latter was easily the longest route reprieved, but other lengthy loss-making lines also survived, including the 88-mile Central Wales line, and the 73 miles of the magnificent Settle & Carlisle line – saved by dogged campaigning and inventive promotion in the face of closure threats and a management run-down which began with Beeching and were not finally lifted until 1989.

The biggest geographical concentrations of reprieves – most of which were given between 1964 and 1970, while Labour was in power – were in the Highlands (always sensitive political territory), west Central Scotland, Merseyside, Yorkshire, Cornwall and eastern East Anglia. The 44-mile East Suffolk line from Ipswich to Lowestoft is an interesting case study. Following strong public protests – there were 1,916 written objections to the Transport Users' Consultative Committee which considered the hardship implications of all closure proposals – the railway secured a Ministerial reprieve. Its costs of operation were then transformed through the application of emerging new methods of operation, in the 'Basic Railway' concept pioneered by BR Eastern Region General Manager, Gerry Fiennes – with tickets sold on board trains ('Paytrains'); destaffing of stations; singling of sections of double track; radio signalling; automatic level crossings; and mechanized track maintenance.

The map also illustrates a little-appreciated story – the surprisingly large number of lines which the report had *not* proposed for closure, but which were in practice axed by the Labour Governments of 1964–70, concerned about railway finances continuing to deteriorate despite the implementation of much of the Beeching closure programme. As the map demonstrates, the pattern of post-Beeching closures was far from evenly spread across the country.

The loss of most of the important cross-country line from Oxford to Cambridge was explored on page 214; not much further north, the late 1960s saw the axing of a collection of routes well beyond the surgery originally proposed by Beeching. The cross-country lines from March and Dereham to King's Lynn lost their passenger services in 1968, while 1969 saw the loss of the King's Lynn–Hunstanton branch (immortalized in a classic TV documentary by the Poet Laureate, John Betjeman) and Dereham's last rail link, to Wymondham. Fortunately the latter has survived as a preserved line, the Mid Norfolk Railway. In a strange sequence of events, Spalding lost its trains to Peterborough and Boston in 1970 as part of the Beeching plan, but then saw the service to Peterborough reinstated less than a year later; to add a further negative twist to the story, the line from March to Spalding was closed as late as 1982.

The other notable concentration of post-Beeching closures was in Fife and along the Forth Estuary. Most of the scenic Fife Coast line was lost in 1965, with the remaining short stub lines to Tayport, St Andrews and Leven not originally proposed for closure – but by the end of 1969 all three had gone. Further up the estuary, Alloa lost all three of its remaining passenger rail routes in 1968 – but happily regained a passenger service to Stirling in 2008, quickly proving to be one of the most successful rail re-openings of modern times. Grangemouth, just across the Forth, lost its passenger train service in 1968, while the last and most serious closure in this part of the world came in 1970 when the direct Edinburgh to Perth main line via Kinross was closed by the London-based Ministry of Transport, to make way for the construction of the M90 motorway by the Edinburgh-based Scottish Office!

BR management's theory was that most passengers from closed branch lines would jump into a car or hop on a bus to the nearest station and continue their journey by train, with BR retaining most of the revenue but ridding itself of many of the costs. In practice, given the lack of integrated timetables and tickets, many passengers simply switched to car or bus for the whole journey.

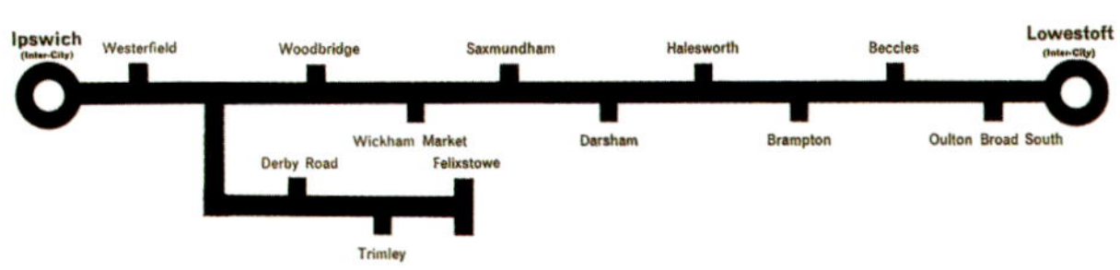

East Suffolk line pocket timetable and map.

Title	Gradient Profiles E1 (part) and M1
From	BR Main Line Gradient Profiles
Publisher	Ian Allan Publishing Ltd
Date	2003 impression of 1966 publication

Gradient profiles have long been used by rail staff and railway enthusiasts to understand the challenge posed to train operation by the vertical topography of specific stretches of route. Unlike the typical map, gradient profiles do not offer a 'bird's eye view' of the horizontal disposition of railways in the landscape, but rather a cross-section along the route of railway, with mileages and gradients numbered beside. Signal boxes traditionally had gradient diagrams for the stretches of track they controlled. These two extracts come from a volume which brought together on a route-by-route basis the gradient profiles of all the main lines and many cross-country routes as they were in 1966, not long before the end of steam traction.

The two profiles here provide an insight to the contrasting challenges posed to locomotive drivers and train service planners by two stretches of the East and West Coast Main Lines (ECML and WCML), both of similar length and at similar latitude, but featuring very different gradients.

These routes were host to the famous – but short-lived – 'railway races' of 1888 and 1895. Prior to 1888 the East Coast route from London to Edinburgh (operated by the Great Northern and North Eastern companies) was much faster than the West Coast route (operated by the London & North Western and Caledonian) – nine hours against ten. The 1887 decision by the East Coast companies to permit 3rd class passengers on the 10 a.m. departure from King's Cross led to a drastic loss of traffic for the West Coast companies. They responded at the start of the 1888 tourist season in early June by matching the East Coast timing with the 10 a.m. departure from Euston. Schedules got faster and faster over the next two months, until the East Coast train did the journey in just 7 hours 27 minutes on 13 August. The next day both sets of companies agreed to minimum train times, and the races stopped.

By 1895, the focus had moved further north with the opening of the Forth Bridge, and Aberdeen was the goal. The mid-

LONDON – NEWCASTLE – EDINBURGH (Continued) E1

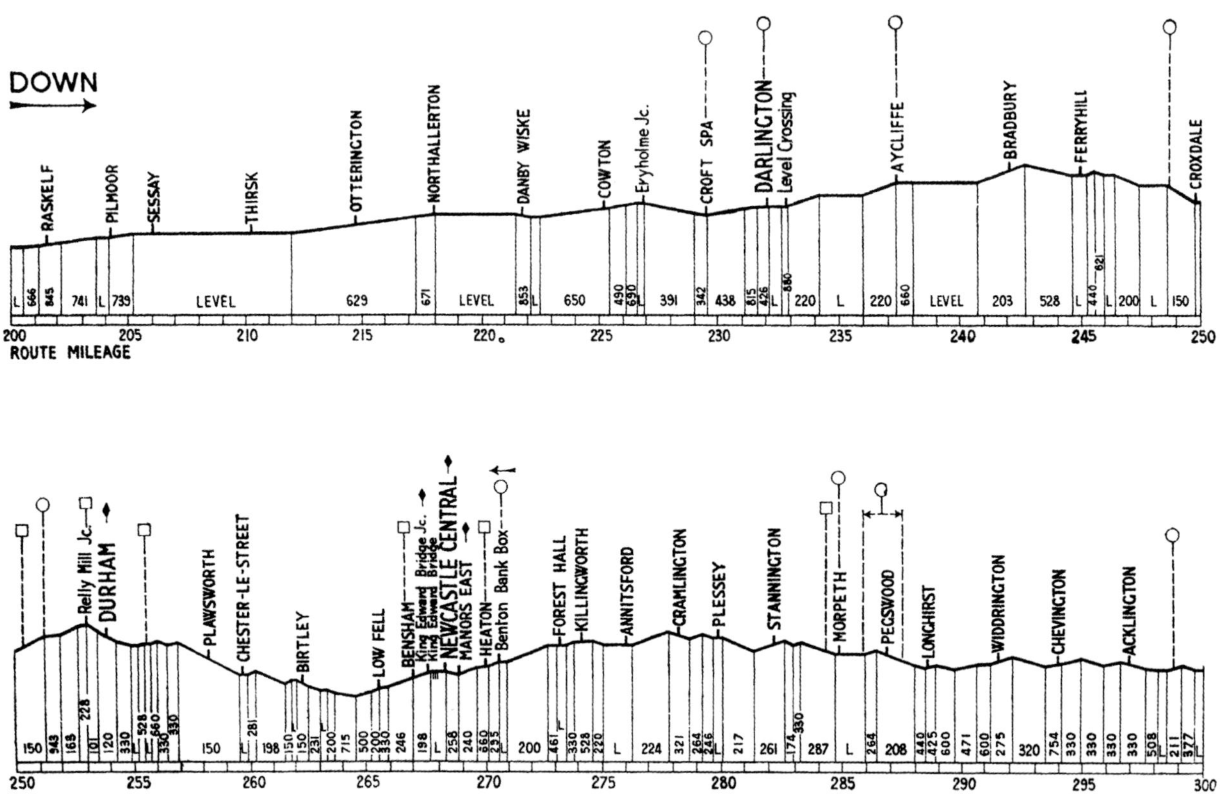

summer races were precipitated by the West Coast companies' concern that the arrival time of the overnight train from Euston did not give enough leeway for the important connection to the Deeside line and the consequent acceleration of their train by 10 minutes. This led to conflicts with the North British at Kinnaber Junction near Montrose, where the two routes converged, and the overnight trains from King's Cross and Euston were now due to pass within five minutes of each other. This started a tit-for-tat slashing of the timings (as in 1888) but their value probably lay more in the resulting publicity than any benefit for passengers, who ended up being deposited in Aberdeen at an unearthly hour in the morning.

The gradient profiles demonstrate just how much more steeply inclined is the WCML, in particular with its four-mile northbound climb at 1 in 75 to Shap summit (and the 10-mile climb to Beattock summit north of Carlisle). On the ECML, much of the route between York and Newcastle is at or near level, with nothing more severe than a very short stretch of 1 in 101 just south of Durham. Throughout its existence, rail companies have responded to the WCML challenge in a variety of ways – by providing 'banking' engines at the rear of trains; double-heading; and latterly with the completion of electrification in 1974. When British Rail completed ECML electrification in 1991, the fastest timing from Edinburgh to London was 3 hours 59 minutes, but the Glasgow to London service was taking over 5 hours. Now the fastest East Coast journey is 4 hours, but the West Coast has seen a radical speed-up – to as little as 4 hours 8 minutes for the 8-mile longer journey – with the introduction of tilting Pendolino trains, designed to overcome the WCML's additional challenge of generally more severe curvature than the ECML.

KEY

SPEED RESTRICTION – SEVERE (30 M.P.H. OR LESS)
" " – MODERATE (35-55 M.P.H.)
" " – SLIGHT (60 M.P.H. OR OVER)

(EUSTON)–LEYLAND–LANCASTER–CARLISLE **M1**

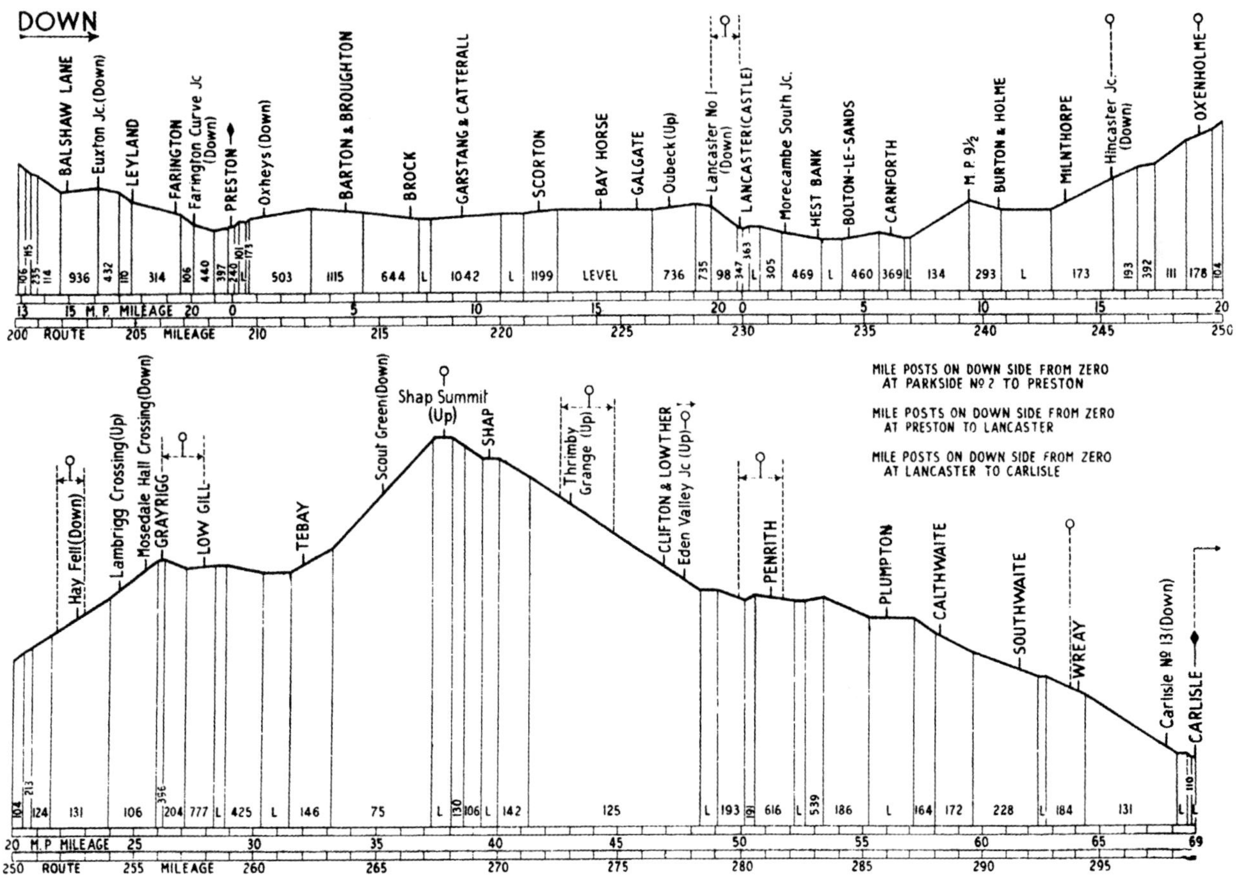

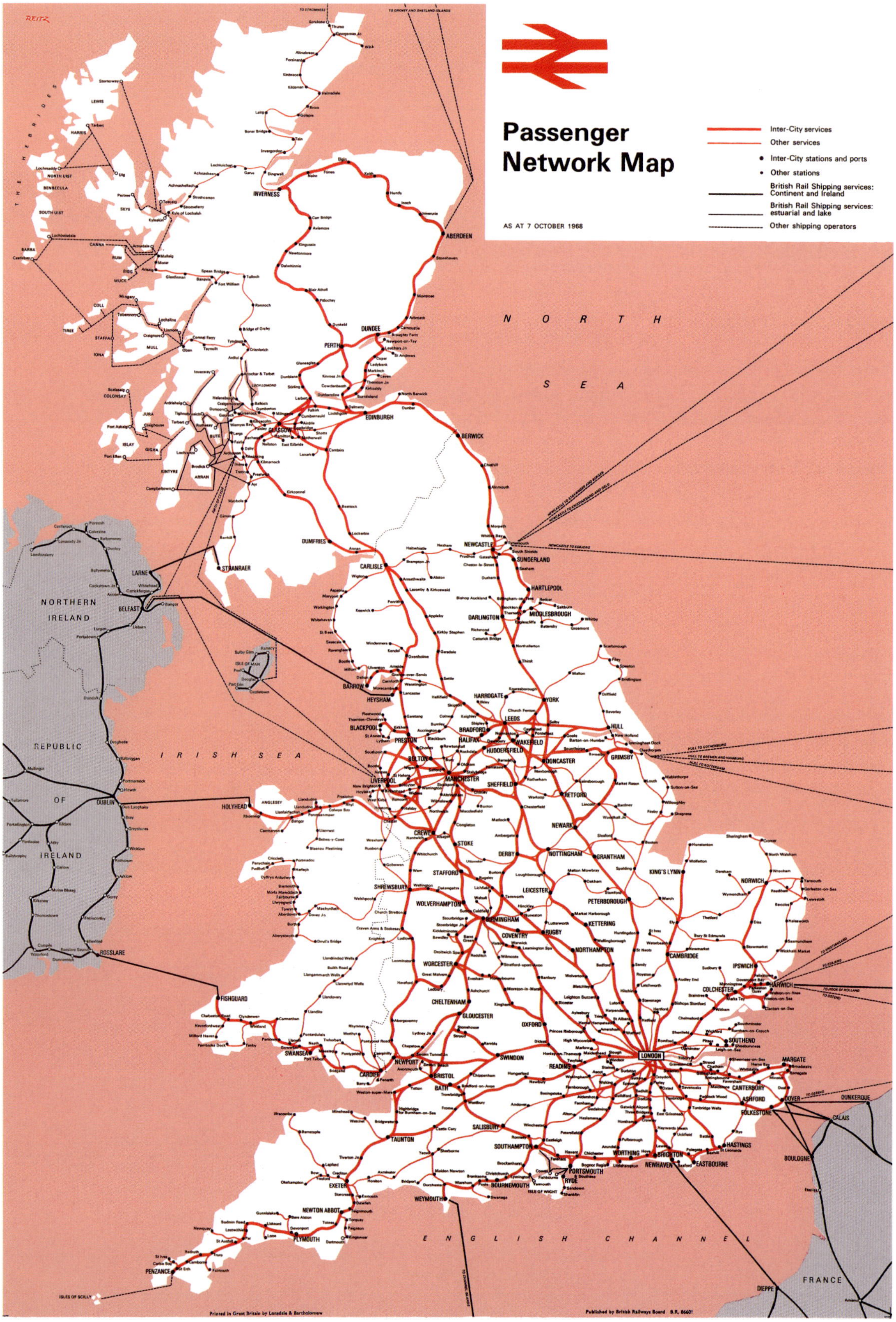
Passenger Network Map
AS AT 7 OCTOBER 1968
Inter-City services
Other services
Inter-City stations and ports
Other stations
British Rail Shipping services: Continent and Ireland
British Rail Shipping services: estuarial and lake
Other shipping operators
N O R T H
S E A
I R I S H S E A
E N G L I S H C H A N N E L
NORTHERN IRELAND
REPUBLIC
OF
IRELAND
FRANCE
INVERNESS
ABERDEEN
DUNDEE
PERTH
GLASGOW
EDINBURGH
BERWICK
STRANRAER
DUMFRIES
CARLISLE
NEWCASTLE
SUNDERLAND
HARTLEPOOL
MIDDLESBROUGH
DARLINGTON
BARROW
HEYSHAM
HARROGATE
YORK
LEEDS
BRADFORD
BLACKPOOL
PRESTON
HALIFAX
WAKEFIELD
HULL
HUDDERSFIELD
DONCASTER
GRIMSBY
BOLTON
LIVERPOOL
MANCHESTER
SHEFFIELD
RETFORD
HOLYHEAD
CREWE
STOKE
NEWARK
DERBY
NOTTINGHAM
GRANTHAM
KING'S LYNN
NORWICH
STAFFORD
SHREWSBURY
LEICESTER
PETERBOROUGH
WOLVERHAMPTON
BIRMINGHAM
COVENTRY
RUGBY
KETTERING
NORTHAMPTON
CAMBRIDGE
IPSWICH
HARWICH
COLCHESTER
WORCESTER
CHELTENHAM
GLOUCESTER
OXFORD
FISHGUARD
SWANSEA
CARDIFF
NEWPORT
BRISTOL
BATH
SWINDON
READING
LONDON
SOUTHEND
MARGATE
CANTERBURY
ASHFORD
DOVER
FOLKESTONE
TAUNTON
SALISBURY
SOUTHAMPTON
PORTSMOUTH
RYDE
WORTHING
BRIGHTON
NEWHAVEN
EASTBOURNE
HASTINGS
BOURNEMOUTH
WEYMOUTH
EXETER
NEWTON ABBOT
PLYMOUTH
PENZANCE
ISLES OF SCILLY
LARNE
BELFAST
DUBLIN
ROSSLARE
DUNKERQUE
CALAIS
BOULOGNE
DIEPPE
Printed in Great Britain by Lonsdale & Bartholomew
Published by British Railways Board B.R. 86601

Title	Passenger Network Map
Publisher	British Railways Board
Date	1968

By the time British Rail published this map in October 1968 (issued with its Western Region timetable), the vast majority of passenger services threatened by Beeching and subsequently withdrawn or reprieved had met their fate. However, the Waverley Route from Edinburgh to Carlisle through the Borders still had three months to run before closure – but the mapmakers had clearly decided that no last-minute campaign would save it, and the line was excised from timetable geography before the trains stopped running.

Route-specific problems with adequate replacement bus services and/or road improvements briefly delayed the end for a few more cross-country routes and branch lines proposed for closure in the 1963 Beeching Report, notably in east Lincolnshire. The map also shows two main lines which had still to close, neither of which were Beeching proposals: the direct Edinburgh–Perth route via Kinross, and the Manchester–Woodhead–Sheffield route. The latter had only been electrified in 1954 but lost its passenger service in 1970 as part of an initiative to eliminate passenger rail route 'duplication' between the two cities, leaving the Woodhead line to concentrate on its core coal traffic – ironically, this eventually went into severe decline, and with its non-standard overhead electrification equipment facing renewal, the line closed completely in 1981.

The reverse of the timetable map is an area map of Western Region (extract shown overleaf) with the appropriate table numbers listed beside each line. The southwest had still to feel the full pain of closures, with the Bridport, Ilfracombe and Minehead branch lines lingering on. The Okehampton branch was a 'post-Beeching' closure in 1972, but remained open for freight and ultimately saw the seasonal reinstatement of passenger services. Most of the Minehead branch was revived as a volunteer-led 'preserved railway', while the Paignton–Kingswear branch moved seamlessly from British Rail control to private ownership at the turn of 1972/1973.

The extracts from the Western Region map on this page show the local routes which survived in the Bristol and Cardiff areas in late 1968. Bristol originally had a reasonable suburban rail network, but, with Temple Meads station some distance from the city centre, it lost the route south through the Mendips to Frome before Beeching; the branch to Portbury on the Bristol Channel opposite Avonmouth during the Beeching period; and the first stretch of the former Midland Railway main line northeast from Temple Meads after the publication of this map. The Severn Beach branch was the only Beeching line reprieve in the Bristol area.

All the Beeching-proposed line closures in the South Wales Valleys were implemented except Cardiff–Coryton, although Bridgend to Treherbert did not succumb until 1970 – subsequently to re-open between Bridgend and Maesteg in 1993, accompanied by other pre-devolution re-openings over the Cardiff 'City' line in 1985 and the Aberdare–Abercynon link in 1988. The Welsh Assembly Government subsequently helped to secure the re-opening of the Barry–Bridgend 'Vale of Glamorgan' line in 2005 and the Cardiff–Ebbw Vale Parkway branch in 2008. All have been successful additions to the rail network.

This national map design had just superseded the early British Railways map style, as illustrated on pages 214–217, but it maintains the latter's twin route categories, in this case distinguishing between 'Inter-City and 'other' services. The number of station types represented has dropped from three in the dense network map of the early British Railways period to just two in 1968 – 'Inter-City stations' and 'other stations'. Given that only selected stations are shown – the complete inventory of Western Region stations is marked on the map on the reverse – the overall impact is a clear and pleasing representation of the national network (plus connecting shipping services) in a familiar geographic style.

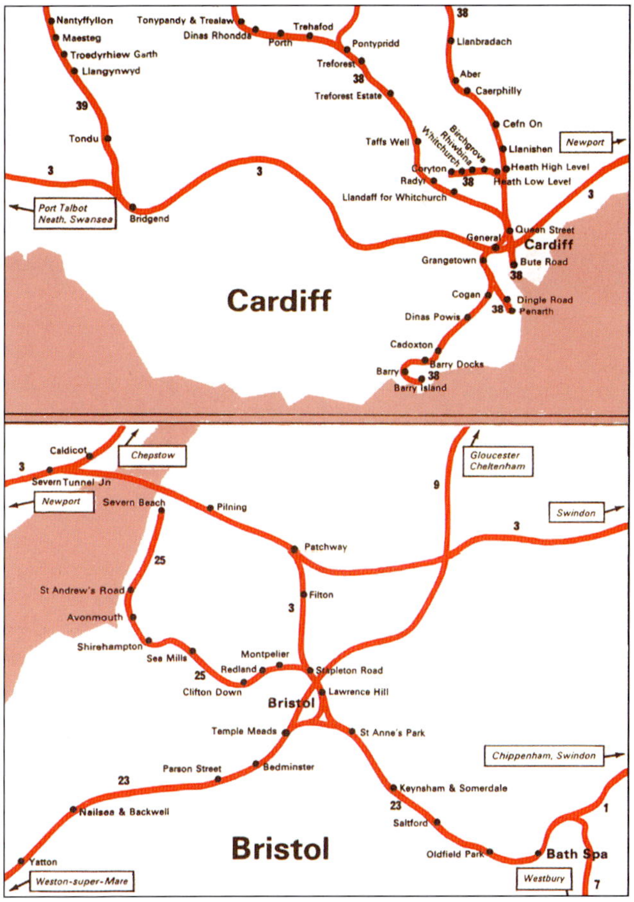

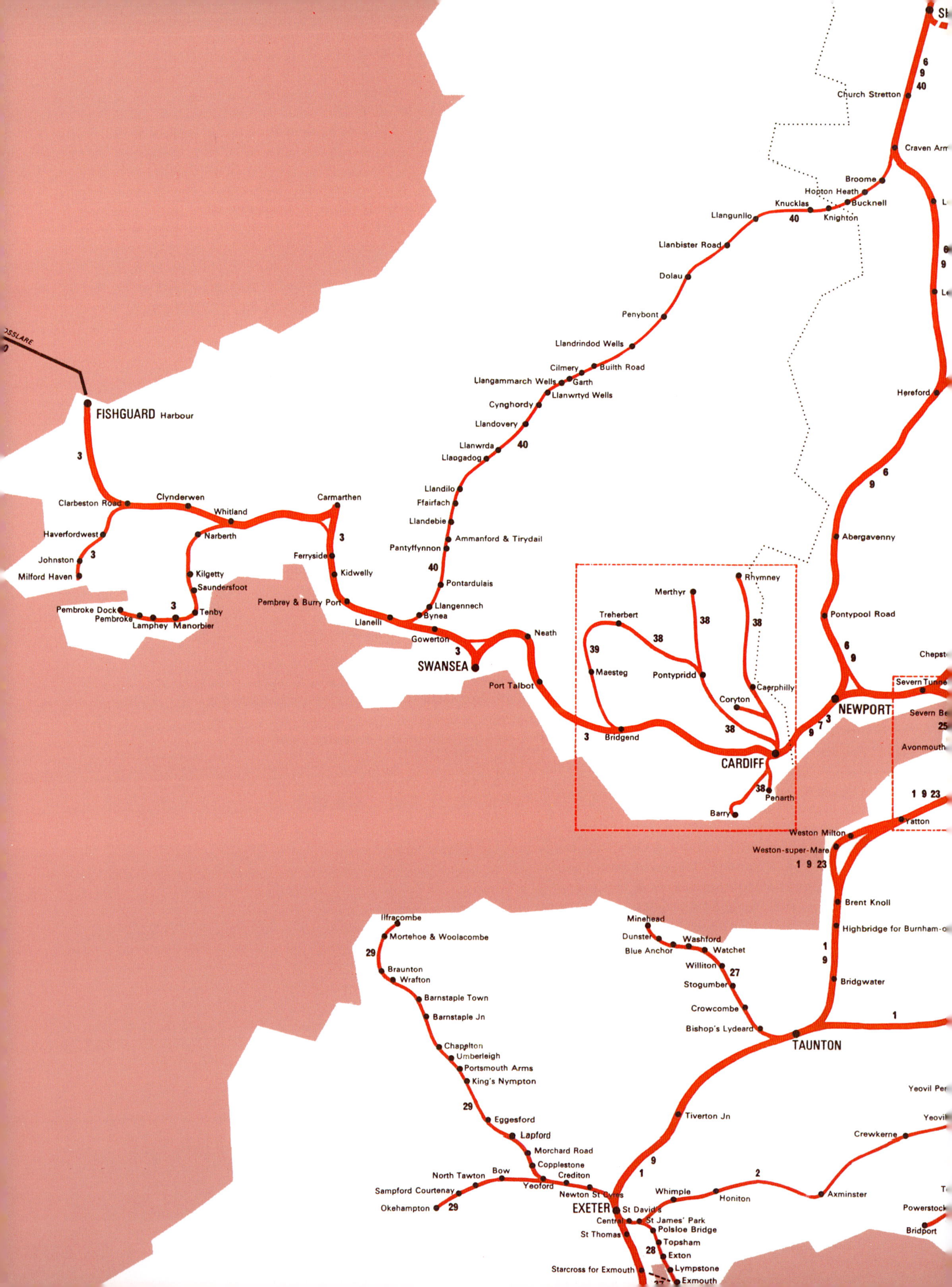

FISHGUARD Harbour
Clarbeston Road
Clynderwen
Whitland
Carmarthen
Haverfordwest
Johnston
Milford Haven
Narberth
Kilgetty
Saundersfoot
Tenby
Pembroke Dock
Pembroke
Lamphey
Manorbier
Ferryside
Kidwelly
Pembrey & Burry Port
Llanelli
Bynea
Llangennech
Pontardulais
Pantyffynnon
Ammanford & Tirydail
Llandebie
Ffairfach
Llandilo
Llangadog
Llanwrda
Llandovery
Cynghordy
Llanwrtyd Wells
Llangammarch Wells
Garth
Cilmery
Builth Road
Llandrindod Wells
Penybont
Dolau
Llanbister Road
Llangunllo
Knucklas
Knighton
Bucknell
Hopton Heath
Broome
Church Stretton
Hereford
Abergavenny
Pontypool Road
NEWPORT
Gowerton
SWANSEA
Neath
Port Talbot
Bridgend
Maesteg
Treherbert
Merthyr
Pontypridd
Rhymney
Caerphilly
Coryton
CARDIFF
Penarth
Barry
Avonmouth
Yatton
Weston Milton
Weston-super-Mare
Brent Knoll
Highbridge for Burnham-o
Bridgwater
TAUNTON
Minehead
Dunster
Blue Anchor
Washford
Watchet
Williton
Stogumber
Crowcombe
Bishop's Lydeard
Ilfracombe
Mortehoe & Woolacombe
Braunton
Wrafton
Barnstaple Town
Barnstaple Jn
Chapelton
Umberleigh
Portsmouth Arms
King's Nympton
Eggesford
Lapford
Morchard Road
Copplestone
Yeoford
Crediton
Newton St Cyres
Bow
North Tawton
Sampford Courtenay
Okehampton
Tiverton Jn
EXETER
St David's
Central
St James' Park
Polsloe Bridge
St Thomas
Topsham
Exton
Lympstone
Exmouth
Starcross for Exmouth
Whimple
Honiton
Axminster
Crewkerne
Powerstock
Bridport
3
40
38
39
38
6
9
7
1
23
25
27
28
29
2

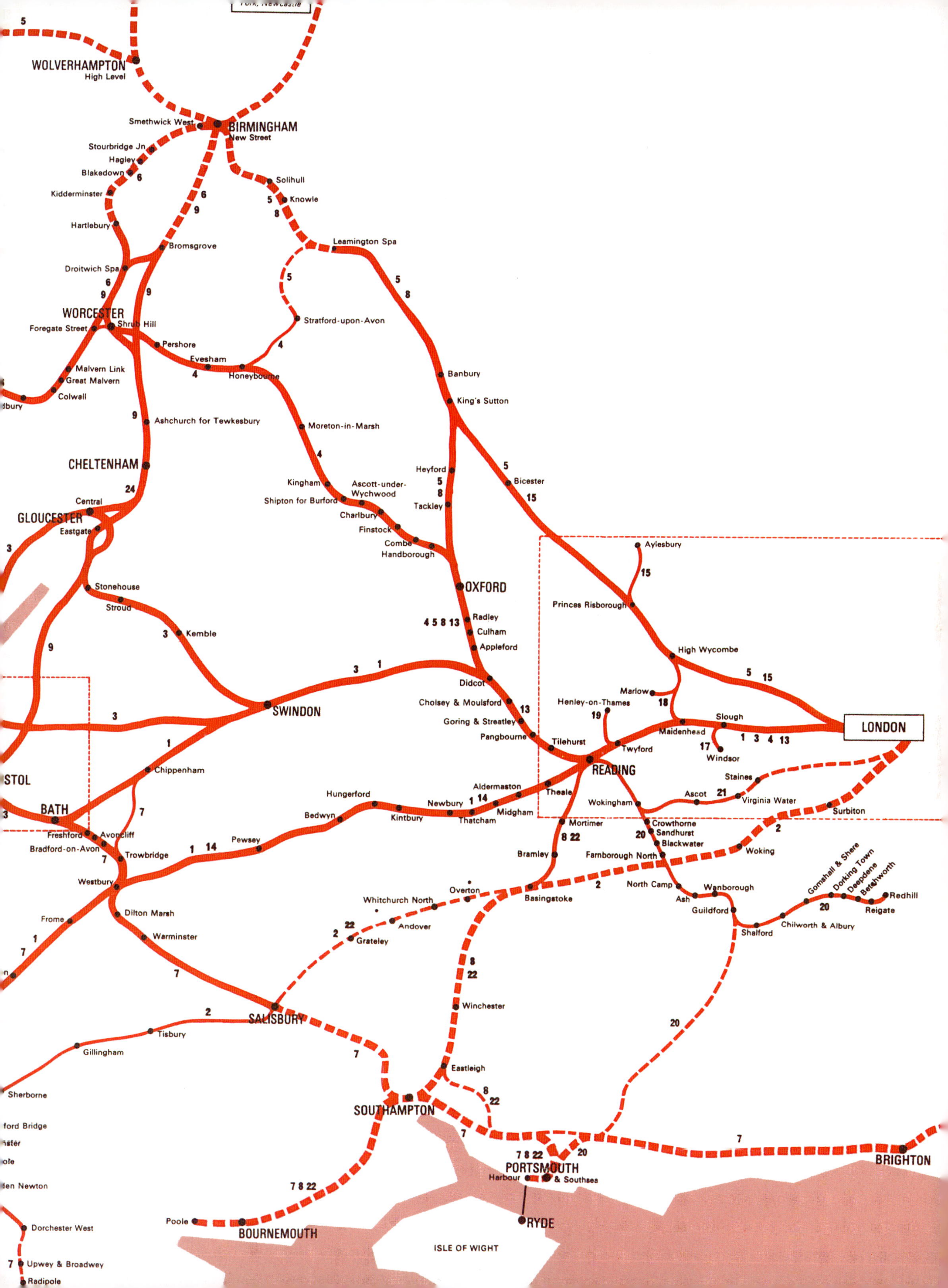
WOLVERHAMPTON
High Level
BIRMINGHAM
New Street
Smethwick West
Stourbridge Jn
Hagley
Blakedown
Kidderminster
Hartlebury
Solihull
Knowle
Leamington Spa
Bromsgrove
Droitwich Spa
WORCESTER
Foregate Street
Shrub Hill
Stratford-upon-Avon
Pershore
Evesham
Honeybourne
Malvern Link
Great Malvern
Colwall
Banbury
King's Sutton
Ashchurch for Tewkesbury
Moreton-in-Marsh
CHELTENHAM
Heyford
Bicester
Kingham
Ascott-under-Wychwood
Shipton for Burford
Charlbury
Finstock
Tackley
Combe
Handborough
Central
GLOUCESTER
Eastgate
Aylesbury
OXFORD
Stonehouse
Stroud
Princes Risborough
Radley
Culham
Appleford
Kemble
High Wycombe
Didcot
Cholsey & Moulsford
Goring & Streatley
Pangbourne
Marlow
Henley-on-Thames
SWINDON
Slough
Maidenhead
LONDON
Tilehurst
Twyford
Windsor
READING
Chippenham
Aldermaston
Theale
Staines
Hungerford
Newbury
Midgham
Thatcham
Kintbury
Bedwyn
Wokingham
Ascot
Virginia Water
Surbiton
BATH
Freshford
Avoncliff
Bradford-on-Avon
Trowbridge
Pewsey
Mortimer
Crowthorne
Sandhurst
Blackwater
Bramley
Farnborough North
Woking
Westbury
Overton
Whitchurch North
Basingstoke
North Camp
Ash
Wanborough
Guildford
Gomshall & Shere
Dorking Town
Deepdene
Betchworth
Redhill
Reigate
Dilton Marsh
Frome
Andover
Grateley
Shalford
Chilworth & Albury
Warminster
Winchester
SALISBURY
Tisbury
Gillingham
Eastleigh
Sherborne
SOUTHAMPTON
PORTSMOUTH
Harbour
& Southsea
BRIGHTON
Poole
BOURNEMOUTH
RYDE
ISLE OF WIGHT
Dorchester West
Upwey & Broadwey
Radipole

THE REVIVAL OF THE PASSENGER RAILWAY

With growing concerns about the environment and oil security – and line closures largely a thing of the past – British Rail continued the development of a rail system geared up to fight road and air competition. The national network of InterCity services became a successful world-leading brand.

MILEPOST 21 Inter-City Makes the Going Easier 258

1971/85 Inter-City Sleeper and InterCity Networks 260

1974 Corby Steel Works and Iron Ore Workings 262

1977/2010 Nottinghamshire Coalfield 264

1970/72/85 North Highland Lines: Tourist Brochures 266

1935/85 Bartholomew's Exmoor/North Devon 270

MILEPOST 22 Going International: Train Ferries and the Channel Tunnel 274

Bideford
East-the-Water
Abbotsham
Alverdiscott
Landcross
Littleham
Huntshaw
Weare Giffard
Buckland Brewer
Monkleigh
Frithelstock
Great Torrington
Little Torrington
Langtree
Bulkworthy
Peters Marland
Merton
Huish
Petrockstow
Meeth
Shebbear
Thornbury
Milton Damerel
Bradford
Sheepwash
Black Torrington
Highampton
Hatherleigh
Cookbury
Beaford
Yarnscombe
St Giles in the Wood
Newton St Petrock
Abbots Bickington
Alwington
Parkham
Goldworthy
R. Torridge
A 386
A 388
A 3072
B 3227
B 3232
B 3220

The term 'Inter-City' was first used by the Western Region of British Railways in the early 1950s when it gave the name to an express train that ran between London Paddington, Birmingham (Snow Hill) and Wolverhampton (Low Level). Leaving Paddington at 9 a.m. on weekdays and returning from Snow Hill at 5 p.m., it provided a fast two-hour journey between Britain's two largest cities. The train was discontinued in 1967 when the newly electrified West Coast Main Line between Euston and Birmingham (New Street) began to provide a faster service between the two cities.

With steam on the way out British Railways made an effort to change its outdated image. Preceded by the experimental XP64 livery, the standardized livery colour of 'Rail Blue' was introduced in 1965 (along with the shortened name of British Rail) while the brand name of 'Inter-City' was first introduced in 1966 for its long-distance expresses – coaches on these routes were given a new livery of two-tone blue and pale grey with 'Inter-City' picked out in white.

The introduction of Inter-City 125 High Speed Train (HST) sets on the Western Region main line from Paddington to Bristol, Cardiff and Swansea in 1976 was a great success. With a maximum speed of 125 m.p.h. (compared to earlier diesels' 100 m.p.h. limit) and vastly improved passenger comfort, they were soon attracting passengers back to Britain's much-maligned railways – helped along the way by innovative marketing such as Jimmy Saville's famous TV advert exhortation to 'let the train take the strain'. Within a few years the HSTs were also operating on the East Coast Main Line from King's Cross, to Edinburgh, Aberdeen and Inverness; the Midland main line out of St Pancras; as far west as Swansea and Carmarthen; and to the southwestern tip of England at Penzance. The completion of West Coast Main Line electrification in 1974 also saw vastly improved services on this vital Anglo-Scottish railway artery.

In 1986, British Rail was divided up into different business sectors, with 'InterCity' (rather than 'Inter-City') as the flagship business. The livery for InterCity coaches and locomotives was changed to dark grey on the upper parts, light grey on the lower, separated by red and white stripes (the famous 'Swallow' logo was added in 1989). The new InterCity sector was divided into seven operating divisions: East Coast; West Coast; Midland; Great Western; Great Eastern; Cross-Country; and Gatwick Express. The sector was also responsible for sleeping car and Motorail car-carrying services. It was a very successful and profitable formula and within a few years British Rail was running more 100 m.p.h. services each day than any other country in the world.

Following the completion of East Coast Main Line electrification in June 1991, locomotive-hauled electric InterCity 225 trains entered service between King's Cross and Edinburgh – their brand name refers to their potential top speed of 225 km/h so it is rather misleading! In fact they 'only' operate up to a maximum speed up 125 m.p.h.

The success of the rebranding of Britain's InterCity network was not lost on others, with the brand name being launched

'Castle' Class No. 5053 'Earl Cairns' heads 'The Inter City Express' into Birmingham Snow Hill in the 1950s.

in West Germany in 1971 to replace the former Trans Europe Express trains – Deutsche Bundesbahn even paid a royalty for using the name to British Rail for many years. Introduced in 1991, Intercity-Express (ICE) high-speed trains now operate throughout Germany while the EuroCity brand name is used by international train services within Europe. InterCity is also the brand name used by Iarnród Eireann for its principal services in the Republic of Ireland.

In Britain the Railways Act 1993 heralded the privatization of Britain's railways. Taking effect on 1 April 1994, British Rail was fragmented into numerous companies, with many of the new train operators taking on franchises not dissimilar to the old InterCity sector operating divisions. Following privatization the term 'InterCity' soon fell into disuse, and is no longer used by train operating companies, reflecting the absence of a single centrally branded network of main line services.

Sporting the InterCity 'Swallow' logo a Class 90 electric locomotive pauses at Crewe with the 'Sussex Scot' in September 1989.

Poster produced for British Rail advertising the new HST 125 Inter-City expresses from Paddington to Bristol and South Wales in 1977.

TV personality Jimmy Saville promoted the highly successful British Rail campaign 'This is the age of the train' during the early 1980s.

In 1956, in response to criticism that new post-nationalization railway rolling stock looked out of date, the British Transport Commission (of which British Railways was part) set up its own Design Panel to produce a co-ordinated design policy. While locomotives and passenger coaches were the primary focus of its initial work, eventually every important aspect of the railway business, its operations and supporting activities came within the sphere of design policy – and this had an impact far beyond Britain. Writing in *British Rail Design* (1986), the Director of Design at Danish State Railways commented:

> *The impact of British Rail Design was immediate, and an encouragement to all those in the international railway community who wanted to improve their ware. British Rail became the leader of a new wave in public design. During the following decade it hit first The Netherlands, then Denmark, Austria, France, Federal Germany, Norway and Switzerland where the national railways followed the British example. All the way, British Rail was the guiding spirit and midwife in the creation of better design standards in public service.*

British Rail's enviable corporate identity incorporated, after 1965, a single new font ('Rail Alphabet') and standardized layouts for posters, leaflets, pocket timetables and maps. In *British Rail 1948–83: A Journey by Design* (1983), Brian Haresnape reflects that, 'this simplicity and constancy provides a strong visual contact with the user amidst the general clamour and confusion created by the surrounding commercial advertising.'

The 1971 Inter-City Sleeper network map comes from an information leaflet for the internal Scottish service from Glasgow and Edinburgh (combined at Perth) to Inverness, which ceased operation in 1990. This is an interesting example of a map produced in a clearly 'geographic' style, but using 'diagrammatic' elements to allow the reader to understand the starting point, main intermediate calls, and destination of individual sleeper services. The accompanying text explaining the amenities on sleeping cars has a delightfully old-fashioned service touch – 'Hot-water bottles are supplied on request.'

As a result of increased day-time train speeds, competition from low-priced airlines, and rail privatization, today's sleeper network is much reduced from the wide spread of towns and cities served in 1971, with just two core routes left – from London to Scotland (Aberdeen, Edinburgh, Fort William, Glasgow and Inverness) and from London to Penzance.

By 1985 the branding had changed from Inter-City to InterCity, and the map of (day time) InterCity Routes is far more diagrammatic, with distances distorted to simplify the presentation, and the British coastline no longer shown. This is certainly an easy map to understand – and does show key connecting cross-country routes – but unlike the 1971 map does not provide information on which towns and cities are directly linked by through train services.

Title	Your journey by Inter-City Sleeper
Publisher	British Railways Board
Date	1971

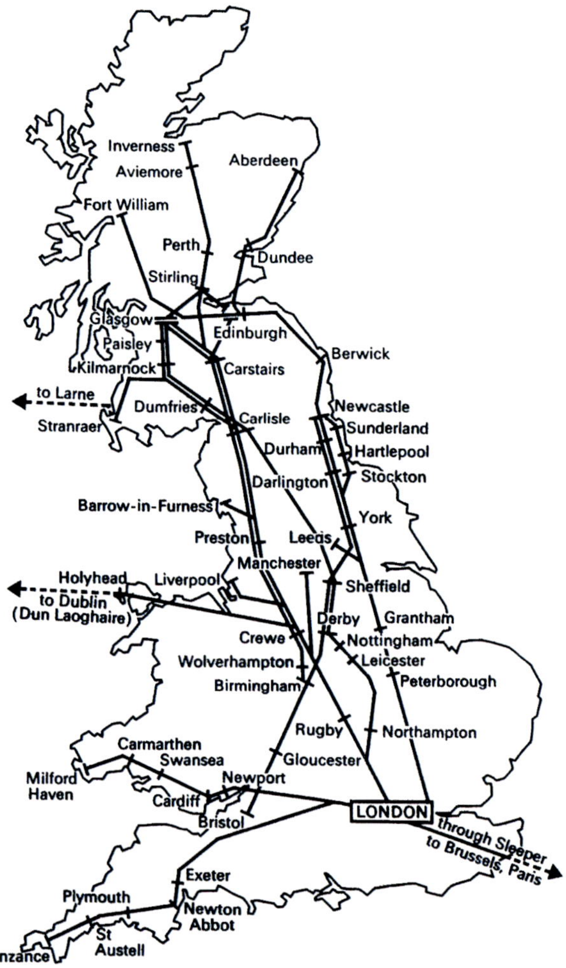

Title	InterCity Routes
Publisher	British Railways Board
Date	1985

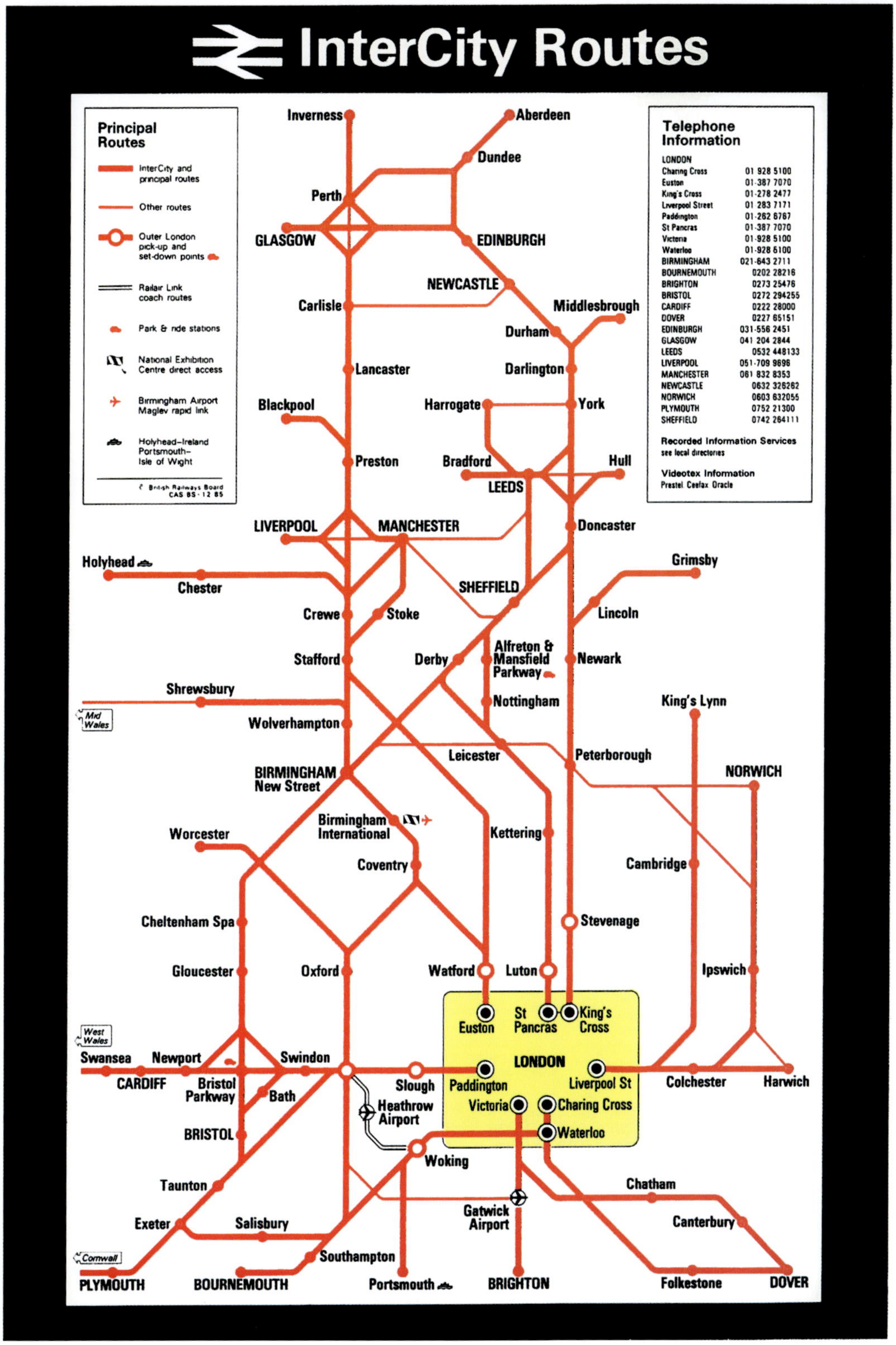

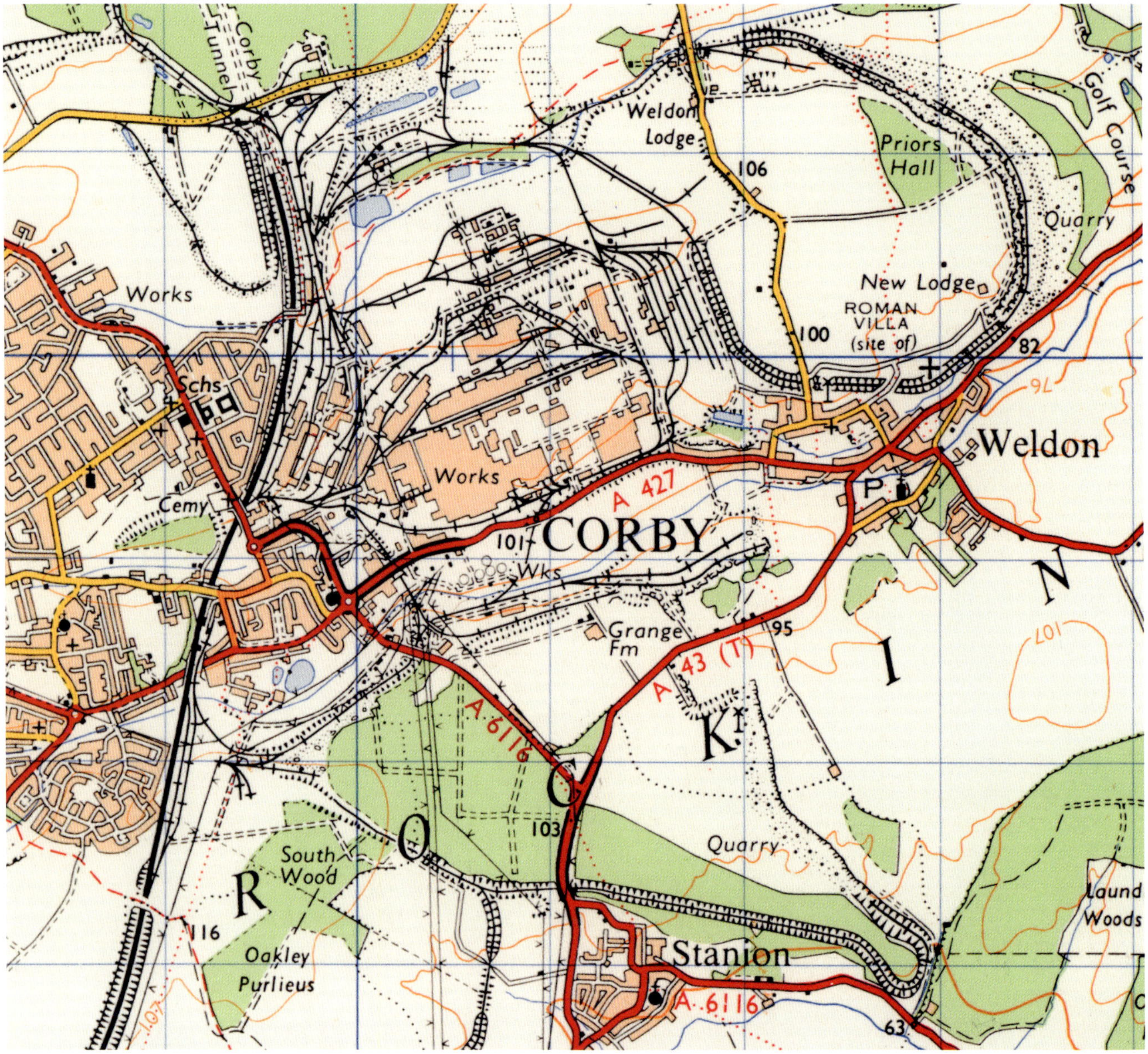

Series	Ordnance Survey 1:50,000 First Series
Sheet	141 (Kettering & Corby) (enlarged)
Date	1974

The principal theme of this map extract is the proximity of a major industry to its source of raw materials, a model that was the norm before the era of globalization.

Local supplies of iron ore – part of a band of ironstone stretching from Lincolnshire to Oxfordshire – led to the transformation of the Northamptonshire village of Corby into one of Britain's major centres of steel production. The Corby blast furnaces, built between 1910 and 1917, were acquired in 1920 by the Glasgow steel manufacturer Stewarts & Lloyds which began construction of a new iron, steel and tube-making plant on the site in 1933. This became the most advanced integrated steel production plant in England, attracting a huge influx of unemployed men and their families to the town, not least from the Glasgow and Lanarkshire areas, where the steel industry was undergoing a downturn. At one point, it was estimated that over 70 per cent of the population of Corby were of Scottish descent, and it is still known as 'Little Scotland'.

As recently as 1959, new railway lines were being built to access ore pits southeast (the quarry shown north of Stanion) and southwest of the steel works, to complement nine

RAILWAYS ON ORDNANCE SURVEY MAPS

From their earliest years, railways have been an important and distinctive feature of Ordnance Survey maps, not least in the popular 1 in. to 1 mile series, which was ultimately replaced by the metric 1:50,000 scale maps from 1974 onwards. The first 1 in. map was published in 1801 and in the next 70 years the series was extended to cover the whole country. At the start of the 20th century colour was introduced, and the 1 in. map reached its final form in the Seventh Series published from 1952.

The Corby map extract is an example of the First Series of the 1:50,000, produced by photographic enlargement of the existing 1 inch map. The new sheets incorporated a number of design and colour changes – for example motorways were coloured blue instead of red, with B roads changing from brown to orange – but the representation of railways was left entirely untouched. However, the subsequent continuing growth of the road system at the expense of the railways ultimately led to changes in rail feature representation, as shown in the contrasting key extracts of the 1974 Corby map and the current edition.

Where a distinction was once made between single track and multiple track rail routes, there is now just one category. A single-track railway takes up a substantially narrower land corridor than a single-carriageway road, but this is not evident from the OS map – neither of the line symbols for road or rail is strictly to scale. Cutting and embankment symbols for rail (and road) remain unchanged, while level crossings are now marked 'LC' rather than by a narrowing of the road line.

In a reflection of rail development nationally, the distinctive ticked line for freight sidings is now used instead for 'light rapid transit system, narrow gauge or tramway', and the hard-to-identify 'ladder' symbol for narrow gauge lines has been dropped. A further sign of rail expansion is the new pecked line symbol for 'track under construction'. The white circle symbol for 'station closed to passengers' – crucial in the immediate post-Beeching era – became superfluous and was eventually dropped, but a new yellow circle symbol has appeared more recently to identify 'light rapid transit system station'. In many ways, the Ordnance Survey has successfully reflected the changing nature and importance of railways in the landscape in the early 21st century – and doubtless there will be further changes to come.

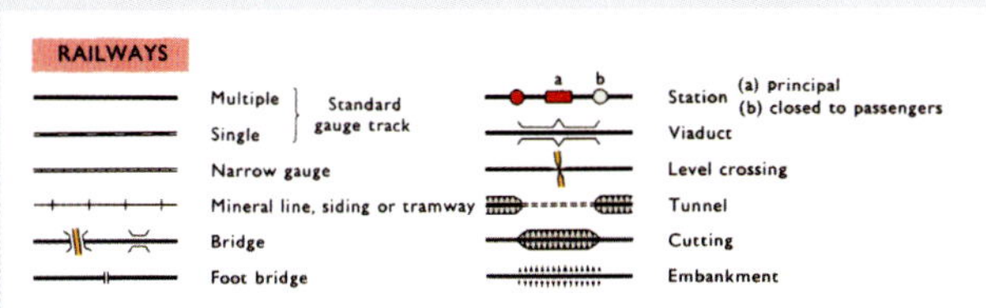

Ordnance Survey railway symbols, 1974

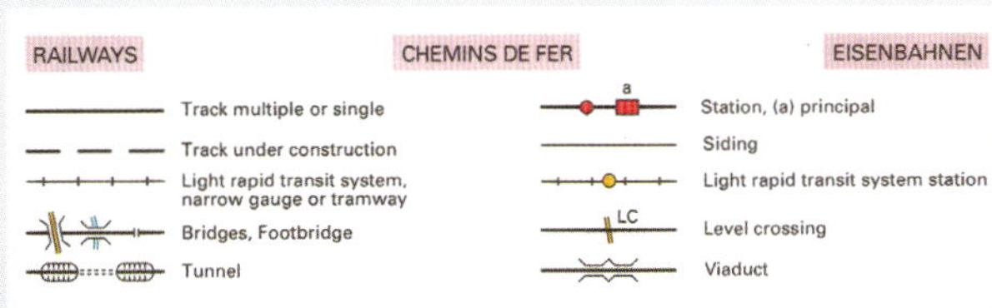

Ordnance Survey railway symbols, 2004

existing short branch lines to the east, northeast and northwest. The map captures the steel works and nearby iron ore workings just a year before the Government agreed a programme that would lead to the phasing-out of steel making at Corby, in favour of plants supplied with higher grade imported ore. The last load of home-produced ore was despatched to the Corby steel works in 1980, although tube manufacturing continued on site, supplied with steel by rail from the remaining plants.

The map shows sidings leading to every part of the works, reflecting the strong hold rail still had on the movement inwards of raw materials and fuel and outwards of finished products and by-products. The internal sidings were controlled by the British Steel Corporation, which operated its own shunting locomotives and wagons to handle the short hauls of iron ore and other traffic movements within the site – 'external' traffic being received from or passed on to British Rail at exchange sidings to the west of the works. The cross-country railway from Kettering on the Midland Main Line through Corby to Oakham and Leicester had lost its passenger services in 1959, and so all railways shown on the map were at the time freight-only.

An experimental Corby–Kettering rail passenger service began in 1987, but suffered from an irregular timetable and unreliable older rolling stock, and was withdrawn in 1990. However, a new station with a bus and taxi interchange opened in 2009, adjacent to the site of the old station, and enjoys an hourly service to Kettering and London. Steel still arrives at the tube works (now owned by Tata Steel) by rail from the Margam plant in South Wales, while the railway has also benefited from Corby's relatively central position in Britain with the establishment of a rail-connected distribution hub on former industrial land to the south of the tube works.

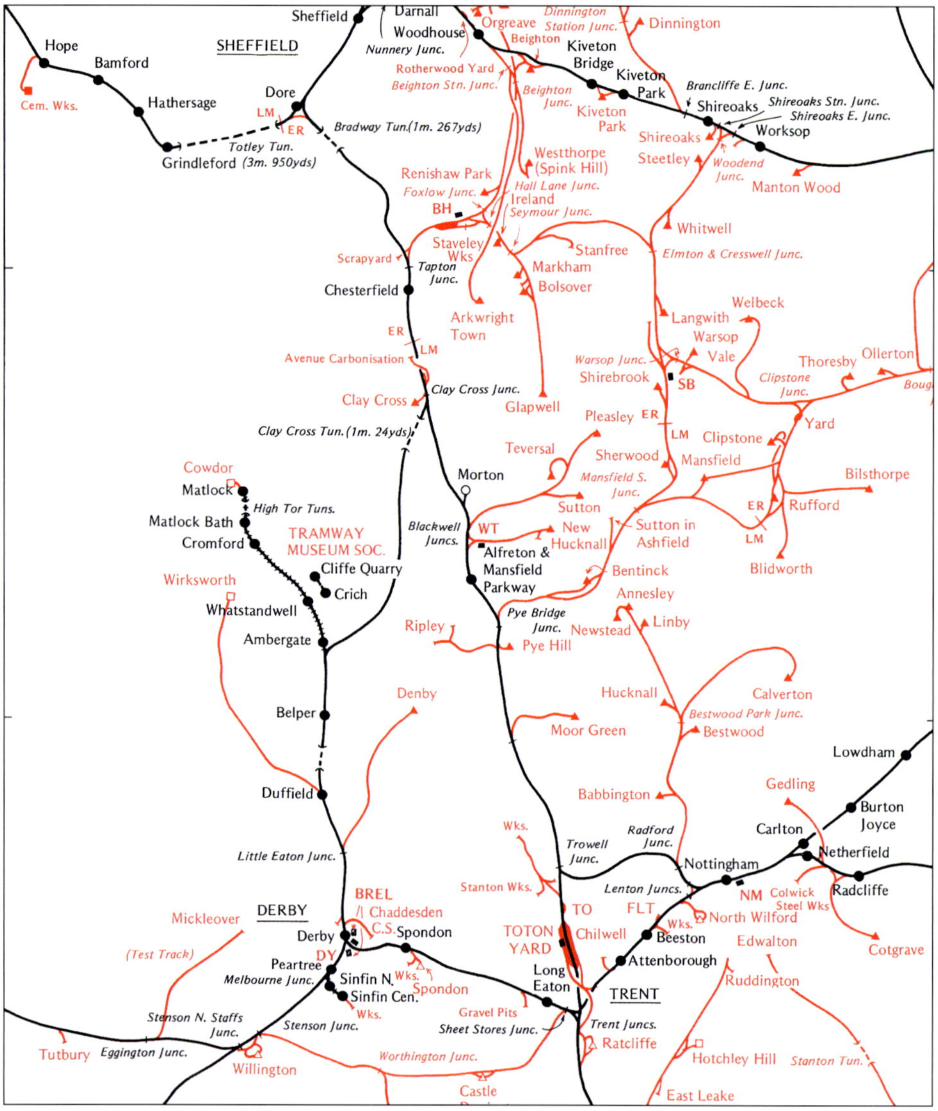

Extract	Page 47
From	*Rail Atlas Great Britain and Ireland – 1st Edition*
Author	S K Baker
Publisher	Oxford Publishing Company
Date	1977

These maps, from the popular 'Baker atlas' used extensively by railway professionals and enthusiasts, illustrate the enormity of change in the Nottinghamshire rail network – and the associated coal industry – between 1977 and 2010. This regional change is a microcosm of the national switch from a freight-led railway system to a passenger-led one.

The most striking feature of the 1977 map is the 43 collieries shown, the majority served by an intricate network of freight-only branch lines over a swathe of country bounded by the Midland Main Line through Alfreton & Mansfield Parkway and the cross-country passenger routes linking Nottingham and Sheffield with Lincoln. The coal mined by the National Coal Board was primarily destined for power stations (four of which are marked on the map), and also supplied to industrial and domestic markets in full trainloads, as well as traditional wagonload traffic on local 'trip' trains which Beeching had identified as a key loss maker – but which survived until shortly before the 1984–5 miners' strike. Many other rail freight terminals and private sidings are shown on the map, including cement works, steel works and a Freightliner terminal ('FLT') west of Nottingham.

By 2010 an extraordinary transformation had taken place. The enormous contraction of the coal industry is immediately apparent, with only four collieries or opencast coal loading points marked – although by 2011 just one of Britain's last three deep mines was actually operational in Nottinghamshire, at Thoresby. Rail freight's fortunes have closely mirrored the coal industry, with the closure or mothballing of many of the freight-only branches. Nottingham Freightliner terminal had also gone, axed as part of a nationwide rationalization of the container-carrying network in 1987.

Extract	Page 53
From	*Rail Atlas Great Britain and Ireland* – 12th Edition
Author	S K Baker
Publisher	Oxford Publishing Company
Date	2010

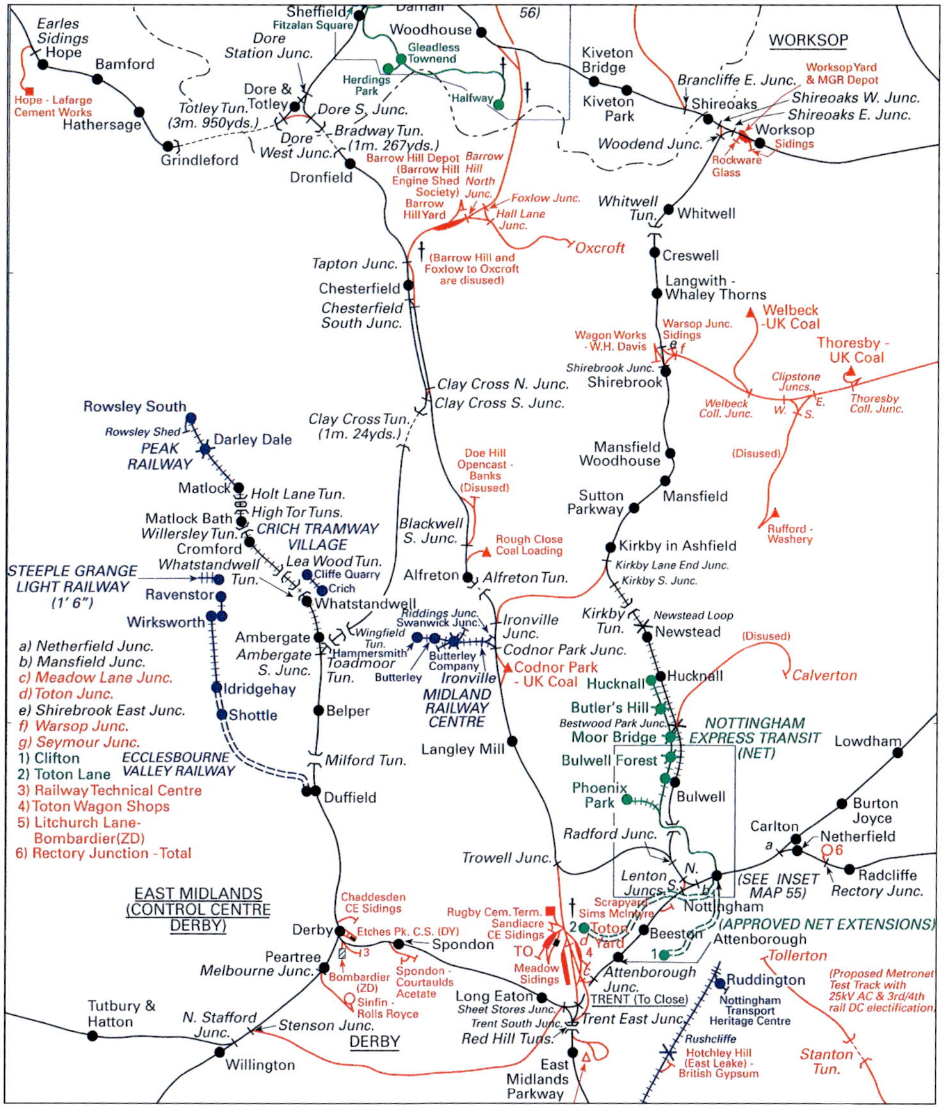

More positively, two of the longest freight routes had been incorporated into the new passenger service on the 'Robin Hood' line from Nottingham to Worksop, which was opened in stages between 1993 and 1998. This heavily populated area, including large towns such as Mansfield, had lost the last of its passenger trains in 1964, and the reinstated service has proved to be very popular. At 29 miles in length, it is by far the longest passenger route re-opening of modern times in Britain.

The Robin Hood line has interchanges with another rail development illustrated on the map – the new Nottingham Express Transit tram system. In the far north can also be seen the southern arms of one more new tram network – the Sheffield-focused South Yorkshire Supertram. A final noteworthy contrast between the 1977 and 2010 maps is the latter's profusion of preserved railways, with no less than six such operations in a relatively small area.

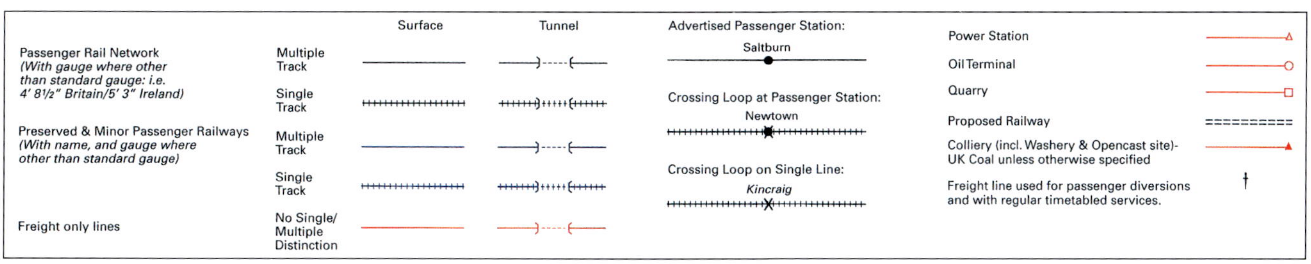

There is a very long tradition of railway companies producing descriptive line leaflets, brochures and booklets for rail passengers – particularly leisure travellers and tourists – and these have typically been accompanied by maps. From Victorian times, such maps have simplified and/or straightened the representation of rail routes in order to ease understanding and fit a convenient printed format for use on trains.

Over its nearly 50-year life, British Rail produced many maps of holiday and scenic routes throughout England, Scotland and Wales – with the apotheosis of this genre perhaps being in the Scottish Highlands, with so many outstanding routes through dramatic 'unspoilt' landscapes.

The three map leaflets here come from a 15–year period and show different styles of simplification designed to suit the fashion or purpose of the time. The 1970 leaflet shown on these pages was produced before the Inverness–Kyle of Lochalsh line faced its second threat of closure – it was first reprieved in 1964 – and its illustration is predominantly abstract. The text evokes traditional attractions for the tourist – scenery, history and folk tales – but also draws the reader's attention to the role of the railway in modern times, including the Ness Viaduct which 'carried the tremendous military traffic of two world wars to the major military installations at Invergordon and to Scapa Flow – a vast achievement for a single-track line.'

By 1972, the Government was planning to withdraw grant aid from the Kyle passenger service – parallel roads had been improved, and the line faced losing most of its freight traffic when the Stornoway ferry switched from Kyle to Ullapool. This time British Rail did not have to face accusations of a run-down – a number of its managers based in Glasgow, shocked by the Ministerial consent to close the Waverley Route through the Borders and the direct Edinburgh–Perth line, were determined to do what they could to help ensure that no more major line closures would take place in Scotland. Rail supporters were surprised by the publication of the 1972 promotional leaflet (shown overleaf), coming in the midst of a campaign to save the railway, and enthusiastically assisted British Rail in its distribution to potential rail users at hotels and guest houses in Inverness and Dingwall. The text of the leaflet unapologetically concludes with reference to the *realpolitik* of the planned fate of this grant-dependent railway:

> *The grants are to be discontinued and as a result the British Railways Board has no alternative but to close the line. The closure will not take place before the end of 1973.*

In practice, a determined campaign of opposition – and the prospects of increased freight traffic from siting an oil production platform yard near the railway – led to a second reprieve, and the Kyle line has gone on to be a major summer attraction for tourists, not just in timetabled

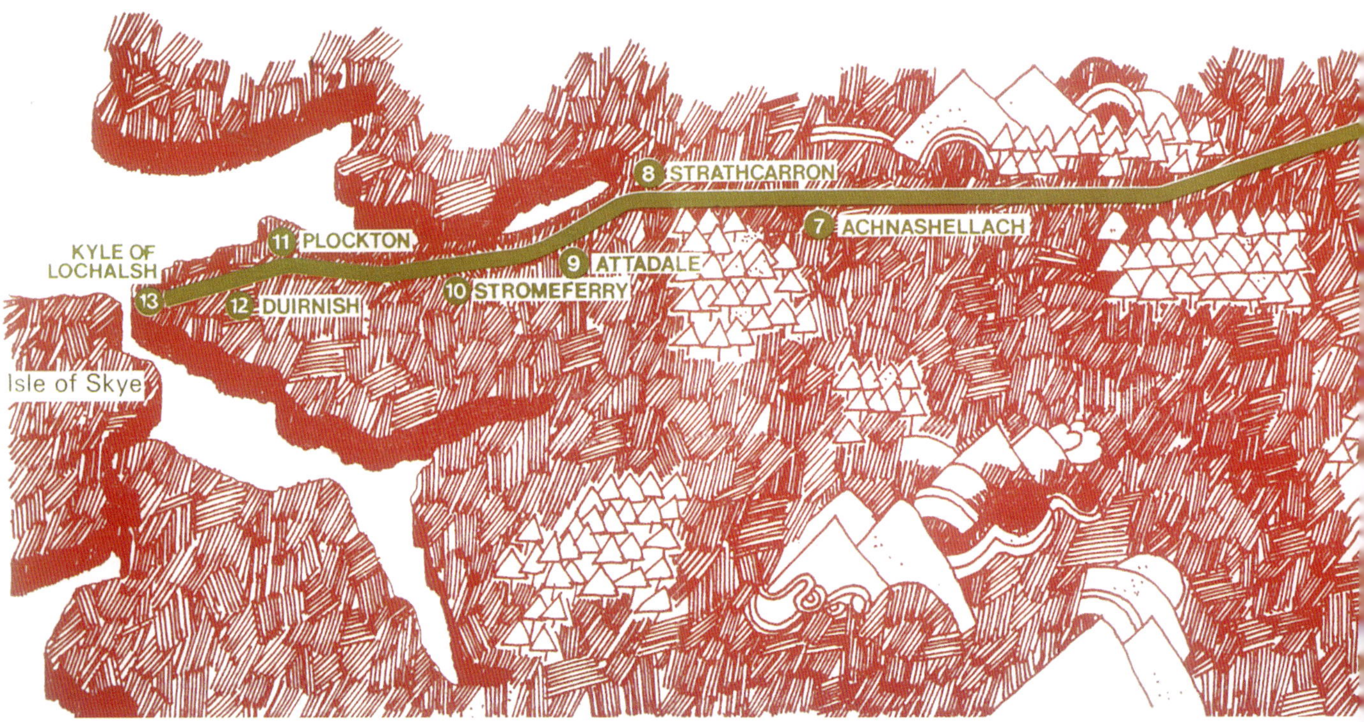

trains but also in charter services from as far away as the south of England.

The 1985 brochure shown overleaf was part of a joint promotion by British Rail and the Government agency, the Highlands and Islands Development Board, which comprised a folder of four different line guides under the branding 'Great Railway Explorations of the Scottish Highlands'. With a lengthy pictorial-style map and a detailed text, it extols the attractions of the Far North line to Wick and Thurso, delves into railway history, and also gives a nod to the modern world served by the railway:

> *Formerly a naval base – and centre of a notorious mutiny in 1931 – Invergordon has a fine harbour, facing Udale Bay and Cromarty Bay on the Black Isle. As you leave Invergordon, the distillery emerges on your left, along with the now closed aluminium factory. To the right lies Saltburn and Nigg Bay, the cradle of vast North Sea oil rigs, and site of the British National Oil Corporation's Beatrice Terminal, where some of Britain's oil comes ashore.*

Title	Rail route. Inverness/Kyle of Lochalsh
Publisher	British Rail
Date	1970

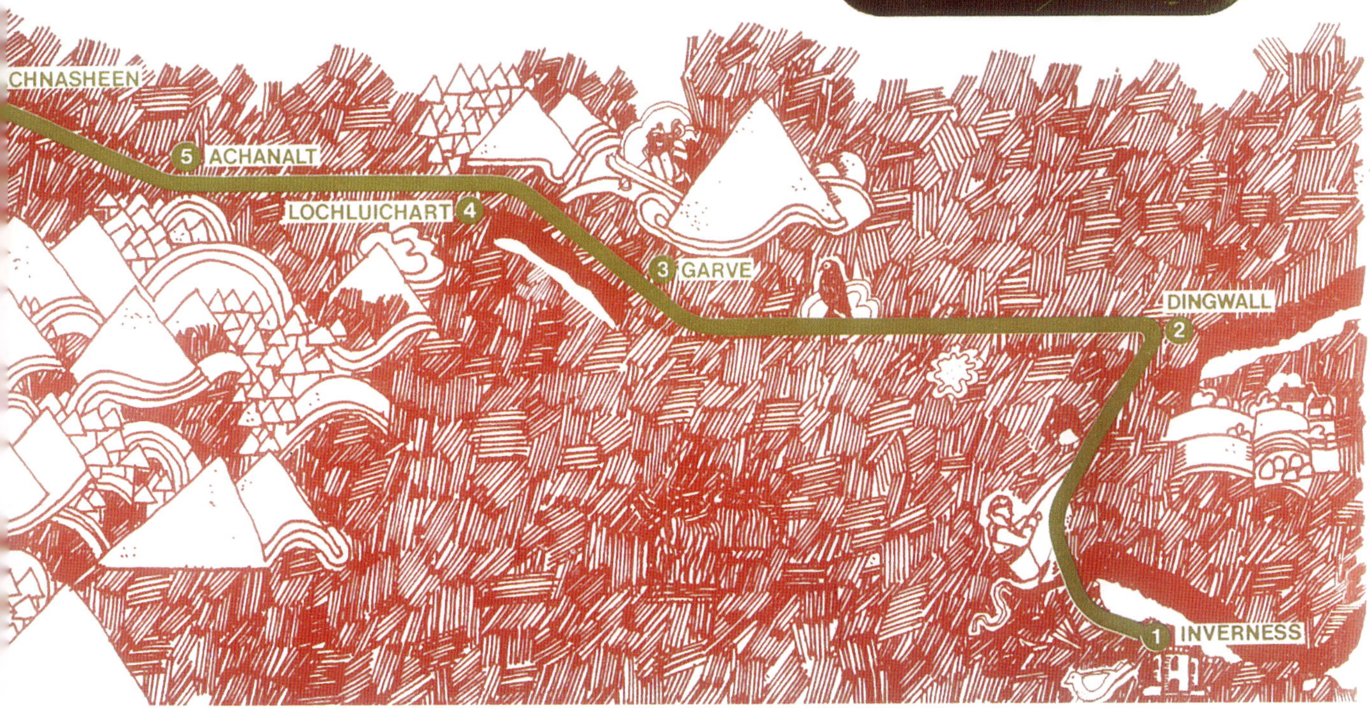

Title	Inverness–Kyle of Lochalsh
Publisher	British Railways Board
Date	1972

Title
Great Railway Explorations of the Scottish Highlands – Inverness to Wick and Thurso
Publisher
Highlands and Islands Development Board/ British Rail
Date 1985

INVERNESS has its roots deep in ancient times, and it provides a splendid starting point for today's Highland journey.

About 563 AD, Inverness was the headquarters of King Brude, monarch of the Picts, the ancient people of Scotland. Local history has it that Brude was converted to Christianity by followers of Saint Columba, on top of Craig Phadraig, a hill overlooking Inverness. Craig Phadraig is a remarkable example of a vitrified fort, and it offers superb views.

View from Inverness Castle, with the River Ness sparkling below.

In modern Inverness you'll find a wealth of shops, bars, hotels and guest houses. You can stroll along the banks of the Ness. Visit Eden Court Theatre. Take a trip to historic Culloden Moor, to such lovely glens as Affric and Moriston, and to Loch Ness with its elusive monster. Shortly after pulling out of Inverness, you'll cross the Ness Viaduct, built in 1862. To your right lies the harbour. To your left stands Tomnahurich Hill, shaped like an upturned boat with trees on top. Known as the hill of the yew trees, or even Fairy Hill, Tomnahurich is the local cemetery.

THE TRAIN SLOWS TO CROSS THE SWING BRIDGE OVER THE CALEDONIAN CANAL.

The Canal is a splendid affair, completed in 1822 by one of Britain's greatest engineers, Thomas Telford. Telford's creation links the North Sea with the Atlantic, joining Inverness to Fort William via Lochs Ness, Oich and Lqchy. If you enjoy cruising and sailing through spectacular scenery, this Canal is made for you. The £160,000 lavished on it by Telford was money well spent.

On through Clachnaharry – originally 'Clachnafhaire', it means Stone of Watching, a lookout point. Glance left up to the rocks. A statue used to stand on the pedestal there, commemorating a clan battle of 1454, fought between the Munros and the McIntoshes. Some say that a Grey Lady has haunted the pedestal since the statue was blown down in a storm a few years ago.

To your right the Kessock Bridge sweeps in spectacular style over the firth to the Black Isle – which is neither black nor an island.

Six miles along the firth you'll see to your right two castles close to each other. The nearer, now a ruin, is Redcastle, built in 1179 by King William the Lion. Beyond is the 16th century Kilcoy Castle. For hundreds of years both were occupied by the MacKenzies of Kintail.

Groam Farm, one of the largest farms in the area, goes by on the left before you go through Clunes.

Watch for Wardlaw Church, for many years the repository of the famous Wardlaw Manuscripts, which record the history of the Clan Fraser from 916 to 1674. You can see the Manuscripts in Inverness Museum.

Approaching Beauly, look right. The ruins of Beauly Priory, founded by John Bisset Lord Lovat, have stood there since 1230.

Over the River Beauly. On your left is Beaufort Castle, seat of the Clan Fraser. The castle was built on the site of the old Castle Downie, which was destroyed by the Duke of Cumberland after the Battle of Culloden in 1746. Culloden was the last pitched battle fought in Britain, and marked the end of the Stuart family's claim to the Crown.

Now you reach your first stop. Some 80 years ago Muir of Ord was the cattle centre of the north of Scotland, and its name derives from 'Moor of the Hammer' – the hammer being the cattle auctioneer's gavel. You'll journey over a golf course which was the site of the cattle market. On your left stood Carn nan Chlarsair, 'the Cairn of the Harper' where Lord MacDonald of the Isles was killed by an Irish harper centuries ago.

AS YOU LEAVE MUIR OF ORD, BEN WYVIS LOOMS UP. FROM ITS SUMMIT, 3,433 FEET UP, YOU CAN ENJOY MAGNIFICENT VIEWS OF THE HIGHLANDS. MIND YOU, IT'S A LONG WALK WHICH CAN TAKE ALL DAY.

Don't be tricked by the rounded bulk of Ben Wyvis – it's one of the highest mountains in Easter Ross. This is rewarding country for walkers, birdwatchers – and anyone who likes exploring.

Conon Bridge goes past. Once renowned for the colour and variety of its cottage gardens, the village is on the River Conon, said to have been the haunt of the dog otter.

After Maryburgh comes Dingwall. The line forks. The Kyle Line wanders left toward mountainous Wester Ross and ends in Kyle of Lochalsh, in the shadow of the Isle of Skye.

Rosemarkie Beach on the Black Isle

like a house surrounded by a wall. It's said to have been built on the site of – and with stone from – an even older castle, which was the stronghold of the Mormaers, or Great Men, who ruled the area.

FINLAEC, FATHER OF MACBETH, WHO INSPIRED SHAKESPEARE'S PLAY, IS SAID TO HAVE BEEN A MORMAER. AND MACBETH HIMSELF WAS BORN IN THE CASTLE.

(The Highlands and Islands are studded with an abundance of castles – you could easily spend a whole holiday castle hunting.)

Along the shores of the Cromarty Firth. Salmon and sea trout can be caught by fly in the River Conon at low tide, so shallow are the waters.

Soon you see the causeway which carries the road across the firth.

Foulis Castle passes on your left. The 18th century seat of the Clan Munro, it was designed in Dutch courtyard style, and replaced an earlier building which had burnt down.

In the background you can still see Ben Wyvis. The ruins of Castle Craig, once home of the Bishops of Ross, are on your right on the foreshore of the Black Isle.

Past Evanton on the Allt Grand River. It used to be known as Novar, after a local estate.

The Black Isle is a peninsula, and a fair, green country of woods, meadows and cornfields, with friendly, pretty villages. The name may derive from the area's lack of snow in winter (it lies low and near the sea). Or it may be because until the 18th century it was uncultivated 'black' moorland.

Dingwall originally was called Inverpeffer, but when the Vikings occupied these lands, the name became 'Thing Vollr', the Valley of the Court.

True to its past, Dingwall remains the administrative centre of the area. You're in a town with a Royal Charter, which was granted by King Alexander II in 1226, when the Vikings were driven out.

Gaze to your left and you'll see a monument to Sir Hector MacDonald. A local crofter's son, he fought in Britain's South African Wars, and rose from the rank of private to general.

Northwards out of Dingwall. On the left look for the castle which is more

Reminder of distant days: The Gate of Negropatam.

On the left, look out for one of the most unusual monuments in the Highlands. Crowning Fyrish Hill is a copy of the Gate of Negropatam, a reminder of Britain's days in India.

This exotic creation was commissioned in the 18th century by General Sir Hector Munro, to relieve local unemployment.

Watch for whisky distilleries on your right as you reach Alness. The Teaninich and Dalmore distilleries are two of many in the Highlands and Islands. While all of

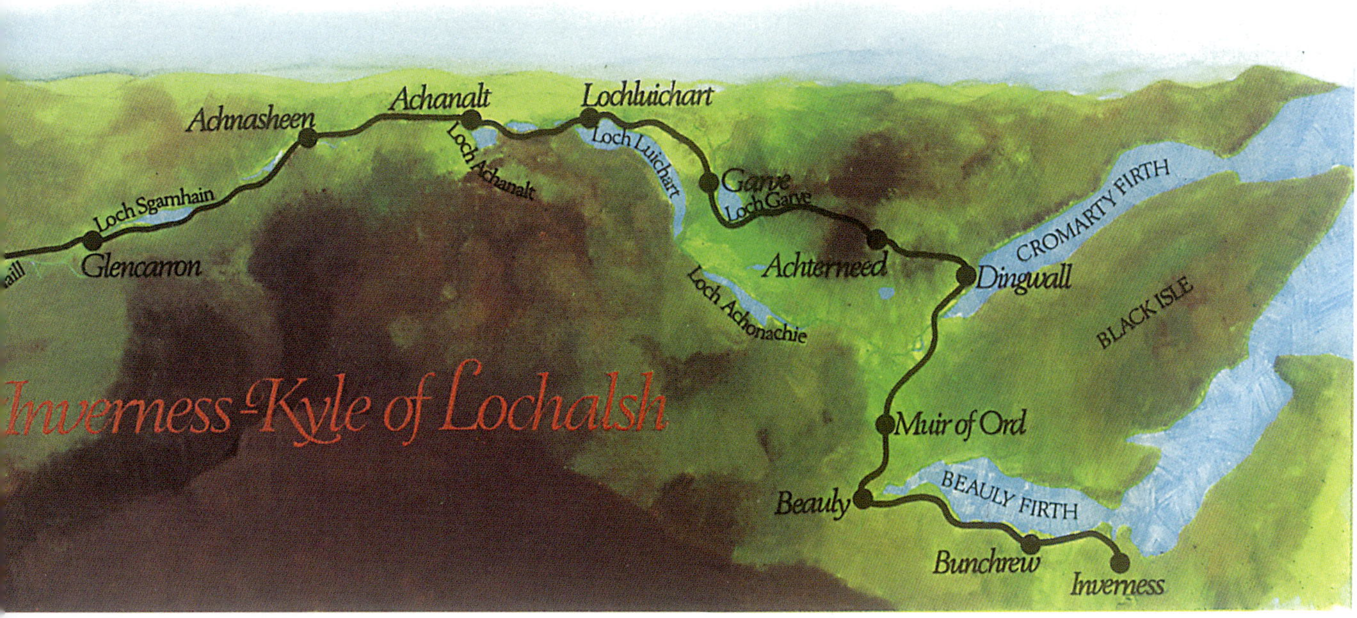

of 'the water of life',
tours – ask at the
etails.
coming a town, and
f nearby Invergordon
prophecy will have
your left, just before

ES FROM A
LLIAM GORDON.
, HE
E TOWN AFTER

Hugh Miller's Cottage, Cromarty.

ore.
romarty, the small
u're interested in
he little museum at
of the most
Past Kildary now,
Fearn, amid fertile
Fearn there's an old
ich was moved there
m Edderton in 1338.
h Eye, one of the few
lochs around here,
gleams on the
right.
North-west
to Tain, ancient
Royal Burgh on
the Dornoch
Firth. Near the

IMPORTANT
ND MANY
KING JAMES IV,
INT DUTHUS IS

nother Royal Burgh,
herland, Dornoch
lfers. If you ever
ch course – one
d his round so much
er had playing golf.'
ed in 1224, and as
ied there. On a less
ne' marks where
s burned to death

A mile north of Tain, look left to see Morangie Distillery. The track skirts the shore of the firth. You pass Meikle Ferry, where at one time you could have a boatman ferry you to Dornoch.

In years to come a bridge may vault the firth, making it even easier to explore the constant delights of this winding coastline.

On through Edderton, with Balblair Distillery on your left. On your right, on a clear day, you'll see Skibo Castle, former home of the late Andrew Carnegie, the Scot who became a millionaire in the United States.

Further on and to the right is the ruin of Spinningdale's old mill. To the left of the mill is a large, bungalow-style house, once home of the film star James Robertson Justice.

The forested slopes of Struie Hill gather above you. You go by the old parish kirk of Kincardine. Ardgay arrives.

THE DISTINCTIVE WHITE KINCARDINE STONE, A FOCAL POINT OF THE WINTER MARKET, STANDS IN THE SQUARE.

To the west, Strath Carron runs deep into glorious mountain scenery. The roads twist off towards the rugged west coast, some 40 miles distant as the eagle flies. One day you might tour over the high passes to the beaches, lochs, sea-angling harbours and islands. For the moment, though, you head north.

The road bridge crossing the channel on the right goes to Bonar Bridge. The channel links the Dornoch Firth with the Kyle of Sutherland.

Over the Carron River you go, to Culrain, the halt for Carbisdale Castle. Now a Youth Hostel, Carbisdale is the only Scottish castle built this century. Beyond is Lamentation Hill, scene of a fierce battle in 1650 between the Marquis of Montrose and the Covenanters, who were fighting to establish their own kind of religious freedom.

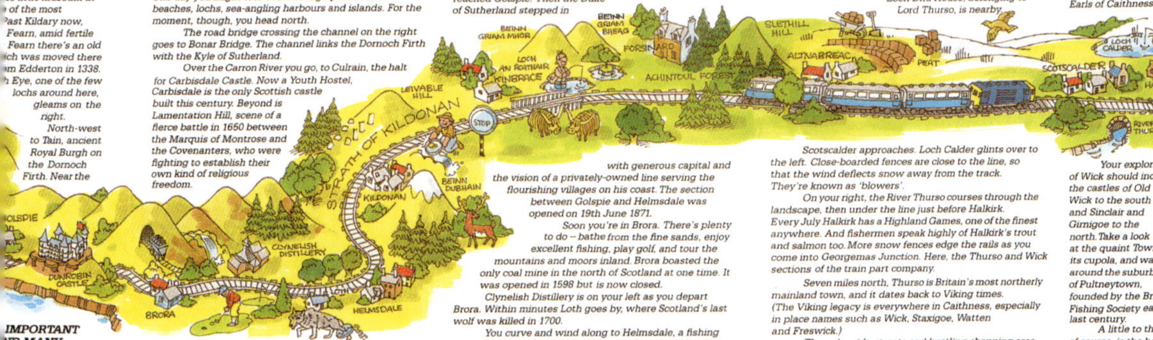

Invershin's request stop goes by. Rocky cuttings interrupt your view. On the left, on a good clear day, you can see all the way to north-west Sutherland, where the mountains cut a ragged edge along the sky.

You veer right and you're in Lairg. Set at the southernmost tip of long, fish-filled Loch Shin, Lairg is the hub of the scattered communities of the north and west. On sheep days Lairg is thronged and noisy and a most enjoyable place indeed. The local buses serve such far-flung romantically named destinations as Lochinver, Elphin, Kinlochbervie and Inchnadamph.

A steep climb heralds Lairg Summit, then it's downhill through Strath Fleet to Rogart. On the way Tressady Lodge appears on the left, to be followed by a stone cairn raised in memory of the Canadian Prime Minister MacDonald, who was a Sutherland man.

Rogart, another important sheep-market centre, precedes The Mound. From here a railway used to run to Dornoch. On you go along the shores of Loch Fleet.

Beinn a Bhragaidh rises on your left. On its top is an unmissable statue, of the first Duke of Sutherland.

Dunrobin Castle, Sutherland

Golspie is next, a busy fishing town and holiday resort. The town provides all the modern services required by the farming and crofting communities which are sprinkled around Sutherland's immense reaches of mountains, glens, lochs, and moors. There's a stone on the old bridge in Golspie, which in chiselled Gaelic, provides the rallying focus for the Clan Sutherland.

Dunrobin Bank means a steep clamber for the train. On the right the North Sea spreads out. Clear days bring views of the mountains of distant Banffshire, down towards Aberdeen.

Straight out of a fairytale, Dunrobin Castle appears on the right, amid its great park. For centuries the seat of the chiefs of the Clan Sutherland, turreted Dunrobin commands superb sea views, and welcomes visitors in the summer.

Finance for the line actually petered out when it reached Golspie. Then the Duke of Sutherland stepped in with generous capital and the vision of a privately-owned line serving the flourishing villages on his coast. The section between Golspie and Helmsdale was opened on 19th June 1871.

Soon you're in Brora. There's plenty to do – bathe from the fine sands, enjoy excellent fishing, play golf, and tour the mountains and moors inland. Brora boasted the only coal mine in the north of Scotland at one time. It was opened in 1598 but is now closed.

Clynelish Distillery is on your left as you depart Brora. Within minutes Loth goes by, where Scotland's last wolf was killed in 1700.

You curve and wind along to Helmsdale, a fishing village nestling between high moorland ridges. Note the new road which runs near the mouth of the river. It sweeps over the site of a castle where the Earl and Countess of Sutherland were poisoned by Isabel Sinclair – just one of the deep and dark incidents which cast shadows upon the realm of Mary Queen of Scots. On a more appetising note, the river is also known for salmon fishing.

You veer from the sea and follow the River Helmsdale up through the Strath of Kildonan. This strath is visited by Prince Charles for fishing and hunting. Beinn Dubhain towers above you.

Kildonan village arrives, Scotland's Klondyke. This was the scene of a gold rush in 1868, when prospectors came from far and near to pan for gold in the Kildonan and Suisgill streams.

The golden dream came to an end in 1911 when an expert pronounced the dust too fine for commercial purposes. Mind you, people still pan the streams . . .

The north of Scotland is rich in prehistoric remains, and you'll find strange, parallel rows of stones around Kildonan, particularly on Leivable Hill.

On past Kildonan Lodge and Suisgill Lodge on your right, then Borrobol Estate House and the River Frithe on your left. Watch for the snow fences on either side of the line. Winters here can bring unbelievably deep drifts.

Beinn Griam Mhor and Beinn Griam Bheag emerge on the left, after the Kinbrace request stop. Loch an Ruathair and Loch Lucy are next, Lucy with the little island at its heart. Between Strath Halladale and Achentoul Forest lies Forsinard. The hotel is popular with fishermen, and no wonder – they have eight lochs to choose from, and the River Halladale.

Slethill Hill is to the left, as you leave Forsinard. You're climbing now, up on to wild, open moorland beneath a big sky. Caithness is wide and flat and enchanting. The air is astonishingly clear up here.

Altnabreac punctuates the wilderness. (All around is peat – the Hydro Electricity Board experimented with the idea of using milled peat for the generation of electricity.)

Loch Dhu House, belonging to Lord Thurso, is nearby.

Scotscalder approaches. Loch Calder glints over to the left. Close-boarded fences are close to the line, so that the wind deflects snow away from the track. They're known as 'blowers'.

On your right, the River Thurso courses through the landscape, then under the line just before Halkirk. Every July Halkirk has a Highland Games, one of the finest anywhere. And fishermen speak highly of Halkirk's trout and salmon too. More snow fences edge the rails as you come into Georgemas Junction. Here, the Thurso and Wick sections of the train part company.

Seven miles north, Thurso is Britain's most northerly mainland town, and it dates back to Viking times. (The Viking legacy is everywhere in Caithness, especially in place names such as Wick, Staxigoe, Watten and Freswick.)

Thurso's wide streets and bustling shopping area, and its full range of accommodation, help make it an ideal touring centre. You'll also discover indoor swimming, golf, parks, the excellent Folk Museum and refreshing walks along the river and the coast. One of the town's most famous sons was Sir William Smith, founder of the Boys' Brigade.

Old Man O'Wick Castle, Caithness.

Thurso needn't be the end of your journey. Modern ferries ply daily out of nearby Scrabster to the Orkney Isles. The cliffscapes and beaches are rich in seabirds and seals. You can go sea angling out of Thurso, into record-breaking waters. Take a walk around Holborn Head and Dunnet Head, for the views over the Pentland Firth. Try surfing – the waves can be world class!

THE ALTERNATIVE FORK IN THE LINE TAKES YOU TO WICK, THROUGH SCENERY WHICH BECOMES LESS AND LESS HIGHLAND.

Notice the way the fields are divided, with great slabs of slate, unlike the drystone dykes of Sutherland and Easter Ross.

The rewarding fishing waters of Loch Scarmclett appear on the left. Loch Watten and Wick Water follow on the right. Around here a battle was fought over the Earldom of Caithness, between Lord Breadalbane and Sinclair of Keiss. Wick rolls in. King James VI made Wick a Royal Burgh in 1589, and it was at one time county town of Caithness. Pay a visit to the Heritage Museum, and the old Sinclair Aisle in the parish church, which is the burial place of the Earls of Caithness.

Your exploration of Wick should include the castles of Old Wick to the south and Sinclair and Girnigoe to the north. Take a look at the quaint Town Hall with its cupola, and wander around the suburb of Pultneytown, founded by the British Fishing Society early last century.

A little to the north, of course, is the best known of all Caithness towns – John O' Groats, last village on mainland Britain.

You have travelled Britain's most northerly railway, and have travelled through an area which is unique in all Europe – the landscapes are unspoiled, the wildlife is teeming and the long summer days give you time to fully enjoy yourself. Let your railway exploration be just a beginning.

This pair of Bartholomew maps, separated by 50 years, illustrates how the rail network was decimated across north Devon. The largest town in the area, Barnstaple, had been second only to Exeter and Plymouth as a West Country focus for competition between the Great Western Railway (GWR) and the London & South Western Railway (LSWR). The GWR provided the most direct route from the rest of the country, over the east–west line from Taunton to its Victoria Road station, but the LSWR ran from the main regional centre at Exeter, and also had north Devon branches from Barnstaple Junction through Barnstaple Town to the coastal resort of Ilfracombe, and eastwards to Bideford on a meandering cross-country route to Okehampton, and to Wadebridge and Padstow in distant north Cornwall. The north Devon network was completed by the independently owned narrow-gauge Lynton & Barnstaple Railway, but this had a short life, closing to passengers (as part of the Southern Railway) in the year of publication of the 1935 map.

Some of Bartholomew's charming idiosyncrasies of description of road and rail features – as seen on the 1921 map of the Swansea area on pages 170–173 – had disappeared by 1935, with station refreshment rooms no longer identified, but the distinctive layer colouring is again seen to advantage in this region of steeply undulating hills and the wide expanse of Exmoor. The map extract overleaf shows the separate company stations in Barnstaple, including the GWR's Victoria Road, with the other stations being part of – post Grouping – a distant outlier of the Southern Railway. Victoria Road closed in 1960, with trains diverted to Barnstaple Junction, but this was to prove to be a very minor cut compared to the surgery which was soon to follow.

The Beeching Report made grim reading for north Devon, and for adjoining west Somerset and north Cornwall – everything was to close, other than Exeter to Barnstaple Junction station (on the southwest side of the town) and the Newquay branch in Cornwall. Despite widespread objections, there was to be no reprieve for any line in north Devon. The line west to Bideford and beyond was the first to lose its passenger services, in 1965, followed by the Taunton route in 1966; the Ilfracombe branch (and Barnstaple Town station) finally closed in 1970, the cost of maintaining the bridge over the River Taw being cited as justification for closure of the more centrally located station. A long freight branch to Meeth, south of Bideford survived, carrying milk and china clay traffic until the early 1980s, after which its route was converted into a recreational path and cycleway.

The track of the old line through Bideford is marked correctly on the 1985 1:100,000 map, but strangely the railway from Barnstaple Junction to Barnstaple Town is still shown as being open, despite having closed in 1970. Perhaps this – and the compromising of the layer colouring principle by the introduction of a separate bright green colouring for 'woods' – was a sign that Bartholomew's was losing its way in the face of intense competition from the Ordnance Survey and specialist mapping companies. Just two years later, this series of maps ceased production.

The one surviving railway in north Devon, from Exeter to Barnstaple, is a shadow of its former self, but has enjoyed something of a revival in the last decade, with a local 'community rail partnership' promoting the route as the 'Tarka Line', named after the animal hero in Henry Williamson's book *Tarka the Otter*.

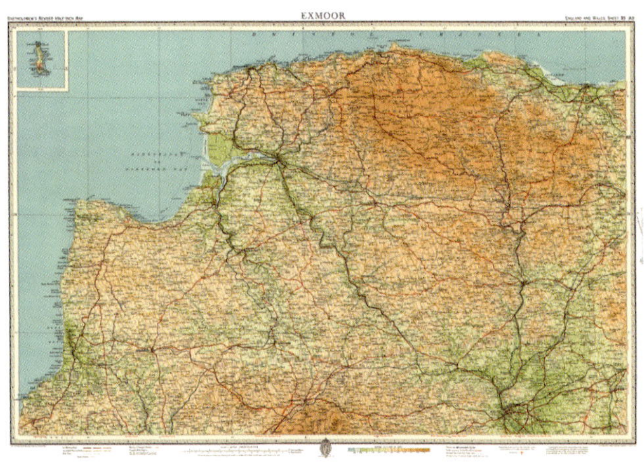

Title	Bartholomew's England & Wales Sheet 35 Exmoor
Mapmaker	J Bartholomew, Edinburgh
Date	1935

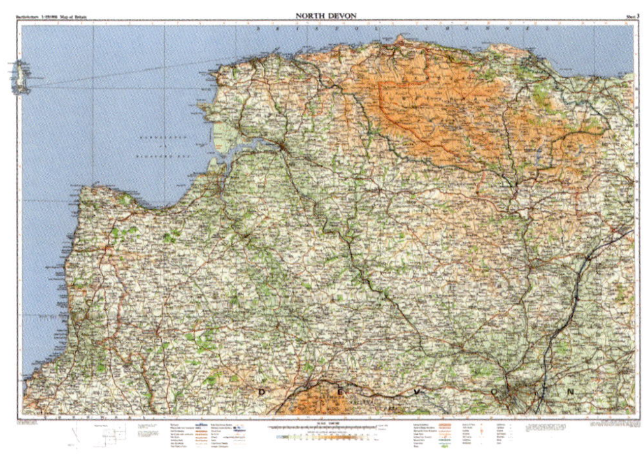

Title	Bartholomew's National Map Series Sheet 3 North Devon
Mapmaker	J Bartholomew, Edinburgh
Date	1985

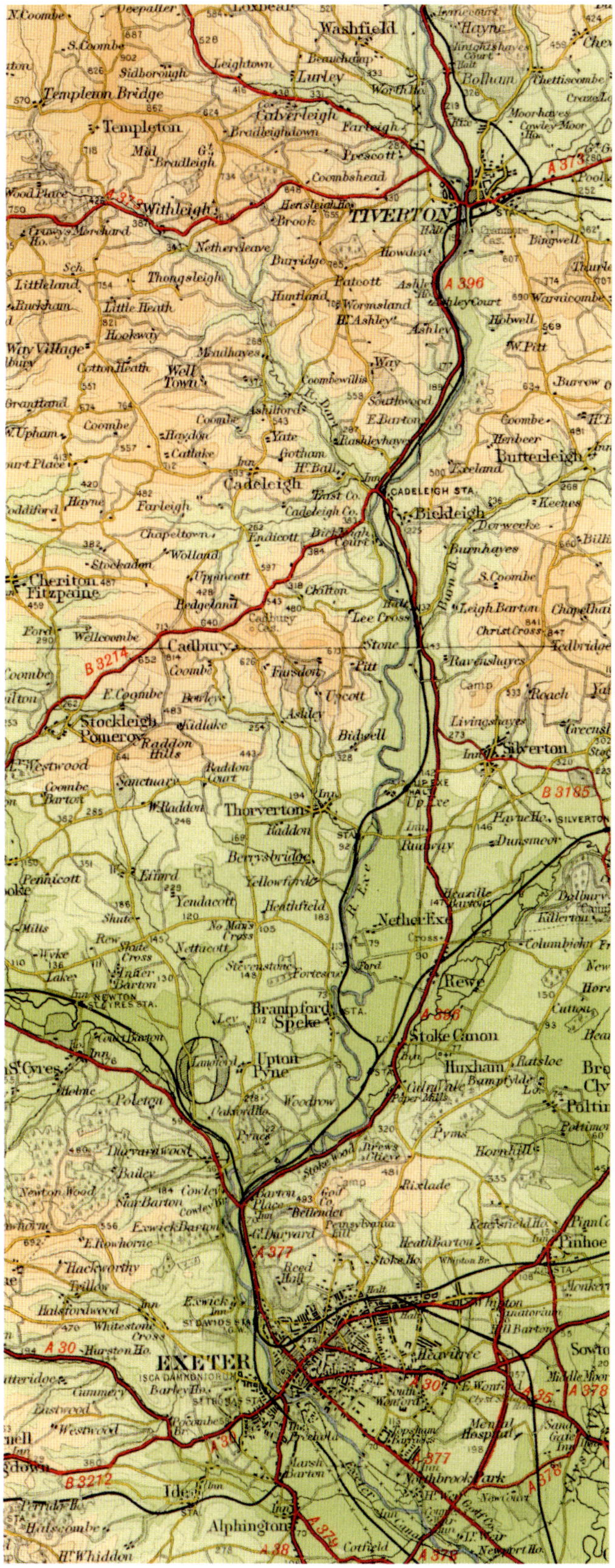

In 1935 train services from seven different directions converged at the county town of Exeter, and three at Tiverton.

By 1985 Exeter, as an important regional centre, was still the junction of five rail routes, but Tiverton had lost all its railways.

In 1935 Barnstaple was the junction of five different rail routes – the Great Western Railway from Taunton; the Southern Railway from Bideford and beyond, from Exeter and from Ilfracombe, and the formerly independent narrow-gauge line from Lynton.

By 1985, the north Devon rail network had shrunk back to a single route, from Barnstaple Junction (by then described as just 'Barnstaple' by British Rail) to Exeter. The map incorrectly shows Barnstaple Town station as still being open, despite having closed 15 years earlier.

Until the opening of the Forth Railway Bridge in 1890, both rail passengers and freight heading north for Perth, Dundee and Aberdeen were forced to break their journey and make the five-mile ferry crossing of the Firth of Forth between Granton, north of Edinburgh, and Burntisland, in Fife. Designed to carry goods wagons on rails, Britain's first train ferry, the 'Leviathan', was introduced on this route by the Edinburgh & Northern Railway in 1850. Another train ferry was introduced across the Firth of Tay a year later. These ferries were meant to be a stop-gap measure but stayed in service for up to 40 years due to delays caused by the Tay Bridge disaster, its subsequent rebuilding and the building of the Forth Bridge.

In the south of England the London, Brighton & South Coast Railway introduced a short-lived train ferry – a converted cargo ship – in the 1880s to carry goods wagons between Langstone Harbour, near Hayling Island, and Bembridge on the Isle of Wight. The ferry service proved unprofitable and ceased running in 1898.

However, as an island nation, Britain depended entirely on sea trade and, by the beginning of the 20th century, the main railway companies all had large vested interests in docks and freight-handling facilities. As we have already seen, carrying goods wagon on rails attached to ships was nothing new, but in 1917 three purpose-built train ferries were introduced to transport war material and munitions to our armies fighting in Belgium and northern France. Running between Richborough Military Port in Kent and Dunkirk the ferries became redundant at the end of the war.

Seeing a golden opportunity for cross-channel rail traffic, the newly formed London & North Eastern Railway and a Belgian company refitted the three ferries and commenced operating a goods-only train ferry service between Harwich and Zeebrugge in 1924. Cross-Channel freight-carrying train ferry services, albeit with newer purpose-built ships, continued operating until the opening of the Channel Tunnel in 1994.

The first passenger train ferry was introduced between Dover and Dunkirk by the Southern Railway and the French national railways (SNCF) in 1936. Operating as the 'Night Ferry' it carried *wagons-lits* (sleeping cars) and baggage vans specially built for the smaller British loading gauge. The overnight train carried passengers in comfort between London's Victoria station and the Gare du Nord in Paris and, later, also to Brussels. Apart from the Second World War years, the train continued in service until 1980.

Plans to build a tunnel under the English Channel go back to the beginning of the 19th century, but the strained relationships between France and Britain put a stop to all of

Produced for a London & North Eastern Railway poster, Frank Mason's painting shows the Zeebrugge train ferry on its way across the North Sea in the 1930s.

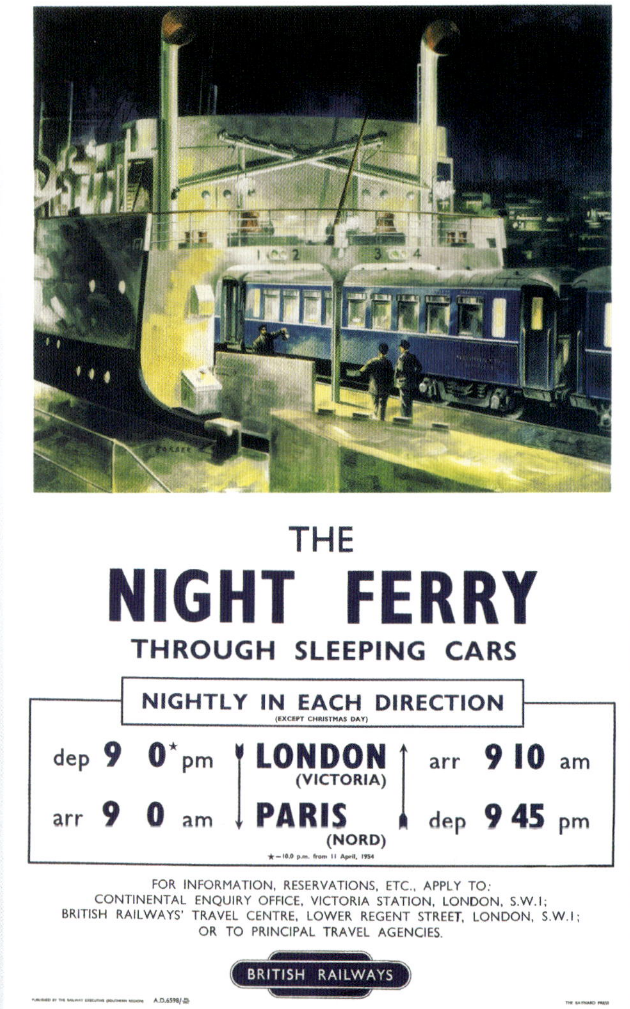

This 1953 British Railways poster advertises the overnight 'Night Ferry' sleeping car service between London Victoria and Paris Nord.

Terence Cuneo's painting of Eurotunnel's 'Le Shuttle' car and lorry carrying train was commissioned for the opening of the Channel Tunnel in 1994. His trademark mice are seen holding the French and British flags in the foreground.

these proposals. Later in the century Sir William Watkin, the enterprising chairman of the Manchester, Sheffield & Lincolnshire Railway, planned a grand continental railway linking northern England to mainland Europe via a Channel Tunnel. Despite exploratory work on both sides of the Channel the tunnel project was abandoned in 1882, although Watkin's Great Central 'London Extension' from the north to Marylebone, which opened in 1899, was still built to the continental loading gauge.

Further proposals for a Channel Tunnel were abandoned after the First World War, and even a 1975 UK–France Government-backed scheme was cancelled (by the British) shortly after tunnelling operations had begun. Finally a privately funded scheme was given the green light at the Treaty of Canterbury in 1986. Responsibility for the scheme was split between British Channel Tunnel Group (two British banks and five construction companies) and the French 'France-Manche (three French banks and five construction companies) and the £2.6 billion raised by private finance was unprecedented for such a project at that time. Tunnelling began in 1988 and was completed, with a cost over-run of 80 per cent, in 1994. The tunnel is used by shuttle trains that carry road vehicles between terminals at Cheriton, near Folkestone, and Coquelles, near Calais; Eurostar services between Paris, Brussels and London; and through international freight services. From 2007 the Eurostar services were rerouted from their previous London terminus of Waterloo International and now travel over the High Speed 1 line to St Pancras International.

A Eurostar passenger train speeds along the new HS1 rail link in Kent. This is the longest new main line to be built in Britain since 1899.

PRIVATIZATION AND BEYOND

The unified railway was split into more than a hundred private companies in an attempt to stimulate competition, improve service and increase efficiency. A number of important new routes opened for passenger and freight business, but the system as a whole now costs the taxpayer far more than when under public ownership.

MILEPOST 23	A Proliferation of Train Operating Companies	278
1964/97	Southampton Docks	280
MILEPOST 24	Recycled Railways: Heritage Railways, Footpaths and Cycleways	282
1997	Bodmin & Wenford Railway	284
2006/08	Track Diagrams of Victoria, Waterloo & York Stations	286
1969–2010	Rail Route Re-openings	290
2008/11	Britain's Rail Network Today	292
2010	High Speed Rail	296

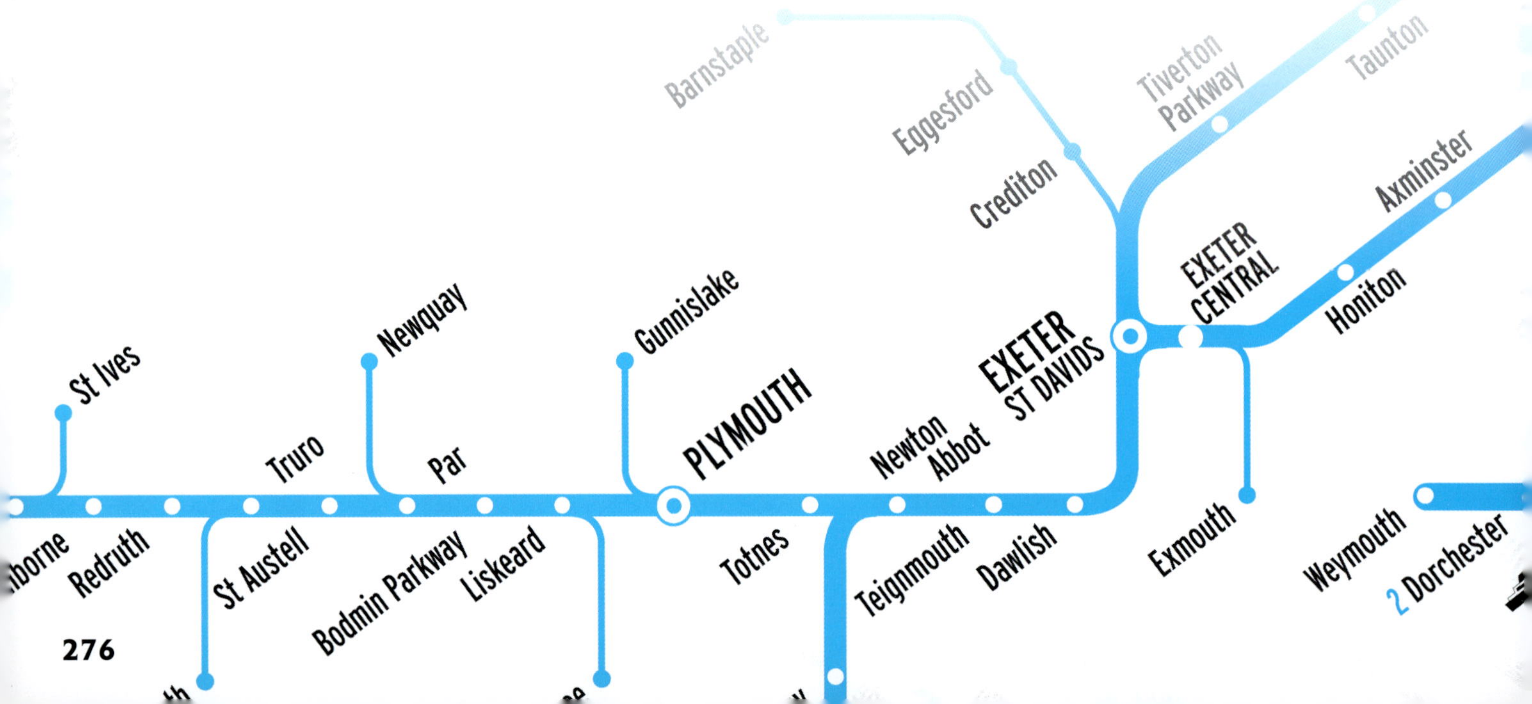

1993 ONWARDS

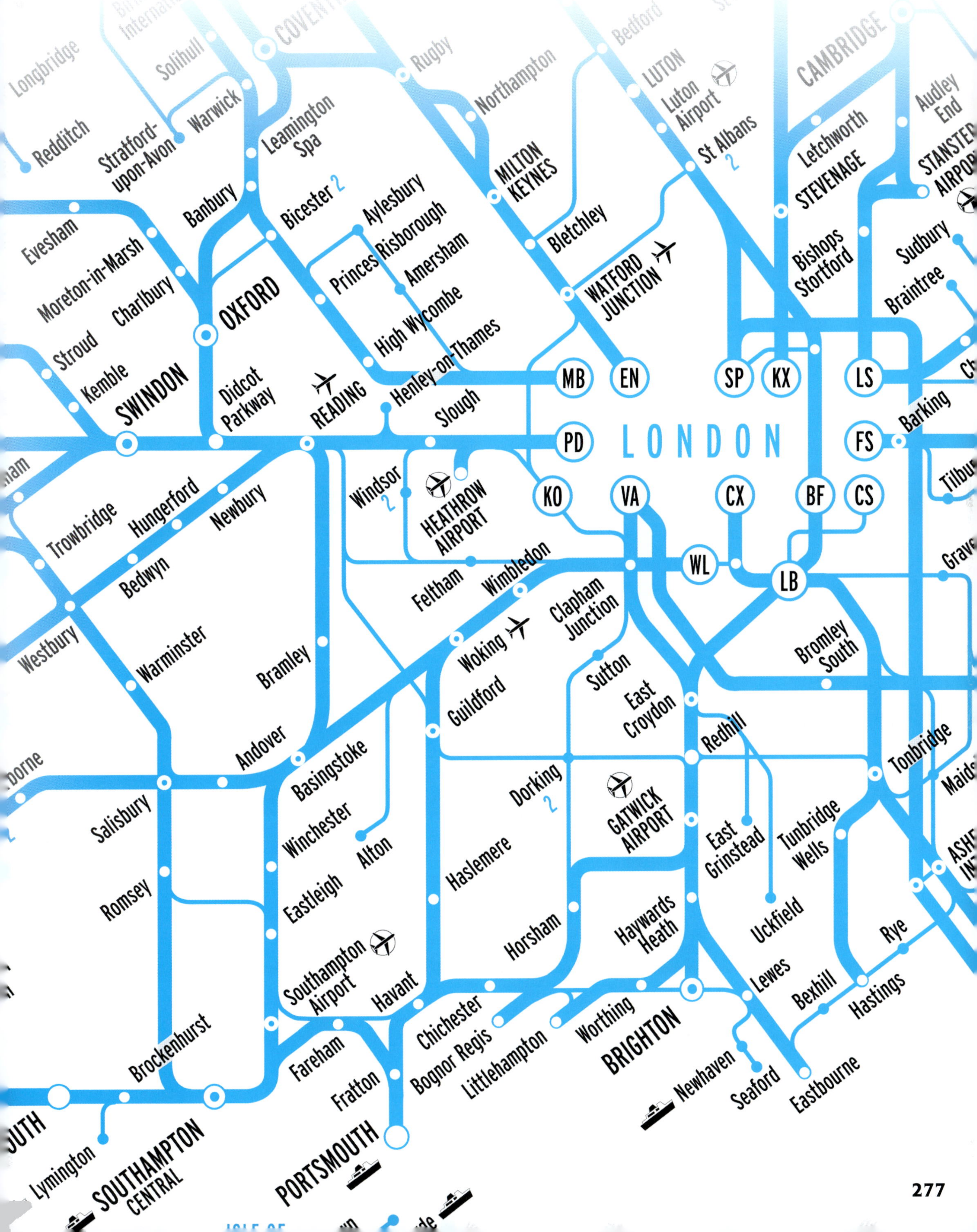

Although selling off nationalized industries into private ownership had been an important part of Margaret Thatcher's policies during the 1980s the privatization of railways was then considered a step too far. Despite the relatively minor sell-offs of British Transport Hotels, Sealink and the railway engineering works, along with the restructuring of railway operations into business sectors such as Railfreight and InterCity, it took until 1993 (during John Major's term as Prime Minister) and the passing of the Railways Act before the 45-year-old state industry was eventually hived off. As with their opposition to the Beeching rail closures 30 years before, the Labour party was against privatization and pledged to overturn it if they came to power – but in practice did not do so.

Under the controversial Act, British Rail was broken up into over 100 separate companies. The Office of the Rail Regulator was set up to regulate all of the main elements of the industry with the Director of Passenger Rail Franchising responsible for the awarding of franchises to 25 passenger train operators. The track, signalling and stations were taken over by Railtrack who, in turn, let out most of the stations to the newly franchised passenger train operators, and contracted out the maintenance and renewal of the infrastructure. Freight services were sold off to two Freight Operating Companies while three Rolling Stock Leasing Companies (ROSCOs) purchased British Rail's locomotives, multiple units and coaches.

The 25 Train Operating Companies were awarded fixed term leases to run services on specific routes – they owned virtually nothing, hiring both locomotives and rolling stock from the ROSCOs, paying Railtrack for the use of the track and infrastructure and contracting out train maintenance and on-board catering.

The contrasting liveries of two Train Operating Companies – Virgin and Central – are very apparent in this photo taken at Birmingham New Street in 2004.

Stobart Rail's 'Tesco Express' from Grangemouth to Inverness speeds along the Highland Main Line near Druimuachdar Summit in 2010.

While privatization of Britain's railways officially came into being on 1 April, 1994 the full effects took some years to filter through. There were also further changes on the way – under a new Labour Government the Transport Act 2000 established the Strategic Rail Authority, a non-departmental public body designed to provide strategic direction for the privatized rail industry. Following the passing of the Railways Act 2005 the SRA was wound up in December 2006 with its duties being split between the Government's Department for Transport, Network Rail (the successor to Railtrack) and the Office of Rail Regulation. Some powers were also devolved to the then Scottish Executive, the Welsh Assembly and the Greater London Authority – and the Scottish Government now has almost complete (non-regulatory) control of the internal Scottish rail network, as well as the power to award the ScotRail passenger train franchise.

By far the most controversial element of the original Act, the story of Railtrack was one of the saddest affairs of the rail privatization saga. Following the Ladbroke Grove accident in 1999, where 31 people were killed and 523 injured, the company was severely criticized for its maintenance of the railway infrastructure and its safety record. With the relationship between Railtrack and the Rail Regulator falling apart, the next major accident, at Hatfield in 2000, was the final straw for the company. Repairs and compensation for the accident coupled with out-of-control costs for rebuilding the West Coast Main Line led the company to go cap in hand to the Government. Some of the tax-payer bail-out was then used to pay a £137 million dividend to its shareholders! Not surprisingly Railtrack was declared bankrupt in 2001 and while it was in administration the Government was forced to plough in a further £3.5 billion in order to keep the railways running. The end of this particular story came in October 2002 when the newly formed 'not for profit' government-created company of Network Rail bought the assets of Railtrack. Today Network Rail has a workforce of 35,000 and a revenue from the passenger and freight train operating companies of £6 billion.

There has been a great deal of debate about the overall impact of rail privatization. Supporters argue that customer service has improved, while opponents respond that quality remains patchy across the network. Many more cheap advance fares are now available, but standard fares are amongst the highest in Europe. Passenger traffic has certainly grown substantially, but sceptics attribute this as much to a period of predominantly good economic fortune and to growing road congestion.

Rail freight was fortunate in being privatized in a much simpler way, with no franchises, and has grown significantly. Rail freight revenues have also benefited from a buoyant economy and from the demise of most of Britain's deep coal mines, which were typically close to key power stations – with the switch to imported coal often resulting in rail movements over hundreds rather than scores of miles.

Reaching an overall verdict on privatization can therefore be difficult, but as Christian Wolmar notes in *Fire and Steam: A New History of the Railways in Britain* (2007):

> *The biggest irony, and indeed failure, of privatization was that far from reducing the cost of the railway to taxpayers, it soared to unprecedented heights. Privatization also brought with it a lack of transparency about the financial affairs of the railways, making comparisons difficult, but a reasonable estimate of the cost to taxpayers since the creation of Network Rail stands at around £5 bn annually.*

As noted in 2011 by Roger Ford, the Industry & Technology Editor of *Modern Railways* magazine, 'By the first decade of the 21st century the privatized railway was costing the tax-payer five times, in real terms, what it cost under BR in 1989/90, and it is still now between four and five times that cost'.

MODERN NETWORK MAPS

Three contrasting styles of modern company network maps are shown here. CrossCountry Trains use a geographic design to show which parts of Britain are served by their trains; East Coast introduces some geographic elements to an essentially diagrammatic design, while Virgin Trains deploy a fully diagrammatic approach, showing which locations are directly linked by their train services. Of these three maps, only Cross Country's shows routes over which the company does not itself operate trains.

CrossCountry Trains

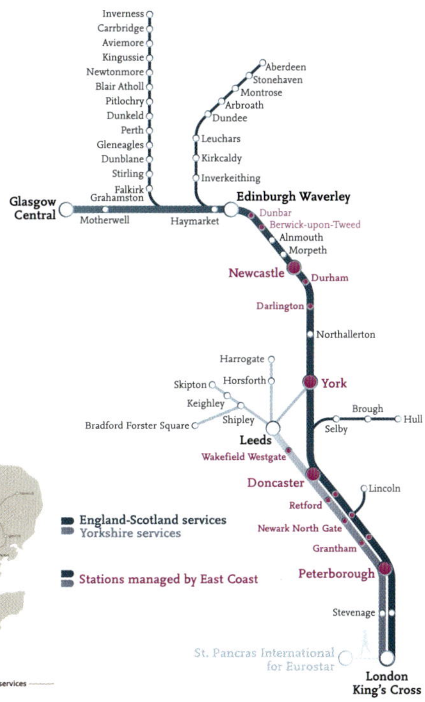

East Coast

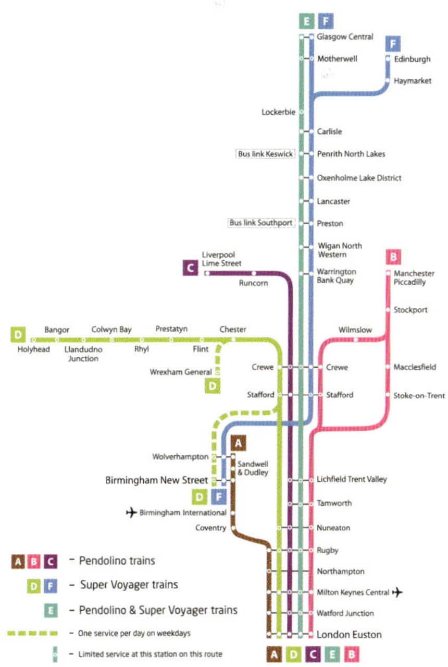

Virgin Trains

As we saw in Milepost 10, the modern docks built in Southampton by the London & South Western Railway (LSWR) were of vital strategic importance during both World Wars. After the Second World War, continental and global shipping services for passengers and freight continued to thrive, but by the time the 1964 map was published transatlantic passenger liners were becoming increasingly unprofitable in the face of air competition – and just four years later the Cunard Line withdrew its year-round service from Southampton to New York.

The 1964 map extract of Southampton Eastern Docks shows freight access lines running to warehouses and transit sheds on every quay, with adjacent access roads evocatively named – Brazil Road, Empress Road and Ocean Road, to name just three. These docks also had their own passenger station – 'Terminus Station', appropriately near the junction of Terminus Terrace and Platform Road. This was the original Southampton terminus of the LSWR, built in 1839, and later replaced as the main passenger terminus for the city by Southampton Central (named Southampton West until 1935) on the through line to Bournemouth. Many of the RMS *Titanic*'s wealthy First Class passengers stayed in the South Western Hotel next to Southampton Terminus before they boarded the ship for their fateful journey in 1912. The station closed to passengers in 1966, prior to rail electrification between Waterloo, Southampton Central and Bournemouth, but the building and the adjacent hotel (now apartments and a casino) have survived.

The original passenger service to Southampton Terminus was later extended into the Ocean Dock Terminal to allow direct 'boat trains' from London Waterloo to terminate at the quayside. Amongst the most famous of these trains was 'The

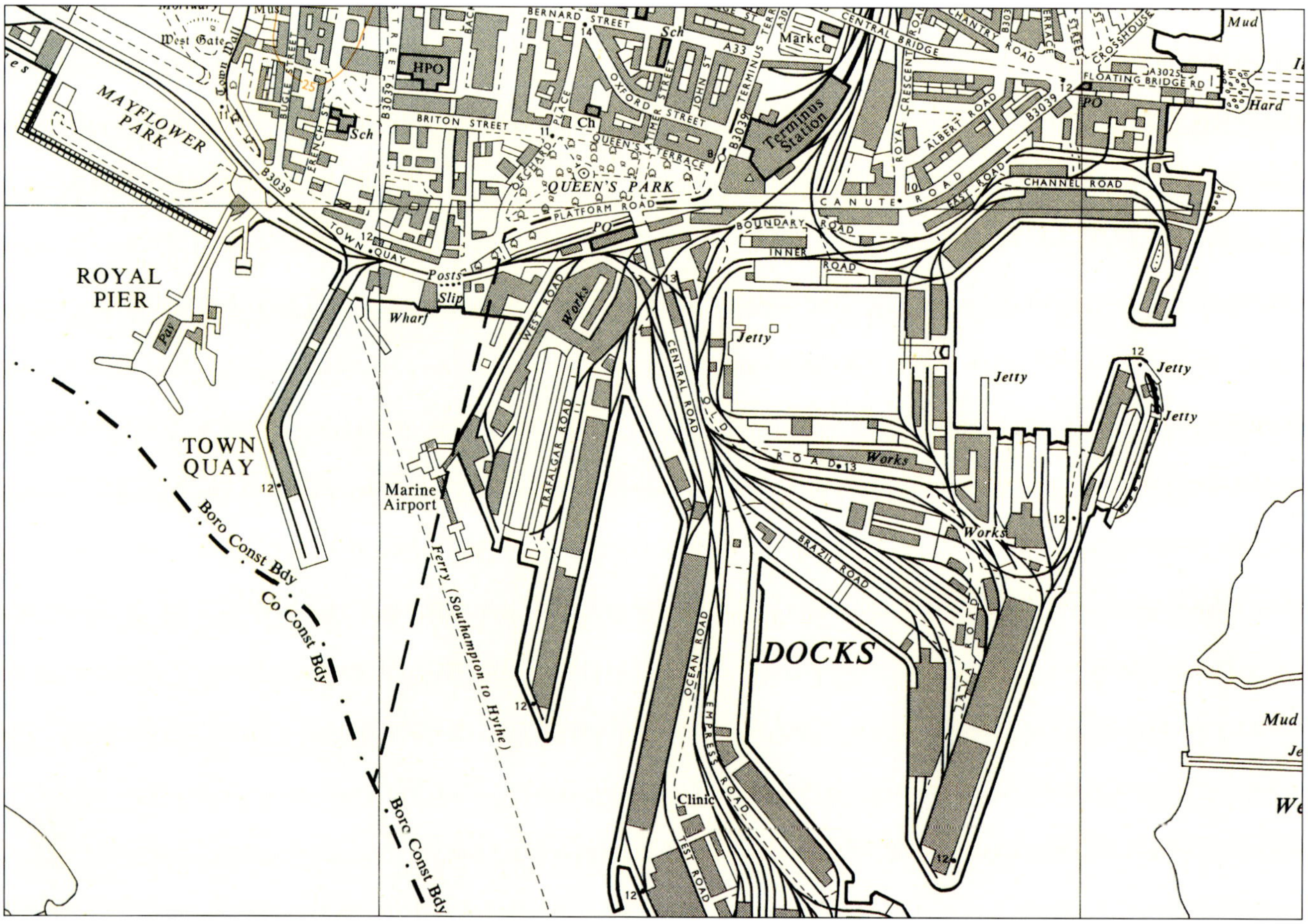

Series	Ordnance Survey 6 in. to 1 mile
Sheet	SU 41 SW (extract)
Date	1964

Cunarder' connecting with the RMS *Queen Mary* and RMS *Queen Elizabeth* transatlantic services to New York. Traditional boat trains ceased operation when liner services finished, but in recent years there has been a revival of chartered trains running from Edinburgh, Glasgow and north of England cities direct to Southampton Western and Eastern Docks, to connect with cruise liners.

Today, as shown in the 1997 Ordnance Survey 1:50,000 sheet extract, rail sidings penetrate the dock complex primarily for freight traffic – notably Britain's second largest container shipping terminal (after Felixstowe). Rail has played a key role in the container revolution since the 1960s, and Southampton has two Freightliner terminals – 'Maritime' in the container terminal itself, and 'Millbrook' just outside the dock boundary – with train services conveying imports and exports to and from inland terminals in Birmingham, Cardiff, Coatbridge, Liverpool, Leeds and Manchester. The map illustrates just how few buildings are involved in modern dock handling – large expanses of open ground are the key requirement for container storage and movement, and in a map of this scale even the mammoth travelling ship-to-shore cranes are not shown.

Directly opposite the Port of Southampton is Marchwood Military Port, built in 1943 to support the D-Day assault on Normandy in 1944, and since used during the Falklands War and other military ventures. The port has its own extensive internal rail system connected to Network Rail's freight-only branch line from Totton to the Fawley oil refinery on the south side of Southampton Water.

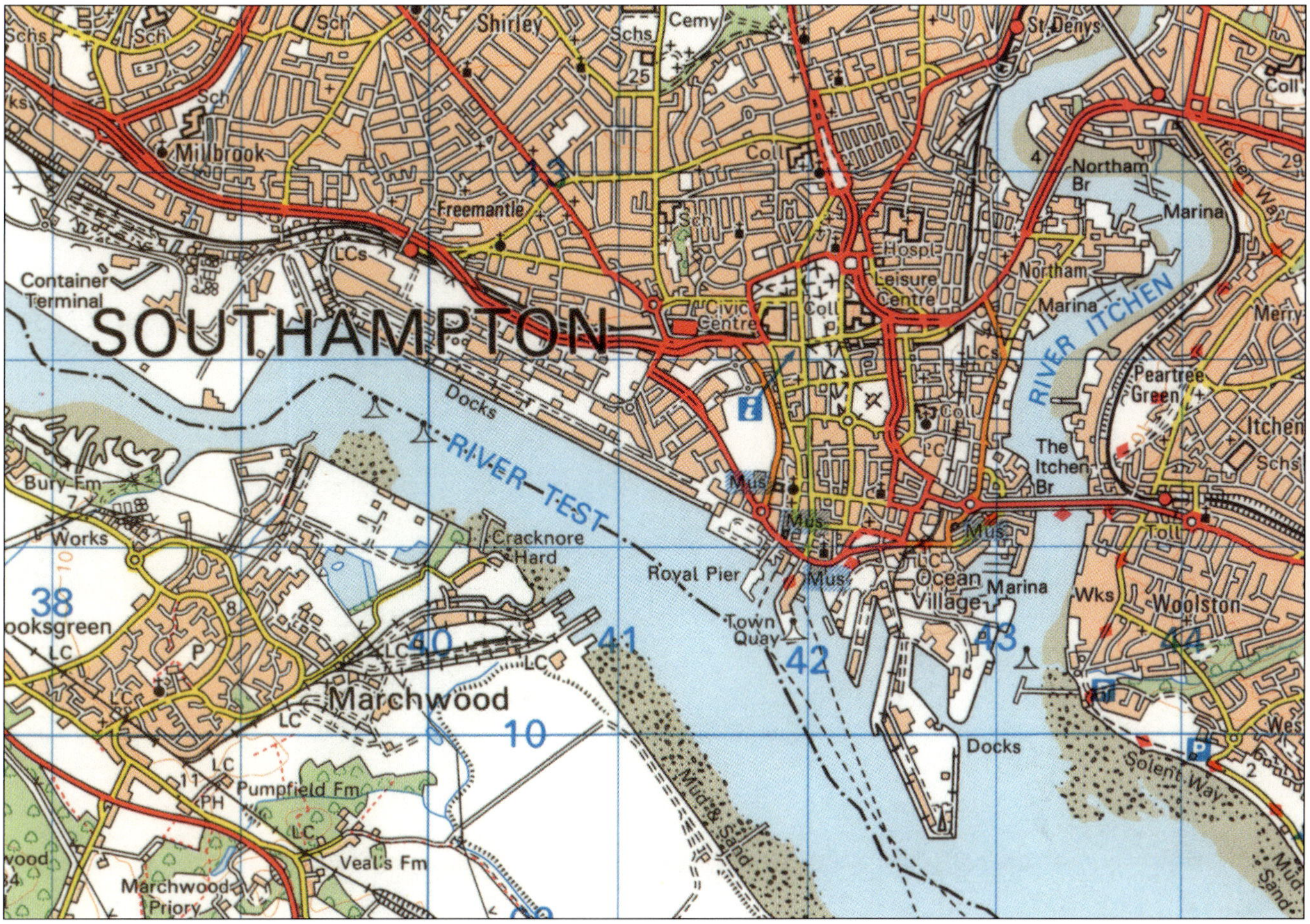

Series	Ordnance Survey 1:50,000 Landranger Series
Sheet	196 (The Solent & Isle of Wight) (extract, enlarged)
Date	1997

Thousands of miles of loss-making railways had already been closed in Britain long before Dr Beeching arrived on the scene. By the outbreak of the Second World War in 1939, over 1,000 route miles had already been closed, not only by the Big Four railway companies but also by many independent companies whose narrow gauge lines had fallen prey to increased competition from road transport.

Following the nationalization of Britain's railways in 1948, the British Transport Commission lost no time in wielding the axe to more loss-making railways. By 1951 the Commission's Branch Line Committee had drawn up a list of a further 3,300 route miles recommended for closure and within 12 years these had been ruthlessly implemented. But worse was to follow.

Enter Conservative Transport Minister, Ernest Marples, and Chairman of British Railways, Dr Richard Beeching. Between them, and also under the following Labour administration, another 4,000 miles of route closures (and 2,000 Beeching-inspired station closures) were implemented, as we saw in Milepost 20. Closures continued until the early 1970s by which time many rural parts of Britain had no rail services at all.

Since then, in what is surely a British phenomenon, the vast network of closed railways across the landscape has seen a new lease of life not only as heritage lines but also as footpaths and cycleways, collectively making a major contribution to the country's tourist industry. To take just one example, the Severn Valley Railway brings in over 200,000 additional visitors to the local economy every year.

While the railway preservation scene took some years to build up steam, the idea of using closed railway lines for recreational purposes goes back to 1937 when the London, Midland & Scottish Railway gave the eight-mile trackbed of the then recently closed Leek & Manifold Valley Light Railway to Staffordshire County Council for use as a footpath and bridleway. Known as the Manifold Way, it is now enjoyed by thousands of visitors each year, contributing much to the local economy. Nearly 50 years later a Bristol-based group of environmentalists, Sustrans, opened its first traffic-free railway path between Bath and Bristol. Today, Sustrans' National Cycle Network extends to 10,000 miles across the length and breadth of Britain; many of their routes incorporate closed railway lines that run through some of the country's most beautiful scenery.

While most of these footpaths and cycleways can be found in the countryside, a few have also found favour with city-dwellers. Notable among these is the Deeside Way on the western outskirts of Aberdeen, the Bristol and Bath Railway Path and the 4½-mile Parkland Walk between Finsbury Park, Highgate and Alexandra Palace in north London. The latter railway was closed in stages between 1954 and 1972 before the trackbed fell into disuse for another 12 years. Reopened as an urban green walkway in 1984, the Parkland Walk is now very popular with weekend strollers, joggers and dog-walkers – unusually, cyclists are banned from using it.

The railway preservation scene had its beginnings in 1951 when a group of enthusiasts led by the writer Tom Rolt saved the narrow gauge Talyllyn Railway in west Wales from closure – the first railway preservation scheme in the world – while in Sussex the Bluebell Railway became the first preserved standard gauge steam-operated passenger railway in the world when it opened in 1960. From these early beginnings over 50 years ago, the preservation movement has grown from strength to strength with currently around 130 heritage railways and museums dotted around Britain. Ranging from major players such as the West Somerset Railway and the North Yorkshire Moors Railway to fledgling projects such as the Aln Valley Railway in Northumberland

A Highgate to Alexandra Palace branch line train disappears away from Cranley Gardens station, shortly before closure in 1954.

The trackbed of the former Alexandra Palace line is now a popular traffic-free route for walkers and joggers in north London.

The narrow gauge Talyllyn Railway in west Wales became the world's first preserved railway when it was saved from closure in 1951.

and the Eden Valley Railway in Cumbria, they all depend for their survival upon an enormous amount of freely given labour by bands of volunteers.

The railway preservation movement was also greatly assisted by a Welsh scrap merchant, Dai Woodham, who, during the 1960s, bought a total of 297 withdrawn steam locomotives. While eventually scrapping 84 of these locomotives at his Barry scrapyard, he concentrated on the easier task of dismantling enormous numbers of withdrawn wooden freight wagons. This stay of execution for the rusting hulks at Barry gave time for the growing railway preservation movement to start a rescue operation – between September 1968 and October 1987 213 of these locomotives were bought by preservationists for use on the growing number of preserved lines across the country. After years of painstaking restoration, many of these steam locomotives can now be seen at work not only on heritage railways but also on a growing number of main-line charter operations.

The culmination of the steam preservation movement in Britain came in 2008 when the brand-new 'A1' Class 4-6-2 No. 60163 'Tornado' emerged from Darlington Locomotive Works. A culmination of 18 years work at a cost of £3 million, this superb machine has not only put in many fine performances on the main line but also draws enormous crowds wherever it appears. Other new-build locomotives of classes not saved from the scrapheap are currently in the pipeline.

The Bluebell Railway in Sussex became the world's first preserved standard gauge railway to operate a public service when it reopened in 1960.

Series	Ordnance Survey 1:50,000 Landranger Series
Sheet	200 (Newquay & Bodmin) (enlarged)
Date	1997

This 1997 sheet of the Ordnance Survey 1:50,000 series illustrates three very different uses of long established railway routes – a main line; a preserved (or 'heritage') railway; and a cycleway and footpath using the solum (trackbed) of an abandoned rail route.

The first line to be opened (as early as 1834) was the Bodmin & Wadebridge Railway – the route entering the map extract from the west – which extended as far as a terminus (later known as Bodmin North) on the north side of the town centre. Parts of its route can be traced from the pecked line running eastwards from Boscarne Junction, the western terminus of the current railway. This was the first steam-worked railway in Cornwall, and one of the first in Britain to carry passengers.

The London & South Western Railway (LSWR) purchased the Bodmin & Wadebridge Railway in 1846, but it was not until 1895 that this isolated branch connected with the rest of its empire fanning out across southwest England from London's Waterloo station. The LSWR's great rival, the Great Western Railway (GWR) opened a branch line from its Plymouth to Penzance main line (at 'Bodmin Road' station, now called Bodmin Parkway) to Bodmin General station in 1887, and a year later opened a connecting line from Bodmin General to Boscarne Junction (the railway running west from the Bodmin terminus on the current map). In 1923, the LSWR lines became part of the Southern Railway and the rivalry with the GWR continued. Particularly in wartime, the local lines formed part of a diversionary route for main-line traffic around Plymouth. In 1944, Field Marshal Montgomery and General Eisenhower arrived at Bodmin by train when visiting the nearby Duke of Cornwall's Light Infantry barracks.

Following the Beeching Report, both branch lines lost their passenger services in 1967, replaced by a bus link to Bodmin Road station. China clay traffic continued thereafter via Bodmin and the Boscarne Junction to Wenfordbridge branch line until 1983, when a need to invest in new track forced closure of the line – its former route can be traced on the eastern bank of the River Camel north of Dunmere.

During the 1980s North Cornwall District Council converted the solum of the former Bodmin & Wadebridge Railway into a cycle- and footpath from Padstow, skirting the western outskirts of Bodmin, to Wenfordbridge – and the picturesque 18-mile length of the 'Camel Trail' now attracts 400,000 visitors a year. The volunteer-led Bodmin & Wenford Railway opened between Bodmin Parkway and Bodmin General in 1990, and was extended to Boscarne Junction in 1996.

In the late 1990s it was proposed to re-open the Boscarne Junction to Wenfordbridge section as a freight railway conveying china clay from the Wenford works by train instead of lorry. More recently there has been a proposal to extend the Bodmin & Wenford Railway westwards to Wadebridge, but objections have stopped both schemes, reflecting the high popularity and economic benefits of the Camel Trail – and the fact that railways are not necessarily seen as environmentally friendly in people's own backyards.

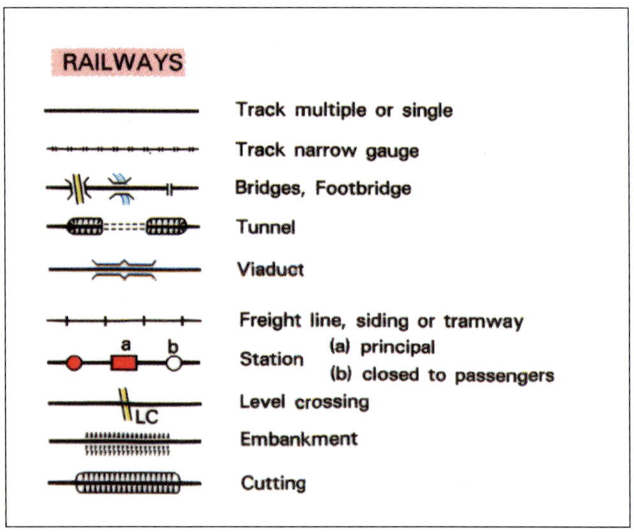

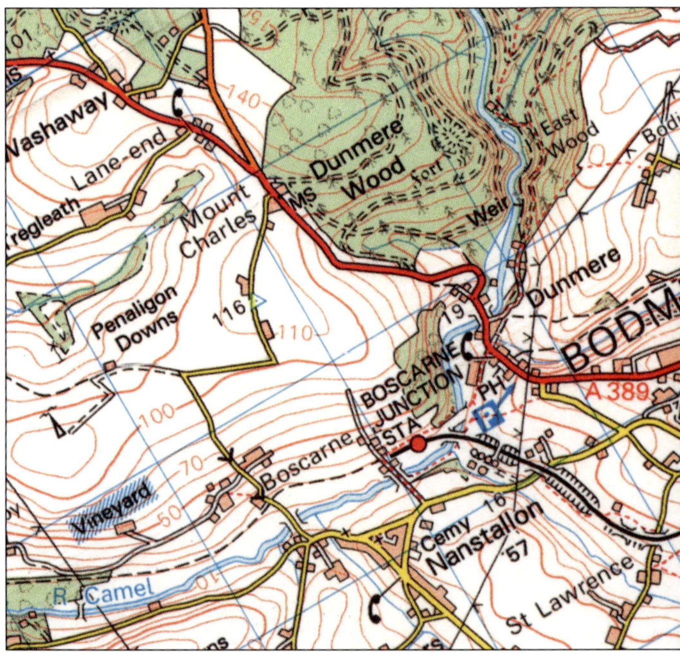

Hauled by preserved Beattie Well Tank No. 30587, a demonstration freight train of vintage china clay wagons heads along the Bodmin & Wenford Railway near Boscarne Junction in Cornwall.

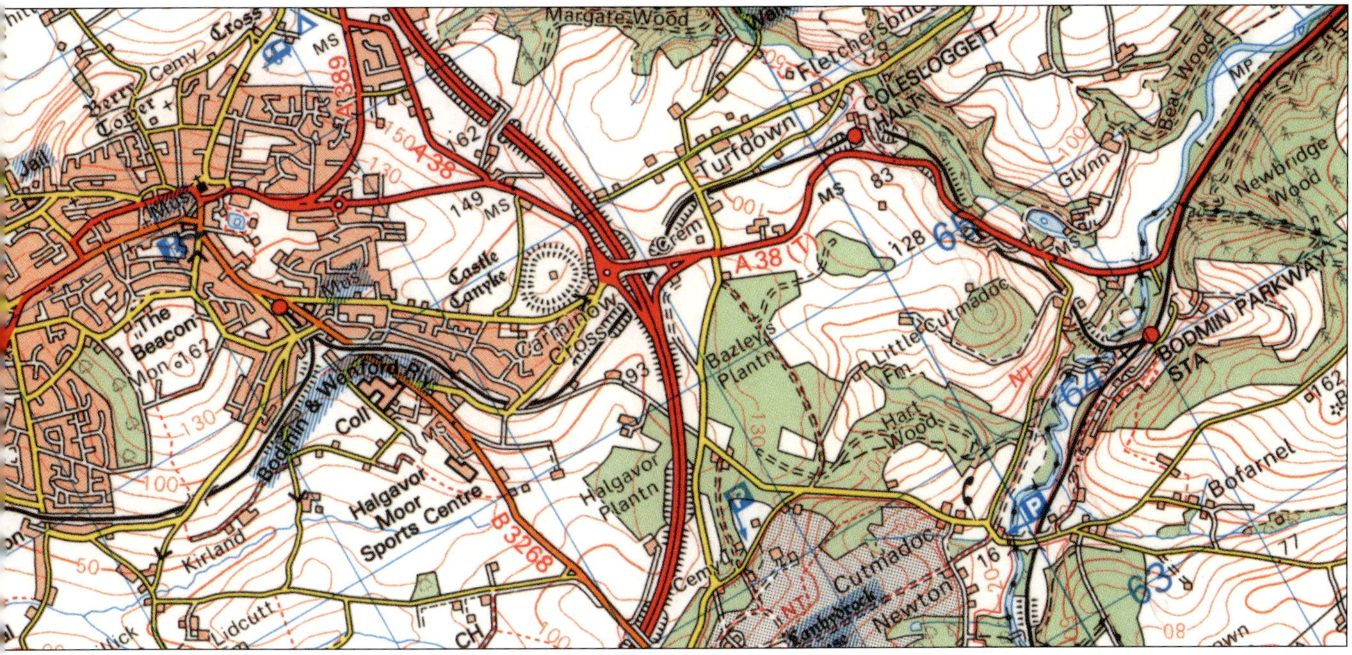

Railway track diagrams have a long history. In *The Railways of Britain: An Unstudied Map Corpus* (2009), the cartographic and railway historians David Challis and Andy Rush record that as the Victorian railway infrastructure became more complicated:

> *Staff needed to know what was where for operational purposes, but they did not require the full topographical layout shown on large, unwieldy regulatory plans, only specific details derived from it. . . . In general terms, operational plans were used for the day-to-day operation of the system and engineering work. Given the linear nature of a railway as well as the nature of the information required, an economical and simple presentation was called for, and increasingly a diagrammatic format was adopted.*

The classic modern inheritor of that tradition has been 'the Quail' – industry shorthand for the *Track Diagrams* – first published in 1998 when career railwayman Gerald Jacobs joined forces with John Yonge of the Quail Map Company as cartographer and publisher, to produce what has now become the standard reference work for infrastructure managers, train planners, transport consultants, railway enthusiasts and modellers alike. In 2004 Mike Bridge at TRACKmaps took over publication of the five volumes covering Britain (based on the old British Rail regions), and subsequently digitized and updated the information.

One of the most dramatic of over 200 pages of diagrams in the Quail series is that showing London's Victoria and Waterloo stations and their approaches. Waterloo, on the south bank of the Thames, is by far the busiest railway station in Britain in terms of passenger throughput, with 86 million passengers a year. It currently has 20 operational terminus platforms – similar to nearby Victoria – the four former Eurostar platforms being out of use since international train services switched to St Pancras in 2007. The station (as well as the track infrastructure) is managed by Network Rail, and all normal train services are operated by South West Trains over an inner and outer suburban network to Berkshire, Hampshire and Surrey, as well as longer distance services to Bournemouth, Exeter, Portsmouth and Southampton.

The rail route approaches to Waterloo are intertwined with the infrastructure of nearby Victoria station, on the north bank of the Thames – the second busiest station in Britain after Waterloo, with 70 million passengers annually. Victoria has 19 platforms, and is managed by Network Rail – with timetabled train services operated by the Gatwick Express, Southern and Southeastern train operating companies to Kent, Surrey and Sussex. The complexity of railway engineering and train operations can be appreciated by

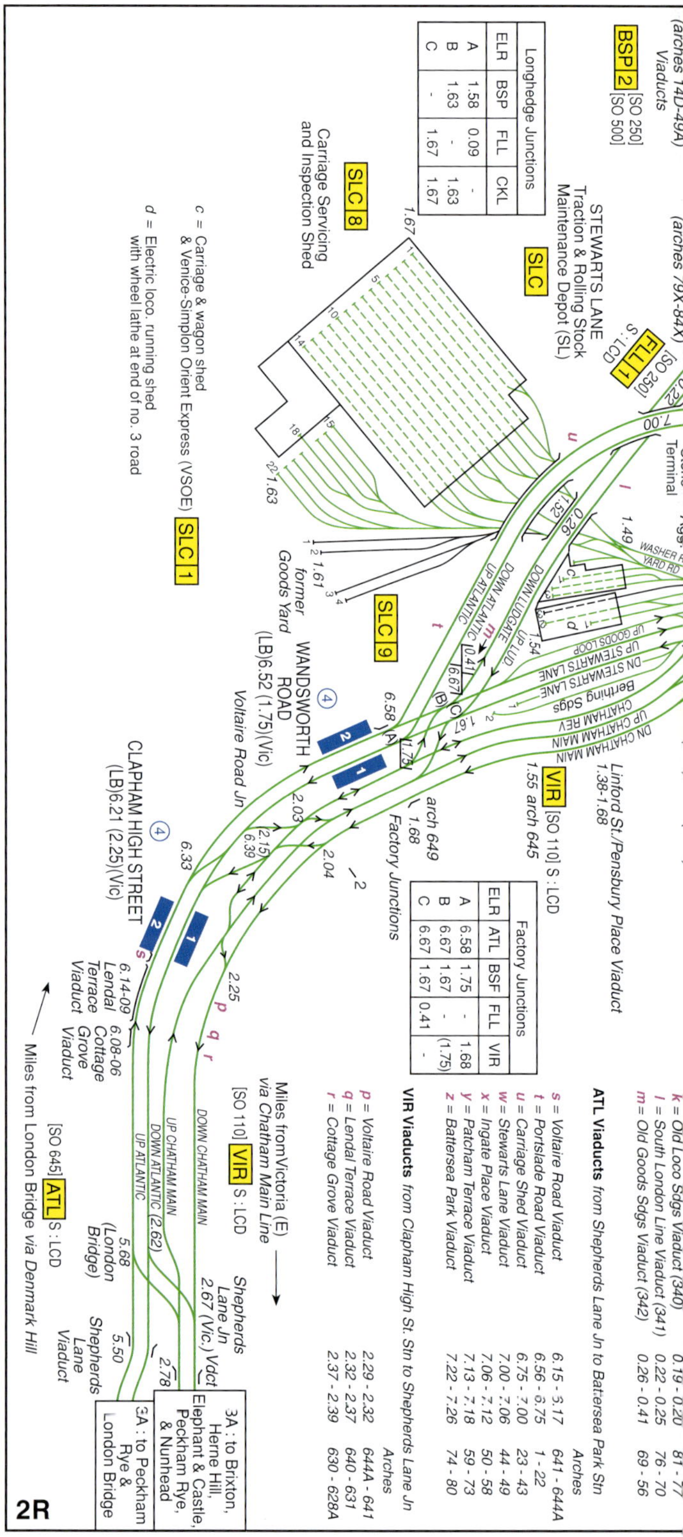

studying the large number of 'crossovers' (from one track to another), route junctions, fly-overs and fly-unders in the area bounded by the two termini and the Stewart's Lane traction and rolling stock maintenance depot.

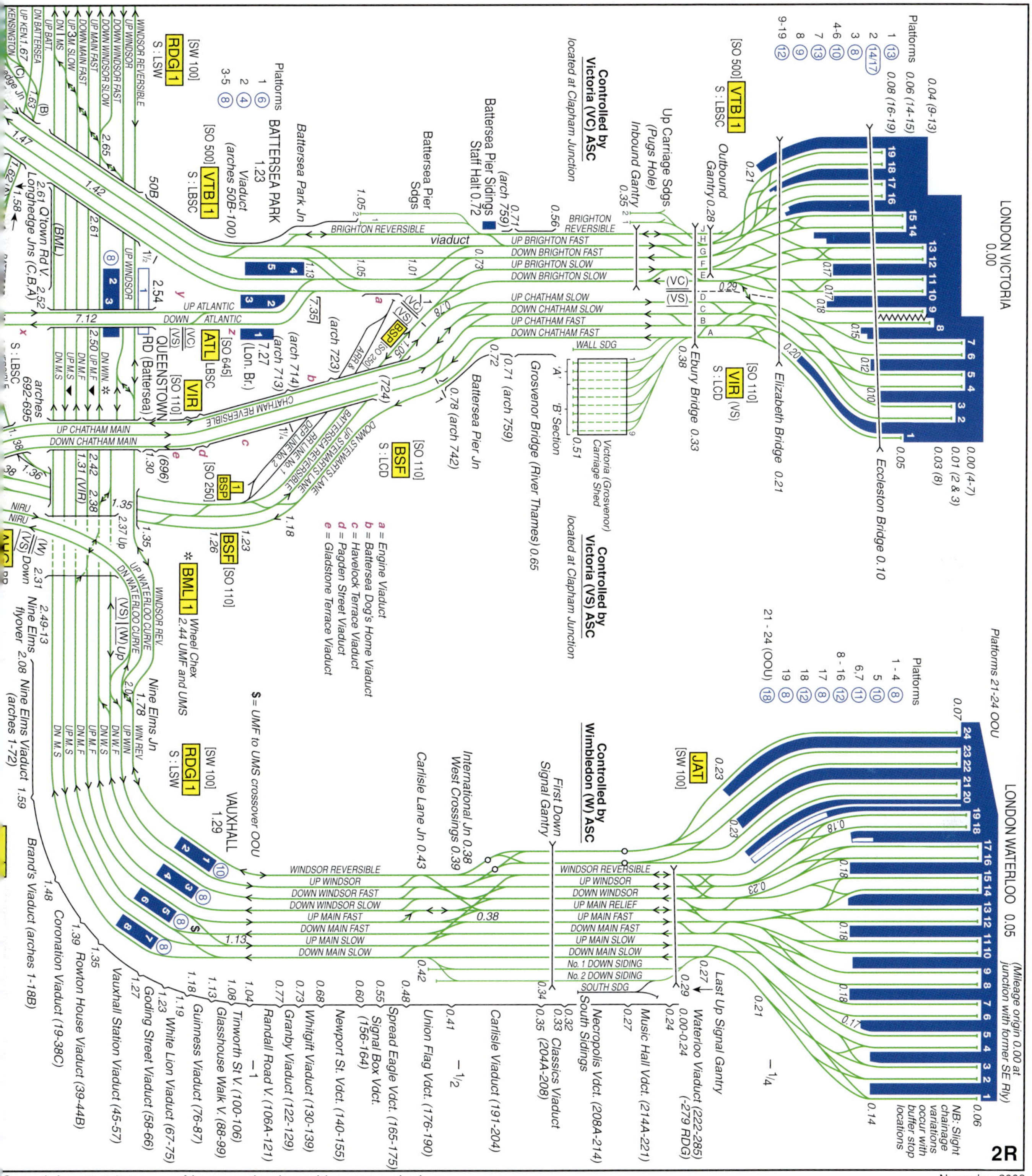

November 2008

Title	Railway Track Diagrams, Book 5, page 2R
Publisher	TRACKmaps
Date	2008

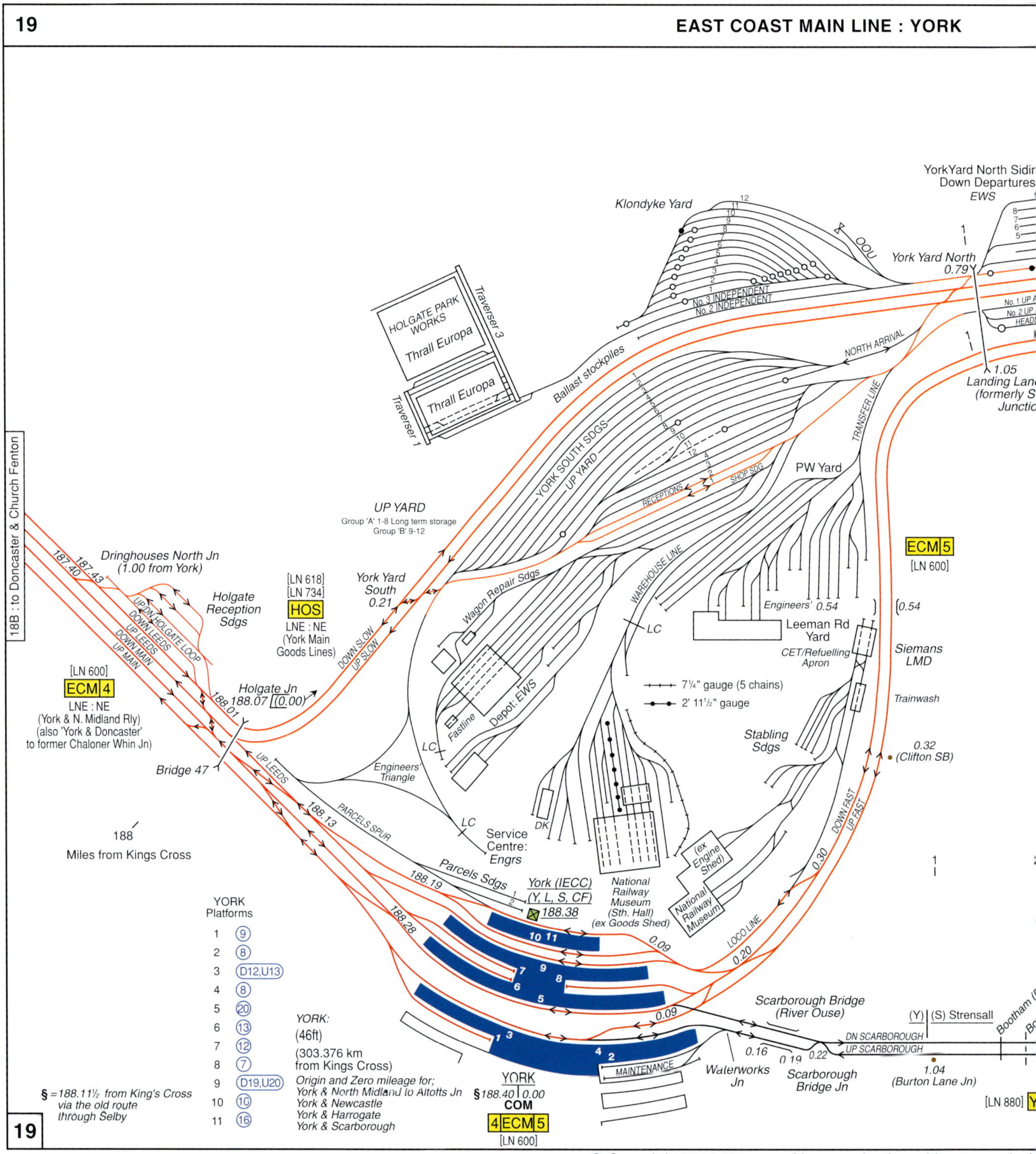
19
EAST COAST MAIN LINE : YORK
18B : to Doncaster & Church Fenton
Klondyke Yard
No. 3 INDEPENDENT
No. 2 INDEPENDENT
York Yard North
0.79
HOLGATE PARK WORKS
Thrall Europa
Traverser 3
Traverser 1
Thrall Europa
Ballast stockpiles
NORTH ARRIVAL
1.05
YORK SOUTH SDGS
UP YARD
RECEPTIONS
SHOP SDG
PW Yard
TRANSFER LINE
UP YARD
Group 'A' 1-8 Long term storage
Group 'B' 9-12
ECM 5
[LN 600]
Dringhouses North Jn
(1.00 from York)
187.40
187.43
Holgate Reception Sdgs
UP ON HOLGATE LOOP
DOWN LEEDS
UP LEEDS
DOWN MAIN
UP MAIN
[LN 618]
[LN 734]
HOS
LNE : NE
(York Main Goods Lines)
York Yard South
0.21
DOWN SLOW
UP SLOW
Wagon Repair Sdgs
WAREHOUSE LINE
LC
Engineers' 0.54
Leeman Rd Yard
CET/Refuelling Apron
(0.54
Siemans LMD
Trainwash
7¼" gauge (5 chains)
2' 11½" gauge
[LN 600]
ECM 4
LNE : NE
(York & N. Midland Rly)
(also 'York & Doncaster' to former Chaloner Whin Jn)
Holgate Jn
188.07 (0.00)
188.01
Bridge 47
Fastline
Depot: EWS
LC
Engineers' Triangle
Stabling Sdgs
0.32
(Clifton SB)
UP LEEDS
188.13
PARCELS SPUR
LC
DK
Service Centre: Engrs
(ex Engine Shed)
DOWN FAST
UP FAST
0.30
188
Miles from Kings Cross
Parcels Sdgs
188.19
York (IECC)
(Y, L, S, CF)
188.38
National Railway Museum
(Sth. Hall)
(ex Goods Shed)
National Railway Museum
LOCO LINE
0.20
0.09
188.28
YORK
Platforms
1 (9)
2 (8)
3 (D12,U13)
4 (8)
5 (20)
6 (13)
7 (12)
8 (7)
9 (D19,U20)
10 (10)
11 (16)
10 11
7 9 8
6 5
1 3
4 2
0.09
MAINTENANCE
Scarborough Bridge
(River Ouse)
(Y) (S) Strensall
DN SCARBOROUGH
UP SCARBOROUGH
0.16
0.19
0.22
Waterworks Jn
Scarborough Bridge Jn
1.04
(Burton Lane Jn)
[LN 880]
YORK:
(46ft)
(303.376 km
from Kings Cross)
Origin and Zero mileage for;
York & North Midland to Altofts Jn
York & Newcastle
York & Harrogate
York & Scarborough
YORK
§188.40 0.00
COM
4 ECM 5
[LN 600]
§ = 188.11½ from King's Cross via the old route through Selby
19

September 2006

Title	Railway Track Diagrams, Book 2, page 19
Publisher	TRACKmaps
Date	2006

The track diagram of York illustrates a very different kind of layout at this important through station on the East Coast Main Line. York has been a major railway centre since Victorian times, and is still a key junction for routes to Edinburgh, Harrogate, Leeds, London and Scarborough. The station is operated by East Coast on behalf of Network Rail and provides a wide range of regional and longer distance services operated by no less than six different train operating companies. Adjacent to the station is the rail-connected National Railway Museum, while the wider railway complex includes freight sidings, engineers' yards and maintenance depots. An unusual feature is the 'Engineers' Triangle' near the station, where track maintenance machines – and steam locomotives – can be turned.

The tracks shown in red are those which have overhead electrified wires – provided as part of the East Coast Main Line electrification completed in 1991 – while only diesel services can operate over the non-electrified lines shown in black. York's signalling centre – shown in green adjacent to Platform 11 – controls all train operations not just in York but also as far away as Northallerton, the fringes of Doncaster and the Leeds area. The track layout at York is still relatively complex, but is much simplified from what it used to be, following a rationalization and resignalling scheme in the late 1980s – one example being the restricted access to the Scarborough line east of the station. In contrast, the infrastructure around Victoria and Waterloo reflects fewer changes in travel and train service patterns over the years.

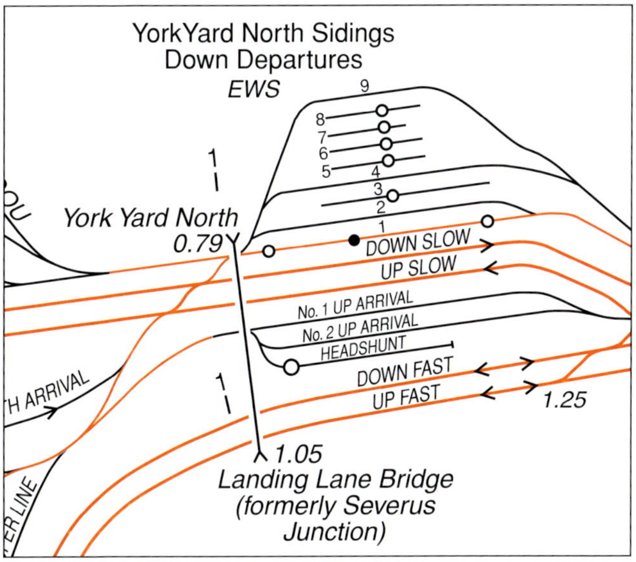

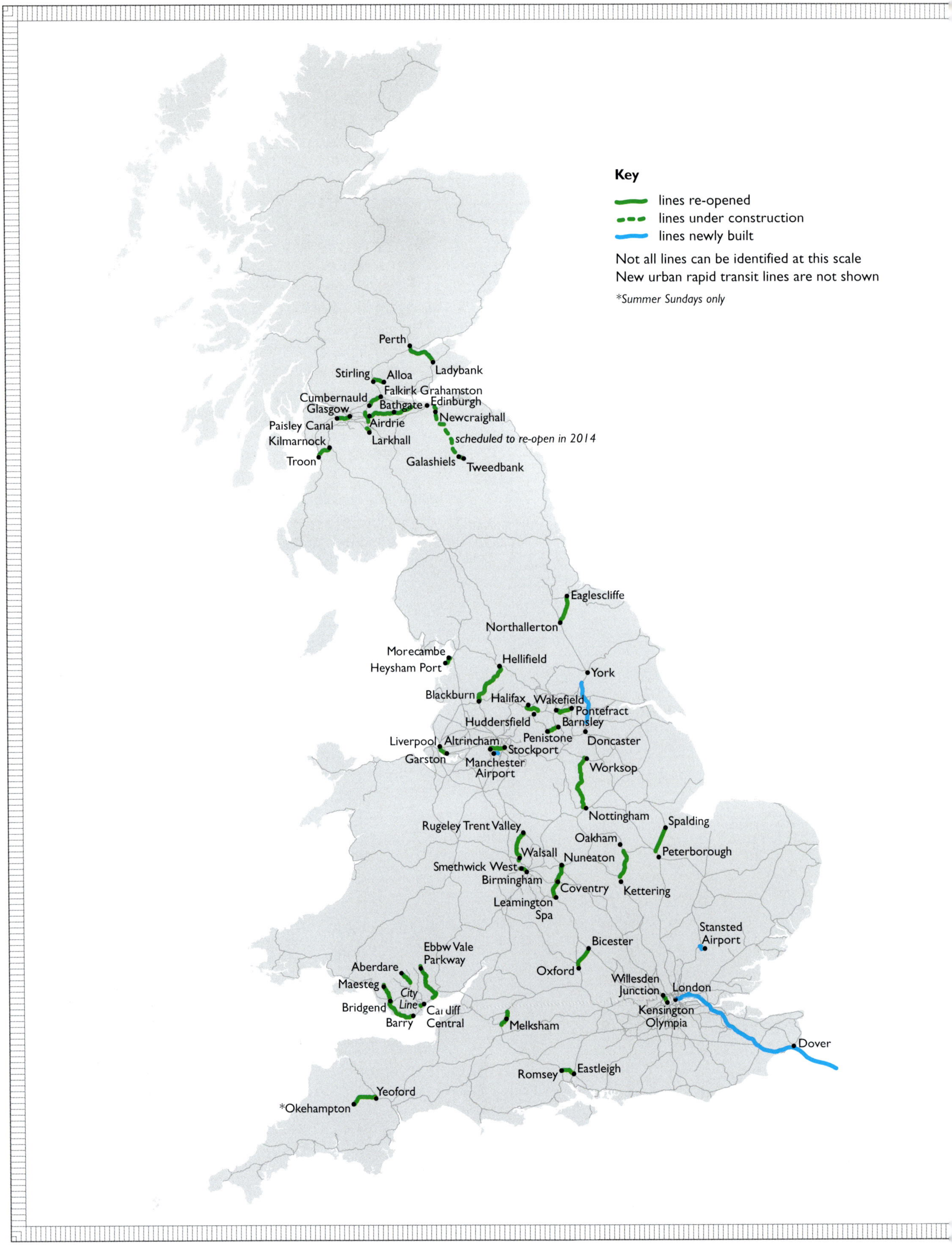
Key
lines re-opened
lines under construction
lines newly built
Not all lines can be identified at this scale
New urban rapid transit lines are not shown
*Summer Sundays only
Perth
Ladybank
Stirling
Alloa
Falkirk Grahamston
Cumbernauld
Glasgow
Bathgate
Edinburgh
Paisley Canal
Airdrie
Newcraighall
Kilmarnock
Larkhall
scheduled to re-open in 2014
Troon
Galashiels
Tweedbank
Eaglescliffe
Northallerton
Morecambe
Heysham Port
Hellifield
York
Blackburn
Halifax
Wakefield
Pontefract
Huddersfield
Barnsley
Penistone
Doncaster
Liverpool
Altrincham
Stockport
Garston
Manchester Airport
Worksop
Nottingham
Spalding
Rugeley Trent Valley
Oakham
Walsall
Nuneaton
Peterborough
Smethwick West
Birmingham
Coventry
Kettering
Leamington Spa
Stansted Airport
Bicester
Ebbw Vale Parkway
Aberdare
Oxford
Maesteg
City Line
Willesden Junction
London
Bridgend
Cardiff Central
Barry
Melksham
Kensington Olympia
Dover
Romsey
Eastleigh
Yeoford
*Okehampton

Title	Rail route re-openings since 1969
By	Collins Bartholomew (using data published by Railfuture in 2010)
Date	2011

After the retrenchment of the Beeching Era, it was some time before the concept of re-opening stations or building new stations became a priority for the railways. However, by the late 1970s, concerns about the environment, road congestion, and changing patterns of urban development were such that a growing number of new names began to appear on the rail network map. This has been a gradual and uncoordinated process, but with a very significant cumulative impact – over 350 new stations and more than 500 miles of new passenger routes opened since 1960 – *Britain's Growing Railway* by Railfuture (2010) noting that:

> *There has been no central government or rail industry plan to achieve this but progress has come usually as a result of local or regional initiatives by rail managers, local authorities and rail campaigners.*

One of the first routes to re-open came in response to the suggestion of a campaigning group, when passenger services were reinstated over the freight line between Perth and Ladybank (15 miles) in 1975 – this followed the closure of the direct line from Perth to Edinburgh via Kinross in 1970 and the consequent lengthy detour via Stirling.

Most station re-openings have been intermediate stops on lines which still had a passenger service – so, typically, no new track or no new trains would be required, just the station itself – but by the late 1980s a growing number of routes which were freight-only, or had been completely abandoned, were being brought back into passenger service.

The peak year for route re-openings was 1987 (seven routes) and 1988 for station re-openings (26 locations) – both under devolved British Rail management – but after a hiatus caused by rail privatization it has increasingly been longer routes which have returned to the passenger network, including Barry to Bridgend (19 miles) and Cardiff to Ebbw Vale Parkway (18 miles of re-opened route), both in South Wales. As noted in *Britain's Growing Railway*, analysis of the re-opened station locations shows:

> *that since 2000, most of the new or reopened stations are either in Scotland or South Wales, where the new national governments have taken a much more determined, pro-active and independent transport role.*

The greatest concentration of re-opened lines has been in Central Scotland, where no less than 15 sections of route have seen the re-introduction of passenger services, latterly funded by the Scottish Government, under which three completely abandoned railways have been re-opened to passenger traffic – Hamilton to Larkhall, Stirling to Alloa and Drumgelloch (Airdrie) to Bathgate.

As well as re-openings, some very significant stretches of entirely new passenger railway have been built in modern times – notably the 12-mile 'Selby diversion' of the East Coast Main Line in 1983 (to avoid future mining subsidence); links to Heathrow, Stansted and Manchester airports; and the 102 miles of the Channel Tunnel and associated 'HS1' rail route from Folkestone to London St Pancras, finally completed in 2007.

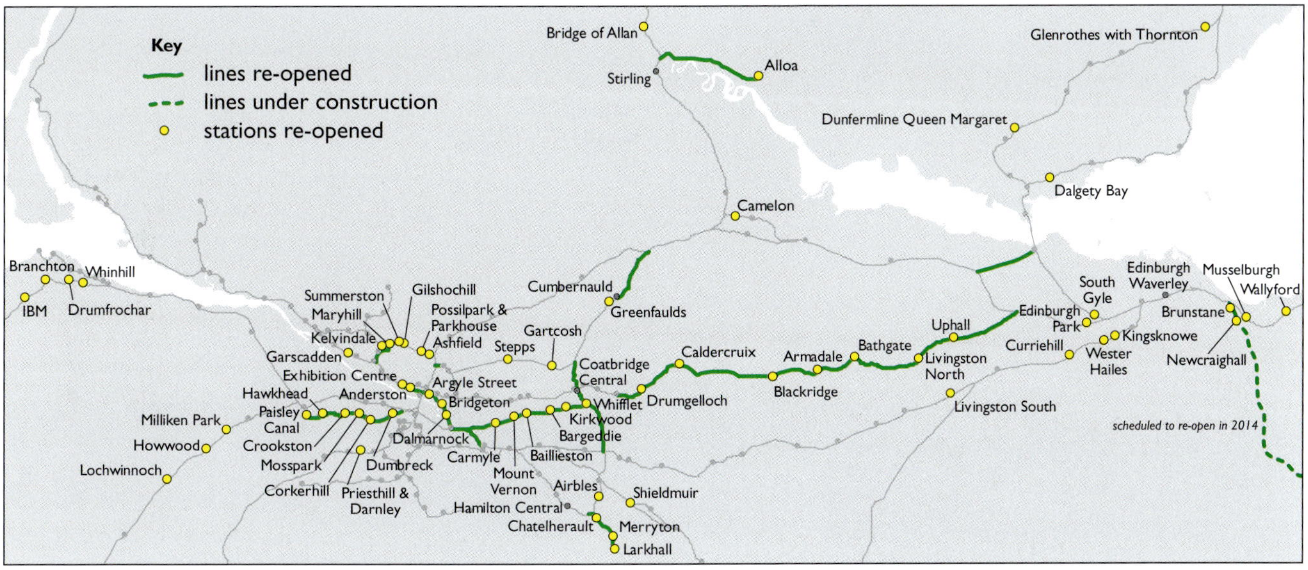

Rail companies have produced or commissioned maps of their networks since the earliest days of the railway. Once the entire British network came into public ownership in 1948, British Railways developed an initial style which lasted until 1967, as illustrated on pages 214–217. In its fresh identity as British Rail a new design was created, as shown on pages 252–255, and elements of this have survived to the present day. Railtrack took over responsibility for the entire infrastructure (the track, as opposed to the trains) from British Rail, and after Railtrack went into administration, control was vested with Network Rail – officially a private-sector organization, but with its debts underwritten by the UK Government, and partial funding of its investment programme by the UK Government, Scottish Government and Welsh Assembly Government.

In its capacity as guardian of the network, Network Rail publishes a geographic network map to illustrate the on-line national timetable which it produces twice a year. Its web site also presents a diagrammatic network map by 'National Rail' – which is not a company, but effectively a brand promoted by the Association of Train Operating Companies (ATOC) – and a 'National Rail Network Diagram', which is an amalgam of both styles, produced by an independent company called Project Mapping. All three are shown in this feature.

Network Rail's map opposite has the merit of showing the recognizable shape of Britain's land mass, together with an essentially geographic representation of every railway line (with adjacent table number) – and every station is named, except in the conurbations where detailed insert maps are shown separately to a different scale. Selected bus and ferry links are also shown, the former including some routes which were 1960s' replacements for rail closures – such as those through the Scottish Borders, between Oxford and Cambridge, and in north Cornwall. The map is a recognizable descendant of the 1968 British Rail map on page 252, with twin categorization of stations – in this case 'principal places' and 'other places' – and use of a similar upper case font for the names of principal places. However, while accommodating a large amount of detail for a national map, the 'hierarchy' of information is not well developed – all lines are of an equal width. The principle of showing every station results in one's eye being drawn to the corridors with greater density of marking, such as rural routes with very frequent stops – as along the Cambrian Coast and the Cumbrian Coast – rather than other rural routes with fewer intermediate stations, like that from Ayr to Stranraer.

The ATOC map overleaf is unashamedly diagrammatic and, interestingly, follows the 'Beck rules' dating back to his 1933 London Underground map – only horizontal, vertical or 45 degree angle lines being used. To provide a clear overview of the main national network, two categories of route are presented – identified by distinctly different line widths – but inevitably the map's simplification prevents every station being shown, even in rural areas. Britain's familiar coastline is understandably omitted, as it would have an uneasy fit with a diagrammatic map, and distances are in some cases very distorted to accommodate to the simplified design – for example, the 60-mile Carlisle–Newcastle line appears to be four times the length of the 59-mile Ayr–Stranraer line. All names (except London) are tilted, when it might have been visually easier to re-orientate the whole map and have horizontal station names.

The Project Mapping design overleaf seeks a compromise between the geographic and diagrammatic approaches, its designer Andrew Smithers endeavouring, 'to show the individuality and personality of the network being depicted, rather than force it to follow traditional abstract rules'. The coastline is shown – albeit in very abstract form in places – and there is a useful hierarchy of visual distinction between four types of route: principal, regional, local and 'limited service'.

In some parts of the country – such as Devon and Cornwall – using only straight lines allows clarity of presentation within a recognizable regional geography. However, in other areas, such as the Cumbrian Coast and northeast Scotland, gently curving lines are introduced, Smithers commenting that this, 'has become an important aspect of the map's design, with softer curves leading to a more pleasing appearance'. Of course, beauty is in the eye of the beholder, railway author and mapmaker Alan Young arguing that the juxtaposition 'creates an unsettling tension between straight lines and self-conscious curves that don't really reflect the true shape of the network'. A large variety of angles are used in the Project Mapping design, with no Beck-type rules to simplify the visual impact, and some distances are severely distorted – with the five-mile Edinburgh–Newcraighall line shown as of equal length to the 59-mile Ayr–Stranraer line. Certain parts of the rail network seem fated to be minimized in importance in modern mapping, irrespective of the technique used.

Map design is of course a matter of personal taste, as well as utility, and it is difficult to achieve comprehensive coverage and ease of understanding in a single design. However, it can be argued that while the novel Project Mapping approach may show the network more clearly than its two contemporaries, the classic 1968 British Rail design succeeded in achieving a pleasing clarity and appropriate hierarchy of information without losing familiar geography – in a way which its modern counterparts have not emulated.

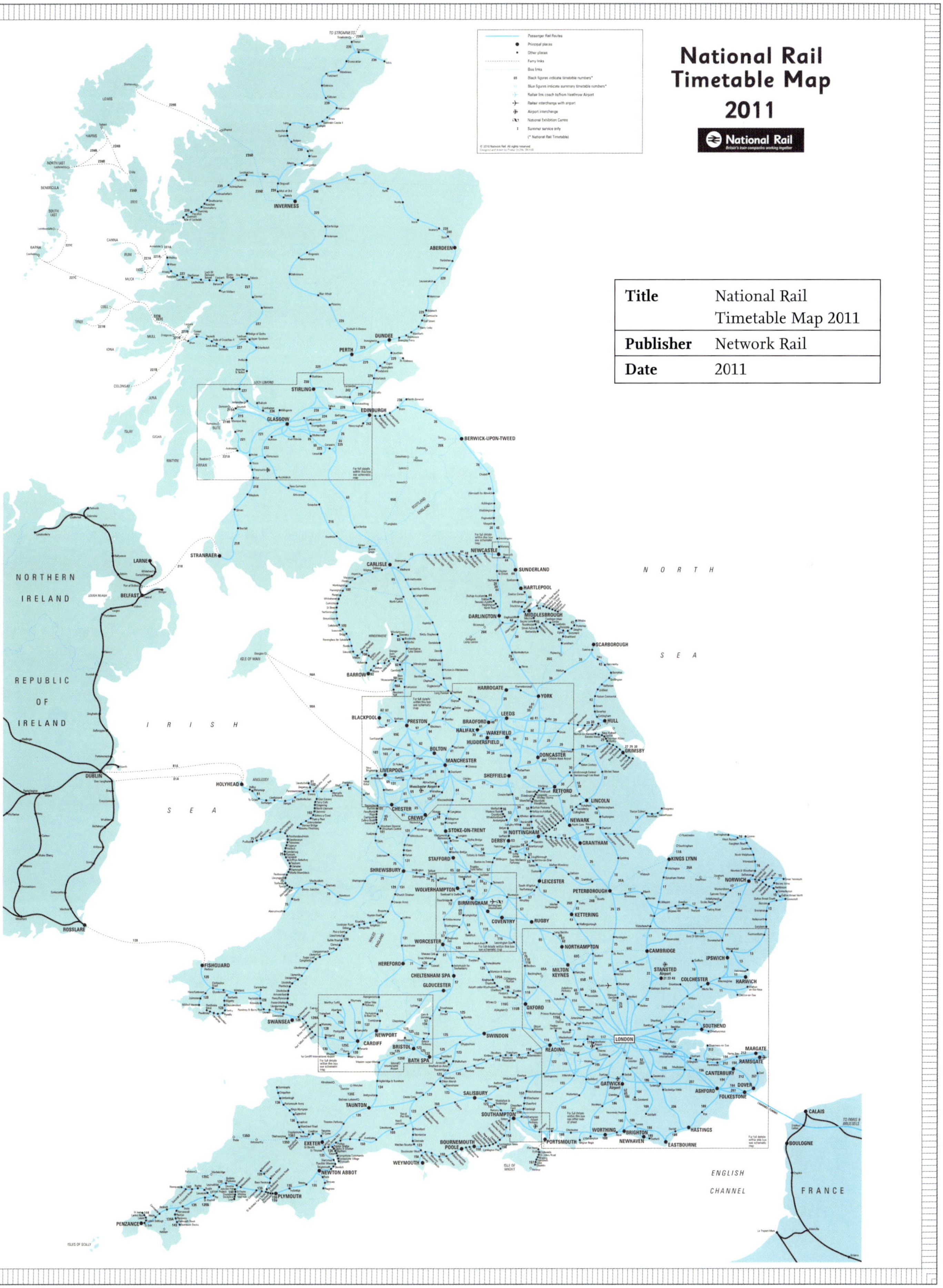

Title	National Rail Timetable Map 2011
Publisher	Network Rail
Date	2011

Title
National Rail Map – schematic
Publisher
Association of Train Operating Companies (ATOC)
Date 2008

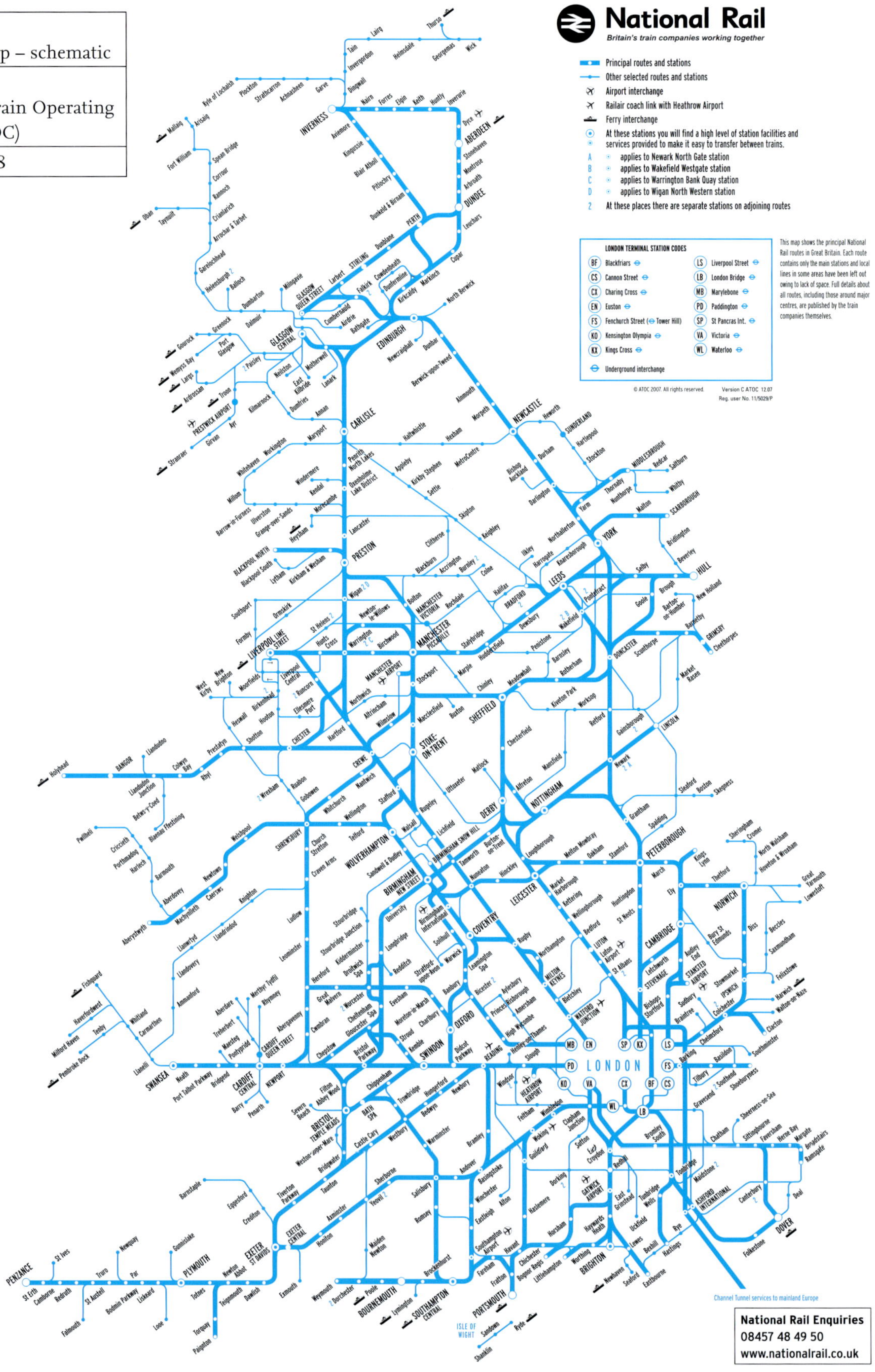

Title	National Rail Network Diagram
Publisher	Project Mapping
Date	2011

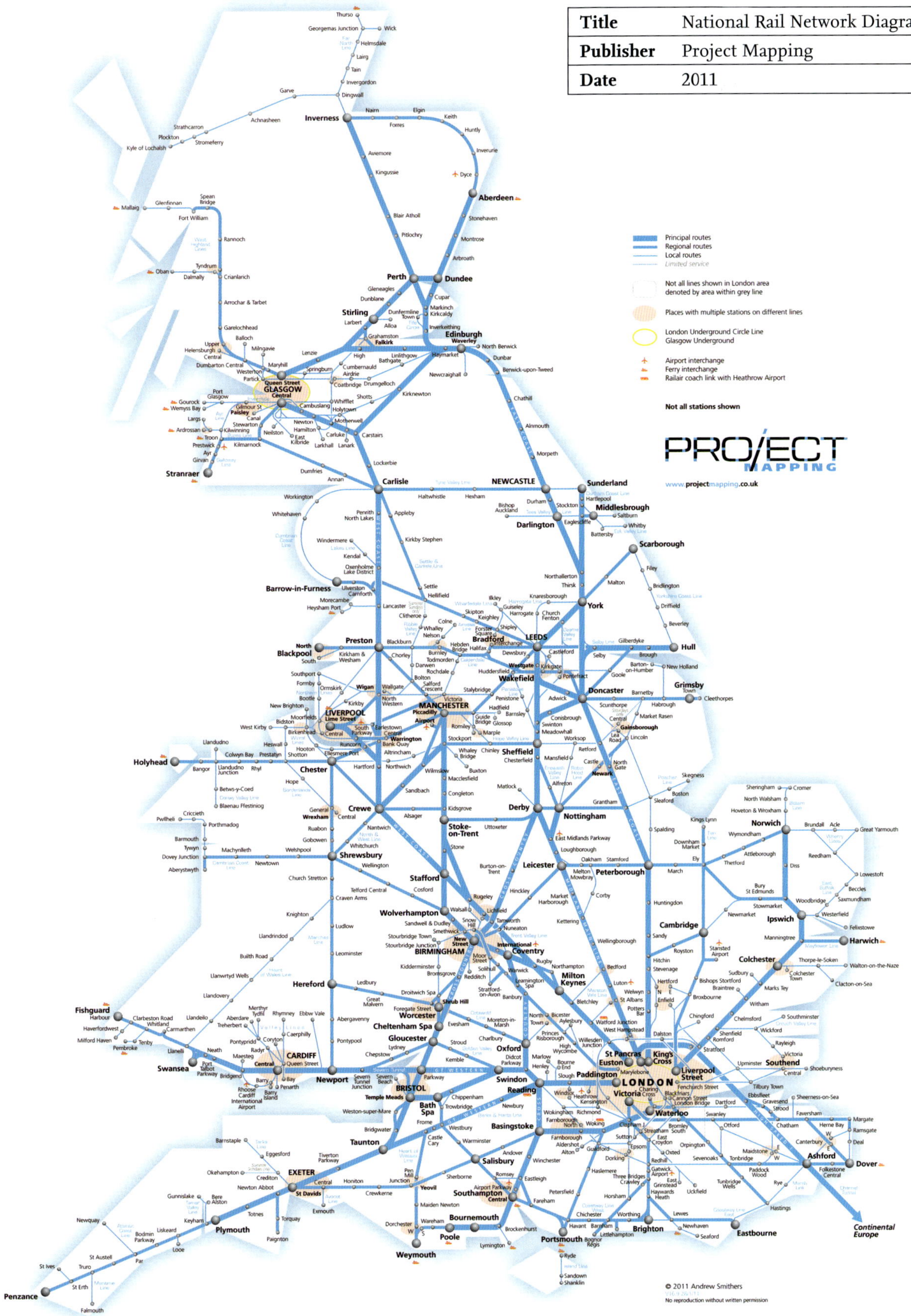

Title	A high-speed rail strategy for Britain
Publisher	Greengauge21
Date	2010

Encouraged by the success of new high speed railways in mainland Europe and the completion of the Channel Tunnel Rail Link (now known as HS1) in 2007, the case for building additional high-speed lines in Britain has gathered pace in recent years. In early 2012, the UK Government confirmed its 'in principle' support for a new ('HS2') route from London to Birmingham, potentially extended to the north of England and central Scotland, while the lobby group Greengauge21 – which published the map on this page – argues the case for a comprehensive British network on the grounds that, 'high-speed rail is an essential part of our national infrastructure and will be needed to accommodate the growth that will come with economic recovery'.

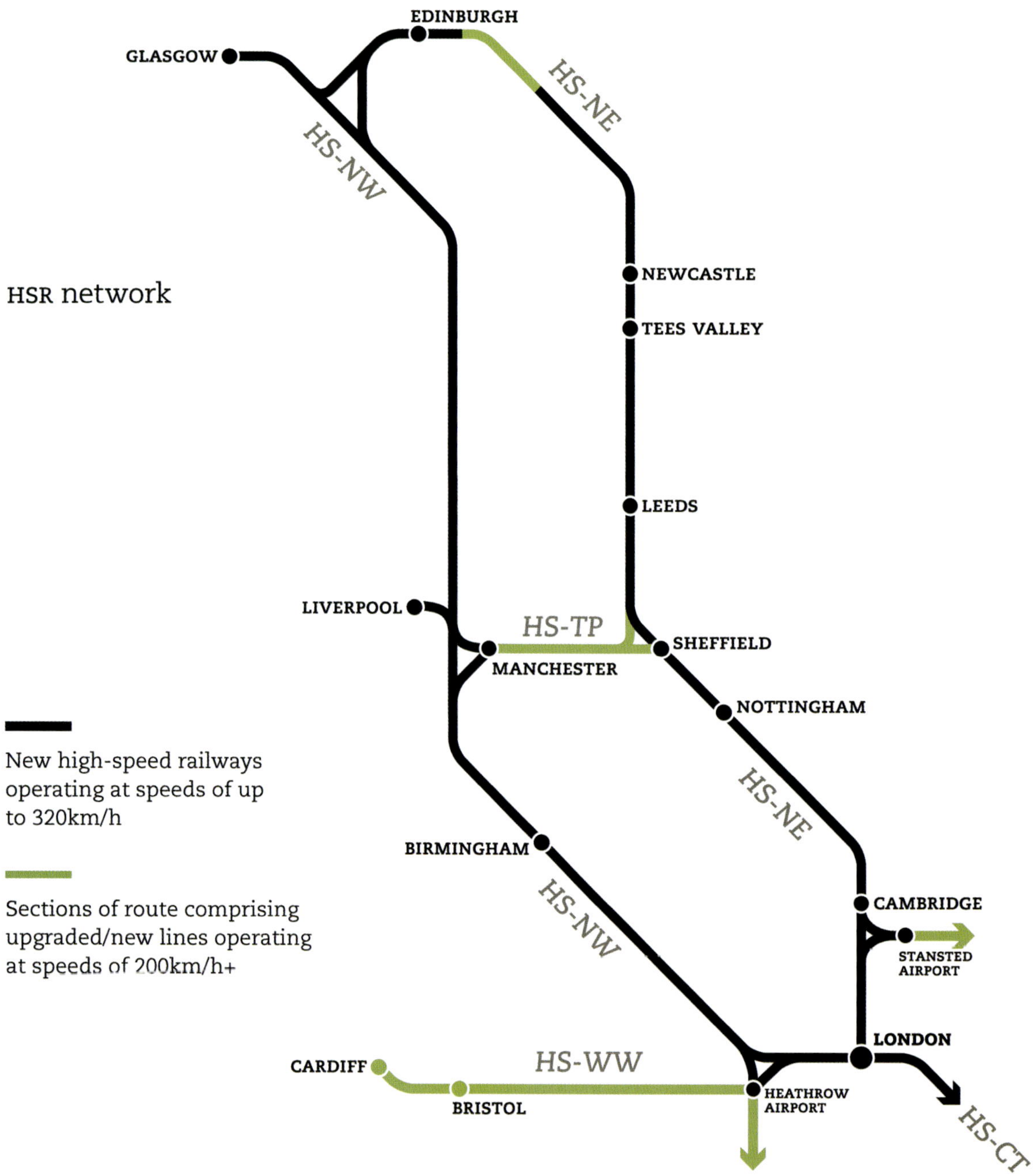

A classic view of Chilterns countryside in the Misbourne Valley between Great Missenden and Wendover – if the HS2 high speed rail route from London to Birmingham is built, it will pass through the middle of this scene.

Not surprisingly, the prospect of noise and visual intrusion from a new 250 m.p.h. railway – with up to 14 trains an hour in each direction – has not been well received along the planned corridor between London and Birmingham. Objectors along the route, which would cut through some beautiful countryside and many affluent communities, have formed an opposition group, StopHS2, whose stance is that:

> *We oppose the HS2 High Speed Rail link, because the business case is based on unrealistic assumptions, the environmental impact has not been assessed, it is not green, the strategic benefits are questionable, and the money could be better spent on other things.*

Supporters of the new line, including the Department for Transport, argue that the impact on the landscape and the local environment would be minimized, for example, by following 'existing rail or road transport corridors, using deep cuttings and tunnels, and avoiding sensitive sites wherever possible'. Wider environmental benefits are claimed by Greengauge21 which has stated that, 'high speed rail is the most sustainable of transport investments', with a projected saving of a million tonnes of carbon annually by 2055. Sceptics point out that Network Rail's 2009 report on an Anglo-Scottish High Speed Railway (HSR) forecast that just 6 per cent of HSR passengers would switch from air transport and 9 per cent from cars, with the vast majority generating additional carbon – the 53 per cent switching from more energy-efficient conventional rail services, and the 32 per cent who would be making entirely new trips.

Arguments for HSR do not, however, rest just on speed – the case is also based on a big release of capacity on existing main lines that, in turn, would allow an expansion of commuter services and freight on rail. Critics have responded that more can be done to accommodate extra traffic on the current network – for example by lengthening existing trains, and reducing the number of intermediate calls on long-distance trains – while an alternative proposal has been for a freight-prioritized rail route from the Channel Tunnel to Scotland, predominantly using existing rail lines and closed track beds, with just four miles of entirely new construction.

HSR would be expensive – Greengauge21 has estimated a full network cost of £69 billion – and critics have argued that it would choke off quicker investment in the existing network, such as more electrification and new trains. In contrast, HSR would involve a long lead time, and its justification depends on difficult assumptions about the state of our economy and environment 40 and more years ahead.

It has also been pointed out that the successful high-speed rail corridors at the heart of mainland Europe have typically involved new sections which secured very dramatic improvements in journey times through by-passing relatively slow conventional routes along river valleys like the Rhine or across significant mountain regions. The geography of Britain's trunk rail routes is very different, and we do not have the same pattern of heavily populated metropolitan regions separated by very long HSR-suited distances. Ultimately, in an age of financial austerity, a national High Speed Rail network – whatever its merits and demerits – may prove to be an idea that came just too late to Britain.

In 1819, in the very first map of this journey through the history and geography of Britain's railways, the engineer Robert Stevenson was uncertain whether he would be building a canal or railway. Our exploration has illustrated how quickly and overwhelmingly railways became *the* dominant mode of transport across the country, remaining so until well after the middle of the 20th century. We have also seen how the railways recovered from the massive retrenchment of the 1960s – caused by car, bus and lorry competition – to carve out a continuing and expanding role in Britain's transport system. New routes and stations have opened in almost every part of the country, and more are due to arrive – including the return of train services to the Scottish Borders in 2014, after an absence of 45 years, and London's new Crossrail route in 2018.

Not only has the density and pattern of the rail network across Britain changed enormously over nearly two centuries, so too have styles of mapping. Where double-pecked lines were good enough to show waggonways in the county maps of 1820, during much of the 20th century the Ordnance Survey used no less than four line symbols to distinguish different types of rail route. As a whole, railways are now a far less dominant part of the transport system, but their character and role have also changed, and new map symbols have been designed to reflect this. Yet, while the technology and techniques for mapping railways have been transformed, there are some reassuring constancies – the black lines and red circles denoting rail routes and stations respectively in *Reid's Railway Rides* of 1847 are still to be seen in the latest Ordnance Survey Landranger maps.

All the best explorations lead to new understanding and appreciation of the country travelled. While hopefully every reader will have learnt something new about railways and maps, each is likely to have taken something different from the history, the geography and the cartography. For the authors, a particular fascination has been the research and analysis which underpins the specially commissioned maps of Beeching reprieves, post-Beeching closures, and railway re-openings since 1969 – seeing these patterns on maps for the first time raises new questions about the politics and economics of railways. Other maps from the most modern era illustrate how local circumstances can change the conventional view of railways as 'environmentally friendly' to an altogether less benign perspective – a standpoint which, as we have seen, can be traced back to the late Victorian era in London.

As we have noted earlier, maps are not just a matter of utility – they also have the capacity to please and delight. The age, background, education and motivation of the reader will to some extent determine his or her view as to whether they are looking at a 'good map' or a 'bad map', irrespective of its theoretical technical merit. Our selection has allowed a very wide variety of maps to be showcased, and different readers will have different views of the most attractive, the most unusual and the most useful.

A recurrent theme of this book has been change, and today we live in an era of constant adjustment to new technology and new fashions stimulated by a highly competitive market place. This produces both opportunities and threats for the

The distinctive 'Double Arrow' platform brickwork logo of the re-opened Robin Hood Line is in evidence at Bulwell train/tram interchange in Nottinghamshire, as an NET tram pulls away.

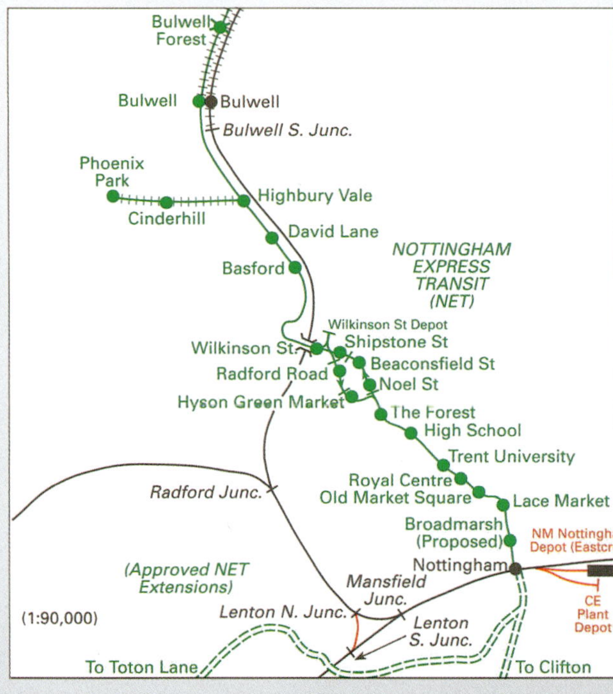

Right: Nottingham area extract from the 'Baker atlas' – see page 265.

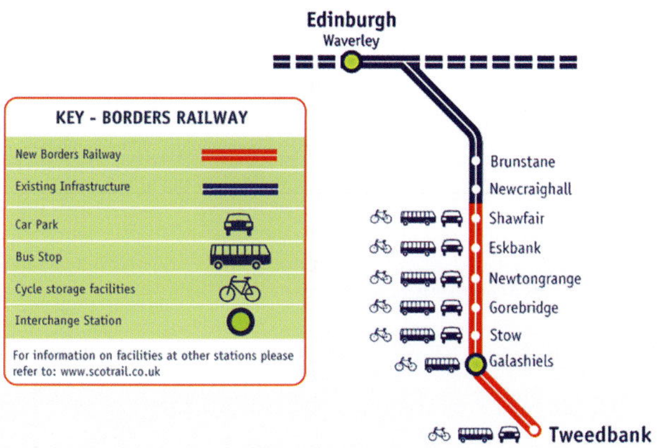

A map of the new Borders Railway by Transport Scotland, the Scottish Government agency which is delivering the project in partnership with local authorities and the rail industry.

Advance engineering works on the new Borders Railway in 2011 – the first stretch of track approaches the main line connection at Millerhill, southeast of Edinburgh.

mapping of railways. So, for example, interactive maps on mobile phones may effectively change what some of us understand by the term 'map', while fast-changing design trends (and the lack of a unified national system of rail services) mean that we are never again likely to see a consistent method of representing railways on publicity maps in the way that British Rail achieved from the 1950s to the 1990s.

Mapping the Railways has inevitably displayed only a very small sample from nearly 200 years of mapping railways in Britain. Beyond this material, which is largely in the public domain, a large corpus of unsung and unstudied private maps of the railways has survived, but as the cartographic and railway historians Challis and Rush have noted, these:

> *were produced by the railway companies for internal or business use and were regarded as private documents from the outset. The vast majority of these remain the preserve of the companies' modern institutional successors. Unsurprisingly these private maps tend to be less, if at all, available.*

It is hoped that this introduction to the story of railways and mapping will stimulate further interest in a neglected aspect of our history and geography, and will encourage the rail industry to make more of its map legacy available to the public for exploration, analysis – and sheer enjoyment.

INDEX

A

Abbeyhill 119
Aberayron 103
Abercynon 253
Aberdare 151, 153, 253
Aberdeen 60, 124, 127, 250, 251, 258, 274, 282
Aberdour 60
Abergavenny 103
Aberystwyth 103
Acton Town 197
Adams, William 34
Adventures of Sherlock Holmes, The 34
Airdrie 223, 291
Airey, John 80, 81, 83
Aldwych 197
Alexandra (Newport & South Wales) Railway 150
Alexandra Palace 136, 282
Alfreton & Mansfield Parkway 264
Alloa 249, 291
Aln Valley Railway 282
Arbroath 15
Armagh 212
Around the World in Eighty Days 34
Arran, Isle of 163
Arrol, William 125
Ashford 37
Ashover Light Railway 162
Aspatria 87
Association of Train Operating Companies 292, 294
Automatic Warning System 213
Aviemore 121, 145
Avonmouth 151, 253
Aylesbury 241
Ayr 234, 292

B

Badminton 95
Baker, Benjamin 125
Baker, S K 264, 265
Bakerloo Line 197
Bala Junction 218
Ballantrae 167
Ballater 121
Banker Hill 17
Bargoed 153
Barnard Castle 45
Barnsley 189
Barnstaple 270
Barnstaple Junction 270
Barnstaple Town 270
Barnstaple Victoria Road 270
Barry 95, 150, 151, 153, 253, 283, 291
Barry, Sir John Wolfe 151
Barry Docks 133
Barry Docks & Railway Company 150
Barry Railway 133, 150
Bartholomew, J (Publishers) 7, 87, 99, 104, 110, 121, 126, 136, 140, 145, 155, 170, 185, 270
Bartholomew, John 95
Bartholomew, John (Jnr) 170
Bartholomew, John George 170
Basford Hall 206
Basingstoke 44
Bath 44, 66, 282
Bath Extension 95
Bathgate 223, 291
Beale, J 35
Beattock 251
Beccles 236
Beck, Henry C (Harry) 9, 194, 196, 197
Bedford 214
Beeching, Dr Richard 7, 229, 232, 233, 234, 236, 245, 247, 253, 264, 278, 282
Beeching Report 7, 8, 90, 161, 151, 178, 201, 214, 229, 234, 236, 241, 249, 253, 270, 284
Belfast 109
Belmont 241
Bembridge 274
Berney Arms 201
Berwick 147
Bethlem Hospital 100
Betjeman, John 249
Beverley 147
Beyer-Peacock 222
Bicester 214
Bideford 270
'Big Four' 34, 185, 191, 204
'Big Four Grouping' 8, 81, 161, 182
Birkenhead 81
Birmingham 28, 90, 176, 206, 281, 296
Birmingham & Derby Junction Railway 43
Bishopsgate 99, 100, 140
Blaenau Ffestiniog 218
Blaenavon 153
Blagdon 95
Blair Atholl 145
Blea Moor 124
Blenkinsop, John 12, 21
Bletchley 214
Bletchley Park 214
Blodwel 179
'Blücher' 21
Bluebell Railway 282, 283
Blyth 83
Bo'ness & Kinneil Railway 60
Board of Trade 167
Bodiam 165
Bodmin & Wadebridge Railway 284
Bodmin & Wenford Railway 284, 285
Bodmin General 284
Bodmin North 284
Bodmin Parkway 284
Bodmin Road 284
Bombardier Transportation 177, 206
Borders Railway 227, 299
Boscarne Junction 284, 285
Boston 249
Boston Lodge 218
Bouch, Thomas 124, 125
Boulogne 132
Bournemouth 223, 280
Bow Road 99
Box Tunnel 66
Bradford 189
Bradshaw & Blacklock 35
Bradshaw, George 23, 34, 35, 37
Bradshaw's Air Guide 34
Bradshaw's Maps of Inland Navigation 35
Bradshaw's Monthly Railway Guide 34, 35
Bradshaw's Railway Companion 34, 35
Bradshaw's Railway Time Tables and Assistant to Railway Travelling 34
Braemar 121
Branch Line Committee 282
Brayton Junction 87
Brecon & Merthyr Railway 153
Brecon Beacons 153
Bricklayers Arms 140
Bridge, Mike 286
Bridgend 95, 253, 291
Bridgeton 110
Bridgnorth 90
Bridport 253
Brigham 87
Brighouse 189
Brighton 37
Bristol 44, 66, 100, 176, 233, 253, 258, 282
Bristol and Bath Railway Path 282
Bristol & Exeter Railway 66
British Channel Tunnel Group 275
British Expeditionary Force (BEF) 159
British National Oil Corporation 267
British Rail 245, 251, 253, 258, 263, 266, 267, 268, 278, 291, 292, 299
British Railways 35, 81, 83, 133, 162, 176, 178, 192, 193, 194, 195, 205, 210, 214, 218, 222, 223, 225, 229, 232, 233, 236, 241, 245, 247, 249, 253, 258, 274, 282
British Railways Board 233, 236, 241, 245, 247, 253, 260, 261
British Steel Corporation 263
British Transport Commission 233, 260, 282
British Transport Hotels 278
Broad Street 100, 136
Brunel, Henry Marc 151
Brunel, Isambard Kingdom 66, 69, 80, 100, 151, 193
Brush Traction 222, 223
Brynmawr 153
'Bubble Act' 42
Buckie 185
Bulleid, Oliver 222
Bullgill 87
Bulwell 298
'Burlington Zephyr' 183
Burntisland 60, 119

C

Cabot, John 66
Cairnryan 167
Calais 275
Caledonian Railway 110, 126, 207, 250
Cambrian Coast line 201, 218
Cambrian Railways 178
Cambrian Railways Society 179
Cambridge 37, 214, 229, 249
Camden 27, 140
Camel, River 284
Camel Trail 284
Campbeltown 15
Campbeltown & Machrihanish Railway 162
Canal Street 119
Cannon Street 100, 213
Canterbury & Whitstable Railway 169
Canterbury East 169
Canterbury Road 169
Canterbury South 169
Canterbury West 169
Cardiff 150, 153, 170, 253, 258, 281, 291
Cardiff City line 253
Cardiff Docks 150, 151
Cardiff Railway 150
Carlisle 15, 147, 189, 212, 226, 234, 236, 251, 253, 292
Carmarthen 258
Carne Point 134
Carr Bridge 121
Castle Cary 95
'Catch-Me-Who-Can' 12, 13
Cathcart 110
Central Line 197
Central Wales line 103, 171, 234, 236, 249
Chad Valley 190
Challis, David 286, 299
Channel Islands 132
Channel Tunnel 101, 133, 274, 275, 291, 296, 297
Charing Cross 100
Chat Moss 26
Cheffins, Charles F 28, 44
Cheltenham 176
'Cheltenham Flyer' 182, 190
Cheriton 275
Cheshire Lines Committee 155
Chester & Holyhead Railway 132
Chesterfield 49
Churchward, G J 204
Cirencester 44
City & South London Railway 194
Clapham Junction 80
Clark, Estra 210
Cleator & Workington Junction Railway 87
Cleator Moor 87
Clyde, River 110, 132
Coalbrookdale 12
Coatbridge 281
Cob 218
Collett, Charles 204
Collins Publishers 170
Continental Railway Guide 34
Coquelles (France) 275
Corby 262, 263
Corkerhill 177
'Cornish Riviera Express' 192
Cornwall 134
Cornwall Minerals Railway 134
Cornwall Railway 134
'Coronation' 183, 191, 192
'Coronation Scot' 192, 183
Corrour 201
Cowdenbeath 127
Craven Dunnill & Co 147
Crew Junction (Scotland) 207
Crewe Toll (Scotland) 207
Crewe 104, 176, 183, 206, 259
Crewe Hall 176
Crewe Junction 28
Cromarty & Dingwall Railway 162
Cromer 185
CrossCountry Trains 279
Crossgates 127
Crossrail 298
Croxley Green 241
Croydon 20
Cruchley, George 73
Cruchley's Travelling Companion 73
Crystal Palace (High Level) 136
Cunard Line 280
'Cunarder, The' 281
Cuneo, Terence 193, 275
Cwmbran 153
Cyclists' Touring Club 170

D

Dalmellington 163
Dalmeny 60

Dalnaspidal 145
Dalston Junction 136
Dalwhinnie 145
Danish State Railways 260
Darlington 21, 177
Darlington Locomotive Works 283
Davies, D 22
Day & Son 69
Deeside Way 282
Delano-Smith, Catherine 8
'Deltic' 222, 225
Department for Transport 278
Derby 49
Dereham 229, 249
Deutsche Bundesbahn 259
Development of the Major Railway Trunk Routes, The 234, 247
'Devon Belle' 266
Dickens, Charles 77
Dingwall 266
Director of Passenger Rail Franchising 278
Docklands Light Railway 197
Doncaster 147, 155, 177, 225
Doric Arch (Euston) 100
Dornoch Light Railway 162
Dover 37, 69, 132
Dovey Junction 201
Dow, Andrew 194, 196
Dow, George 194, 195
Dowlais 12
Druimuachdar Summit 145
Drummore 167
'Duchess of Hamilton' 183
Dumfries 121
Dundee 124, 274
Dunkirk (France) 159
Dunmere 284
Dunoon 163
Durham 41, 251

E
'Earl Cairns' 258
Earlestown 28
East Coast Main Line 7, 45, 49, 99, 223, 225, 227, 250, 251, 258, 289, 291
East Kent Light Railway 162, 169
East Lincolnshire line 241
East Suffolk line 201, 229, 236, 249
Eastern Counties Railway 100
Eastleigh 177
Ebbw Vale 103, 153
Ebbw Vale Parkway 253, 291
Eden Valley Railway 283
Edinburgh 60, 95, 110, 119, 124, 126, 136, 183, 207, 234, 236, 249, 253, 258, 260, 266, 274, 292
Edinburgh & Glasgow Railway 27
Edinburgh & Northern Railway 274
Edinburgh Geographical Institute 126
Edinburgh Waverley 119, 121, 225
Edmondson, Thomas 81
Elham Valley Railway 169
'Elizabethan, The' 183, 225
Elvanfoot 167
English Channel 132, 158
English Electric 222, 223
English Heritage 176
Epping 197
Eurostar 275, 286
Euston 27, 49, 80, 104, 183, 206, 213, 250, 251
Euston Hotel 104
'Evening Star' 176, 205
Exeter 95
'Experiment, The' 21

F
Fakenham 229
Falklands War 281
Far North Line 145, 236, 249
Fawley 281
Felixstowe 245
Ffestiniog Railway 162, 165, 218
Fiennes, Gerry 233, 234
Fife 60, 127
Fife Coast line 249
Finchley Road 77
Finsbury Park 282
First World War 158, 182, 189, 190, 212, 275
Firth of Forth 60, 119, 124, 125, 274
Firth of Tay 124, 125, 274
'Flying Hamburger' 183
'Flying Scotsman, The' 225
Folkestone 37, 132, 169, 275, 291
Ford, Roger 279
Fordell Railway 127
Fort Augustus 145
Fort William 114, 145
Forth Bridge 60, 119, 126, 250, 274
Fowey 134
Fowler, John 125
France-Manche 275
Fraserburgh 241
Freight Operating Companies 278
Freightliner 206, 245, 264, 281
Frome 253
Furness Railway 190

G
Galashiels 227, 236, 241
Gare du Nord (Paris) 274
Garland, Ken 197
Gatwick Express 286
General Electric 223
General Motors 223
Girvan 95
Gladstone, William 247
Glasgow 15, 60, 110, 114, 183, 206, 222, 251, 260, 262
Glasgow & South Western Railway 121, 126, 177
Glasgow 'Blue Trains' 223
Glasgow Central 104
Glasgow Central Low Level 110
Glasgow Queen Street 110
Glasgow Underground 110
Glastonbury 241
Gloucester 44, 124, 176, 233
Gobowen 178, 179
Gowerton South 171
Grand Junction Railway 26, 27, 28, 42, 176
Grangemouth 249
Grantham 183
Granton 119, 126, 207, 274
Granton Harbour 119
Great Central Railway 73, 77, 101, 133, 155, 158, 194, 241, 275
Great Eastern Railway 99, 100, 185, 201
Great Grimsby & Sheffield Junction Railway 132
Great Hall (Euston) 100
Great North of England Railway 45
Great North of Scotland Railway 121
Great Northern Railway 100, 177, 189, 190, 250
Great Western Railway 34, 43, 44, 57, 66, 69, 73, 80, 81, 95, 100, 103, 124, 134, 140, 158, 150, 151, 153, 171, 176, 178, 182, 185, 190, 206, 222, 270, 284
Greater London Authority 278
Greenburn Light Railway 163
Greengauge 21 296
Greenock 132
Greenwich Mean Time 81
Greenwich Park 136
Greenwood, Christopher 16
Greenwood, Pringle & Co 16
Gresley, Nigel 182, 183
'Gresley, Sir Nigel' 6
Grimsby 133, 155, 241

H
Halifax 189, 210
Halwill Junction 95
Hamilton 126, 291
Hampstead Tunnel 77
Hansard, Henry 57
Hardwick, Philip 100, 104
Hardy, R H N 233
Haresnape, Brian 260
HarperCollins Publishers 170
Harrogate 225
Harrow & Wealdstone 213, 241
Hartlepool 83, 147
Harwich 133, 274
Hassall, John 190
Hastings 164
Hatfield 212, 278
Haverhill 229
Hawick 226, 236, 241
Hayling Island 274
Haymarket (Edinburgh) 27
Headcorn 164
Heathrow 291
Hebden Bridge 189
Hedley, William 13
Hellifield 189
Hengoed 153
Her Majesty's Stationery Office 233
Hereford 103
Hetton Colliery Railway 21
Hexham 32
Heysham 132
High Speed 1 7, 9, 124, 275, 291, 296, 297
Highgate 196, 282
Highland Railway 121, 145, 167, 185
Highlands and Islands Development Board 267, 268
Hindle, Brian 73
Hobson, William Colling 41
Holborn Viaduct 101
Holyhead 132
Hornsea 45
Horwich 177
Horwich Railway Mechanics Institute 177
Huddersfield 6, 189, 210
Hudson, George 43
Hull 132, 161
Hull & Barnsley Railway 73, 161
Hull Paragon 161
Humber, River 132
Hungerford 44, 69
Hungerford Market 100
Hunstanton 249
Hunterston 151
Huskisson, William 26, 212

I
Ian Allan Publishing 250
Iarnród Eireann 259
Ilfracombe 73, 253
Immingham 133, 151
'Inter City, The' 258
Inter-City 258, 260
Inter-City 125 (HST), 213, 247, 253, 258, 259
'Inter-City Express, The' 193
Inter-City Sleeper 260
InterCity 260, 261, 278
Invergordon 267
Inverkeithing 60
Inverkeithing Bay 127
Inverness 60, 121, 127, 234, 249, 258, 260, 266, 268
Ipswich 229, 234, 236, 249
Ireland 132
Irish Sea 108, 132
Ironbridge Gorge Museum Trust 147
Isle of Man 132
Isle of Wight 69, 132, 274

J
Jacobs, Gerald 286
Jowett, Alan 160, 161, 153, 169, 189, 201
Jowett's Railway Atlas 8, 160, 153, 189
Jowett's Railway Centres 160, 161, 169, 201
Jubilee Line 197
Junction Diagrams 80, 81

K
Kain, Roger 8
Keighley & Worth Valley Railway 189
Keighley 189
Keith 185
Kensington Olympia 136
Kent & East Sussex Railway 162, 165
Kettering 263
Killingworth colliery railway 21, 22
Kilmacolm 126
Kilmarnock & Troon Railway 27
Kilsby Tunnel 27
King George Dock (Hull) 161
King's Cross 100, 140, 147, 183, 192, 225, 250, 251, 258
King's Lynn 234, 249
Kinghorn 60
Kingstown (Dún Laoghaire) 132
Kingswear 253
Kingussie 145
Kinnaber Junction 251
Kinross 249, 253, 291
Kirkcaldy 60
Kirkhill 110
Kirkliston 60
Kyle of Lochalsh 234, 236, 266, 267, 268

L
Ladbroke Grove 212, 213, 278
Ladybank 291
Lanarkshire & Dumbarton Railway 110
Lancashire & Yorkshire Railway 177, 189
Lancaster 189
Langstone Harbour 274
Largs 126
Larkhall 223, 291
Larne (NI) 132
Launceston 95
'Launceston Castle' 182
Leadhills & Wanlockhead Light Railway 162, 167
Leeds 12, 183, 189, 225, 281
Leek & Manifold Valley Light Railway 282
Leicester 49, 263
Leith 119, 126, 207
Leith Docks 119
Leith Walk 119
Letts & Co 77
Leven 249
Lewis, Isle of 163
Leyburn 45
Light Railway Syndicate 162
Light Railways Act 162
Lincoln 37, 236, 264
Linefoot 87
Liverpool 27, 132, 176, 206, 245, 281
Liverpool & Manchester Railway 21, 26, 27, 28, 32, 42, 212
Liverpool Street 100, 136, 205, 223
Llandilo 103
Llandudno Junction 218
Llanelli 103, 150, 171
Llanidloes & Newtown Railway 178
Lochinver 163
Locke, Joseph 26
Locke, Robert 27
Lockington 45
'Locomotion' 21
London 37, 66, 73, 77, 95, 99, 136, 140, 196, 197, 247, 296
London & Birmingham Railway 26, 27, 43, 80, 100, 104
London & Brighton Railway 37
London & Greenwich Railway 100
London & North Eastern Railway 182, 183, 185, 191, 192, 195, 196, 201, 194, 195, 222, 274
London & North Western Railway 27, 28, 49, 87, 103, 104, 124, 140, 155, 150, 153, 171, 176, 189, 190, 214, 250
London & South Western Railway 95, 100, 133, 158, 177, 270, 280, 284
London Bridge 100
London, Brighton & South Coast Railway 100, 140, 274
London, Chatham & Dover Railway 100, 101, 169
London, Midland & Scottish Railway 176, 182, 185, 191, 192, 194, 195, 204, 205, 206, 282
London Passenger Transport Board 196
London Transport 196, 197
London Underground 9, 196, 292
Lord's Cricket Ground 77
Lords Tunnel 77
Lostwithiel 134
Louth 241
Lowestoft 201, 229, 234, 236, 249
Lybster 162
Lyme Regis 95
Lynton & Barnstaple Railway 270

M
Macassey, L Livingston 108
Macauley, Zachary 81
Machrihanish 15
MacMillan, Harold 232
'MacPuff' campaign 236
Maesteg 253
Maidenhead 66, 158, 182
Mais, Stuart Petre Brodie 191
Major, John 278
Maldon 229
Mallaig Extension 114, 121, 145
'Mallard' 183, 192
Malton 45
Manchester 28, 35, 155, 176, 189, 206, 234, 253, 281, 291
Manchester & Leeds Railway 124
Manchester Journal 35
Manchester, Sheffield & Lincolnshire Railway 275
Manifold Way 282
Mansfield 265
March 201, 236, 249
Marchwood Military Port 281
Margam 263
Markinch 14
Marples, Ernest 232, 233, 282
Maryhill 110
Marylebone 77, 101, 275
Maryport & Carlisle Railway 87
Maryport 87
Mayo, Charles 190
Meeth 270
Melton Constable 229
Menzies, John 95
Mersey, River 28
Merthyr Tydfil 103, 153
Metropolitan Railway 77
Meyers Lexicon 119
Mid-Norfolk Railway 249
Middlesbrough 83, 147
Middleton colliery railway 12
Middleton Railway 21
Midland & Great Northern Joint Railway 185, 201, 232
Midland Counties Railway 43
Midland Railway 43, 77, 101, 124, 132, 189, 253
Millerhill 226, 227, 299
Milnsbridge Viaduct 6
Minehead 73, 253
Minffordd 218
Ministry of Supply 205
Mirfield 189
Modern Railways 279
Modernisation and Re-Equipment of British Railways 205, 206, 222
Modernisation Plan 178, 214, 226, 232
Monks Coppenhall 176
Monktonhall 227
Montrose 251
Morningside 60
Morpeth 147
Mundesley-next-the-Sea 229

N
Nantwich 206
Napoleonic Wars 12
National Coal Board 264
National Cycle Network 282
National Railway Museum 147, 183, 289
Neilston 110
Ness Viaduct 266
Network Rail 35, 278, 279, 281, 286, 289, 292, 293
New Holland 132
New Street (Birmingham) 206, 258
Newbottle Waggonway 17, 41
Newcastle & Carlisle Railway 32
Newcastle-upon-Tyne 12, 13, 21, 83, 182, 182, 192, 194, 251, 292
Newcastleton 235
Newcraighall 227, 292
Newport (Gwent) 150, 153
Newport, Abergavenny & Hereford Railway 57
Newquay 73, 134, 270
Newton (Lancs) 26, 28
Newton (Scotland) 110
Newtonmore 145
Newtown 178
Newtown & Machynlleth Railway 178
'Night Ferry' 274
Nine Elms 140
Nine Mile Point 103
Norfolk & Suffolk Joint Railway 201
Normanton 49
North British Locomotive Company 222, 223
North British Railway 60, 110, 121, 126, 155, 190, 251
North Channel 108
North Cornwall District Council 284
North Eastern Railway 83, 133, 147, 161, 177, 250
North Eastern Tile Company 147
North Frodingham 147
North London Railway 100
North Midland Railway 43
North Queensferry 60
North Staffordshire Railway 190
North Woolwich 136
North Yorkshire Moors Railway 45, 282
Northallerton 225, 289
Northampton 214
Northern Ireland 108, 121, 167, 236
Norwich 201, 223, 247
Nottingham 101, 264, 265

O
Oakham 263
Office of the Rail Regulator 278
Okehampton 253, 270
'Old Worse and Worse' 57
Olive Mount Cutting 26, 27
Ongar 197
Ordnance Survey 7, 9, 16, 41, 49, 52, 60, 134, 164, 165, 170, 178, 206, 218, 219, 220, 221, 226, 227, 262, 263, 270, 284, 298
Oslo 35
Oswestry 178
Outram, Benjamin 12
Overton, George 22
Oxenhope 189
Oxford 44, 214, 249
Oxford Publishing Company 264, 265
Oxford, Worcester & Wolverhampton Railway 57

P
Paddington 44, 66, 95, 100, 101, 158, 176, 183, 213, 258
Padstow 73, 95, 270
Paignton 253
Paisley 15
Palace Gates (Wood Green) 136, 241
'Papyrus' 183
Par 134
Parkhead 110
Parkland Walk 282
Parkmore Waggonway 12
Parkside Junction 28
Partick 110
Parton 163
Pease, Edward 22
Pendleton (Lancs) 35
Pendolino trains 251
Penistone 189
Penrhyn, Lord 12
Penydarren plateway 12
Penzance 258, 260
Perth 60, 127, 145, 213, 249, 253, 260, 266, 274, 291
Peterborough 201, 234, 249
Peterhead 241
Pewsey 73
Philadelphia 17
Pick, Frank 196
Plymouth 44, 132, 284
Pontardulais 171
Pontypool 153
Port Carlisle 15, 185
Port Glasgow 126
Port Talbot 150
Portbury 253
Porthmadog 218
Portpatrick & Wigtownshire Joint Railway 163
Potters Bar 212
Preston 189
Primrose Hill Tunnel 27
Project Mapping 292, 295
'Puffing Billy' 13

Q
Quail Map Company 286
Queen Elizabeth, RMS 281
Queen Mary, RMS 281
'Queen of Scots, The' 225
Queensbury 189
Quintinshill 212

R
Railfreight 278
Railfuture 291
Railtrack 278, 279, 292
Railway Children, The 189
Railway Clearing House 80, 81, 87, 94, 99, 150
Railway Executive, The 214
Railway Gazette, The 167
Railway Mania 42, 43
Railway Operating Division (ROD) 158
Railway Shareholders' Guide 34
Railways Act 1993 259
Railways Act 2005 278
Rainhill Trials 21, 26, 32
Ramsgate 213
Ratcliffe-on-Soar 151
Readers Digest 170
Reading 66
Redhill 37
Reid, Andrew 49
Reid's Railway Ride 48, 49, 50, 51, 225, 298
Rennie, George 26
Rennie, John 26
Reshaping of British Railways, The 232, 233, 236, 241, 245
Reynolds, Richard 12
Rhins of Galloway 167
Rhins of Galloway Light Railway 163
Rhondda 151
Rhymney Railway 150, 153
Ribblehead Viaduct 124
Richborough Military Port 274
Richmond 45
Richmond, Leonard 190
Riddles, Robert 204, 205
Ripon 225
Robertsbridge 164, 165

Robin Hood line 265, 298
'Rocket' 21, 26, 212
Rolling Stock Leasing Companies 278
Rolt, Tom 282
Rolvenden 164
Rother Valley Railway 164
Rotherham 49
Royal Albert Bridge 193
Royal Engineers 158
Royal Exchange & London Assurance Corporation Act 42
Royal Scots Regiment 213
Rugby 49
Rush, Andy 286, 299

S

Saffron Walden 229
St Albans City 241
St Andrews 249
St David's Harbour 127
St Ives (Cornwall) 73, 249
St John's (Lewisham) 213
St John's Wood Tunnel 77
St Pancras 9, 77, 101, 258, 286
St Pancras International 275, 291
Saltash 193
Saltburn 147
Saltend 161
Sandbach 28
Sanderstead 241
Saville, Jimmy 258, 259
Saxmundham 236
Scapa Flow 266
Scarborough 147, 210
Science Museum 13
Scotland Street 119
Scotland Street Tunnel 119
ScotRail 278
Scott, George Gilbert 100
Scott's Railway Companion 32
Scottish Executive 163, 278
Scottish Government 292
Scrabster 167
Seaham Harbour 41
Sealink 133, 278
Seaspeed 133
Second Boer War 158
Second World War 158, 159, 183, 189, 192, 193, 204, 214, 232, 274, 280, 282
Secretan, Murray 190
Selby 291
Settle 189
Settle & Carlisle line 73, 124, 234, 249
Sevenoaks 37
Severn, River 124, 132
Severn Beach 253
Severn Tunnel 124
Severn Valley Railway 90, 282
Shap 191, 251
Sheffield 49, 155, 189, 241, 253, 264
Sheffield, Ashton-under-Lyne & Manchester Railway 124
Sheringham 185
Shildon 21
Shrewsbury & Chester Railway 178
Shrewsbury 90, 103, 171, 206, 236
Shropshire & Montgomeryshire Railway 162
'Shuttle, Le' 275
Siddick 87
'Silver Jubilee' 182, 183, 192
'Silver Link' 183
Sir Joseph Causton & Sons 114
Skegness 190
Skipton 210
Skye, Isle of 163
Smith, W H 95
Smithers, Andrew 292
Snailbeach District Railway 162
Snow Hill (Birmingham) 258
Solent, The 132
Solway Firth Viaduct 185
Solway Junction Railway 87
Somerset & Dorset Joint Railway 185, 241
Sonning Cutting 66
South Acton 197
South Eastern & Chatham Railway 100, 177
South Eastern Railway 100, 132, 169
South Queensferry 60
'South Sea Bubble' 42
South Sea Company 42
South Shields 194
South Yorkshire Supertram 265
Southampton 132, 133, 158, 245, 280
Southampton Central 280
Southampton Eastern Docks 280
Southampton Ocean Dock Terminal 280
Southampton Terminus 280
Southampton West 280
Southern Railway 182, 185, 191, 222, 270, 274, 284
Sowerby Bridge 189
Spalding 201, 236, 241, 249
Springburn Park 110
Staffordshire County Council 282
Staines 241
Standedge Tunnel 124, 189
Stanford, Edward 77
Stanier, William 183, 204, 205
Stansted 291
Stedeford Committee 232
Stephens, Colonel H F 162, 165
Stephens, Patrick 160, 153, 169, 189, 201
Stephenson, George 13, 20, 21, 22, 23, 24, 26, 27, 43, 212
Stephenson, Robert 21, 26, 27, 37
Stevenson, Robert 14, 298
Stewart's Lane 286
Stewarts & Lloyds 262
Stirling 249, 291, 291
Stobcross 110
Stocksmoor 189
Stockton 21, 23
Stockton & Darlington Railway 21, 26
Stockton Quay 21
Stoke Bank 183
StopHS2 297
Strachur 163
Stranraer 95, 121, 132, 167, 234, 236, 292
Strategic Rail Authority 278
Stratford Market 99
Strathearn 15
Strongitharm, A H 87
Struan 145
Sudbury 229
Summit Tunnel 124
Sunderland 16, 17, 83
Surrey Iron Railway 20
'Sussex Scot' 259
Sustrans 282
Swaffham 229
Swansea 150, 151, 170, 171, 236, 247, 258, 270
Swansea & Mumbles Railway 20, 171
Swansea High Street 171
Swansea Victoria 103, 171
Swindon 95, 176, 183, 205, 222, 223
Swinton 49, 147

T

Taff Vale Railway 150
Talyllyn Railway 282, 283
'Tarka Line' 270
Tarka the Otter 270
Tata Steel 263
Taunton 73, 95, 270
Taw, River 270
Tay Bridge 124, 125
Taylor, Fred 101
Tayport 249
Tees, River 21
Teesside 83
Temple Meads (Bristol) 66, 95, 253
Tenterden 165
Tenterden Town 164, 165
Thames, River 99
Thamesport 245
Thatcher, Margaret 278
Thetford 201, 229
Thirsk 45
Thomas, John 114
Thomson, Dr James 14
Thurso 167, 234, 236, 236, 249, 267, 268
Tilbury 245
Titanic, RMS 280
Tonbridge 37
'Tornado' 283
Torrington 95
Totley Tunnel 124
Totton 281
Tower Bridge 125
TRACKmaps 286, 287, 289
Train Operating Companies 278
Trans Europe Express 259
Transport Act 247
Transport Act 2000 278
Transport Users Consultative Committee 234, 249
Treaty of Canterbury 275
'Tregenna Castle' 182
Treherbert 253
Trevithick, Richard 12, 13
Trinity 119
Tweedbank 227
Tyne, River 12, 41
Tyne & Wear Metro 83
Tyne Dock 133
Tynemouth 83, 147
Tyneside 83, 194, 195

U

Ullapool 163, 266
Underground Group 197
US Army Transportation Corps 159
Uxbridge 241

V

Vacher & Sons 77
Vale of Glamorgan line 95
Vale of Glamorgan Railway 253
Victoria (London) 100, 274, 286
Victoria & Albert Docks 99
Victoria Line 197
Vignoles, Charles 26
Virgin Trains 279

W

Wadebridge 95, 270
Wandsworth 20
Warrington & Newton Railway 28
Warrington 28
Washington 41
Washington Staithe 16
Waterloo 100, 280, 284, 286
Waterloo International 275
Watford Junction 241
Watkin, Sir William 275
Watt, James 12
Waverley Route 226, 227, 234, 235, 236, 237, 241, 253, 266
Wear, River 16, 41
Weaver Junction 206
Welch, Vic 195
Wellington 90
Wellington, Duke of 26, 43
Wells-next-the-Sea 229
Welsh Assembly Government 253, 278, 292
Wemyss Bay 126
Wenfordbridge 284
West Auckland 23
West Coast Main Line 27, 28, 222, 223, 250, 251, 258, 278
West Drayton 241
West Herrington 17
West Highland line 145, 201
West Highland Railway 114
West Midland Railway 57
West Somerset Railway 282
Westbury (Wilts) 44, 95
Wetherby 45
Whitby 45, 147, 210
Whitby & Pickering Railway 45
Whitchurch 178
Wick 234, 249, 267, 268
Wick & Lybster Railway 162
Wigan 28
Wigton 87
Wilkinson, Norman 191
Williamson, Henry 270
Wilson, Harold 233
Winchburgh Junction 127
Wisbech 201
Withernsea 147
Witton Park 21
Wolmar, Christian 245, 279
Wolverhampton 90
Wolverhampton (Low Level) 258
Wood, John 87
Woodham, Dai 283
Woodhead route 222, 253
Woodhead Tunnel 124
Woodside 241
Worcester 233
Worcester & Hereford Railway 57
Workington 87
Worksop 265
Wylam 21
Wylam colliery railway 13, 21
Wymondham 249

X

XP64 258

Y

Yarmouth 201
Yonge, John 286
York 43, 45, 49, 52, 147, 225, 251, 289
York & Newcastle Railway 43
York & North Midland Railway 43, 52
York Castle 43
York Union Banking Company 43
Young, Alan 292
Ystrad Mynach 153

Z

Zeebrugge (Belgium) 133, 274

ACKNOWLEDGEMENTS

t = top; b = bottom; r = right; l = left; m = middle

Maps and additional material courtesy of:
Ian Allan Publishing Ltd: 250–251, 264–265, 298br
ATOC: 276–277, 294
BRB (Residuary) Ltd: 228–231, 236–244, 246, 247, 249, 252–255, 260, 261, 266–269
© Collins Bartholomew: 248, 290–291
Andrew Dow: 194–195
East Coast Main Line Company Ltd: 279bm
Freightliner Ltd: 245
© Greengauge 21: 296
Julian Holland collection: 80t, 150t, 166, 202–203, 214–217
Bill Jamieson collection: 252–255
Alan Jowett: 152–153, 160–161, 168–169, 188–189, 200–201
Mapseeker: 4–5, 10–11, 23t, 36–39, 56–59, 64–65, 72–79, 82–85, 110–113, 118–119, 130–131, 136–143, 174–175, 184, 185tr, 186–187, front cover, back cover
Bruce McCartney collection: 185b
National Library of Scotland: endpapers, 9br, 14–19, 22, 23b, 24–25, 28–33, 40–41, 44–55, 60–63, 68–71, 86–99, 102–109, 120–123, 126–129, 134–135, 144–145, 154–157, 164–165, 170–173, 178, 180–181, 206, 207tl, 207b, 208–209, 218–221, 226–227, 256–257, 262, 270–273, 280–281, 284–285
National Railway Museum / Science & Society Picture Library: 210–211
Network Rail: 293
Ordnance Survey ® map data licensed with the permission of the Controller of Her Majesty's Stationery Office. © Crown copyright. Licence number 100018598: 227, 262–263, 280–281, 284–285
Andrew Smithers: 295
David Spaven collection: 114–117, 224–225, 228–231, 236–244, 246–247, 249br, 260–261, 266–269
© TfL from the London Transport Museum collection: 196t, 197bl
TRACKmaps, reproduced by kind permission: 286–289
Transport for London: 198–199
Transport Scotland, © Crown copyright: 299tl
Virgin Trains: 279br
XC Trains Ltd: 279bl
Alan Young: 179b

Photo credits:
Ben Ashworth: 233tl
Henry Casserley: 162/163b, 167b, 212b, 282bl
Colour-Rail: 179tl, 192ml, 258b, 259tr, 278bl, 299b
Paul Francis: 297t
John Goss: 283b, 285t
Julian Holland: 163tr, 190ml, 190bl, 190br, 191tr, 192br, 193tr, 212tr, 213b, 222tr, 232tr, 278tr, 282br, 283tl
M C Kemp: 204
Bruce McCartney: 235t, 235b
Gavin Morrison: 6/7b
National Railway Museum / Science & Society Picture Library: 12b, 13tl, 13b, 20tl, 20b, 21bl, 26t, 27tr, 27b, 34b, 35tr, 35bl, 42tl, 42b, 43bl, 66/67b, 67t, 80bl, 81b, 100t, 101tr, 101b, 124t, 125tr, 125b, 132t, 133t, 133br, 151t, 158t, 159b, 176t, 177t, 177b, 182tr, 182b, 183b, 191b, 192tl, 193b, 205tr, 205b, 223b, 232tl, 234bl, 259bl, 259br, 274bl, 274br, 275t
© railphotolibrary.com: 8/9t, 151br, 192bl, 275br, 298bl
Bill Roberton: 299tr
Gordon Sharp: 146, 147tr, 147mr, 147br, 148tl, 148ml, 148bl, 148–149
David Spaven: 119br
Frank Spaven: 207mr
Alan Young: 179tr

Cover image:
Now a listed structure and nominated as a UNESCO World Heritage Site, the iconic Forth Railway Bridge across the Firth of Forth near Edinburgh was opened in 1890 and has the second-longest single cantilever bridge span in the world. © Anna Kucherova/Shutterstock.com.

With thanks for research assistance/advice to:
Andrew Dow, Chris Fleet, Roger Ford, Michael Jowett, Neil MacDonald, Allan McLean, David Prescott, Doug Rose, Gordon Sharp, John Yellowlees, Alan Young.

With special thanks to Bill Jamieson for historical and mapping research.

BIBLIOGRAPHY

Alycidon Rail http://www.alycidon.com – the official website of Roger Ford's *Informed Sources*
Beeching: Champion of the Railway? RHN Hardy, Ian Allan Ltd 1989
Britain's Growing Railway, Railfuture (the Railway Development Society Ltd) 2010
British Rail Design, James Cousins, Danish Design Council 1986
British Rail 1948-83: A Journey by Design, Brian Haresnape, Ian Allan Ltd 1983
English Maps: A History, Catherine Delano-Smith and Roger JP Kain, The British Library 1999
Fire & Steam: A New History of the Railways in Britain, Christian Wolmar, Atlantic Books 2007
Henry Beck invented what? http://www.dougrose.co.uk/index_henry_beck.htm
I Tried to Run a Railway, GF Fiennes, Ian Allan Ltd 1967
Maps and Plans for the Local Historian and Collector, David Smith, Batsford Books 1988
Maps for Local History, Brian Hindle, Batsford Books 1988
Mr Beck's Underground Map, Ken Garland, Capital Press 1994
New ideas in the interpretation of complex rail networks, Andrew Smithers in 'The Bulletin of the Society of Cartographers', Vol 43, Issues 1&2, SoC 2009
The Oxford Companion to British Railway History, edited by Jack Simmons & Gordon Biddle, Oxford University Press 1999
The Railways of Britain: An Unstudied Map Corpus, David Millbank Challis and Andy Rush, in 'Imago Mundi' (the international journal for the history of cartography) Vol. 61, Part 2, Informa Ltd 2009
Telling the Passenger Where to Get Off: George Dow and the Development of the Diagrammatic Railway Map, Andrew Dow, Fastline Books 2005.

CHESHIRE
Chester
Mold
Denbigh
Rathin
DENBIGH SH.
CONWAY & LLANWRST
Congleton
Sandbach
Crewe
LON. & NOR. WES.
W. CHESHIRE
LONDON & NORTH WESTERN
NORTH STAFFORD
Newcastle under Lyme
Stoke
Stone
Stafford
Wrexham
Nantwich
Ruabon
Llangollen
Corwen
L. & C.
L. RY.
Bala
C. & B.
B. & D.
Dolgelly
Whitchurch
Drayton
DRAYTON JUNCTION
WELLINGTON & DRAYTON
Ellismere
Wem
Oswestry
W. SHROPSHIRE
Llangynnog
Lanfyllan
O. & N.
SHREWSBURY & CHESTER
Shrewsbury
Newport
SHROPS. UNN.
SHREWSBURY & BIRMINGHAM
Wellington
Welshpool
Minstrelsey
Machynlleth
MONTGOMERY SH.
NEWTOWN & MACHYNLLETH
OSWESTRY & NEWTOWN
Newtown
BISHOPSCASTLE
Montgomery
Bishopscastle
SHROP SH.
Wenlock
WENLOCK
SEVERN VALLEY
Wolverhampton
W. & B.
Bridgenorth
Dudley
MIDLAND COUNTIES
Bushb
Llanidloes
MILFORD
M. W.
RADNOR SH.
KNIGHTON RY.
Knighton
Ludlow
SHREWSBURY & HEREFORD
L. & C. H.
WELCH
Stourport
BEWDLEY
Tenbury
TENBURY
Kidderm
Kingston
L. & K.
K. & E.
Leominster
WORCESTER
W. B. & L.
HEREFORD SH.
CENTRAL WALES
MID WALES
Hay
HAY
HEREFORD
BRECON
Hereford
MIDLAND
Ledbury
Malvern
WORCESTER
T. & M.
Tewkesbury
Llandovery
Talgath
Brecon
WEST
HEREFORD ROSS & GLOUC.
DEAN FOREST
Newent
GLOUCESTER
G. & D. F. RY.